Embedded Systems Design
with 8051 Microcontrollers

ELECTRICAL ENGINEERING AND ELECTRONICS
A Series of Reference Books and Textbooks

1. Rational Fault Analysis, *edited by Richard Saeks and S. R. Liberty*
2. Nonparametric Methods in Communications, *edited by P. Papantoni-Kazakos and Dimitri Kazakos*
3. Interactive Pattern Recognition, *Yi-tzuu Chien*
4. Solid-State Electronics, *Lawrence E. Murr*
5. Electronic, Magnetic, and Thermal Properties of Solid Materials, *Klaus Schröder*
6. Magnetic-Bubble Memory Technology, *Hsu Chang*
7. Transformer and Inductor Design Handbook, *Colonel Wm. T. McLyman*
8. Electromagnetics: Classical and Modern Theory and Applications, *Samuel Seely and Alexander D. Poularikas*
9. One-Dimensional Digital Signal Processing, *Chi-Tsong Chen*
10. Interconnected Dynamical Systems, *Raymond A. DeCarlo and Richard Saeks*
11. Modern Digital Control Systems, *Raymond G. Jacquot*
12. Hybrid Circuit Design and Manufacture, *Roydn D. Jones*
13. Magnetic Core Selection for Transformers and Inductors: A User's Guide to Practice and Specification, *Colonel Wm. T. McLyman*
14. Static and Rotating Electromagnetic Devices, *Richard H. Engelmann*
15. Energy-Efficient Electric Motors: Selection and Application, *John C. Andreas*
16. Electromagnetic Compossibility, *Heinz M. Schlicke*
17. Electronics: Models, Analysis, and Systems, *James G. Gottling*
18. Digital Filter Design Handbook, *Fred J. Taylor*
19. Multivariable Control: An Introduction, *P. K. Sinha*

Additional Volumes in Preparation

Embedded Systems Design with 8051 Microcontrollers

Hardware and Software

Zdravko Karakehayov
Technical University of Sofia
Sofia, Bulgaria

Knud Smed Christensen
Technical University of Denmark
Lyngby, Denmark

Ole Winther
Technical University of Denmark
Lyngby, Denmark

MARCEL DEKKER, INC.   NEW YORK · BASEL

ISBN: 0-8247-7696-8

Great care has been taken in preparing and testing the hardware and software examples. However, Marcel Dekker, Inc., and the authors make no warranty with regard to the documentation or programs contained in this book. The accompanying software is licensed solely on an "as is" basis. The only warranty made with respect to the accompanying software is that the diskette medium on which the software is recorded is free of defects. Marcel Dekker, Inc., will replace a diskette found to be defective if such defect is not attributable to misuse by the purchaser or his agent. The defective diskette must be returned within 10 days to: Customer Service, Marcel Dekker, Inc., P.O. Box 5005, Cimarron Road, Monticello, NY 12701, (914) 796-1919.

This book is printed on acid-free paper.

Headquarters
Marcel Dekker, Inc.
270 Madison Avenue, New York, NY 10016
tel: 212-696-9000; fax: 212-685-4540

Eastern Hemisphere Distribution
Marcel Dekker AG
Hutgasse 4, Postfach 812, CH-4001 Basel, Switzerland
tel: 41-61-261-8482; fax: 41-61-261-8896

World Wide Web
http://www.dekker.com

The publisher offers discounts on this book when ordered in bulk quantities. For more information, write to Special Sales/Professional Marketing at the headquarters address above.

PREFACE

The field of embedded computers is growing with high production volumes. Some typical examples are consumer electronics, computer peripherals and control systems in cars and aircraft. Built-in computers significantly improve the parameters of the devices under control.

This book reflects our vision of how small-scale, control-dominated embedded systems should be taught to students and professionals. The intended audience includes students from computer engineering, computer science, electrical engineering, applied electronics and applied physics departments. The reader is expected to have some basic knowledge of digital design and programming with a high level language C. The hardware and software concepts discussed in the book are oriented to courses, such as embedded systems, microcontrollers and industrial microcomputer systems. The design examples which are based on 8051 microcontrollers are readily transferable to other devices.

The text is organized to cover several topics that are imperative to the system design of embedded computers:

Microcontrollers viewed as parallel running processors and embedded peripherals

We study the most popular 8-bit microcontroller, the 8051, in Chapter 2. We emphasize the interaction between the software and the hardware subsystems. In that regard, the flags taxonomy and symbols introduced in Chapter 1 require close attention. A microcontroller with sophisticated embedded peripherals, the 83C552, is focused on in Chapter 7.

Assembly language programming for 8051 microcontrollers

We present the basis of this subject in Chapter 3 and add more examples throughout the book. The electromechanical clock design, Chapter 4, and the digital clock, Chapter 7, illustrate how different applications are driven by software written in assembly language.

Digital interfacing

This theme is addressed in Chapter 4 at two hierarchy levels. First, we discuss interfacing to typical embedded systems components. Second, we investigate the digital interface between embedded systems and the outside world. Chapter 7 provides examples of industrial interfaces. The interface between microcontrollers and field programmable logic devices is discussed in the EPROM emulator case study in Chapter 11.

Analog interfacing

The analog interface design begins in Chapter 5. The basic concepts are followed by hardware and software examples. The analog subsystems of the 83C552 microcontroller are focused on in Chapter 7. Chapter 9 introduces an extra example for interplay between the digital and analog domains: the speech machine.

Serial interfaces for distributed embedded systems

This topic is first introduced in Chapter 4 by two serial interfaces, RS-232 and RS-485. Chapter 6 covers the interaction between personal computers and microcontrollers. Chapter 8 adds two more specific serial interfaces: the I^2C bus and the CAN bus.

A systematic design methodology for embedded systems

The alternative to writing code in a high level language is discussed in Chapter 9. This chapter includes a set of examples that put together with the assembly language programs indicate the trade-offs between the different levels of abstraction. Chapter 10 summarises previously used design techniques and provides a discussion for appropriate debugging tools. Essentially, Chapter 10 outlines the main motivations behind the hardware-software co-design. Some embedded applications will demand concurrent design of hardware and software; some will not. Chapter 11 demonstrates the design of two debugging tools, a simplified EPROM programmer and an EPROM emulator. While the EPROM programmer project can be split into hardware design and software design, a successful design of an EPROM emulator requires a co-design approach. The processor plus ASIC architecture of the EPROM emulator is a typical example of linking microcontrollers and FPGAs to improve the performance.

All of the example programs in the book which have a file name are included on the accompanying disk.

Many people deserve special mention for offering us important advice. First, the students of the courses Embedded Systems 4532 at Technical University of Denmark and Applied Microcomputer Systems at Technical University of Sofia guided us by their comments and questions. Profs. Bent Nielsen and Hani Kamel used drafts in their classes in the Department of Applied Electronics. We improved the book through their constructive criticism. We thank the reviewers Dr. P. J. Miller of Aston University and Dr. Russ Hersch of Silicom for their many useful comments. Valuable feedback was also received from anonymous reviewers. We thank our teaching colleague Emil Saramov of Technical University of Sofia for his significant contribution to the ZEBRA and EPROM emulator projects. Profs. Thorkild Larsen, Hans-Kurt Andersen, Christen Monberg and Olav Jensen of Technical University of Denmark gave us detailed information about data acquisition systems and industrial control. We are grateful to Evgeni Liharsky and Dr. Vladimir Alexiev for their hardware and software implementations at the Technical University of Sofia. Finally, we would like to thank B. J. Clark, Linda Schonberg and Lila Harris of Marcel Dekker, Inc. who provided excellent and professional support during the editorial and production phase.

Zdravko Karakehayov, zgk@computer.org

Knud Smed Christensen, ksc@iae.dtu.dk

Ole Winther, ow@iae.dtu.dk

CONTENTS

Chapter 1

BASIC CONCEPTS

The full understanding of microcontrollers and their applications demands knowledge of all aspects of digital systems, from logic gates to microcomputers. In this chapter we outline the basic features of logic gates and memory elements. We use these components to design combinational and sequential logic circuits. Next, we briefly discuss memories and microprocessors to be able to deal with microcomputers. In section 1.8 we explain the basic characteristics of the microcontrollers. Also, we introduce a flag's taxonomy and symbols which are in use throughout the book. Finally, we discuss embedded applications and the impact of the Internet expansion on future designs.

1.1 Logic gates

Logic gates are the essential building blocks of the digital systems. Figure 1.1 shows basic logic functions and the logic gate symbols. Furthermore, the correspondence input - output values are presented in truth tables. Finally, Figure 1.1 depicts the logic equations consistent with Boolean algebra laws.

For example, the first logic gate is an inverter. The inverter has an input A and an output Y. The next gate computes a little more complex function called AND. The output of the AND gate will be 1 when both inputs are 1. Watch out for a small detail: the AND symbol $\wedge$ could be omitted in the formula.

Naturally, our aim is to be prepared for implementation of any logic function. Having in mind that technology will dictate the most suitable function or functions we need to sort through the possibilities and find functionally complete sets. For instance, the set NOT, AND and OR can be used to form every possible Boolean function. The same applies for NAND or NOR functions.

Figure 1.2 shows a transfer characteristic for the inverter. The transfer characteristic presents the gate output voltage as a function of the input voltage. We can obtain the characteristic by connecting an adjustable voltage source to the gate input and using a voltmeter to measure pairs of input and output voltages.

Logic function	Logic gate symbol	Truth table	Equation
NOT	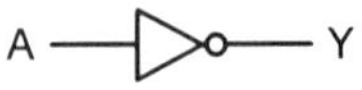 A —▷o— Y	A Y 0 1 1 0	$Y = \overline{A}$
AND	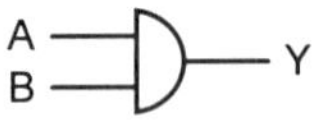 A, B —D— Y	B A Y 0 0 0 0 1 0 1 0 0 1 1 1	$Y = A \wedge B = AB$
NAND	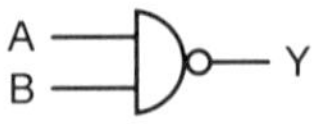 A, B —Do— Y	B A Y 0 0 1 0 1 1 1 0 1 1 1 0	$Y = \overline{A \wedge B} = \overline{AB}$
OR	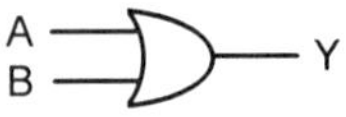 A, B —D— Y	B A Y 0 0 0 0 1 1 1 0 1 1 1 1	$Y = A \vee B$
NOR	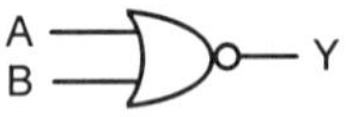 A, B —Do— Y	B A Y 0 0 1 0 1 0 1 0 0 1 1 0	$Y = \overline{A \vee B}$
XOR	 A, B —D— Y	B A Y 0 0 0 0 1 1 1 0 1 1 1 0	$Y = A \oplus B$
XNOR	 A, B —Do— Y	B A Y 0 0 1 0 1 0 1 0 0 1 1 1	$Y = \overline{A \oplus B}$

Figure 1.1 Logic functions and logic gates.

The curve displays all possible output voltages. The grid is 1 volt. The logic 1 (high) voltage set is separated from the logic 0 (low) set by a grey area. In a static state gates are not allowed to produce undefined (grey) output voltages. Inevitably, switching from one output level to another will cause the output voltage to pass the grey area.

Furthermore, Figure 1.2 indicates logic levels by the horizontal border lines of the grey area. There are two other logic levels which are relevant for the inputs and can be seen in Figure 1.2 as dashed lines. The differences between the corresponding logic levels are termed noise margins.

A major parameter which defines the ability of a logic gate to switch fast from one level to another is the delay. The delay is also related to a property called power consumption. Usually, any attempt to decrease the delay leads to increase of the power consumption and vice-versa.

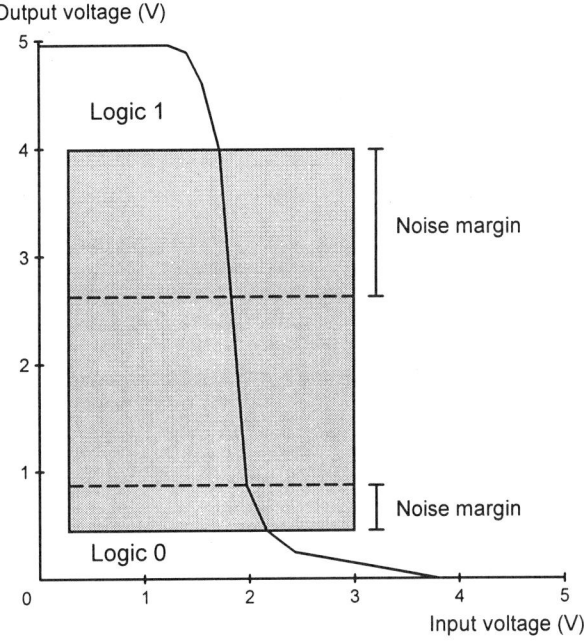

Figure 1.2 Logic levels and the transfer characteristic.

If the logic gates are viewed at a deeper level they can be broken down into two types. First, logic gates built from bipolar transistors are fast, but require relatively more power. Second, logic gates based on MOS (Metal-Oxide Semiconductor) transistors consume less power, however the delays are longer.

The power consumption is crucial in two ways. Obviously, we need to minimize the power used, especially for portable systems. Nevertheless, the power consumption affects the performance indirectly. The logic gates reside in Integrated Circuits (IC). It is imperative for the designers to be able to put as many transistors on a single chip as the application demands. Unfortunately, the ceiling of the heat dissipation limits the number of transistors and therefore the performance.

The density of the integrated circuits can be measured by the number of active elements (transistors) or gates. Figure 1.3 classifies the ICs according to the transistor count. It is certainly implied that the ranges are a general picture rather than strict frames.

Under the MOS technology branch, we distinguish between NMOS and CMOS integrated circuits. The NMOS ICs use n-channel transistors. The CMOS (Complementary MOS) gates are based on pairs of p-channel and n-channel transistors. The key advantage of the CMOS technology is that the circuit consumes practically no power in a static state. The CMOS technology dominates the microcontroller's market.

Beneath the CMOS branch of the taxonomy tree, we distinguish between static and dynamic logic gate circuits. The dynamic logic circuits depend on the charge of internal capacitors which must be refreshed. A certain amount of the microcontrollers are still produced as dynamic logic devices. As a result, the IC will not operate properly if the clock frequency is below the lower limit.

Classification		Transistors
SSI :	Small Scale Integration	< 100
MSI :	Medium Scale Integration	100 - 1000
LSI :	Large Scale Integration	1000 - 100 000
VLSI :	Very Large Scale Integration	100 000 - 1000 000
ULSI :	Ultra Large Scale Integration	> 1000 000

Figure 1.3 An integrated circuits classification.

In contrast, the static CMOS logic, typical for the new microcontrollers, is capable of saving power by decreasing the clock frequency when necessary.

1.2 Combinational logic circuits

Digital systems are built from numerous logic gates. It would be useful for us if we can isolate parts of the system which contain only Combinational Logic Circuits (CLC). The main property of combinational logic is that it has no closed loops or feedback. Breaking down the system into smaller pieces will help us to organize the design process and employ appropriate methods.

Figure 1.4 shows a truth table of an example combinational logic circuit. The circuit is called a decoder. In addition, Figure 1.4 includes the logic equations of this 2-input, 4-output circuit. There are four possible input combinations and each of them activates a single output.

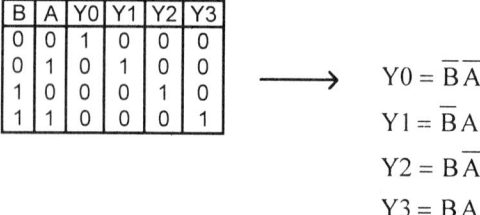

$$Y0 = \overline{B}\,\overline{A}$$

$$Y1 = \overline{B}\,A$$

$$Y2 = B\,\overline{A}$$

$$Y3 = B\,A$$

Figure 1.4 The decoder truth table and output equations.

Once we have prepared the logic equations, it is a simple task to draw the decoder combinational logic circuit (network) which can be seen in Figure 1.5. This example circuit consists of two levels of logic gates. The number of levels and the gate delay define the overall delay of the circuit.

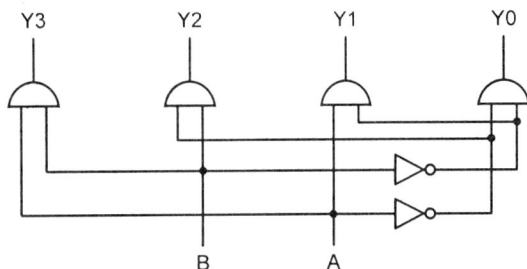

Figure 1.5 The decoder combinational logic network.

A typical application of the decoder is to select memory components according to the current requirements of the system. When a digital system accesses a location in the memory it generates a certain number of signals termed address. The address specifies which memory component is activated and which location in the component is the target. Using the decoder from Figure 1.5 we apply two address signals to the inputs A and B. The outputs Y0 through Y3 go to the select inputs of the memory integrated circuits.

It is certainly implied that actual designs will deal with mass-produced combinational logic blocks. Figure 1.6 presents a popular 3-line to 8-line decoder 74HCT138. Most obviously, the number of inputs is increased to 3 which brings the number of outputs to 8. In contrast to the previous decoder, the outputs are active low (glance the bubbles) which fits better to the commonly used active low select inputs. Finally, the decoder itself has three select inputs $\overline{E1}$, $\overline{E2}$ and E3. If the decoder is not selected all outputs are inactive (high) regardless of the inputs A, B and C. In order to select the 74HCT138, both inputs $\overline{E1}$ and $\overline{E2}$ must be tied to the ground and the input E3 tied to high.

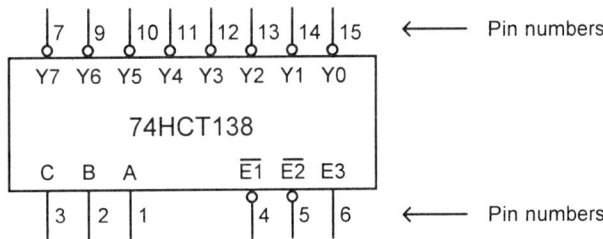

Figure 1.6 The 74HCT138 3-line to 8-line decoder.

You might be surprised when you see the logic symbol of this decoder in different catalogs. In some of them, the output's letters appear as we show them in Figure 1.6. In some others, the output values are negated in the decoder logic symbol ($\overline{Y0}$ through $\overline{Y7}$). What does it mean when the bubbles are still there? The fact is that an asserted output will be at low level.

A similar inconsistency in the graphical presentation applies for the select inputs $\overline{E1}$ and $\overline{E2}$. Again, the decoder is selected by pins #4 and #5 pulled down and pin #6 pulled up.

The 74HCT138 decoder is a representative of a high-speed CMOS family labelled HCT. The logic levels of this family are the same as the logic levels of the famous Transistor Transistor Logic (TTL).

1.3 Latches and flip-flops

Immediately adjacent to the logic gates are the latches and the flip-flops. The combinational logic circuits could result in useful building blocks, however they do not introduce new fundamental features. But, the latches and flip-flops do. They can be used to store logic values. This distinctive characteristic is achieved by feedback or special design techniques [Wolf 1994a].

There are many types of memory elements as far as the functionality is concerned. Figure 1.7 shows three of the most commonly used memory elements, presented by their state transition tables (functional tables).

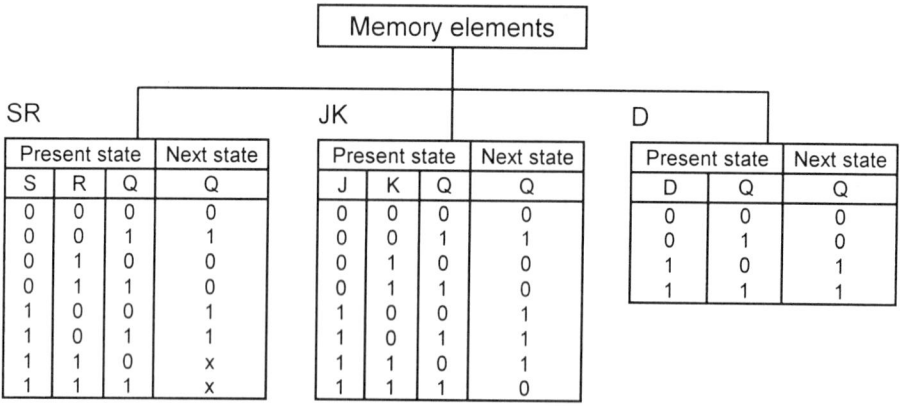

Figure 1.7 Memory elements and functional taxonomy.

• SR memory elements are set by a S input and reset by a R input. When both inputs are not asserted (low) the output Q is not changed. The last two transitions in the table lead to unstable states. Consequently, two active (high) inputs are not allowed.

• JK memory elements are similar to the SR counterparts. As you can see in the table, an input J plays the role of the S input and a K input substitutes for the R input. An advantage of the JK element is that it will be toggled (the state will be changed) if both inputs are asserted high.

• D memory elements have one input. They are also known as delay elements. The output Q copies the input value after some delay.

Furthermore, in Figure 1.8 we look at the memory elements from a different angle. In this taxonomy we focus on the timing parameters. At the highest level, we distinguish between asynchronous and synchronous memory elements.

• The asynchronous memory elements possess only data inputs. When the pattern of the data inputs is changed, the stored value is also changed and appears at the output after some delay. The asynchronous memory elements are termed latches. An example of an IC which contains four latches is 74LS279 [Texa 1989].

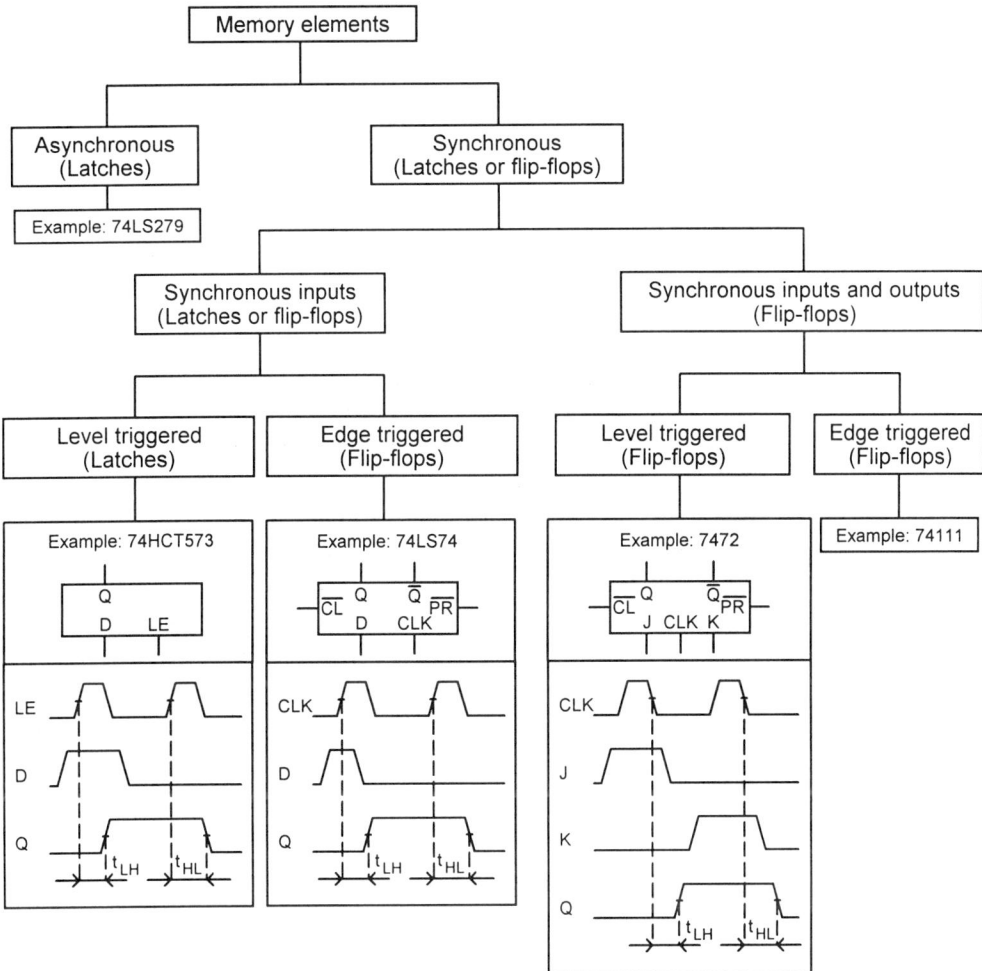

Figure 1.8 Memory elements and timing taxonomy.

• The synchronous memory elements have an extra input called clock input. By means of the clock input two important things can be controlled. First, the access to the memory element is allowed by the clock input. Second, under the synchronous inputs and outputs branch of the taxonomy both inputs and outputs are dependable on the clock input. In other words, we can choose the time when to activate the input circuitry and store the value and when to present the updated value to the output.

The synchronous inputs branch is further classified into level triggered latches and edge trigged flip-flops. These two types of memory elements differ significantly in the nature of the timing. We use two example ICs: 74HCT573 which contains eight D latches and 74LS74 which includes two D flip-flops [Phil 1994a, Texa 1989].

When the 74HCT573 clock input (Latch Enable) becomes high, the logic value applied to the input D is passed to the output Q after some delay. We can expect differences between low-to-high delay (t_{LH}) and high-to-low delay (t_{HL}). As long as the clock input LE is high, the output Q will follow the input D (the latch is transparent).

In contrast, the 74LS74 flip-flop is controlled by the low-to-high transition of the clock signal. Hence, the logic value on the input D must remain stable around the rising edge of the clock signal. The clock input of the 74LS74 is indicated by CLK. Since the flip-flop stores the input value around the low-to-high transition of the clock signal, the delays are measured as shown in Figure 1.8.

Using flip-flops with synchronous inputs and outputs, we are in position to control when the input value is stored and when the stored value is presented at the output. As a level triggered flip-flop example we have chosen 7472 [Texa 1989]. Memory elements such as the 7472 are also termed master-slave flip-flops, because they contain two latches. When the clock input (CLK) of the 7472 is high the master section stores the data. The slave section, which controls the flip-flop's outputs, still maintains the old value. When the clock input becomes low the inputs J and K are blocked and the stored value is moved to the slave section to appear at the output after some delay. In fact, the 7472 has three J inputs (J1, J2 and J3) and three K inputs (K1, K2 and K3). An AND logic function is implemented between the corresponding inputs.

Furthermore, the 74LS74 and 7472 flip-flops have an output $\overline{Q}$ which emits the logic complement of the Q output. Moreover, the mentioned flip-flops possess asynchronous inputs $\overline{CL}$ (Clear) and $\overline{PR}$ (Preset). The $\overline{CL}$ and $\overline{PR}$ inputs are active low and affect the stored value regardless of the state of the other inputs.

Finally, edge triggered master-slave flip-flops, such as 74111, accept the input data around the rising edge of the clock pulse. After this short period, the inputs may be altered while the clock is high without affecting the logic value stored in the master section. When the clock input moves from high to low, the value stored in the master latch is copied to the slave section and appears at the output. As far as the delay measurement is concerned, the timing diagram of the 74111 is identical to the one presented for the 7472.

1.4 Sequential logic circuits

A dramatic step forward in the design of digital systems was the concept of Sequential Logic Circuits (SLC). Other terms used for this model are Finite State Machines (FSM) or sequential machines.

We distinguish between two basic types of sequential machines: Mealy machines and Moore machines. Figure 1.9 compares both structures. The present state of the sequential circuit is kept in the memory elements. The next state is a function of the X inputs and the present state. The Z outputs are defined by the present state and the X inputs for the Mealy machine. The Moore machine outputs are a function of the present state. The states, for example if we use two flip-flops, are 00, 01, 10 and 11. The input combinational logic circuit (Input CLC) takes care of the sequence of the states to be consistent with the specification. The output logic (Output CLC) yields the desired output values.

The presence of feedback in the architecture dictates the use of flip-flops as memory elements. Latches are not capable of cutting off the loop. Therefore, the memory elements timing taxonomy (Figure 1.8) should be used when we design sequential machines. It is certainly implied that we can construct master-slave flip-flops from latches.

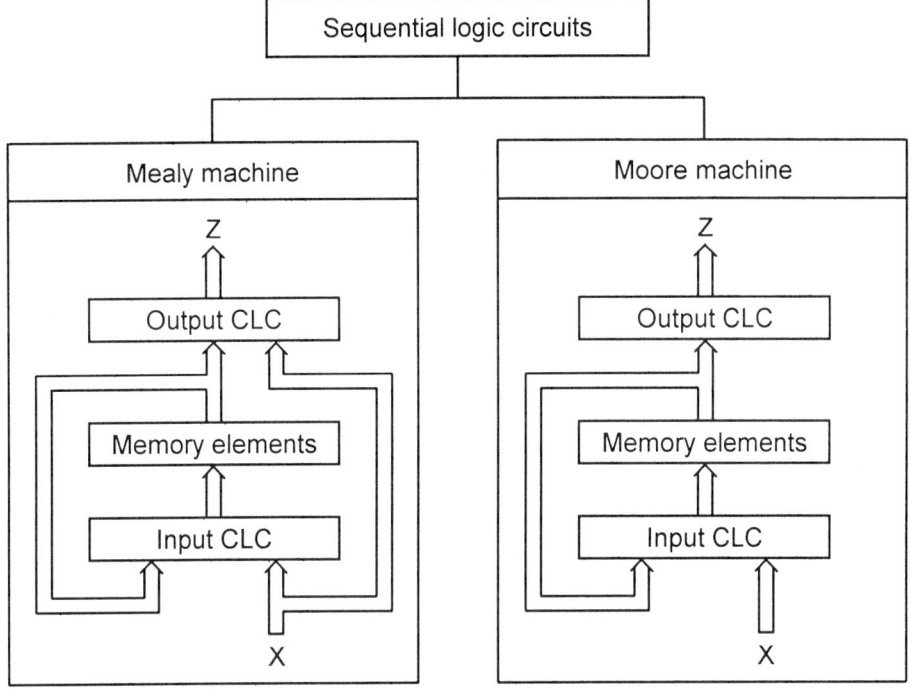

Figure 1.9 Sequential logic circuits and architectural taxonomy.

The Moore machine structure is a simplified version of the Mealy machine. It is often additionally reduced by discarding the block labelled Output CLC. In this case, the SLC outputs are taken from the block "Memory elements". The classical parameters such as speed, power consumption and price apply for the SLC in the same way they do for the logic gates.

As shown in Figure 1.10, the sequential machines can be broken down into asynchronous, synchronous and self-timed.

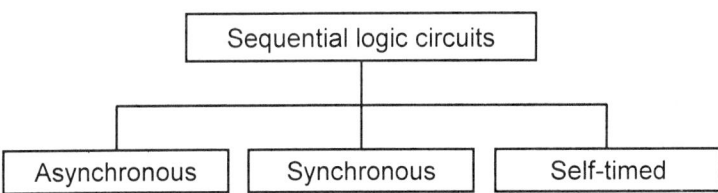

Figure 1.10 Sequential logic circuits and timing taxonomy.

The vast majority of sequential machines are synchronous. The synchronous approach is a good trade-off between speed, complexity and reliable design methods. Theoretically, asynchronous sequential machines could run faster in some cases, however the design is too complicated and the actual behavior might be unpredictable.

Figure 1.11 traces the design of a two-bit counter. We want to build the counter as a synchronous Moore machine. The specification of a FSM can be done either as a state transition table or a state transition graph [Wolf 1994a]. In this example we use a state transition table. Later on, in the EPROM emulator case study (Chapter 11), a FSM is described by a state transition graph. In fact, we combine the specification with the encoding phase in the state transition table. This is justified when the intended machine is simple and the encoding (state assignment) is obvious. A more sophisticated counter could have a start/stop control input which is the case with the microcontroller's embedded counters. For simplicity, we assume now that the sequential circuit counts all the time.

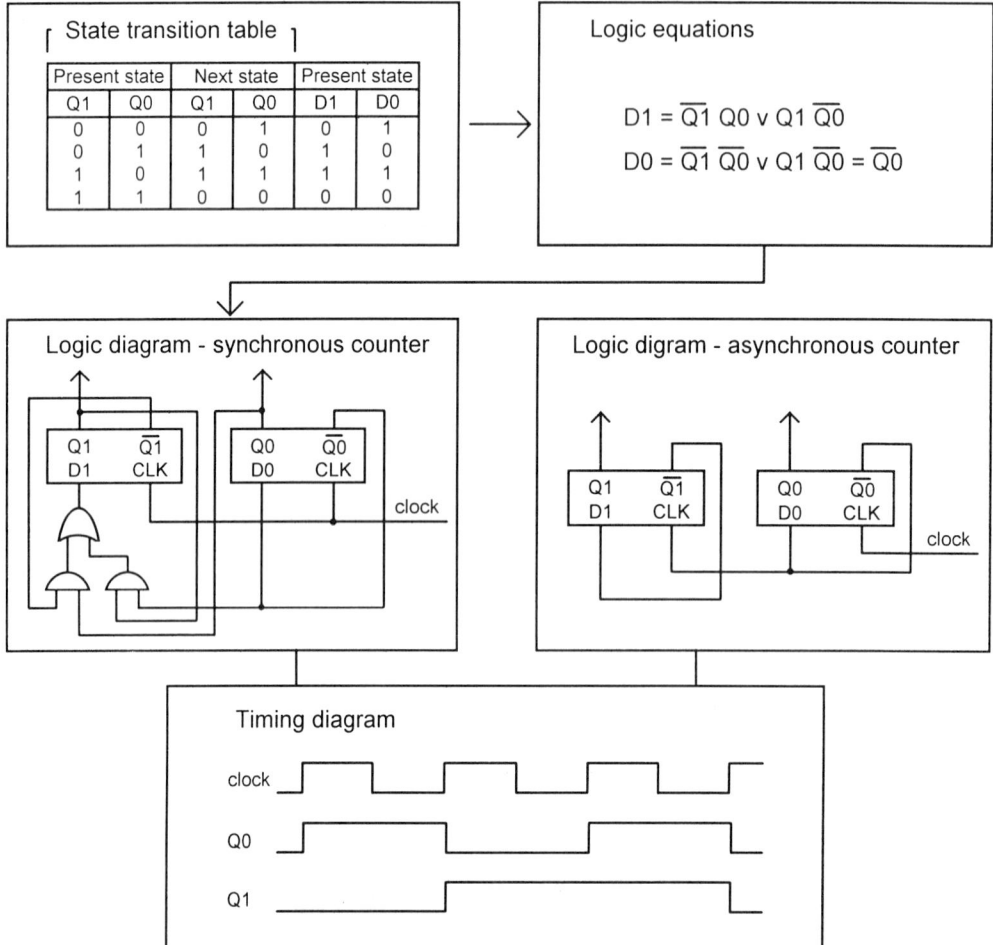

Figure 1.11 Two views of designing counters.

Furthermore, we have allocated D type, edge triggered flip-flops (74LS74) for the counter and attached two extra columns to the encoded state transition table. The columns contain the logic values which must be applied to the inputs of the flip-flops in the present state in order to ensure the desired transition to the next state.

Converting the truth tables for the inputs D1 and D0 into logic equations is done by covering all 1s from the corresponding function. Naturally, complex designs will require simplification methods to be applied manually or through Computer Aided Design (CAD) tools. Once we have the equations, it is routine work to draw the logic diagram of the synchronous counter.

In contrast to the formal synchronous counter design, we could use our intuitive sense to build an asynchronous counter, which is also presented in Figure 1.11. In principle, using heuristics to design even moderate scale sequential circuits is a risky job. On the other hand, we obtain a simpler solution with our asynchronous counter. The downside of the asynchronous counter is that the flip-flops are not switched simultaneously. Increasing the number of flip-flops will lead to proportional delay of the most significant bits and the counter behavior expressed as a timing diagram will diverge from the one laid out in Figure 1.11.

Apart from the other machines, the self-timed circuit's functionality doesn't depend of the propagation delays [Mull 1959, Stau 1994, Kish 1994]. This distinctive feature is achieved by introducing special circuits which indicate the completion of the transient process. The main benefit of this approach is that the sequential machine operation is adapted to the current timing parameters of the circuits. For example, if the power supply or temperature variations increase the propagation delays of the circuits, the machine will run slower, but utilizing completely the timing properties again. Even more, when for instance, a gate sticks its output to a certain logic level, the sequential machine will stop and manifest its loyalty to the user. Inherently, the self-timed designs possess a big potential for fault tolerance.

A drawback of the self-timed sequential circuits is the overhead which will vary from application to application. Another shortcoming is that the design methods are more complicated.

What does the self-timed design mean for the microcontrollers? Surprisingly, the main goal might be reduction of the power consumption rather than increasing the performance. Modern microcontrollers have power management modes. If a microcontroller is faster and completes execution of a certain program in a shorter time, it could stay longer in a power reduction mode and cut down the total energy consumption.

1.5 Memories

The memory elements built in the sequential machines store the current state. When there is a need for computer programs or data to be accommodated, the requirements for memory capacity increase drastically. High-density memories are produced by special technologies. In terms of embedded systems design, the major memory parameters are price, power consumption, speed and the capability to store data when the power is switched off.

Figure 1.12 presents a memory organization and the corresponding memory component. The memory can be accessed by quanta of data called bytes. A byte consists of 8 bits. A bit can be either low (logic 0) or high (logic 1). We lay out the addresses as hexadecimal numbers. The memory capacity for this example is 64K byte (K=1024).

The IC interface can be broken down into address, data and control. The example memory component has 16 address inputs (specified by a slash and 16) and 8 data inputs/outputs. In an attempt to save pins, the data interface is based on bidirectional lines. Naturally, if the functionality is limited only to read operations, the data interface is simplified to pure outputs.

The control inputs are used to select a specific memory component or to define the data flow direction (read or write).

Figure 1.13 shows a classification which is relevant for both memory components and on-chip microcontroller memory. The memories included differ significantly as far as the technology is concerned.

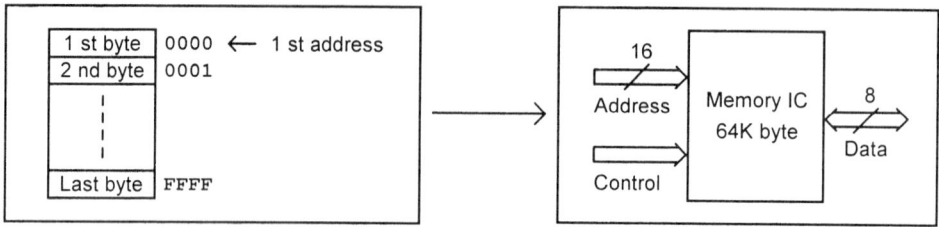

Figure 1.12 A memory organization and the corresponding memory component.

We distinguish between two basic types of memory components [Wolf 1994a, Herb 1996, SGS 1994] :

• Read Write Memory (RWM), which is better known as Random Access Memory (RAM), is used to store temporary data. The information can be written to and read from, however it will be lost when the power is switched off.

• Read Only Memory (ROM) is used to store programs. As the name indicates, the code can be read, but not written. ROM memories are non-volatile devices. Thus, the pattern is available immediately when the power is applied.

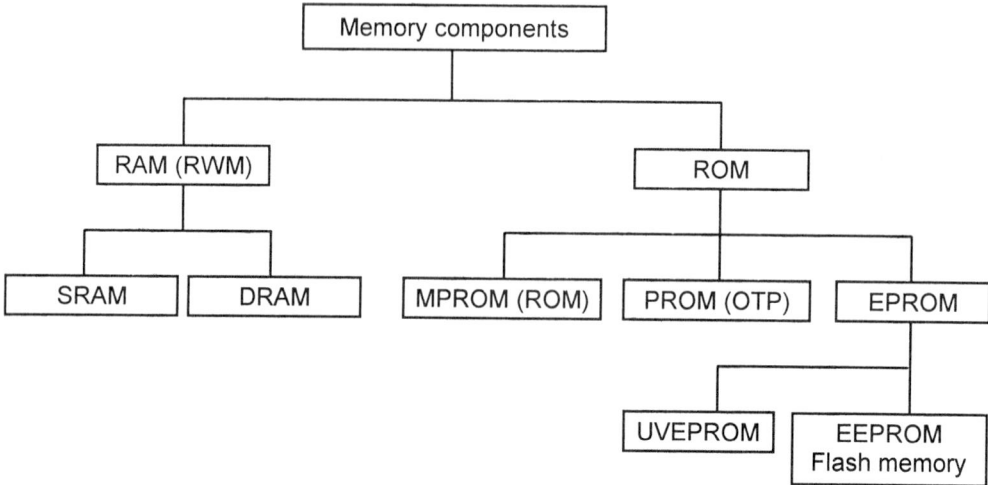

Figure 1.13 A memory classification.

Under the RAM branch of the classification there are two types of RAMs: static (SRAM) and dynamic (DRAM). The SRAM is faster, however the DRAM has a smaller cell layout and is more suitable for bigger sizes. Usually, the microcontrollers expand their memory by SRAM components.

The ROM components are further classified into three essential types:

Mask Programmable ROM (MPROM), which is frequently referred to as ROM, is a factory programmable component. The ROM may be a cost effective solution only if ordered in high volumes.

The Programmable ROM (PROM), as you might expect, is programmed by the user. The PROM contains fuses that once blown can not be recovered. An equivalent term OTP (One Time Programmable) memory is very popular for the microcontroller's internal PROM.

Taking into account the requirements of the design process, we'll see a desperate need for Erasable PROM (EPROM) components. At this level of the taxonomy we distinguish between UVEPROM (erasable by ultraviolet light) and EEPROM (electrically erasable PROM). The UVEPROM devices are similar to the OTP memories, but they have a transparent window in the package for exposure.

The EEPROM is erased by a special electrical mode and there is no need for the device to be exposed with ultraviolet light, which takes a long time. The EEPROMs can be further divided into out-of-circuit programmable and in-circuit programmable (flash memories). Currently, the flash memory technology can offer higher density than other EEPROM processes. Flash memory is fast, reliable and suitable for a large spectrum of applications.

1.6 Microprocessors

The next step toward complex building blocks is a quantum leap. We are discussing a very sophisticated unit termed a microprocessor. As the name implies, it is a solid-state Central Processing Unit (CPU).

The first general purpose microprocessor was introduced by Intel in 1971. The Intel 4004, which is its manufacture's designation, could process four bits. The next year Intel released the first 8-bit microprocessor, the 8008.

Figure 1.14 shows the basic structure of a microprocessor. In essence, the CPU would be workable only if linked to memory. The primary function of a CPU is to execute instructions. The instruction codes should be stored in the memory.

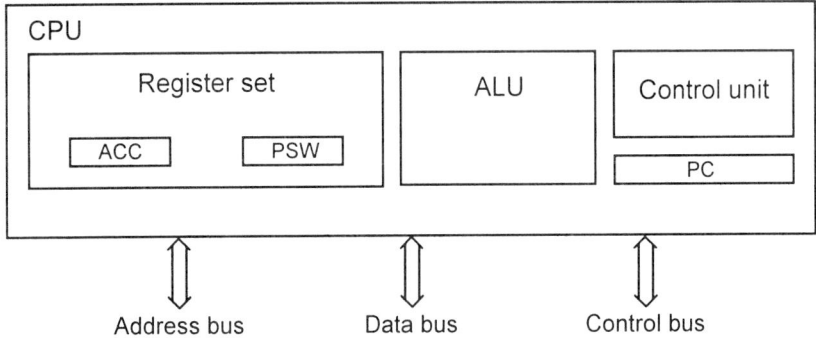

Figure 1.14 The basic structure of a microprocessor.

Similarly to the memory component's interface presented in Figure 1.12, the microprocessor possesses address, data and control lines. In addition, peripheral devices could be linked to the microprocessor via the bus.

The major blocks which are implemented in the microprocessor are a register set, an Arithmetic Logic Unit (ALU) and a control unit. The register set includes program accessible registers such as an accumulator (ACC), Program Status Word register (PSW) and several others. The ALU performs the arithmetic and logical (Boolean) operations. The control unit insures the proper execution and sequencing of instructions. A register called Program Counter (PC) is used to address the instructions in the memory.

The CPU fetches an instruction from the memory, decodes it and executes the specified operation. At the same time, the program counter PC is incremented in order to reach the address of the first byte of the next instruction. The next instruction is then fetched and the execution continues.

The accumulator is a destination register for a big group of operations, especially transfer, arithmetic and logical instructions. Likewise, the accumulator is frequently used as a source register.

Furthermore, the microprocessor modifies specific bits in the PSW register to indicate, for example, carry or overflow. The user could take this opportunity to organize conditional jumps in the program.

The control section of the microprocessor's interface is more complicated compared with the memory control lines. For instance, the memory could be organized in more than one address space which would require extra control signals.

So far, we have assumed that the microprocessor uses only one register to address the memory. This method is termed linear addressing. Figure 1.15 indicates two other approaches. The key point is how to extend the address range of early microprocessors in the modern devices. While using the linear method we simply address the memory by a bigger register. Applying paging or segmentation involves two registers.

Paging implies that the total amount of memory is broken down into equal pages. The memory is addressed by two registers. A page register specifies the page and a normal address register works within the page. The designers of new microprocessors could use paging to implement bigger address spaces. The users might apply paging to expand the memory space of the available computers. Furthermore, paging the memory could be an alternative used to make the code more efficient.

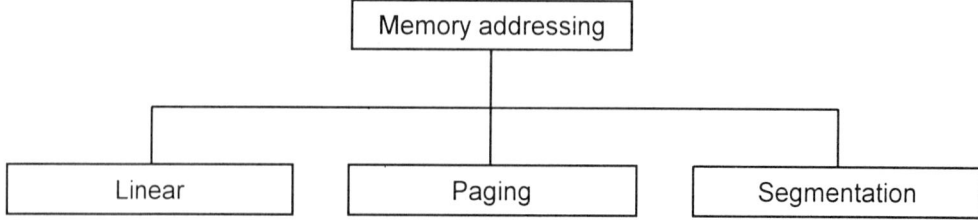

Figure 1.15 Methods for memory addressing.

Finally, segmentation is a method similar to paging. Again, there are two registers which form the address. However, when calculating the address the registers added are partly overlapped. For example, two 16-bit registers generate a 20-bit address. The segment register is shifted four bits to the left before addition.

The performance capability of a microprocessor is closely related to the size of the CPU. Surely no one would dispute that 16-bit microprocessors are faster than the 8-bit counterparts and the 32-bit machines gain additional performance through the larger CPU. Nevertheless, there have been other factors which shape the performance of the microcomputers. For example, the microprocessor instruction set plays a significant role in that regard. There are two basic concepts in the design of microprocessors: Complex Instruction Set Computers (CISC) and Reduced Instruction Set Computers (RISC). As discussed earlier, microprocessors read instruction code from the memory and the total performance depends on how fast the processor is and how fast the memory is. Naturally, if the microprocessor speed is higher, its burden must be heavier. In other words, the processor should execute more complex instructions which will allow less frequent access to the memory. That was the motivation behind the CISC processors developed in the late '60s. Later on, in light of the reduced difference in speed between microprocessors and memories, the RISC processors were designed to execute simple, basic instructions. Regardless of more frequent access to the memory, the performance with a RISC processor was increased two to four times. The simplified architecture brings an additional benefit: the occupied chip area is smaller and extra functions can be implemented.

The real performance will vary from application to application. A rough estimation of the performance is the number of MIPS (Millions Instructions Per Second). However, a program never uses only one instruction. A more precise approach is to test the processor by a program or a set of routines designed especially for this purpose. Unfortunately, the benchmark test will produce reliable results if it is closely related to the specific application.

1.7 Microcomputers

To this point, we have described the basic building blocks of digital computers. Now, we will interface them in order to construct a simple microcomputer. If there is a need we will introduce new circuits to outline a typical system.

Figure 1.16 shows basic microcomputer architecture. There can be no doubt that we need a microprocessor, a ROM and a clock. The microprocessor executes the instructions. The ROM contains the instruction codes. The clock (oscillator) controls the timing. A real system would use a certain amount of RAM locations and a Peripheral Input Output (PIO) adapter. The PIO links the microprocessor and peripheral devices such as keyboards and displays. All blocks are attached to three common buses. Two motivations underlie the bus architecture. First, the number of the microprocessor's pins is kept down. In other words, the microprocessor is connected to the bus rather than to each component individually. Second, the microcomputer is easily expandable. Additional ICs could be linked to the bus according to requirements.

Logically, the microprocessor controls the address bus. The data bus lines are bidirectional. The microprocessor can read data from a memory component (the PIO's registers might be viewed as a part of the memory), or it can send data out to the memory. Memory components such as RAM or ROM take over the data bus when the microprocessor reads. Therefore, the output buffers of the devices must be somewhat different. In addition to the two logic states (high output level and low output level), they possess an extra state termed OFF state (high impedance state). When the three-state buffer is in the OFF state, you could think that the IC and the bus are no longer joined by any physical bond. The microcomputer shown in Figure 1.16 will work properly if only the devices attached to the bus obey the following restriction: only one IC can open its output buffers to drive the bus at a time. The situation when two or more devices activate their outputs is named bus contention. Overriding the bus discipline by bus contention will result in significant currents where two outputs try to establish different levels on a line of the bus.

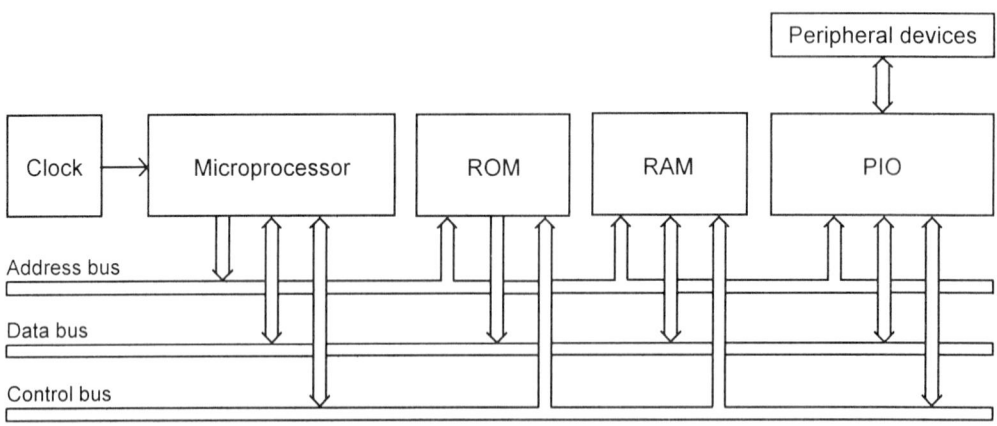

Figure 1.16 Basic microcomputer architecture.

Peripheral devices can influence the system in two ways. The microcomputer can read signals through the PIO. High priority peripheral outputs might interrupt the microprocessor (directly or via the PIO) and activate appropriate subroutines.

Figure 1.17 presents fundamental design philosophies for small scale computers. We distinguish between bus oriented computers, single board computers and mixed designs.

Bus oriented computers are distributed over a few separate boards. The boards are plugged into connectors on an extra board (the bus board) which links them together. The key advantage of this approach is adaptability. For example, if you need more memory, you could just plug in an additional board. Furthermore, such systems are easy to test and maintain.

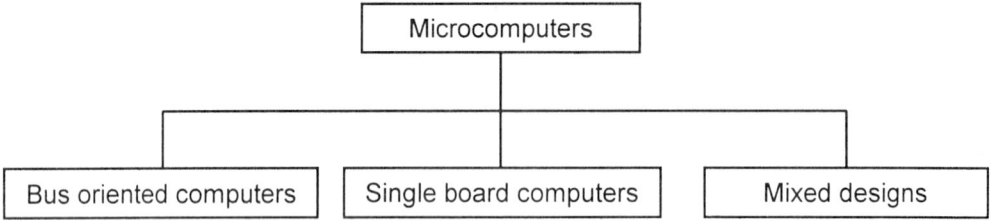

Figure 1.17 Design taxonomy of microcomputers.

In contrast, single board computers are more attractive for the manufacturer from an economical point of view. Another advantage of the single board solution is reliability. Though hardly perfect, this design approach works well enough. We will discuss the complete design of a single board computer in Chapter 4.

In addition, Figure 1.17 indicates a mixed design approach that compromises between the first two design philosophies [Rosc 1994]. Thus, a board, called a system board (or motherboard) contains the essential computer core and connectors for further expansion. Mixed designs have been widely used in the world of Personal Computers (PC).

Along with personal computers, Figure 1.18 shows two other types of computers which are relevant to this text.

Programmable Logic Controllers (PLC) are used for industrial control. Since they interface a vast number of input and output devices, they are designed as bus oriented computers. Other motivations for the bus oriented design of the PLCs are the requirements for adaptability and convenience to maintain.

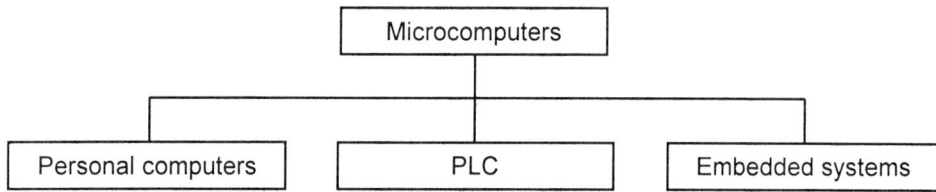

Figure 1.18 Functional taxonomy of microcomputers.

Finally, a branch of the taxonomy tree called embedded systems presents computers which perform a specific function, usually as a part of a bigger system. Small scale, control dominated embedded systems are often designed as single board computers.

Embedded computers are, for example, the TV-set control system, the engine control unit in a car, the laser printer embedded computer and so on. Compared with PLCs, the embedded systems are more specialized both in hardware and software.

The microcomputer software is usually not confined to an application program. As is frequently the case, a program termed an operating system (OS) is placed between the hardware and the application program. There are a few substantial benefits associated with this concept. First of all, the operating system provides standard access to the microcomputer's hardware. This advantage is vital for PCs where application programs developed from different vendors need to work with numerous peripheral devices. In the field of embedded systems, there are also sophisticated operating systems, but in the typical case, they are limited in size or even absent. Indeed, if a single application program utilizes hardware which is never changed, an operating system would be unnecessary. Typically, predictability is a main requirement and in this case we use the term real-time operating system (RTOS).

1.8 Microcontrollers

Discussing the design taxonomy in Figure 1.17 we could add an extra branch: single chip microcomputers. Integrated circuit manufacturing made possible the implementation of complex systems into a single chip. Essentially, a single chip solution for the architecture shown in Figure 1.16 leads to significant advantages. Most obvious, reliability, power consumption, weight and size are improved. A subtle effect of the integration is the smaller parasitic capacitances and therefore the higher speed. Furthermore, the single chip design may be a cheaper solution.

An alternative for the term single chip microcomputer is microcontroller. Both names are considered to be equivalent, even though microcontroller could be misleading to some extent. It is also used for embedded microcomputers based on general purpose microprocessors. Recapitulating briefly, the terms single chip microcomputer and microcontroller are used synonymously in this text.

The vast majority of 8-bit microcontrollers are CISC machines. There are, also, 8-bit microcontrollers based on high speed RISC architecture and two of them can be seen in Appendix A.

The microcontroller's hardware can be efficiently utilized if only combined with appropriate software. Figure 1.19 shows a classification of programming languages. A language, probably hidden for many users, is termed machine language. This is the only form required by the microcontroller. Each program written in other languages must be translated into machine language. Being a sequence of binary codes, the machine language is rarely used for creating software.

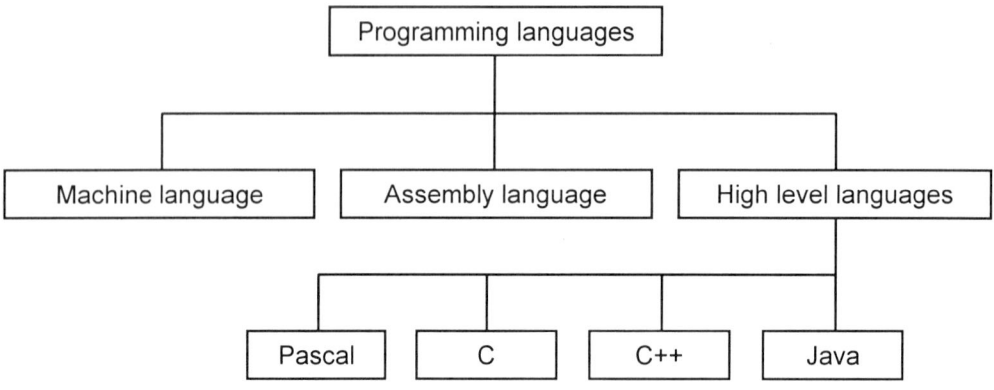

Figure 1.19 Programming languages.

However, replacing the machine code with a few letter long mnemonics substantially improves the situation. The language is called assembly language. The assembly language possesses the same functional capabilities over the microcontroller hardware compared with the machine language. Using assembly language we are in position to design the most efficient programs. It is certainly implied that assembly language programs must be translated into machine language. The process is automated by a program called assembler. Even though the assembly language has the biggest potential for writing useful programs in terms of speed and required memory, the designing process itself might not be very efficient.

Programming in high level languages will help us to keep the software design under control, especially for big programs. The high level language programs consist of statements which are a step forward to a natural language. Software written in a high level language, such as Pascal, C or C++, must be translated into a machine language by means of a program termed compiler. The resulting machine code produced by a compiler is decidedly slower and requires more memory.

The embedded control applications are characterized by high production volumes. Consequently, the component cost is vital for success. In spite of the primary importance of the price, there are several other factors which must be taken into account when a microcontroller is selected.

Naturally, the application will demand a certain degree of performance which should be met. Thus, the price/performance ratio, measured in MIPS/dollar becomes a dominant issue. Two key points are relevant in that regard. First, a parallel embedded system built from two or more microcontrollers could be a cheaper solution compared to a system based on a single, but more expensive processor. Second, it might be misleading if the performance as a parameter is replaced by bus width. There are some other features which contribute to the absolute performance.

First of all, modern microcontrollers are not confined to the architecture presented in Figure 1.16. It is frequently the case, peripheral devices, such as timer/counters, serial ports, analog-to-digital converters (ADC) and digital-to-analog converters (DAC) are integrated on the chip. Those peripherals run concurrently with the CPU. The parallel running hardware blocks can interrupt the CPU and activate appropriate subroutines. Logically, great importance is attached to the interaction between the CPU and hardware blocks. The communication is organized by flags. The CPU modifies flags in order to adjust a hardware block to the current requirements of the system. For example, the program sets a flag in the interrupt block to make an input sensitive to the rising edge of the signal. In addition, the CPU reads flags from the hardware subsystems to obtain status data. For instance, the program can check a flag from the serial port which indicates a received character. As you might expect, the flags are usually grouped in bit addressable registers.

Nevertheless, the embedded control specifications are frequently given as individual algorithms for each output, rather than byte or word oriented procedures which are typical for the classical data processing. The designers of microcontrollers made the right step forward providing the devices with bit-addressable memory. The bit-addressable memory accommodates input and output image tables and bit variables. In addition, bit-addressable memory can store user defined flags which will help the software modules to interact.

Both user defined flags and special function flags (usually placed in bit-addressable registers) form the basis of the taxonomy in Figure 1.20. In an attempt to avoid errors and to make the dialog more efficient, we introduce graphical symbols for the special function flags. The symbols contain no ambiguity and support our graphical approach.

The table in Figure 1.20 provides a detailed description of the flags. The generalized flag placed in the lower left corner of the picture accompanies every figure where flags are involved. Theoretically, there are nine types of special function flags. We have chosen seven examples from the 8051 and 83C552 microcontrollers for illustration. For instance, the flag RI (Receive Interrupt) is set automatically by hardware. The flag must be reset by software. The flag RI can also be set from the program. The flag RI makes possible the interaction between the CPU and the serial port. Likewise, the flag ADCI allows an embedded ADC to inform the program that the conversion is completed. The flag is set by hardware and must be reset by software. Special function flags which are set and reset only by software, such as the flag TR1, are usually shown without the small circles.

Furthermore, releasing new versions of microcontrollers that run at higher clock speeds boosts the performance. However, the temptation to squeeze speed out of the microcontroller might be an unacceptable solution. In contrast with the intuitive sense, some applications will demand lower oscillator frequency due to the limitations for power consumption and radio-frequency interference.

In terms of power consumption, the microcontrollers usually possess three modes: normal operation, idle (standby) and power down (clock-off). In idle mode the CPU is blocked, but the embedded peripherals are still running. In the event of an external or internal interrupt, the idle mode is terminated. The power down mode further declines the consumption, however the microcontroller has to be restarted. Since the power consumption and the performance are closely related, a MIPS/watt ratio will help us to compare microcontrollers.

In many cases, compatibility influences the system-component allocation. If the development process is based on used microcontrollers, a big part of the software could be easily modified for the new requirements. High-level languages, when applied, also alleviate compatibility problems and help designers to keep programming under control. The C language dominates the embedded market and the C compilers memory requirements lead to on-chip memory enhancements in the new versions.

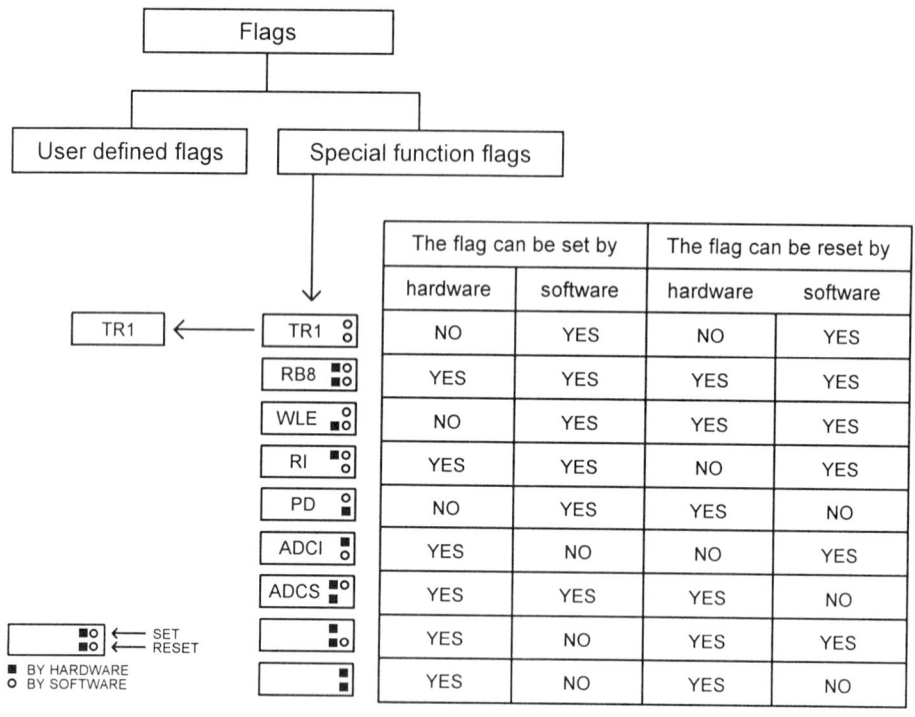

	The flag can be set by		The flag can be reset by	
	hardware	software	hardware	software
TR1	NO	YES	NO	YES
RB8	YES	YES	YES	YES
WLE	NO	YES	YES	YES
RI	YES	YES	NO	YES
PD	NO	YES	YES	NO
ADCI	YES	NO	NO	YES
ADCS	YES	YES	YES	NO
	YES	NO	YES	YES
	YES	NO	YES	NO

Figure 1.20 The flags taxonomy and symbols.

Moreover, the availability and the price of development tools should be taken into account when a microcontroller is selected. Also, considering the possible development tools in the early stages of the project would be the prudent course of action.

Ultimately, the producer of embedded computers needs a guarantee about the future availability of the components. It should be pointed out that second-sourcing might be a problem for some new, high-performance devices.

While the microcontrollers incorporate peripherals for control dominated applications, there is another type of processors termed Digital Signal Processors (DSPs) which have a CPU customized for data-intensive operations, such as digital filtering. In many cases, DSPs have analog-to-digital and digital-to-analog converters as embedded peripherals. There are applications, such as cellular phones, which require a microcontroller and a DSP. Integrating both processors on a single chip seems to be the most logical solution for this case.

Currently, the 8-bit microcontrollers dominate the market. Most experts are predicting that the share of 8-bit devices will be 60 percent of the microcontroller market in 2000.

1.9 Embedded systems

The embedded systems or embedded computers are difficult to introduce by a simple definition. Rather than imposing frames, we lay out typical examples in Figure 1.21.

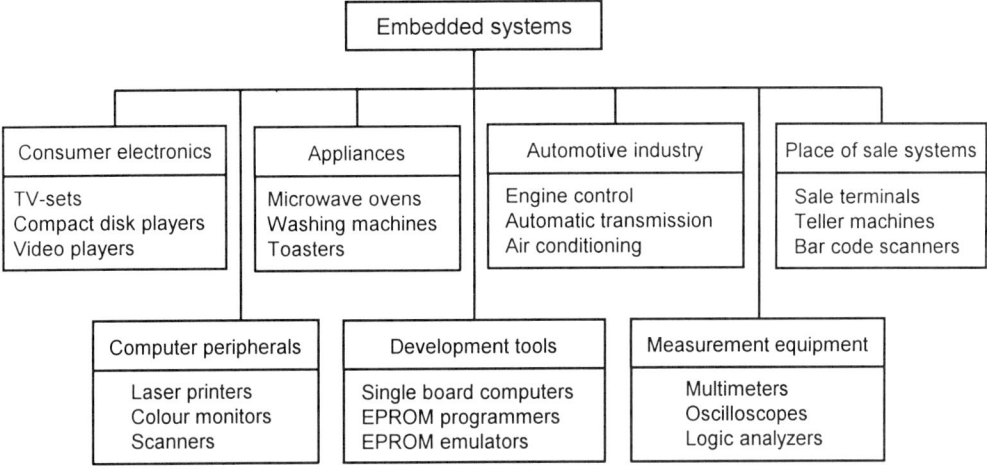

Figure 1.21 Typical embedded applications.

The most important parameters of the embedded systems are price, power consumption, fault-tolerance and user interface. The embedded computers improve the parameters of the devices under control. Furthermore, they provide friendly interface and advanced features.

Embedded systems usually include sensors, a processor and actuators. As is frequently the case, the computing power is brought by microcontrollers, as shown in Figure 1.22. In addition, microcontrollers provide useful peripherals.

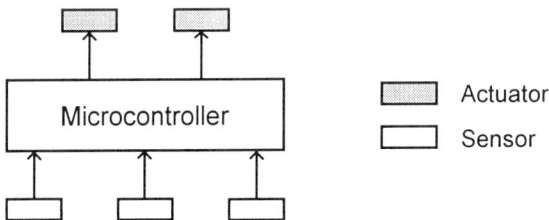

Figure 1.22 Embedded system architecture.

The embedded computer reads input data from sensors, applies a control algorithm and provides output data via the actuators. The output values influence the control object and the effects can be observed through the sensors, thus closing the loop. The sensors can be broken down into two groups: sensors with digital output and sensors with analog output. For example, a digital sensor can indicate the end of the tape in a video recorder. Due to the fact that we live in an analog world, the embedded computers must be adapted to analog signal quantities. The link is done in two steps. First, sensors convert nonelectric physical variables, such as temperature, light, pressure, weight or humidity into voltage. Second, ADCs supply the embedded computer with digital values which correspond to the measured analog voltages.

Typically, the embedded systems are designed and produced under tight deadlines. Moreover, the design teams must cover several disciplines, such as VLSI design, software engineering, microcontrollers, distributed systems and real-time systems.

In the near future, all phases of the embedded computers design process will be supported by matured system-design tools. Up to now, the research has been focusing on three areas:

• First, the desired functionality is implemented by a processor or microcontroller and an Application Specific Integrated Circuit (ASIC). The ASIC is used to improve the timing parameters of the embedded system [Wolf 1994b].

• Second, the microcontroller is produced by a modular method. The process begins with a CPU core and continues with integration of all on-chip blocks like memory and peripherals which cover the requirements of the specific application. This approach is beneficial for the manufacturer due to the fact that the microcontroller is produced as a Standard Product (SP). At the same time, the device is an ASIC from the user point of view.

• Finally, distributed embedded systems, built from several microcontrollers working concurrently, is a promising approach. The objective is improved timing properties and in some cases, cost-effective solutions. In addition, distributed systems are easily expandable, adjustable to changing conditions and convenient to maintain.

1.10 The Internet

The birthday of the Internet is considered January 1, 1983. This date marks the change of the ARPANET to Transmission Control Protocol/Internet Protocol (TCP/IP). The network ARPANET was developed in 1969 to support research and educational organizations. In 1992 the World Wide Web (WWW or Web) was invented as an implementation of hypertext. The basic idea was to organize access to randomly associated documents by linking them together.

Every computer connected to the Internet is termed a host. While some computers have a permanent link to the network, others establish temporary dial-up connections. As far as the information flow is concerned, we distinguish between client programs and server programs. Client programs request information or services. Server programs provide information or services. Each host has a unique 32-bit address named IP address. However, a common practice is a proxy server to provide access to several other computers using a single IP address.

As an information source, the Internet unrolls a succession of magnificent possibilities. We can open Web pages using a client termed a Web browser. Currently, browsers such as Microsoft Internet Explorer and Netscape Navigator are in wide use. Web pages are based on HyperText Markup Language (HTML). The hallmark of the HTML is the ability to link a word or phrase to other documents by means of tags. Resources are accessed through their addresses. The addresses are alternatively labelled Uniform Resource Location (URLs). The interaction between a browser and a Web server is consistent with a protocol named HyperText Transport Protocol (HTTP).

Also, there is another type of file that is stored on Web servers. Portable Document Format (PDF) files are a useful option for the exchange of technical information, such as data sheets. When a PDF file is linked to a HTML file, you can view, copy and print it. A very popular program for this purpose is Adobe Acrobat Reader. If you don't know where to find the information you need, a search tool could help you. Figure 1.23 shows the names of four widespread search engines and their URLs.

In addition to the possibility to provide information, the Internet can be used to link and control embedded systems. Embedded computers designed to be network accessible are referred to as Internet appliances. Inevitably, the Internet embedded systems must include additional hardware and software subsystems. Moreover, the demands for some basic resources, such as memory, are increased as well. A good trade-off, especially for home appliances, is to implement the connection to the Internet through a proxy server. The proxy server would run TCP/IP and route traffic to embedded systems over a serial or wireless link. Currently, there are Web servers designed as single board computers. Rather than disk storage they have a certain amount of flash memory.

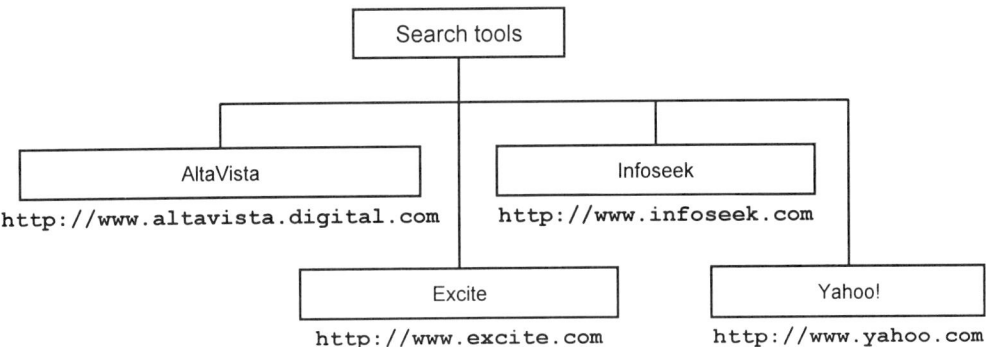

Figure 1.23 Search tools and their URLs.

The design of Internet appliances attracts the attention of an increasing number of companies. Many solutions are based on Java technology. Java is an object-oriented programming language developed at Sun Microsystems, Inc. Interestingly, the original Java team developed software for consumer electronics and started the design of a new language, more suitable for embedded applications. As a result they created Java, a small and reliable language. Subsequently, Java matured as an ideal programming language for Internet applications. Java programs are compiled to Java virtual machine instructions labelled bytecode. The stack-based bytecode instruction set is platform independent. Since Java has no pointer architecture, no direct access to memory, security is an inherent advantage. Once downloaded to a host, the bytecode is converted into native machine language. The conversion process can be accomplished by interpreters/compilers called Java Virtual Machine (JVM). However, even the simplest interpreter could be rather complex for many embedded applications. As always in the computer realm, when software imposes limitations, it is replaced by hardware. Java processors were developed to execute directly the bytecode. After gaining recognition for Internet programming, the Java technology is back in the embedded system domain, thus closing the loop.

In essence, the image of an intelligent embedded system is associated with a device which incorporates a Web server and therefore can be accessed through a Web browser. Practically, a smart device can send an e-mail to order supplies in case of impending shortage. From the other end, manufacturers could download firmware for upgrades when necessary. The existence of more and more Internet appliances will add a certain degree of machine-to-machine communications to the Web's primary person-to-person links.

1.11 References

Robert W. Atherton, "Moving Java to the factory", *IEEE Spectrum*, December 1998, pp. 18-23.

Fredrick M. Cady, *Microcontrollers and Microcomputers*, Oxford University Press, 1997.

Douglas V. Hall, *Microprocessors and Interfacing*, McGraw-Hill, 1992.

Steve Heath, *Embedded Systems Design*, Newnes, 1997.

L. J. Herbst, *Integrated Circuit Engineering*, Oxford University Press, 1996.

Arthur van Hoff, Sami Shaio and Orca Starbuck, *Hooked on Java*™, Addison-Wesley, 1996.

Barry Holmes, *Programming with Java*, Jones and Bartlett , 1998.

Jerry Honeycutt et al., *Using the Internet*, Que, 1998.

Randy H. Katz, *Contemporary Logic Design*, Benjamin/Cummings, 1994.

Michael Kishinevsky, Alex Kondratyev, Alexander Taubin and Victor Varshavsky, *Concurrent Hardware*, John Wiley & Sons, 1994.

Harlan McGhan and Mike O'Connor, "PicoJava: a direct execution engine for Java bytecode", *Computer*, October 1998, pp. 22-30.

Christopher L. Morgan and Mitchell Waite, *8086/8088 16-Bit Microprocessor Primer*, BYTE/McGraw-Hill, 1982.

David E. Muller and W. S. Bartky, A theory of asynchronous circuits. *Proceedings of an International Symposium on the Theory of Switching*, pp. 204-243, Harvard University Press, 1959.

Bill Peisel, "Designing the next step in Internet appliances", *Electronic Design*, March 23, 1998, pp. 50-56.

Philips Semiconductors, *High-speed CMOS Logic family*, *Data Handbook IC06*, 1994a.

David J. Preston, "Internet protocols migrate to silicon for networking devices", *Electronic Design*, April 14, 1997, pp. 87-94.

Winn L. Rosch, *Hardware Bible*, Sams Publishing, 1994.

Kenneth H. Rosen, *Discrete Mathematics and its Applications*, McGraw-Hill, 1995.

Manfred Schlett, "Trends in embedded-microprocessor design", *Computer*, August 1998, pp. 44-49.

SGS-Thomson Microelectronics, *Memory products*, 1994.

Kang G. Shin and Parameswaran Ramanathan, "Real-time computing: a new discipline of computer science and engineering", *Proc. IEEE*, vol. 82, No. 1, January 1994, pp. 6-23.

Jorgen Staunstrup, *A Formal Approach to Hardware Design*, Kluwer, 1994.

Texas Instruments, *The TTL Data Book*, Volume 1, 1989.

John F. Wakerly, *Digital Design, Principles and Practices*, Prentice-Hall, 1994.

Tom Williams, "Tools and protocols link embedded system over the Internet", *Electronic Design*, August 18, 1997, pp. 91-98.

Tom Williams, "Embedded operating systems take on tools, languages, and modules", *Electronic Design*, January 26, 1998, pp. 67-88.

Wayne Wolf, *Modern VLSI Design*, Prentice Hall, 1994a.

Wayne Wolf, "Hardware-software co-design of embedded systems", *Proc. IEEE*, vol. 82, No. 7, July 1994b, pp. 967-989.

Chapter 2

THE 8051 MICROCONTROLLER

2.1 Introduction

In this chapter we first examine the architecture and memory organization of the 8051 microcontroller. Next, we discuss all subsystems which are available in the standard versions.

In 1980, the American chip manufacturer Intel introduced the 8051 microcontroller. Now a huge number of software compatible versions are produced by many firms. The name given to the whole group of microcontrollers is 8051. The main features of the 8051 microcontroller are:

- 8-bit CPU
- Instruction set of 111 operations including multiplication and division
- Boolean processor
- 64K Program Memory address space
- 64K Data Memory address space
- Two 16-bit Timer/Counters
- 32 bidirectional I/O lines
- Full duplex serial port
- On-chip clock oscillator.

The 8051 microcontroller has 4K bytes on-chip Program Memory (factory mask-programmable ROM). The device also possesses 128 bytes of internal Data Memory. Figure 2.1 outlines the major versions released by Intel [Inte 1992, Inte 1996]. Note that the name 8051 when used for a group of software compatible microcontrollers does not indicate the device technology or the size of the on-chip memory. Also, we assume that the versions under this name possess the classical set of peripherals, which are discussed in the current chapter.

The oscillator frequency for most devices can vary between 3.5 and 24 MHz, but some versions can run at 33 MHz. While microcontrollers based on dynamic logic can not operate below a certain clock rate, static devices are able to slow down the oscillator frequency to zero.

If the oscillator frequency is 12 MHz it will take one, two or four microseconds a given instruction to be executed. Assuming that the one microsecond instructions are the most frequent, the microcontroller can be viewed as an engine, capable of delivering a peak performance of one MIPS.

The devices designated with "C" in the middle of the name are CMOS versions and they not only draw less current, but can use two power management modes, Idle and Power Down (PD).

Device	Internal Program Memory	Internal Data Memory	Timer/ Counters 16-bit	Interrupt Sources	Power Reduction Modes	Oscillator Frequency (MHz)
8051AH	4K bytes ROM	128	2	5	---	3.5 - 12
8031AH	---	128	2	5	---	3.5 - 12
8751BH	4K bytes EPROM	128	2	5	---	3.5 - 12
80C51BH	4K bytes ROM	128	2	5	Idle, PD	0.5 - 24
80C31BH	---	128	2	5	Idle, PD	0.5 - 24
87C51	4K bytes EPROM	128	2	5	Idle, PD	0.5 - 24
80C52	8K bytes ROM	256	3	6	Idle, PD	0.5 - 33
80C54	16K bytes ROM	256	3	6	Idle, PD	0.5 - 33
80C58	32K bytes ROM	256	3	6	Idle, PD	0.5 - 33
83C51FA	8K bytes ROM	256	3	7	Idle, PD	0.5 - 33
83C51FB	16K bytes ROM	256	3	7	Idle, PD	0.5 - 33
83C51FC	32K bytes ROM	256	3	7	Idle, PD	0.5 - 33

Figure 2.1 The 8051 microcontrollers produced by Intel.

Being deliberately compressed, the last two sections of the table (Figure 2.1) contain only ROM devices, however ROM-less and EPROM counterparts are also available.

The 8031 version appears to be a very popular single chip microcomputer because of its lower price. Furthermore, in many applications the system requires additional memory and the on-chip program memory is not a necessity.

2.2 Architecture

Using the microcomputer architecture from Figure 1.16 and detailing the block peripheral devices, we obtain the 8051 internal structure shown in Figure 2.2. We assume that the internal Program Memory is 4K bytes. For simplicity, the blocks are connected only by a data bus.

The 8051 is an 8-bit machine. Thus, the register set includes four 8-bit registers. In addition, there is one 16-bit register which is used to address the external Data Memory.

The most frequently used register is an accumulator (ACC). Almost all arithmetic operations use the accumulator. All data transfers with the external Data Memory are implemented through the accumulator. Register B is involved in multiplication and division operations. Program Status Word register (PSW) contains a few flags, such as carry and overflow. Another register termed Stack Pointer (SP) is related to the stack operations. It is incremented before data is stored. The stack may reside anywhere in internal RAM. The Stack Pointer is initialized to 07H after a reset. This causes the stack to begin at address 08H.

The data pointer register (DPTR) is a 16-bit register. It can be viewed as two separate registers: a high byte (DPH) and a low byte (DPL). The most common access to the external Data Memory is based on DPTR.

As usual, the Program Counter (PC) addresses the instructions in the Program Memory. The amount and the type of the Program Memory will vary from version to version (see Figure 2.1). The 8031 microcontroller lacks internal Program Memory. All microcontrollers possess at least 128 bytes of internal Data Memory.

The on-chip peripherals consist of four parallel ports, a serial port, two timer/counters and an interrupt unit (system). All four ports are built from output registers (P0 through P3) and buffers (P0-B through P3-B). A key point here is that Port 0 and 2 interface the external memory. Port 3 is also multifunctional and links peripheral devices with the 8051 interrupt system, serial port and timers/counters.

Finally, an on-chip oscillator circuit is available as well.

2.3 Memory organization

The designers of the 8051 microcontroller organized the memory in two parts: Program Memory and Data Memory. The memory can also be divided into four separate address spaces. Figure 2.3 shows the address spaces and example read instructions oriented to each memory. Furthermore, Figure 2.4 reveals all details associated with the 8051 memory organization. We assume that the on-chip Program Memory is 4K bytes.

Program Memory

Program Memory is the first memory address space, which is 64K bytes long. In other words, the designers can place up to 64K bytes of physical memory in this address space. The lower 4K (8K or more) portion of this memory may reside on-chip.

After reset (RST), the CPU begins execution from location 0000H. As is also shown in Figure 2.4, each interrupt is assigned a fixed location in the Program Memory. Interrupts cause the CPU to jump to a specific location and run a service routine. External interrupt 0 ($\overline{INT0}$), for example, is assigned to location 0003H. If external interrupt 0 is used, its service routine must begin at location 0003H. If an interrupt is not used, its service location is available as general purpose Program Memory. Interrupt service locations are spaced at 8-byte intervals. If an interrupt service routine is short enough, it can reside entirely within that 8-byte interval. Longer service routines can use a jump instruction to skip over some other interrupt locations, if they are in use.

Microcontrollers which possess internal Program Memory can execute instructions either from it or from external memory components. The selection is done by a dedicated input as explained in section 2.4.

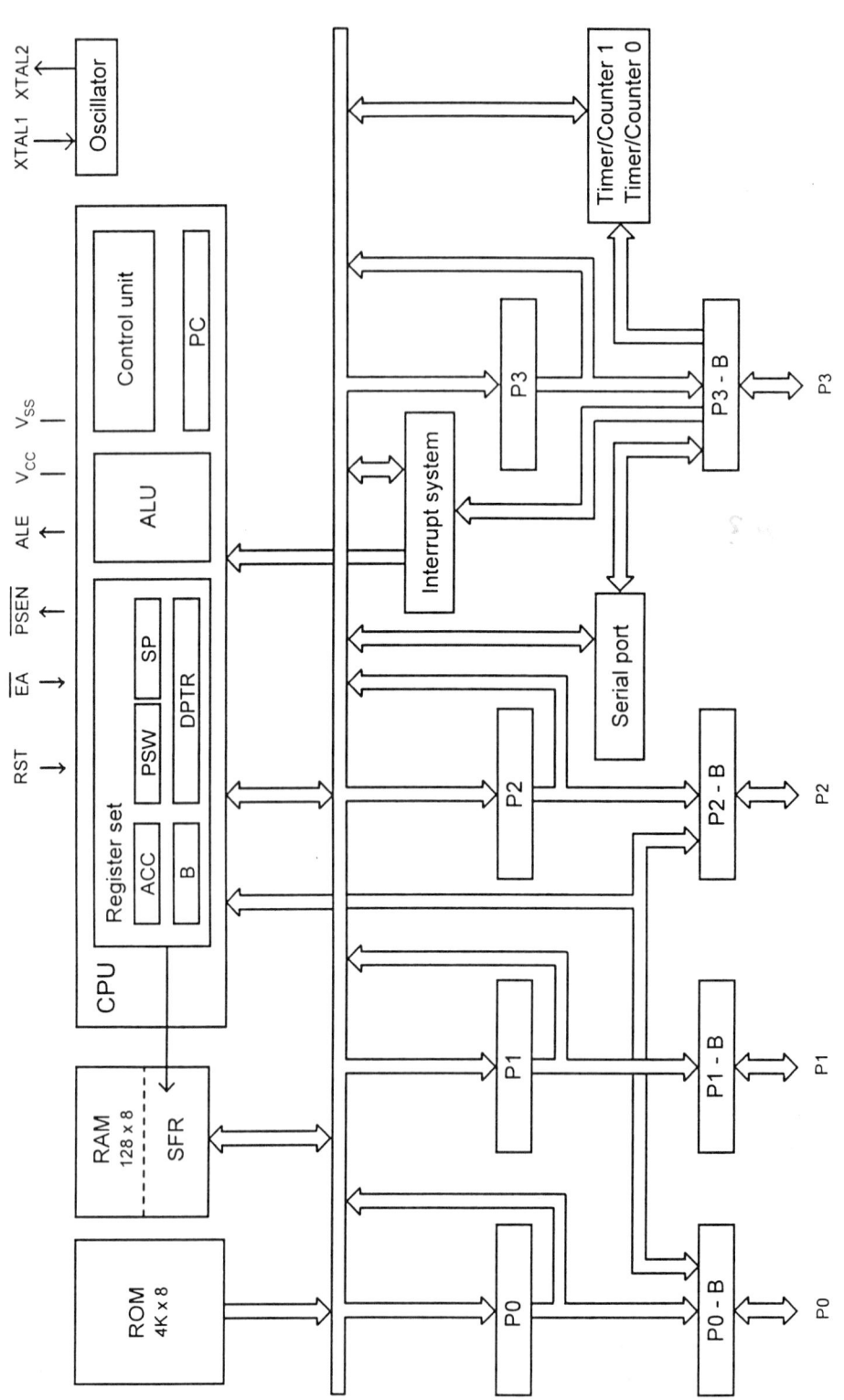

Figure 2.2 The 8051 block diagram.

External Data Memory

The next memory address space is termed external Data Memory (external RAM). The program can access up to 64K bytes of external Data Memory.

Move instructions (**MOVX**) are used to exchange data between the accumulator and the external Data Memory. The external Data Memory is addressed indirectly through registers DPTR, R0 or R1.

In some cases, a memory component can occupy an address area from both Program Memory and external Data Memory and act as RAM memory for programs and data. The benefit of this approach is that code can be downloaded and executed from the external RAM.

Both the Program Memory and the external Data Memory form a total amount of 128K bytes. Some applications might require bigger memories than 64K bytes for programs and/or 64K bytes for data. Working around this problem the embedded systems designers apply appropriate techniques to expand the memory.

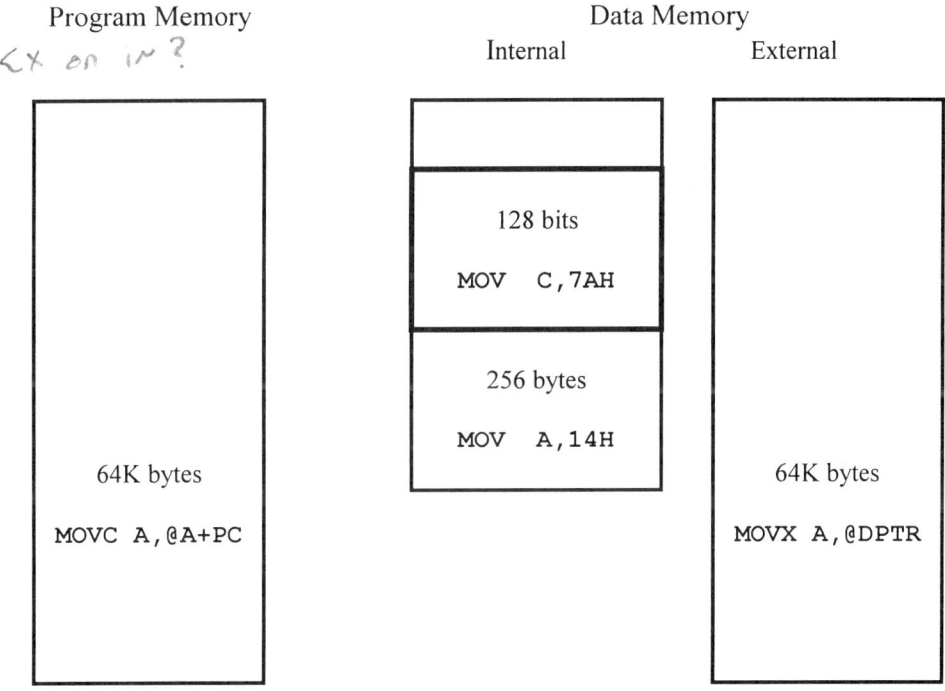

Figure 2.3 The four address spaces accessed by read example instructions.

Internal Data Memory

A memory address space called internal Data Memory has a size of 256 bytes. The internal Data Memory (internal RAM) is divided into two blocks: first 128 bytes for general purpose RAM and a second part starting from address 80H, the Special Function Register (SFR) area. Thus, the registers ACC, B, PSW, SP and DPTR plus all other control and status registers can be viewed as locations in the internal Data Memory.

Program Memory   Data Memory

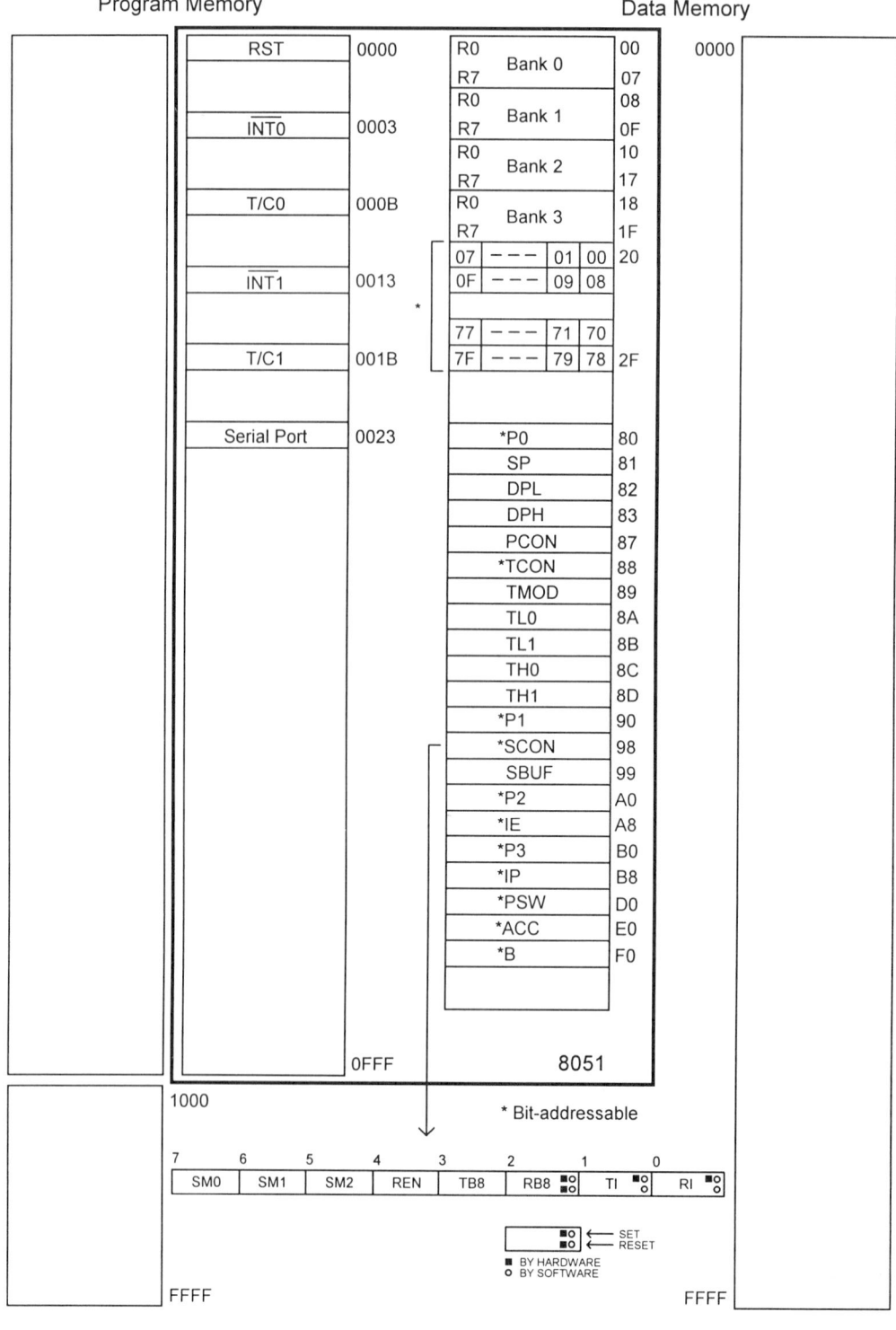

Figure 2.4 The 8051 memory structure.

In other words, the SFRs are organized as an extension of the internal RAM (see the arrow in Figure 2.2). In contrast to a general purpose microprocessor, the 8051 microcontroller integrates peripheral devices and they need control and status registers. Logically, all microcontroller's registers, general purpose and related to embedded peripherals, are placed in a common group.

The lowest 32 bytes of the general purpose RAM are grouped into 4 banks of 8 registers. Program instructions call out these registers as R0 through R7. Two bits in the Program Status Word register (PSW) select which bank is in use. This approach allows more efficient use of code space.

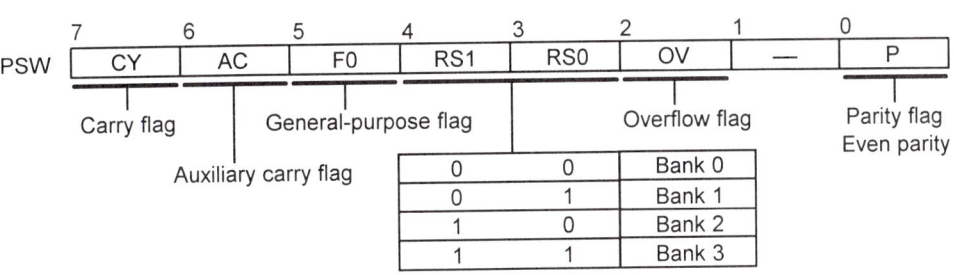

Figure 2.5 The Program Status Word register.

Figure 2.5 shows the register PSW in detail. The register performs both control and status functions. Two flags, RS1 and RS0, determine the current register bank. The other flags provide status information. The parity flag is set if the accumulator contains an odd number of 1s. Examples about the carry flag, auxiliary carry flag and overflow flag are discussed in section 3.6. The flag F0 is available as a general purpose flag.

Bit -addressable memory

As discussed earlier, the microcontroller is capable of addressing individual bits in the memory. Along with the other address spaces, Figure 2.3 displays a bit addressable area in the internal Data Memory. The size of this segment is 128 bits. Obviously, two ways to access the data are possible: by bytes and by bits. Furthermore, Figure 2.3 contains an example instruction MOVC   7AH which moves a bit from address 7A (hexadecimal) to the carry bit in the register PSW.

The specific address organization of the bit addressable space is shown in Figure 2.4. The bit addressable segment itself is marked by *. The least significant bit in byte 20H has a bit address of 00. The most significant bit in byte 2FH is accessed by bit address 7FH.

Moreover, a certain number of SFRs, marked by * in Figure 2.4, are also bit addressable. For example, the serial port control register SCON is a bit addressable register. As shown in Figure 2.4, the register contains eight special function flags. Three of them are processed both by hardware and software. What does it mean for the programmer of the microcontroller? For instance, the receive interrupt flag RI is set automatically by hardware when a character arrives. The symbols placed in the RI field manifest that the flag can be cleared only by software. Thus, the programmer is advised in a readable way that a clear bit instruction must be included in the program.

The register PCON is not a bit-addressable register, but it contains two special function flags: PD and IDL. The flags can be accessed through byte-oriented instructions.

In summary, the bit addressable space of the 8051 microcontroller overlaps the internal Data Memory both in the real RAM and in the SFRs.

2.4 Pin definitions and functions

The capability to interface external memory has an impact on pin organization. In general, there are two types of microcontrollers. First, those microcontrollers which have on-chip Program Memory and are not capable to use external memory and second, microcontrollers, such as the 8051, which can interface external memory. Of course, some of them lack internal Program Memory and must fetch instructions from external memory. In the second case, a large majority of the microcontroller pins are occupied for memory interface. As indicated in Figure 2.2, Port 0 and Port 2 link the microcontroller with the external memory.

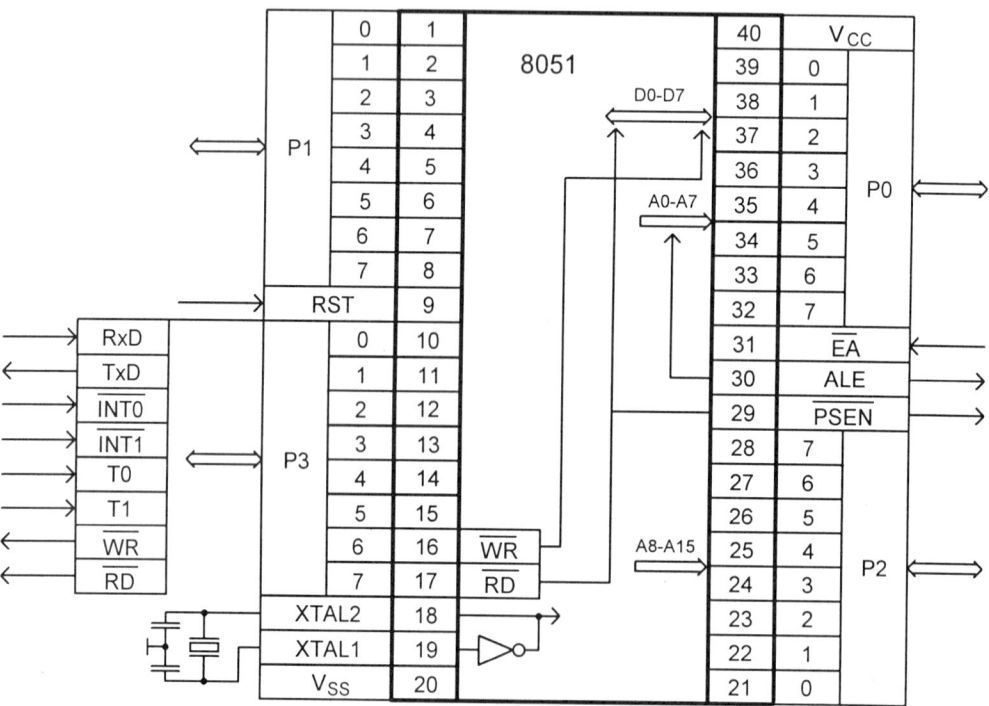

Figure 2.6 The 8051 pin definitions and functions.

In addition, the parallel ports, the serial port, the interrupt system and the timer/counters need pins to interact with the environment.

Figure 2.6 shows a pin configuration of the 8051 microcontroller in a 40-pin dual in-line package. Port 0 can be used either as a general purpose port or to access external memory. When external memory is used, Port 0 forms the low order byte of the external memory address (A0 - A7). Next, the byte being written or read is transferred through Port 0. Port 0 is capable of driving 8 Low power Schottky Transistor Transistor Logic (LS TTL) loads.

Port 1 is a general purpose port and does not possess any other special function. Port 1 can drive 4 LS TTL loads. Port 2 can be used either as a general purpose port or to emit the high order address byte (A8 - A15) when external memory is used. Port 2 is designed to drive 4 LS TTL loads.

Port 3 can be used either as a general purpose port or to perform the following special functions:

P3.0	RxD	Serial port input
P3.1	TxD	Serial port output
P3.2	$\overline{INT0}$	Interrupt 0 input
P3.3	$\overline{INT1}$	Interrupt 1 input
P3.4	T0	Timer 0 input
P3.5	T1	Timer 1 input
P3.6	$\overline{WR}$	External Data Memory write strobe
P3.7	$\overline{RD}$	External Data Memory read strobe

Port 3 can drive 4 LS TTL inputs.

The external Data Memory write strobe $\overline{WR}$ is connected to the write enable inputs of the external Data Memory components. Similarly, the external Data Memory read strobe $\overline{RD}$ controls the output buffers of the external RAMs. In order to illustrate the correspondence between the signals ($\overline{WR}$, $\overline{RD}$ and $\overline{PSEN}$) and the data flow direction, we linked the pins with the corresponding ends of the bi-directional arrow D0-D7 (Figure 2.6).

The reset input (RST or RESET) is an active high input. When pulled high for two machine cycles while the oscillator is running, it resets the microcontroller. All registers are cleared, except the SFRs P0 through P3, which are set to FFH. As yet another exception, the stack pointer SP is set to 07H. The microcontroller jumps to address 0000H and executes the first instruction. Microcontrollers produced by NMOS technology can draw backup power for the internal RAM through the RST pin.

Using external memory, the designer must store the low-order address byte (A0-A7) in a register. The 8051 microcontroller provides an output to clock the register. The output is termed Address Latch Enable (ALE). The output ALE can drive up to 8 LS TTL loads. Furthermore, the device has an output to synchronize the read from the external Program Memory. The output is asserted low and labelled Program Store Enable ($\overline{PSEN}$). Up to 8 LS TTL loads can be connected to this output. An input named External Address ($\overline{EA}$) is used to select external Program Memory when an instruction read from the internal Program Memory is also possible (the microcontroller possesses on-chip Program Memory and the address is within the range). The pin is asserted low.

Two pins, XTAL1 and XTAL2 are related to the built-in oscillator. We usually connect a crystal to use the internal oscillator, but an external source can drive the microcontroller as well. Finally, the power supply pins are V_{CC} and V_{SS}. Once the 8051 microcontrollers were designed for $V_{CC} = 5V$. Now, there are devices capable of working in the range 2.7 V to 6 V.

2.5 Timing

The control unit of a microprocessor or microcontroller is responsible for the proper handling of the stream of instructions. Each instruction includes one or more steps (cycles). At the same time, an instruction can be viewed as a deeper level program which performs a series of specific actions. Executing this program, termed microprogram or microcode, the control unit emerges as a processor within the CPU. There can be no doubt, every processor needs memory to read the instructions. Consequently, the control unit must contain a ROM. Alternatively, the control unit can be designed as a sequential machine. As discussed earlier, CISC computers use microcode. In the case of RISC processors, the control unit can be designed as a FSM. Practically, the users might not be interested in the internal architecture of the microcontroller. However, they should be aware of the timing parameters of the microcomputer and the essential design rules. The topics of this section are the following matters:

- Basic timing: the interaction between the microcontroller's CPU and memory.
- The microcontroller's instructions parameters: execution time and number of bytes.

Figure 2.7 shows the basic timing diagram. A machine cycle includes 12 oscillator periods. The cycle is divided into 6 states - S1 through S6. Each state consists of 2 phases - P1 and P2. As an example of one-byte, one-cycle instruction we use **INC A** (Increment accumulator A). A fetch/execute cycle from external Program Memory includes the following steps:

- The opcode is read through Port 0 (Cycle 1, State 1).

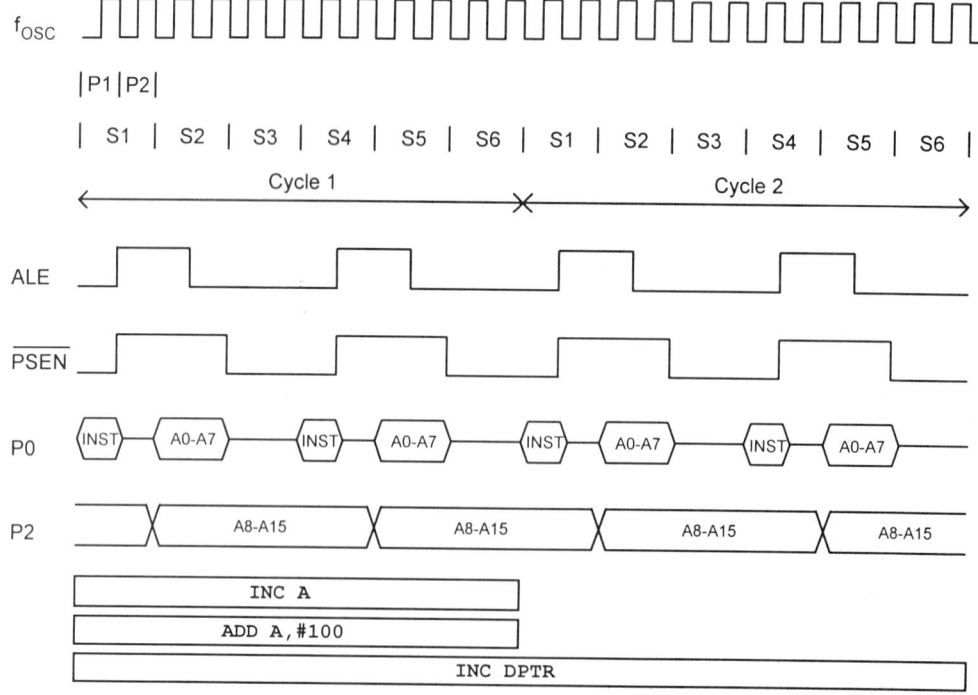

Figure 2.7 The 8051 external Program Memory execution.

• The low address byte (A0 - A7) is latched into the external register buffer on the high-to-low transition of ALE pulse. The high address byte (A8 - A15) is emitted through Port 2.

• There is a fetch at Cycle 1, State 4, but it is a one-byte instruction and the byte read is ignored. The Program Counter is not incremented.

• The low address byte (A0 - A7) of the next opcode is prepared.

The instruction ADD   A,#100 is employed as an example of a two-byte, one-cycle instruction. The only difference with the previous instruction is at Cycle 1, State 4 when a real fetch is performed. The instruction execution times and number of bytes can be found in Appendix D.

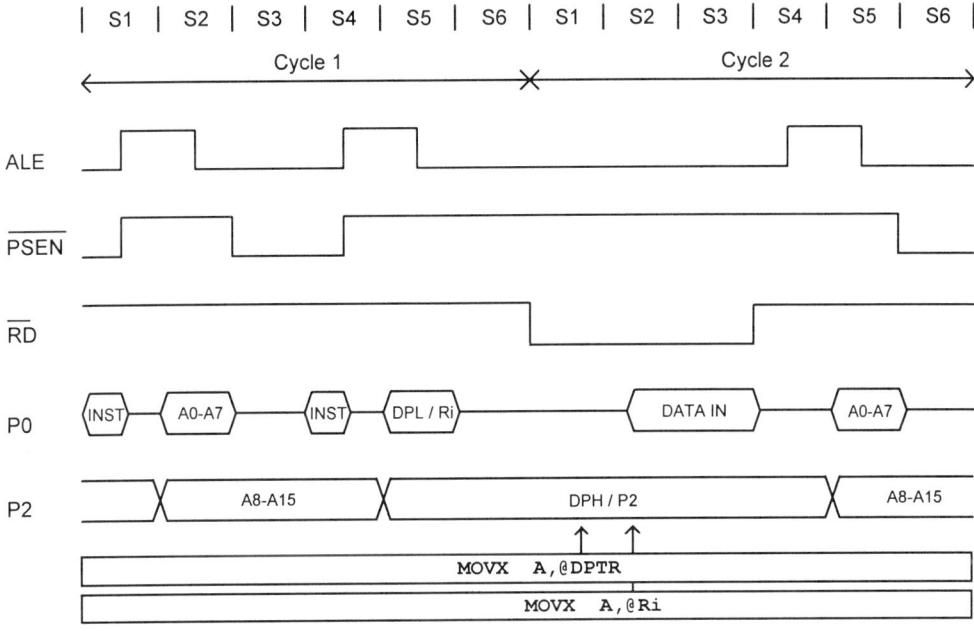

Figure 2.8 The 8051 external Data Memory read operation.

When a one-byte, two-cycle instruction, such as INC DPTR is executed, the opcode is read at Cycle 1, State 1 and the next three bytes at Cycle 1 (State 4) and Cycle 2 (State 1 and 4) are discarded. The type of the accessed memory does not influence the output Address Latch Enable (ALE). It is activated in all cases of instruction codes. The output $\overline{\text{PSEN}}$ is asserted only when the microcontroller reads from external Program Memory.

Another key point concerning the basic timing is the interaction with the external Data Memory. Figure 2.8 depicts the timing when the external Data Memory read operation is performed. An active strobe signal $\overline{\text{RD}}$ enables a read operation. As you might expect, when the strobe signal $\overline{\text{WR}}$ is low, an external Data Memory write operation will be under way.

The external Data Memory access employs the indirect addressing mode either through register R0/R1 (8-bit address) or DPTR (16-bit address). As can be seen from Figure 2.8, the difference between the two read instructions

MOVX   A,@DPTR
MOVX   A,@R$_i$

is in the Port 2 pattern.

In the event of an 8-bit address (MOVX   A, @R$_i$), Port 2 emits the last code written to the output register buffer. We named this approach paging in section 1.6. A peculiar point in this case is that the first Address Latch Enable (ALE) pulse of the second cycle is missing. Therefore if external Data Memory is used, it will be impossible output ALE to clock other devices with a constant rate.

Furthermore, we can achieve an important feature if the external RAM strobe $\overline{RD}$ is replaced by $(\overline{PSEN})\,(\overline{RD})$. In this case, the external RAM receives an additional function: to act as Program Memory. Practically, we use the RAM as external Data Memory and download the code. Next, we employ the RAM as Program Memory and execute the program. Naturally, using the external RAM as a Program Memory component will require consistency between the RAM timing parameters and the microcontroller requirements for external Program Memory.

2.6 Parallel ports

Parallel ports are simple, but efficient channels which link the microcontroller with off-chip peripherals. As a rule, parallel ports are bidirectional and therefore can be used either as input or output. The output circuitry is latched and the logic values remain on the pins as long as required.

There are two questions which the user should ask when studying a new microcontroller:
• How can I program the port to be input or output?
• How big is the driving capability of a port?
Discussing the 8051 port structures and operation we will answer these questions.

The 8051 microcontroller has four parallel input/output ports. Figures 2.9 through 2.12 show the port structures. Each I/O line can be independently programmed to be used either as an input or an output. If an input is required the port bit latch must be set. In that event the output driver transistor turns off.

During the reset function, 1s are written in all port latches and the I/O lines act as inputs. This is a good start for Port 3, which has additional functions, such as, interrupt and counter inputs.

When external memory operations take place, the CPU writes FFH to the Port 0 register automatically (Figure 2.9.)

As outputs, the fan-out of Port 0 lines is 8 LS TTL inputs. When Port 0 is used as a general purpose output, external pull-ups are required, as indicated in Figure 2.9. Port 1 through Port 3 can drive up to 4 LS TTL loads.

When an instruction affects a port output, the new value appears at the output pins during the state S1P1 of the next instruction.

Single chip microcomputers are designed to interface different peripheral devices. A typical application is for direct drive of transistors. If a port bit emits a high level, the external transistor will turn on. Consequently, the pin voltage will be interpreted as low level. If "read-modify-write" instructions are based on the output voltage level, the result will be incorrect. The microcontroller is organized so that the "read-modify-write" group of instructions read the latch rather than the pin. The next examples cover all operations of this type:

```
ANL    P1,A        ; Logical AND
ORL    P1,A        ; Logical OR
XRL    P2,A        ; Logical EX-OR
```

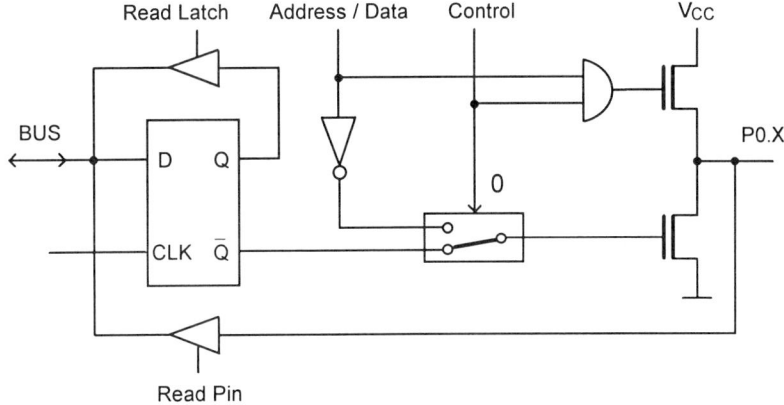

Figure 2.9 The 8051 Port 0 latches and I/O buffers.

Action	Control	Latch		Pin	
Input	0	FF		Input data	
	0	FF (by CPU)		D0 - D7	
Output	0	Output data	1	Output data	OFF (pull-ups required)
			0		0
	1	FF (by CPU)		A0 - A7 / D0 - D7	

```
JBC     P1.3,L12    ; Jump to the label L12 if bit P1.3 is set
                    ; and clear bit P1.3
CPL     P2.4        ; Complement bit P2.4
INC     P1          ; Increment P1
DEC     P2          ; Decrement P2
DJNZ    P1,L3       ; Decrement P1 and jump to label L3 if P1 is
                    ; not zero
MOV     P1.4,C      ; Move carry bit to bit P1.4
CLR     P2.7        ; Clear bit P2.7
SETB    P1.4        ; Set bit P1.4
```

The last three instructions read the port byte, modify the selected bit and then write the new byte back to the output register.

As shown in Figures 2.9 through 2.12, when the CPU reads a latch, the corresponding Read Latch signal is activated. A port pin value is placed on the internal bus in response to a Read Pin signal.

Typically, ports 0 and 2 are occupied by the external memory interface. Port 3 covers the serial channel, the interrupt system and the timer/counter inputs. What is left as a general purpose parallel port is only Port 1. Design methods which aim to expand the microcontroller's parallel interface are discussed in section 4.3. Microcontrollers that possess more ports are presented in Chapter 7.

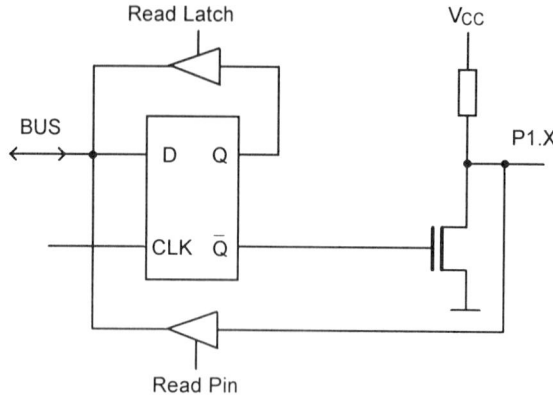

Figure 2.10 The 8051 Port 1 latches and I/O buffers.

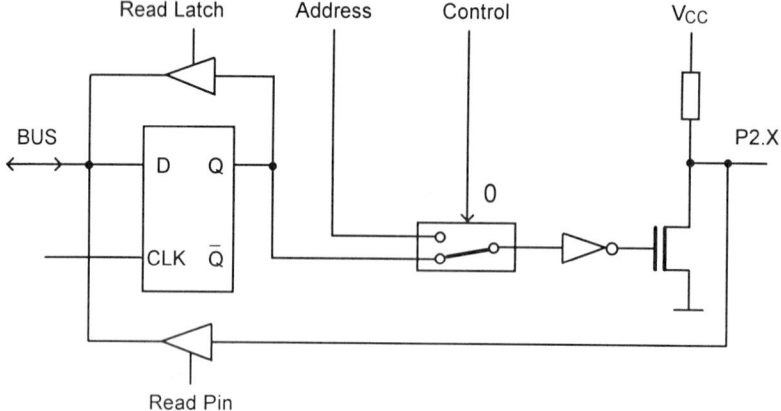

Figure 2.11 The 8051 Port 2 latches and I/O buffers.

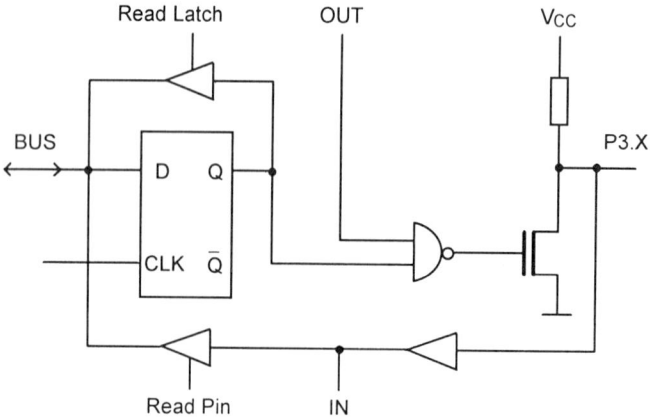

Figure 2.12 The 8051 Port 3 latches and I/O buffers.

2.7 Timer/counters

A turn around the embedded microcomputer applications shows that the following cases are very widespread:

- Counting pulses from sensors.
- Frequency measurements.
- Period measurements.
- Precise time delay generation.

The 8051 microcontroller responds to this demand by two 16-bit timer/counters: Timer/Counter 0 and Timer/Counter 1. Each timer/counter can be individually programmed either to count internal clock pulses (timer function) or to count external pulses (counter function).

It is certainly implied that some tasks, for example time delays, can be done by pure software solution. The first weakness of the software approach is that it works one at a time. The microcontroller can not generate two or more time delays in parallel, only by software.

Another question will arise if you migrate to a new device (from the same family). The new microcontroller may run at a different oscillator frequency (most likely higher) and the program must be modified to maintain the same time delay. The situation will become worse if the new microcontroller speeds up the computing by an instruction queue. The queue would make the precise execution time unpredictable.

When a timer/counter is used as a counter, the code in the 16-bit register (counter) is incremented after each 1-to-0 transition at its corresponding input. The external input is sampled during S5P2 (state 5, phase 2) of every machine cycle. If the input is high in one cycle and low in the next cycle, the count will be incremented. The new code appears in the register at state S3P1 of the cycle following the one in which the active transition was found.

The maximum count input frequency is 1/24 of the oscillator frequency as a result of at least one cycle in high level and one in low level. There are four timer/counter operating modes. Mode 0 is based on a 13-bit counter, mode 1 on a 16-bit and mode 2 assumes an 8-bit counter with automatic reload. Mode 3 can be used only for Timer/Counter 0, which is split into one 8-bit timer/counter and one 8-bit timer.

Mode 0

Figure 2.13 shows the structure of Timer/Counter 1 in mode 0. Both timer/counters can be accessed as a high byte - TH1 (TH0) and a low byte - TL1 (TL0). In mode 0 only 5 bits of register TL1 (TL0) are used. Mode 0 is selected by writing code 00 into mode bits M0 and M1 in Timer/counter MODe control register (TMOD). Both possibilities - timer or counter are expressed by a switch in Figure 2.13. Inspection shows that the timer function is achieved by 0 in a control bit ($C / \overline{T}$). Setting the bit $C / \overline{T}$ allows the counter function. The oscillator frequency is divided by 12.

An important point to note is that when the count rolls over from all 1s to all 0s, it sets the timer interrupt flag TF1 from Timer/counter CONtrol register TCON. In this way a source in the interrupt system is activated. The flag TF1 is cleared automatically by hardware when interrupt is processed. This is illustrated by a connection between the vector address 001BH and the reset input (R) of the flag TF1 (Figure 2.13).

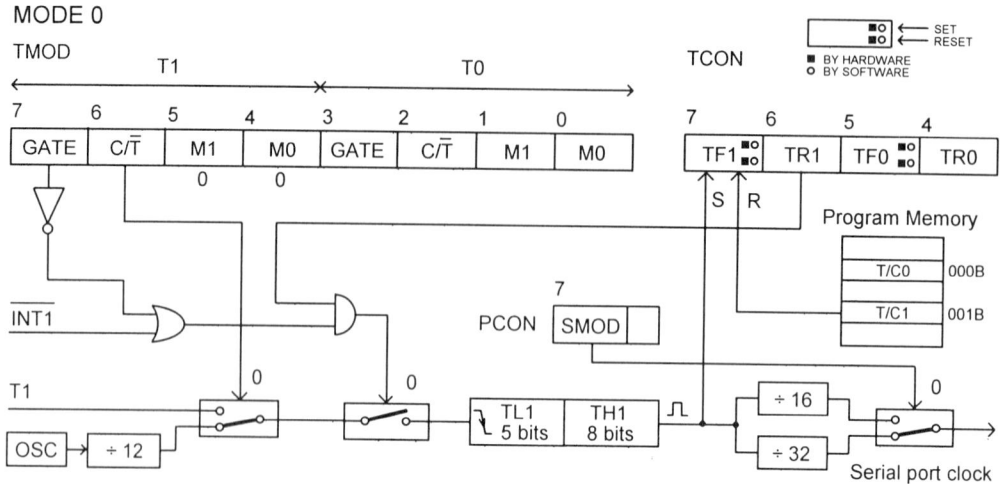

Figure 2.13 The 8051 Timer/Counter 1 mode 0, a 13-bit counter.

Timer/Counter 1 starts counting (the switch is closed) when Timer Run flag (TR1) is set and either a bit called GATE is reset or the input $\overline{INT1}$ is pulled high. On the base of the timer function and a set bit GATE, pulse width measurements can be organized. The pulses are applied to the input $\overline{INT1}$.

An additional function of Timer/Counter 1 is to generate the serial port clock. Every time when the count rolls over from all 1s to all 0s, a pulse is directed through a divider by 16 or 32 to the serial port circuit. The switch in the divider will be in position 0 (division by 32) if a bit termed Serial MODe (SMOD) from the Power CONtrol register (PCON) is low.

Timer/Counter 0 owns the same structure. It uses the four least significant bits from the register TMOD and bits TF0 and TR0 from the register TCON. The only difference is that Timer/Counter 0 can not clock the serial port.

Mode 1

In this mode the full 16-bit length of register TL0 (TL1) is used. All other features are identical to the operation in mode 0.

Mode 2

Figure 2.14 shows the Timer/Counter 1 in mode 2. In this mode, the timer/counters cores are organized as 8-bit counters with automatic reload. Mode 2 is selected by code 10 in bits M1 and M0 from the register TMOD. Mode 2 is based on an 8-bit counter (TL1) with auto-reload by means of the high byte register (TH1). When timer/counter rolls from all ones to all zeros the flag TF1 is set and the code from register TH1 is moved to register TL1 without changing its content. The contents of the register TH1 is preset by software.

The only difference between both timer/counters is that Timer/Counter 0 lacks the possibility to clock the serial port.

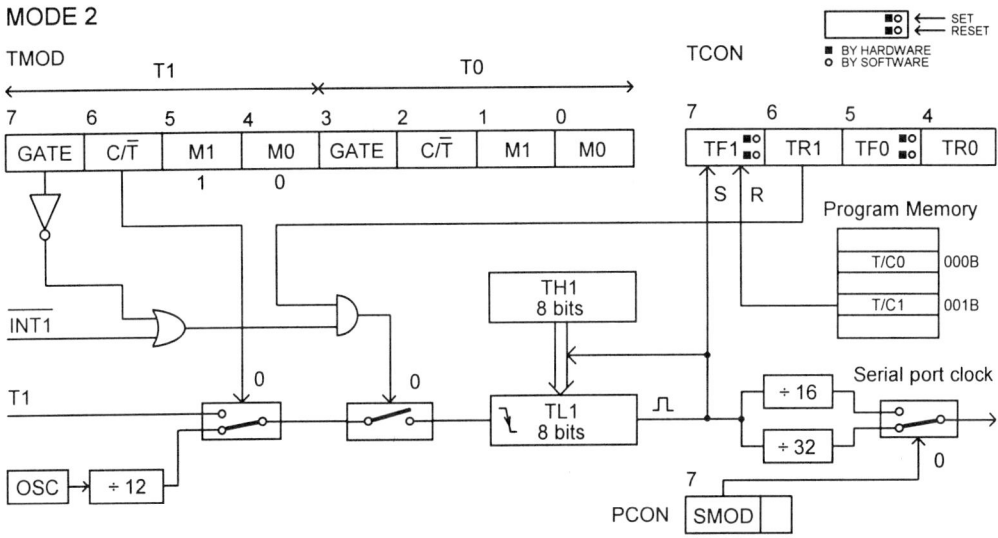

Figure 2.14 The 8051 Timer/Counter 1 mode 2, an 8-bit auto-reload mode.

Mode 3

Timer 1 in mode 3 holds its count. The effect of switching to mode 3 is the same as to clear the bit TR1. Figure 2.15 shows Timer 0 in mode 3. In this mode Timer 0 is decomposed into two separate 8-bit counters TL0 and TH0. Register TL0 can be used either as a timer or as a counter. The other part, TH0, can be employed only as a timer. Timer TH0 needs a start (run) bit and TR1 is devoted to this purpose. In the same way, the bit TF1 is taken from Timer/Counter 1 and is used as an interrupt request flag.

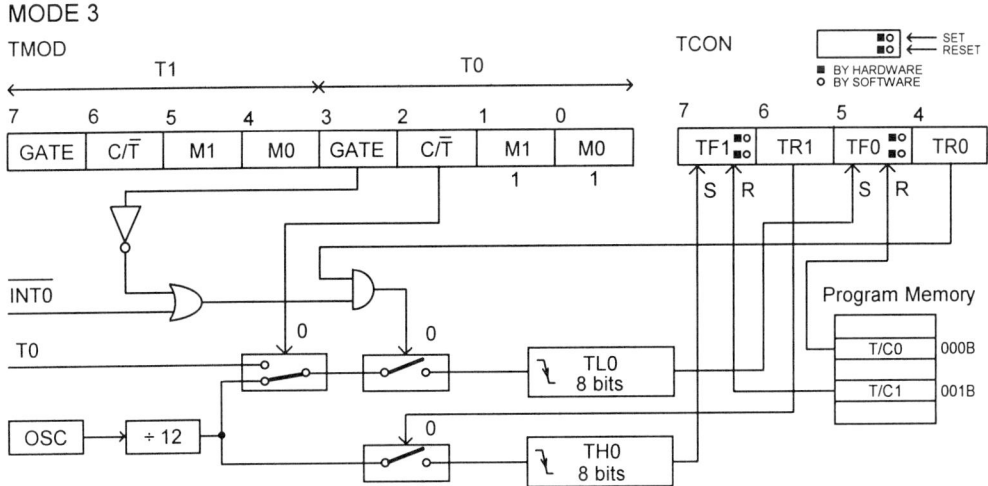

Figure 2.15 The 8051 Timer/Counter 0 mode 3: two 8-bit counters.

When Timer/Counter 0 runs under mode 3, Timer 1 can be turned on and off by switching it out of and into its own mode 3. Therefore, it will be possible to generate serial port clock or may be used for another purpose if the interrupt flag is not required.

The registers TMOD, TCON and PCON are cleared by a reset.

Even though the timer/counters are hardware blocks which run concurrently with the CPU, they depend on the oscillator frequency. Both timer/counters are identical, except for the serial port clock feature and mode 3. Timer/counter application examples and problems are discussed in section 3.10.

2.8 Serial port

The 8051 microcontroller, like almost all other computers, processes data in parallel. At the same time, the communication between computers is often done by serial links. Inevitably, a serial bus would be slower than a parallel bus, but it is cost-effective and very popular for small embedded systems and personal computers. The rate of transmission is called the bit rate. The bit rate is frequently termed baud rate. Apart from the fact that bit rate and baud rate are not always equal, they are used synonymously in this text.

The microcontroller subsystem which converts the parallel data into a serial bit stream and vice versa is called a serial port. The serial port is alternatively labelled Universal Asynchronous Receiver Transmitter (UART). The serial port of the 8051 is a full duplex port and therefore it can transmit and receive simultaneously. The receive register is buffered and it is possible to go on reception of a second byte before a previously received byte has been read. The receive and transmit registers are both named SBUF. The serial port is controlled by a Serial port CONtrol register (SCON). Moreover, the register SCON acts as a status register including a Transmit Interrupt flag (TI) and a Receive Interrupt flag (RI). Figure 2.16 shows the register SCON and its relation to the incoming and outgoing frames.

A set interrupt flag TI indicates that the transmit buffer is empty and can be loaded again. The interrupt flag TI is set by hardware at the end of the 8th bit period in mode 0 or at the beginning of the stop bit in the other modes when the serial port transmits. The flag TI must be cleared by software.

A set receive interrupt flag RI alarms that the receive buffer is full and should be read. The interrupt flag RI is set by hardware at the end of the 8th bit period in mode 0 or through the stop bit period in the other modes, when the serial port receives. Likewise, the flag RI must be cleared by software.

As you might expect, mode 0 is one of four possible modes of operation that can be selected by bits SM0 and SM1 from register SCON. Unlike what you might expect, the bit SM0 is the most significant bit (MSB) when the mode is coded. A bit REN from register SCON enables serial reception for all modes. It can be set by software to enable or cleared to disable the reception.

Mode 0

This synchronous mode is used to interface shift registers. In mode 0 the serial bit stream goes either in or out through the line RxD (P3.0). The shift clock is generated on pin TxD (P3.1). Eight bits are either transmitted or received. The least significant bit (LSB) is transferred first.

The baud rate is

$$BR = \frac{f_{OSC}}{12} \text{ , bps}$$

Transmission is initiated by loading the register SBUF.

To start the reception, the following condition must be satisfied

$$REN \wedge \overline{RI} = 1$$

The logic equation implies that the reception is allowed by setting the bit REN and by clearing the receive interrupt flag RI. The flag REN is cleared in the beginning. The flag RI is reset after each received byte.

Mode 0 is commonly used for fast data transfer in conjunction with parallel-to-serial and serial-to-parallel registers.

Mode 1

Mode 1 is the standard full-duplex mode. A data word includes a start bit (low), 8 data bits and a stop bit (high). The LSB is transferred first. In mode 1, 2 and 3 the bit stream is transmitted through the pin TxD and received through the pin RxD. The baud rate can be programmed by Timer/Counter 1. Transmission is initiated by a move to the register SBUF. Reception is initiated when the line RxD is switched from high to low. If the first bit after the transition is high, the receive circuit will be reset and go back to wait for another high-to-low transition.

Embedded systems very often work in noisy environments. To alleviate this problem, the 8051 serial port estimates the bit which is being received on the base of three samples. The value accepted is the one that appeared at least twice. A 4-bit counter is used to split the bit time into 16 periods. A bit detector samples the three states in the middle (7th, 8th and 9th) to avoid frequency deviation problems.

If the conditions described in the next logic equation are met

$$REN \; \overline{RI} \; (\overline{SM2} \vee STOPBIT) = 1$$

then the following actions will be performed:

- The data received in an input shift register is moved into register SBUF.
- The stop bit from the frame is loaded into the flip-flop RB8 of the register SCON.
- The receive interrupt flag RI is set.

If the described conditions are not satisfied the received frame will be lost.

As can be seen from the logic equation for mode 1, the inverted bit SM2 and the received value for the stop bit are involved in an OR function. Normally, the bit SM2 is set, however you have some leeway to clear the bit SM2 and override the stop bit.

Mode 2

This mode facilitates multiprocessor communications. The frame includes a start bit (low), 8 data bits, an additional bit (D8) and a stop bit (high). The LSB is transferred first.

The baud rate is

$$BR = \frac{f_{OSC}}{64 - 32 PCON.7} \text{ , bps}$$

The register PCON occupies the place between Timer/Counter 1 and the serial port, as shown in Figures 2.13 and 2.14.

Figure 2.16 attempts to illustrate modes 2 and 3. The transmission is initiated again by a move to register SBUF. The additional bit D8 takes its value from register SCON, bit TB8. The transmit interrupt flag TI is set at the beginning of the stop bit (shift register empty).

As usual, the reception will be under way if the input RxD is changed from high to low. Assuming

$$\text{REN } \overline{\text{RI}}\left(\overline{\text{SM2}} \vee \text{D8}\right) = 1$$

the serial port will perform the following actions:

- The data (D0-D7) from the input shift register is moved into register SBUF.
- The received additional bit D8 is stored into flip-flop RB8 of the register SCON.
- The interrupt flag RI is set.

Otherwise the received frame is lost.

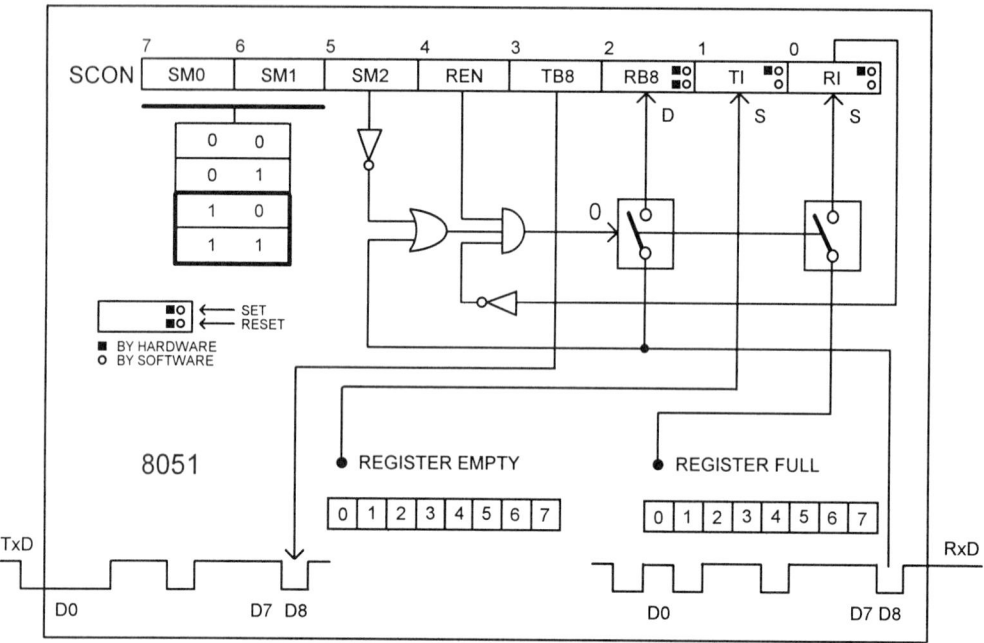

Figure 2.16 The 8051 serial port control register SCON, mode 2 and 3.

Mode 3

Mode 3 differs from mode 2 only in the baud rate control. The Mode 3 baud rate is controlled by Timer/Counter 1.

Figure 2.17 covers all possible combinations between the serial port and Timer/Counter 1 modes. In general, mode 0 and 2 have a fixed baud rate depending on oscillator frequency. Mode 1 and 3 possess a programmable baud rate option based on Timer/Counter 1. If that's the case, the baud rate can be controlled by writing a certain code to register TH1. In this way either oscillator frequency f_{OSC} or external signal with frequency f_{T1} applied to input T1 is divided to form the desired baud rate. If the serial port in mode 1 and 3 is combined with Timer/Counter 1 in counter mode, the ceiling of the input frequency is $1/24\,f_{OSC}$.

Figure 2.18 illustrates several widespread baud rates and examples how they can be organized. On the last line in the table, the baud rate of 110 bps is obtained when Timer 1 in mode 1 is loaded with code FEEBH after interrupt. For all other examples, the column TH1 contains the reload value.

Serial port	Mode 0			$\text{BAUD RATE} = \dfrac{f_{OSC}}{12}$
	Mode 1 Mode 3	Timer 1	Mode 0	$\text{BAUD RATE} = \dfrac{f_{OSC}}{12.2^{13}(32-16\text{PCON.7})}$
			Mode 1	$\text{BAUD RATE} = \dfrac{f_{OSC}}{12.2^{16}(32-16\text{PCON.7})}$
			Mode 2	$\text{BAUD RATE} = \dfrac{f_{OSC}}{12(256-\text{TH1})(32-16\text{PCON.7})}$
		Counter 1	Mode 0	$\text{BAUD RATE} = \dfrac{f_{T1}}{2^{13}(32-16\text{PCON.7})}$
			Mode 1	$\text{BAUD RATE} = \dfrac{f_{T1}}{2^{16}(32-16\text{PCON.7})}$
			Mode 2	$\text{BAUD RATE} = \dfrac{f_{T1}}{(256-\text{TH1})(32-16\text{PCON.7})}$
	Mode 2			$\text{BAUD RATE} = \dfrac{f_{OSC}}{64-32\text{PCON.7}}$

Figure 2.17 Baud rate selection.

Figure 2.18 also shows that if oscillator frequency 11.0592 MHz is used, several baud rates will be available under software control (see also Chapter 6).

Baud rate	Serial port mode	Timer/Counter 1	Timer/Counter 1 mode	f_{OSC} (MHz)	SMOD PCON.7	TH1
1 Mbps	0	x	x	12	x	x
375 Kbps	2	x	x	12	1	x
62.5 Kbps	1, 3	timer	2	12	1	FF
19.2 Kbps	1, 3	timer	2	11.0592	1	FD
9.6 Kbps	1, 3	timer	2	11.0592	0	FD
4.8 Kbps	1, 3	timer	2	11.0592	0	FA
2.4 Kbps	1, 3	timer	2	11.0592	0	F4
1.2 Kbps	1, 3	timer	2	11.0592	0	E8
110 bps	1, 3	timer	1	12	0	FEEB

Figure 2.18 Baud rate examples.

2.9 Interrupt system

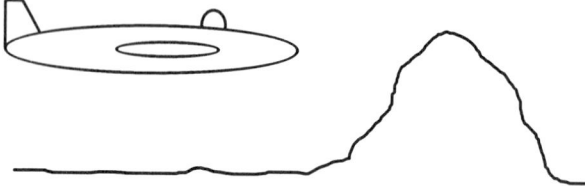

Embedded systems are reactive systems. In many applications they must be viewed as real-time systems. Real-time systems are divided into hard real-time and soft real-time. The picture above illustrates a hard real-time system, for which failure to react by a given deadline can be catastrophic. Soft real-time systems perform tasks which should be completed within specified periods, but the consequences of not meeting the deadlines are not severe.

As discussed earlier, microcontrollers keep track of different events (external or internal) by manipulating flags. There are two methods for the CPU to interact with the events (tasks) through the flags.

• First, the polling approach results in a wait loop. For example, a character must be sent over the serial port and the program checks the flag TI in a loop. When the flag TI is set (serial buffer is empty), the microcontroller leaves the loop and moves a byte to the serial buffer register. Obviously, this method is not perfect. Even for high baud rates, there is big difference in speed between the CPU and the serial port.

• Second, the interrupt method implements a hidden check of the flags. Hidden, because it is done by hardware, concurrently with the program execution. When the hardware finds an asserted flag, an interrupt request is generated. The CPU completes the instruction in progress, saves the Program Counter so that it can return later and jumps to a specific program called interrupt subroutine. When the interrupt subroutine is over, the processor jumps back to the main program by reloading the Program Counter with the saved address. The interrupt method may give us timing parameters which we would not otherwise be able to obtain.

Different applications may demand different parameters of the interrupt system. For example, an embedded system may require several external interrupt inputs. Contrary to that, another application may not use external interrupt inputs at all, but the system performance would rely on a prioritized interrupt scheme. Figure 2.19 shows the interrupt system. The system can be broken down into three parts: interrupt sources, enable circuit and priority circuit.

Interrupt sources

There are six independent events which can generate interrupt requests: conditions on the interrupt inputs, overflows of the timer/counters and completion of the serial port transmission/reception. We marked the sources in Figure 2.19 by thick line - two external interrupt inputs $\overline{INT0}$ and $\overline{INT1}$, two timer/counter interrupt flags TF0 and TF1, and finally the serial port by means of the flags TI and RI. Note that the interrupt flags TI and RI generate a common request. The interrupt flags are active when they are high.

The external inputs can be used as either low level activated or falling edge activated interrupt sources. The input $\overline{INT0}$ is programmed by bit IT0 (Interrupt 0 Type) in register TCON. The input $\overline{INT1}$ depends on bit IT1, register TCON.

When the aim is the low level mode, the control bit IT0 (IT1) is reset. The external interrupt inputs are sampled once each machine cycle. Therefore, to be accepted correctly, an input signal should last at least one machine cycle (12 oscillator periods). However, the proper operation requires the low level to continue more than one cycle. A low level applied to the input $\overline{INT0}$ ($\overline{INT1}$) will keep the interrupt flag IE0 (IE1) high. In low level mode an attempt is made to clear the flag every cycle. As long as the input remains low the attempt will be unsuccessful. The input signal must be hold low until the service routine is under way and must be pulled high before the end of the interrupt subroutine.

An interrupt input will work in the falling edge mode if the corresponding control bit (IT0 or IT1) is set. The input should be pulled high for at least one cycle and then pulled low for at least one cycle. If this condition is satisfied, the corresponding flag IE0 or IE1 (Interrupt 0/1 Edge flag) will be set. The flag is cleared automatically by hardware when the service routine is in progress.

Enable circuit

The second part of the interrupt system is an enable/disable block. Each source needs an individual and a total permission. This phase is controlled by an Interrupt Enable register (IE). Each source is enabled by setting the corresponding bit (closed switch in Figure 2.19). Bits Enable eXternal (EX0/EX1) take care for external interrupt inputs. Flip-flops Enable Timer (ET0/ET1) cover timer/counters and bit Enable Serial (ES) is devoted to the serial port. Finally, a bit called Enable All (EA) controls all sources.

Note that setting an interrupt enable flag may unlock requests which are too old to be taken into account.

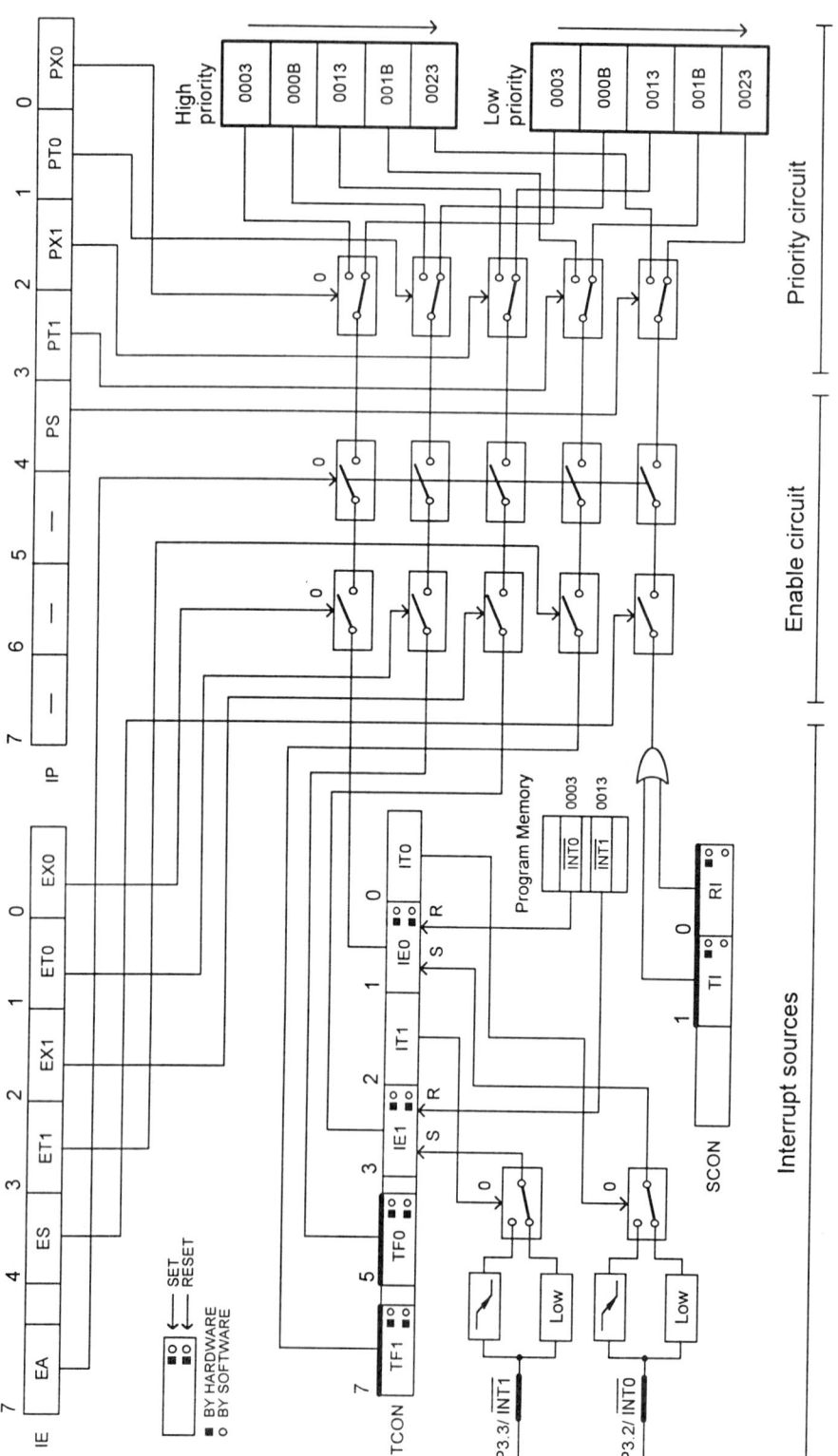

Figure 2.19 The 8051 interrupt system.

Priority circuit

The third part of the interrupt system is a priority level circuit. There are two priority levels: high and low. Each request can be individually directed to a priority level by an Interrupt Priority register (IP). If a bit in register IP is set, the corresponding interrupt will join the high priority group.

Bits Priority eXternal (PX0/PX1) define external request priority. A timer/counter's priority is controlled by bits Priority Timer (PT0/PT1). A flip-flop Priority Serial port (PS) selects the serial port priority.

Furthermore, Figure 2.19 shows the vector locations (from 0003H through 0023H) which are reloaded to the Program Counter in the event of interrupt.

If at a certain moment an external interrupt input has to be enabled and old requests rejected, the flag IE0 (IE1) is cleared first and the corresponding bit in register IE is set afterward.

A low priority servicing subroutine can be interrupted by a high priority interrupt request. A high priority interrupt can not be interrupted. To follow this procedure, the interrupt system includes two non-program accessible flip-flops. One of them shows that a high priority interrupt is being serviced and does not allow further interrupts. The other indicates that a low priority interrupt is under way and prevents all but high priority interrupts.

In case of two or more requests received simultaneously, an internal priority rule is used. As shown in Figure 2.19, external interrupt $\overline{INT0}$ possesses the highest priority level in each group and the serial port interrupt is associated with the lowest.

The interrupt system hardware is ready with the asserted flags (and their priority) until state S6 of every cycle.

The interrupt of highest priority will commence with state S1 of the next cycle if the process is not blocked by any of the following conditions:

• An interrupt of equal or higher priority level is under way.

• The current machine cycle is not the final cycle in the instruction (the instruction in progress must be completed).

• The instruction in progress is RETI or any access to registers IE or IP. An important point here is that at least one other instruction will be executed after the instruction RETI or any read or write concerning registers IE or IP.

Thus, the interrupt procedure includes the following actions:

• One of the non-program accessible flip-flops is set (the priority level is fixed).

• The interrupt flag is cleared. This action does not affect the flags TI and RI.

• The contents of the Program Counter are pushed onto the stack.

• The Program Counter is reloaded with the corresponding vector location. The execution proceeds until the RETI instruction is reached. The RETI instruction clears the priority level flip-flop which is active, pops the Program Counter from the stack and continues the interrupted program.

The interrupt system response time is measured from the interrupt request till the beginning of the first instruction of the interrupt subroutine. The 8051 microcontroller response time is always more than 3 cycles and less than 9 cycles. If the application requires registers to be saved in the stack, the actual response will be additionally delayed.

2.10 Power reduction modes

In the domain of portable systems and remote devices, the power consumption is imperative. The CMOS 8051 microcontrollers have two power reduction modes: Idle and Power Down. Both modes are activated by software.

Idle mode

Figure 2.20 shows the CMOS 8051 microcontrollers clock tree. If an instruction sets the bit IDL in register PCON, it will be the last instruction executed before the Idle mode gates off the clock signal from the CPU. The consumption is declined about 80 - 90%. The interrupt system, serial port and timer/counters continue to run. Internal RAM and SFR save their information.

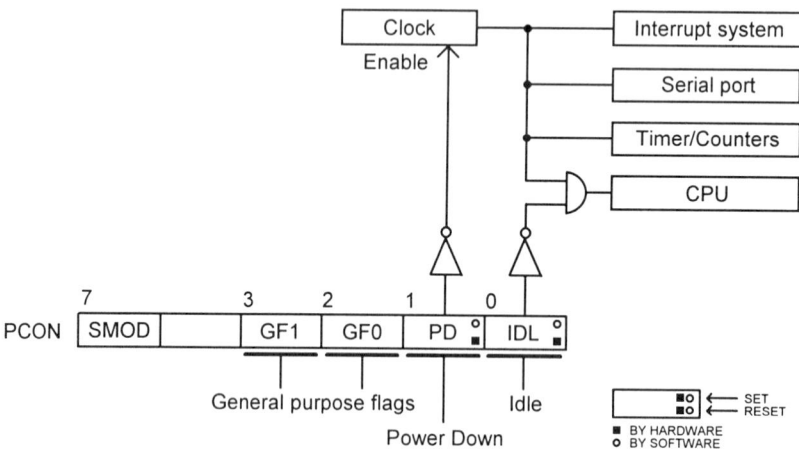

Figure 2.20 The microcontroller's clock tree.

Since the register PCON is not bit addressable, a byte instruction must be used to set the flag. For example,

```
ORL     PCON,#1     ; Enter Idle mode
```

Bits GF0/GF1 (General-purpose Flag) could be set simultaneously with the bit IDL to indicate that the Idle mode has been entered.

Figure 2.22 shows output signals for both power management modes. In Idle mode, outputs ALE and $\overline{\text{PSEN}}$ emit 1's. The intention is for external Program Memory to disable its output buffers. The port pins save their pattern in Idle mode. When the microcontroller executes out of external Program Memory, Port 0 is left in an off state and Port 2 continues to emit the high byte of the address (PCH).

There are two possibilities to terminate Idle mode.

• If an interrupt is activated, the bit IDL in register PCON will be cleared by hardware. Figure 2.21 shows the sequence of actions associated with Idle mode. When Idle mode is terminated by an interrupt, the microcontroller runs a subroutine. Immediately after the instruction RETI which marks the end of the subroutine, the microcontroller jumps to the instruction next to the one that activated the Idle mode.

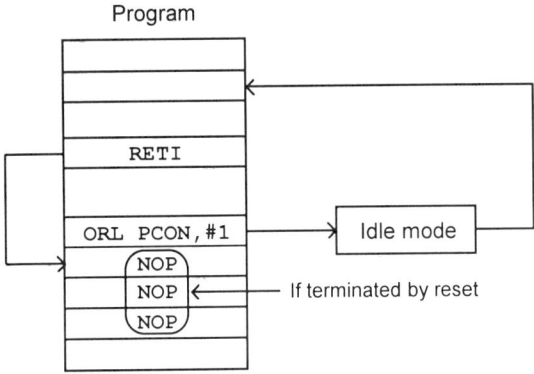

Figure 2.21 Idle mode operation.

• The Idle mode can also be terminated by a reset. In that event the RST input needs to be pulled high for two machine cycles (the clock is still running). Clearly, all SFRs are initialized to their reset values and the program should start from address 0. Unlike what you might expect, two or three machine cycles may take place before the internal reset procedure takes control. During that period, the microcontroller will execute instructions from where it left off. The problem will be solved if three **NOP** instructions follow the instruction that invokes the Idle mode.

Power Down mode

Power Down mode is activated by setting bit PD in the register PCON. As can be seen from Figure 2.20, the clock is frozen and it does not matter if the internal or external oscillator is used. The internal RAM saves its data. The outputs ALE and $\overline{\text{PSEN}}$ emit 0s. The reason is minimal power consumption.

Pin	Idle mode		Power Down mode	
	Internal execution	External execution	Internal execution	External execution
ALE	1	1	0	0
$\overline{\text{PSEN}}$	1	1	0	0
P0	SFR data	OFF	SFR data	OFF
P1	SFR data	SFR data	SFR data	SFR data
P2	SFR data	PCH	SFR data	SFR data
P3	SFR data	SFR data	SFR data	SFR data

Figure 2.22 Output signals in power-saving modes.

The port pins continue to output whatever data was written to them. There is only one exception for Port 0. The only way to terminate Power Down mode is a reset. The input RST must be high at least 2 machine cycles (we assume that the oscillator has already been stabilized, which usually takes 10 ms).

In Power Down mode the supply voltage can be reduced to as low as 2V. The supply voltage, if changed, must be decreased and restored within the Power Down mode.

2.11 Programming the internal Program Memory

Memory technologies were discussed in section 1.5. Figure 2.23 shows a classification for the 8051 microcontroller internal Program Memory. The in-system reprogrammable flash memory is the most convenient option. Using microcontrollers with internal Program Memory may decline the number of components in the system. Along with the code, the designers can program a few lock bits which will help to protect the code.

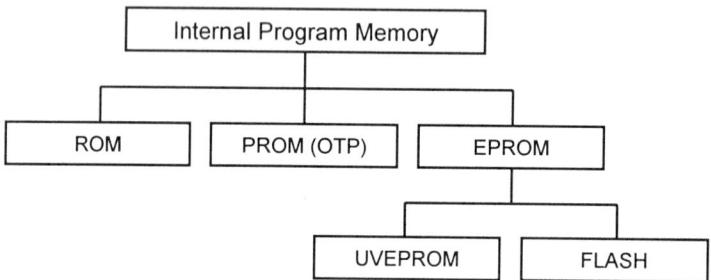

Figure 2.23 Implementations of the internal Program Memory.

Figure 2.24 is a simplified description of the programming procedure. The microcontroller is an 8051 software and pin compatible version produced by Atmel [Atme 1997]. Some of the AT89C52 main features can be seen in Appendix A. Essentially, the microcontroller possesses 8K bytes flash memory which uses a low-voltage (V_{CC}) program enable signal. Furthermore, the Program Memory is erased electrically for 10 ms. When erased, all locations contain FFH.

Figure 2.24 shows three modes: program, read (verify) and erase. The memory is programmed byte-by-byte. Once a bit has been programmed to low, the only way to set it is to erase the entire flash memory array. The programming mode requires the combination of control signals shown in Figure 2.24. The input ALE / $\overline{\text{PROG}}$ is pulsed down to accomplish the programming of one byte. The byte-write cycle is self-timed and the programming pulse can vary between 1 and 110 μs. Typically, one byte is programmed for no more than 1.5 ms. The internal flash memory is erased by pulling down the input ALE / $\overline{\text{PROG}}$ for 10 ms. In the read (verify) mode external pull-up resistors must be used.

The AT89C52 microcontroller has three lock protection bits. Programming the lock bits we can achieve three levels of protection. First, MOVC instructions executed from external Program Memory can not read bytes from the internal Program Memory. Second, on the top of the first level, verify is disabled. Finally, the external execution is disabled as an additional restriction to the previous level.

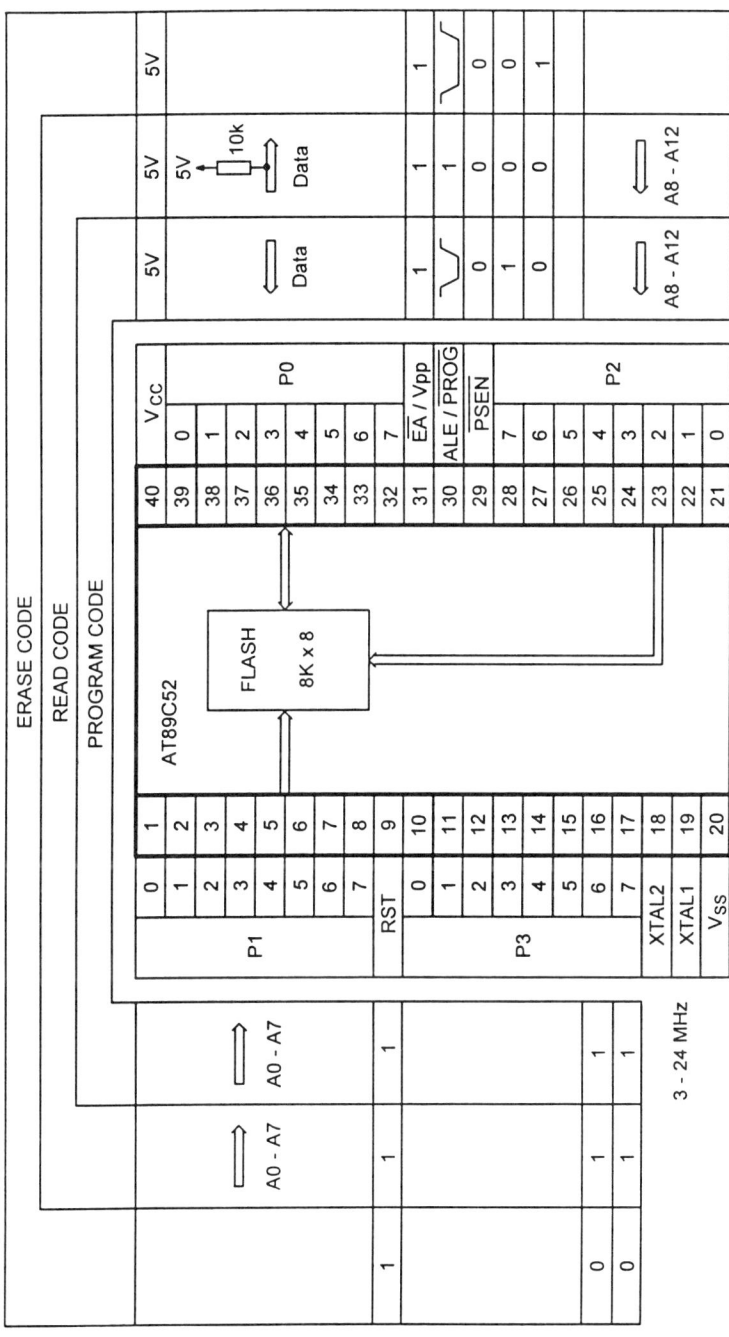

Figure 2.24 Programming the internal flash memory.

2.12 References

Atmel, *Microcontroller Data Book*, 1997.

Kenneth J. Ayala, *The 8051 Microcontroller*, West Publishing Company, 1991.

Intel, *MCS 51 Microcontroller Family User's Manual*, Volume I, 1992.

Intel, *Embedded Microcontrollers*, 1996.

Zdravko Karakehayov, Knud Smed Christensen and Ole Winther, *Embedded Systems*, Technical University of Denmark, Department of Applied Electronics, 1995.

Zdravko Karakehayov and Stanislav Grigorov, *Single Chip Microcomputers*, Technica, Sofia, 1992.

Zdravko Karakehayov and Emil Saramov, *Applied Microcomputer Systems*, Technical University of Sofia, 1995.

Philips Semiconductors, *80C51-Based 8-Bit Microcontrollers, Data Handbook IC20*, 1997a.

James W. Stewart, *The 8051 Microcontroller*, Regents/Prentice Hall, 1993.

Sencer Yeralan and Ashutosh Ahluwalia, *Programming and Interfacing the 8051 Microcontroller*, Addison-Wesley Publishing Company, 1995.

Information about 8051 microcontrollers can be found at:

AMD	http://www.amd.com
Atmel	http://www.atmel.com
Dallas Semiconductor	http://www.dalsemi.com
Integrated Silicon Solution	http://www.issiusa.com
Intel	http://www.intel.com
OKI	http://www.okisemi.com
Philips Semiconductor	http://www.philips.com
Siemens	http://www.siemens.com

An 8051 directory, searchable by product type and company is published by

Miller Freeman, Inc.   http://www.directories.mfi.com/embedded/8051

A large spectrum of 8051 related subjects are discussed by Russ Hersch in

http://www.faqs.org/faqs/microcontroller-faq/8051

Chapter 3

THE 8051 ASSEMBLY LANGUAGE PROGRAMMING

3.1 Introduction

In the last chapter, we discussed the 8051 microcontroller architecture, memory organization and embedded peripherals. In this chapter, we are going to discuss how to build programs which harness the microcontroller's hardware to perform a certain job.

First, forget for a while about the hardware details. Think of the concept address space and get down to the specific implementation presented in Figure 2.3. As discussed then, the 8051 microcontroller is capable of accessing locations in three byte addressable spaces plus a bit addressable space. So, when an instruction is fetched, the first byte (the opcode) specifies not only the operation but, also, the selected address space. Furthermore, the instruction contains the location address within the indicated address space. Therefore, a value in the instruction address field does not mean anything if we do not know the address space.

In essence, instructions specify operations and address spaces combined with corresponding addresses. In an attempt to make the program more readable, we can use the standard register names, such as A, B or DPTR rather than specific addresses. Analogously, we could employ labels attached to addresses, which is a common practice as well. In both cases the assembler (as a program created to help us), will convert all names into specific addresses.

When we split an instruction into operation, address space and address, we should take into account some alternative solutions. In fact, we need rules to distinguish, for example, between actual address and an immediate value defined directly within the instruction. Following a set of rules, we will be able to apply a certain number of addressing modes.

3.2 Addressing modes

The 8051 microcontroller employs eight addressing modes. These are the register addressing mode, the direct addressing mode, the immediate addressing mode, the indirect addressing mode, the based-indexed indirect addressing mode, the relative addressing mode, the extended addressing mode and the implied addressing mode.

Register addressing mode

The idea behind the register addressing mode is to form short instructions, for example, in one byte. The one byte long opcode specifies both operation type and address. As the name implies, the address indicates a certain register. All registers are placed in the internal Data Memory address space. Using one byte we have 256 combinations available. Obviously, there is a trade-off between the number of operations and the number of registers. The designers of the 8051 microcontroller selected accumulator, register B, register DPTR and the registers from the current bank R0 through R7 to be used in the register addressing mode.

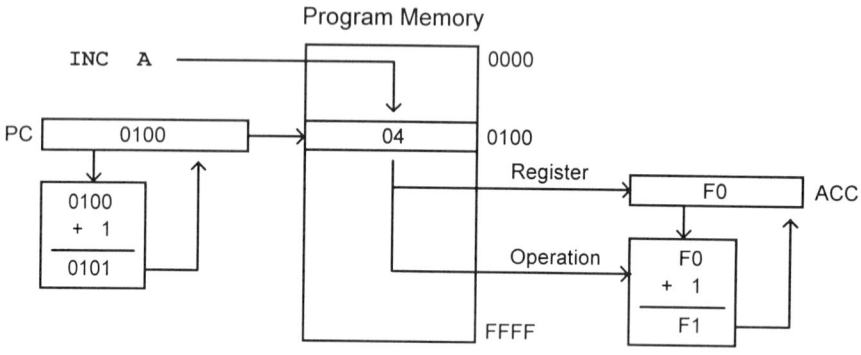

Figure 3.1 Register addressing mode.

Figure 3.1 illustrates the register addressing mode. The example instruction is increment accumulator (`INC A`). The instruction employs encoding 04H. This code, viewed as an operation, is an increment. Along with the operation, the same code specifies as an operand accumulator. In this example, the accumulator contains F0H before and F1H after execution of the instruction. When an instruction is in process of completion, the Program Counter must address the first byte of the following instruction. In this particular example, the instruction is one byte long and the Program Counter is incremented as well.

Direct addressing mode

Direct addressing mode instructions specify operands within the internal Data Memory address space. An example of pure direct addressing mode is shown in Figure 3.2. The instruction

        MOV   10H,20H

moves a byte from address 20H to address 10H. Both locations belong to the internal Data Memory address space (direct address).

The instruction opcode 85H indicates a move operation. Also, it says that the following bytes are a source address and a destination address. In this example, a data byte 22H is moved from direct address 20H to direct address 10H. The instruction is three bytes long and the Program Counter is increased to 1003H.

Immediate addressing mode

In the last example (Figure 3.2), we moved 22H to a certain destination (direct address 10H). Our aim was to move the value stored at direct address 20H. In some cases, the programmer knows the value which should be moved and makes it a part of the instruction.

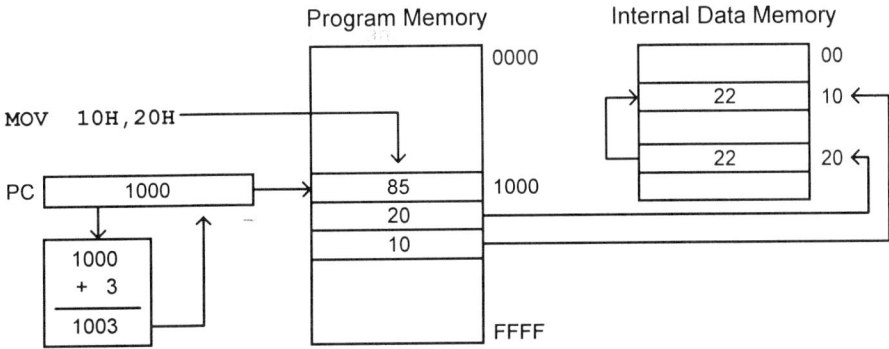

Figure 3.2 Direct addressing mode.

Figure 3.3 shows an example instruction. The opcode is 75H. The second byte defines the destination address (10H). The third byte contains the immediate value (22H), which is copied to the destination address. The pound sign (#) in the instruction mnemonic specifies immediate data. If we miss the #, the assembler will use the direct addressing mode. Consequently, the instruction will move a random value taken from direct address 22H.

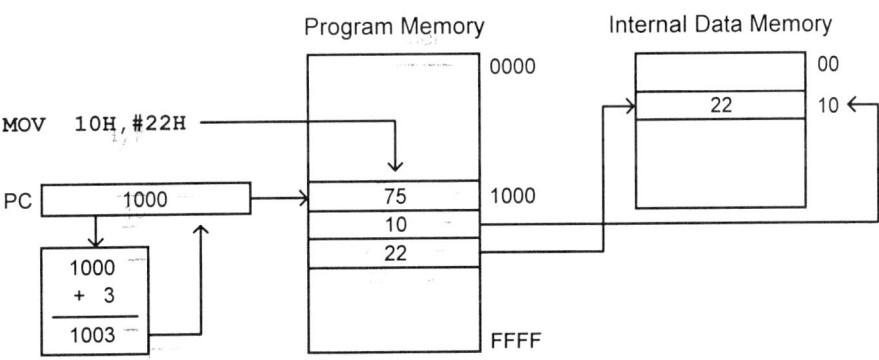

Figure 3.3 Immediate addressing mode.

Indirect addressing mode

The idea of using a two-step procedure to find the actual address was briefly mentioned in section 2.5, Figure 2.8. First, the opcode points out a register which could be R0, R1 or DPTR. Second, the code in the register is used as an operand's address. The indirect method is related to the internal Data Memory address space (via registers R0 and R1) and to the external Data

Memory (through registers R0, R1 and DPTR). As far as the instruction mnemonic is concerned, the symbol @ must precede the register name.

An example of a move instruction based on indirect addressing mode is shown in Figure 3.4. The instruction opcode F6H says that the register R0 holds the actual address (47H). The accumulator code (1EH) is moved to the destination address (47H).

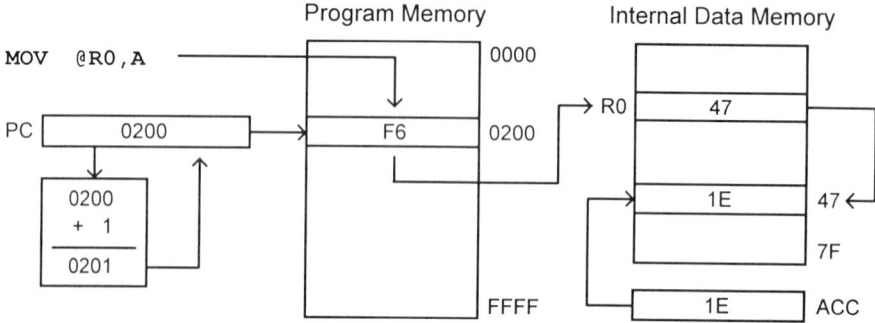

Figure 3.4 Indirect addressing mode.

Based-indexed indirect addressing mode

As the name implies, this is again an indirect addressing mode, however there are two registers involved. The address is the sum of the codes in the registers. Two instructions of this type are available:

```
MOVC    A,@A+DPTR
MOVC    A,@A+PC
```

Both instructions move code from the Program Memory to the accumulator. A typical application is access to look-up tables. A base register can be either DPTR or the PC. The accumulator points to a certain value within the table.

Figure 3.5 shows an example of based-indexed indirect addressing. In the end of the instruction ACC contains E4H.

Relative addressing mode

Along with the data transfer and arithmetic instructions, the code usually employs and program control instructions. These instructions, such as jump, call to subroutine, or return from subroutine, define the actual order of execution. Reduced to its basic principle, the program control instructions modify (or not) the code in the Program Counter. In general, there are two options. First, a new value is moved to the Program Counter. Second, in the case of relative addressing mode, the code in the PC is modified by adding a displacement, which is a part of the instruction.

The displacement code is added to the PC as a signed two's complement number. Consequently, the jump is either forward or backward in the program depending on the offset sign. Figure 3.6 shows an example of a jump instruction based on relative addressing mode. The instruction opcode is 80H. With the assumption that the program has to jump to address 4008H, the offset must be 06H. Remember, the Program Counter is always incremented to the point of the first byte of the following instruction. On the top of this, the microcontroller adds the offset (06H).

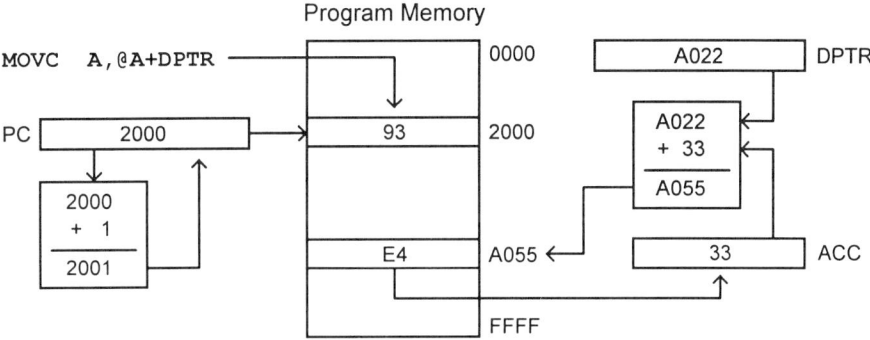

Figure 3.5 Based-indexed indirect addressing mode.

As yet another example, if the displacement were EEH (negative), the program would jump to address 3FF0H (backward in the memory).

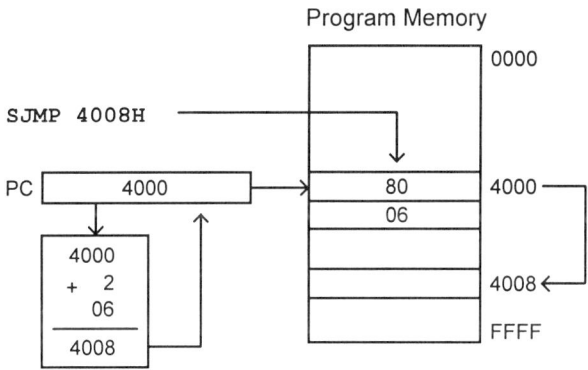

Figure 3.6 Relative addressing mode.

Extended addressing mode

While discussing the indirect addressing mode earlier in this section, we explained how to access the 64K byte external Data Memory. Also, we noticed the use of the based-indexed indirect addressing mode for the Program Memory. Furthermore, we presented an instruction which makes the program to jump as far as a byte offset allows. Of course, a one byte displacement is not enough to reach an address everywhere in a block of 64K bytes. Obviously, we need a method to jump and call subroutines in the full spectrum of the Program Memory. The extended addressing mode satisfies this demand.

Figure 3.7 depicts an example instruction, a long jump to address 60F0H. The second and third bytes, which carry the jump address, are copied to the Program Counter. This example reflects a basic rule for the 8051 microcontroller: the low-order address byte is placed in high-order address location.

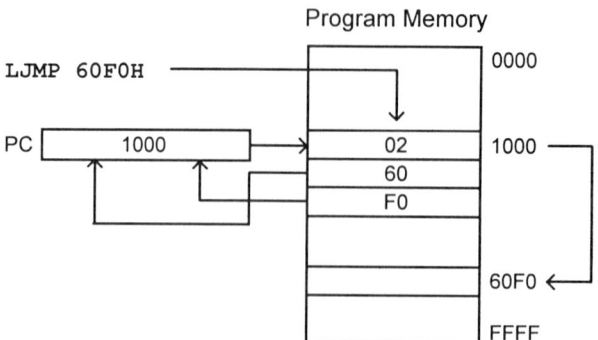

Figure 3.7 Extended addressing mode.

Implied addressing mode

There are cases when the instructions do not specify directly any operands. For example, a no operation instruction (NOP) which does nothing, but to increment the Program Counter, belongs to this group. In fact, every microcontroller needs a NOP instruction. For instance, we might need to introduce a delay in the program.

3.3 Instruction set

The 8051 microcontroller instruction set includes 111 instructions. Most of them (49) are single-byte, 45 are two-bytes and 17 are three-bytes long.

The instructions, organized alphabetically, are presented in Appendix D. The following abbreviations are relevant:

R_n	Register R0 - R7 of the currently selected register bank.
dir	8-bit internal Data Memory address (direct address).
$@R_i$	8-bit internal Data Memory or external Data Memory address defined indirectly through register R0 or R1.
#data	8-bit constant included in instruction.
#data16	16-bit constant included in instruction.
addr16	16-bit destination address.
addr11	11-bit destination address.
rel	Signed (two's complement) 8-bit offset byte.
bit	Bit address.

In Appendix D we present the instruction mnemonics (in bold) and a short description followed by the number of bytes, cycles, encoding (the machine code), the affected bits from the register PSW and detailed description. Appendix E contains a summary of the instruction set.

The 8051 microcontroller instructions can be split according to their functions as follows:
- Data transfer instructions
- Arithmetic instructions
- Logical instructions
- Boolean instructions
- Program control instructions.

Before discussing the instructions in each group we show how assembly language programs are created and debugged.

3.4 Assemblers and linkers

Traditionally, each line of an assembly language program contains one instruction. What is retrieved from the instruction set is the mnemonics of the instructions. What is added to form a readable program are labels and comments. Thus, each line is composed of up to four sections, as can be seen in the following example:

```
Labels       Opcode        Operands         Comments
DEL1:        MOV           A,#2FH           ;Load accumulator
```

At this point, it is natural to ask, how we are going to convert the assembly language program to the 8051 machine language? In the case of a short program, we can hand code it. If we use a program, termed an assembler, to convert our source program into ones and zeros, we could benefit from computer assistance. Naturally, we should be aware of the assembly language mnemonics and the rules imbedded in the assembler.

Practically, the translating program runs on a personal computer and is executed on another CPU. If that is the case, the precise term is cross-assembler.

The instruction fields can be used as follows:

- The label field is used to attach symbolic names to assembly language instructions. In this way, an instruction may refer to the label, rather than to the memory address. Labels must begin in the first column and possess a leading alphabetical character. Labels must be followed by a colon.

- The opcode section indicates a specific operation. For example, MOV for move, INC for increment and so on.

- The operands field states the specific registers or memory locations which are involved in the operation. For instance, the following instruction

```
MOV    R1,5AH
```

specifies for a source operand the value in the internal Data Memory, address 5AH. The destination operand is the register R1. The source operand is always being the one on the right and the destination operand, the one on the left. Operands are separated from the operator by one or more spaces.

As always, there is a rule which is used for the numbers type definition. We have just employed H to indicate hexadecimal. Moreover, binary digits must be followed by a B. The default (no letter) is a decimal number.

Essentially, what you write in the operands field defines the addressing mode.

• The comment field contains information of how the program works. Even though the assembler does not process any of the text after the comment delimiter, the field should not be neglected. Comments are the most reliable source of information about the program logic and design history. The comment field is separated from the operands by a semi-colon. A comment-only line is indicated by a semi-colon character on the first position.

It is, perhaps, worth mentioning at this point, that we have to advise the assembler where to place the binary codes. For example, an interrupt subroutine must be bound to its vector. In this and some other cases we use assembler directives to instruct the assembler. The assembler directives are not translated into machine codes. For example, the directive origin

          ORG   2000H

tells the assembler that the first byte of the following instruction must be placed on address 2000H.

Likewise, we can use an assembler directive EQU (equate) to define a value which the assembler will use to substitute in other instructions. For instance,

          RESET   EQU   30H

So, whenever the word **RESET** appears in an instruction, the assembler substitutes it by 30H. This approach could be applied for both labels and immediate values.

Another assembler directive, called define byte, allocate, define and name bytes in the memory. For instance,

          TEXT1:   DB   'ABCD'

The end of the program is indicated by a directive **END**.

When we write an assembly language program and the application requires a group of instructions to be executed several times, we could organize a subroutine. Consequently, the subroutine code is stored in the memory only once. One call instruction is used to activate the subroutine which saves memory. Inevitably, this approach requires access to the stack and also results in additional execution time.

There is another possibility supported by the assemblers which allows us to write a certain group of instructions only once. It is called a macro. By means of two words, **MACRO** and **ENDM**, the instructions are grouped under a certain name. Every time the assembler meets a macro name in the program it inserts the corresponding group of instructions defined in the beginning. It is certainly implied that when a macro option is used, the program memory and execution rate are not changed.

Figure 3.8 shows the steps we follow to create and debug an assembly language program. Initially, we use a program called editor to type in our program and create a source file. For example, we type **EDIT** and press return. An editor, which comes with DOS, is off and running. We type in the program, save the file under a certain name, for example, **TEST.ASM** and exit the editor.

The next step is to assemble the program. Different people can have different assemblers. The assembly language examples in this book have been processed by a program named **ASM51** [Chri 1995]. Thus, we type **ASM51 TEST** and press return. The assembler translates the instructions into machine code and eventually displays error messages on the screen. The most frequent errors are incorrectly spelled mnemonics and undefined names. Unfortunately, the assembler is not capable of finding instructions (or their parts) which are inconsistent with the logic of the code.

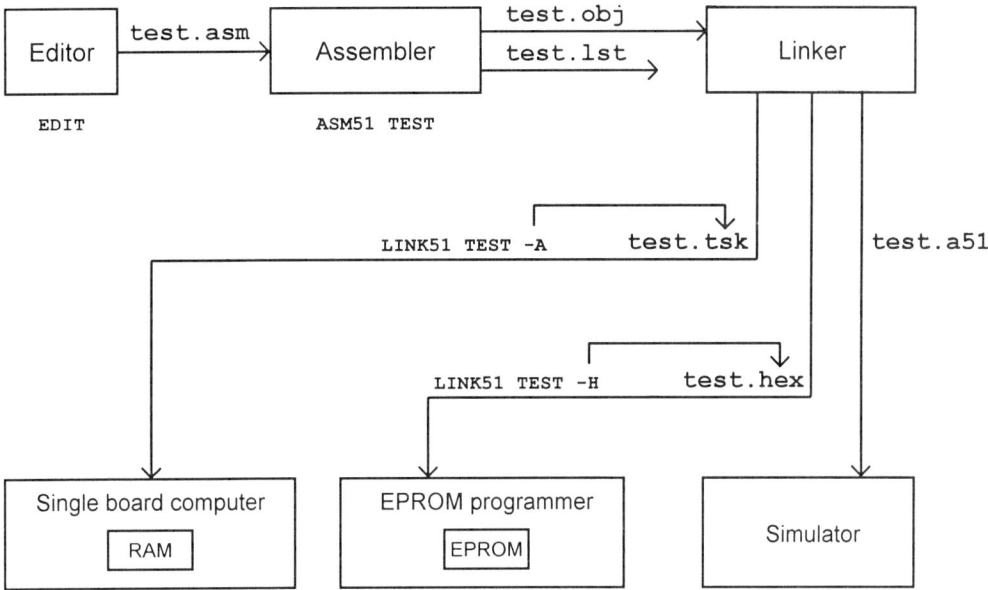

Figure 3.8 Steps in creating and debugging assembly language programs.

The assembler generates a file called object file. The object file contains the program translated into machine language, but it still is not ready to run. The extension of this file is .obj. In addition, the assembler yields a file called list file, extension .lst. The list file combines the assembly language statements, the binary codes and an address part. The list file is generally used to produce a printout of the program in order to make the debugging process more convenient.

Once we have eliminated all errors reported by assembler, we are ready to approach the next phase, to link the program. This is done by a program called linker. Now we take a final decision about the addresses and produce executable code. As the name implies, we can also link together modules which are written and debugged concurrently. However, the linker must be used, even if the program has only one module.

The standard input for a linker are files with extension .obj. We can start the linker by typing LINK51 TEST. As this program is a part of a bigger system, it has been designed to work in conjunction with a simulator [Chri 1995]. The simulator entry utilizes a file test.a51 which is generated by the linker. An alternative solution is to execute and debug the program on a single board computer, such as the one designed in section 4.5. Furthermore, you might want to program an EPROM. An example design of an EPROM programmer can be found in section 11.2. Logically, when the final destination of the code vary between different systems, the file itself has to be different. Practically, you can force the program LINK51 to produce pure code (test.tsk) by typing

```
LINK51 TEST -A
```

The file **test.tsk** can be downloaded into a single board computer's RAM and executed from there (see also section 4.5). Due to the fact that files with extension **.tsk** do not carry the starting address, the single board computer must be advised where to place them in the memory. Furthermore, if the goal is to burn an EPROM, we must use a file format consistent with the EPROM programmer, such as Intel HEX format (see an example in section 3.10). Typing

        LINK51 TEST -H

will instruct the linker to generate an Intel HEX file (**test.hex**). In contrast to the previous option, the Intel HEX file contains both code segments and there starting addresses. Again, the code can be downloaded to a single board computer for debugging purposes. In theory, you could debug the code by burning EPROMs, however this is not very practical. EPROMs will come on the scene when the software has been proved bug-free.

3.5 Data transfer instructions

The discussion about data transfer instructions was begun in section 2.3, when we introduced the microcontroller's memory address spaces. As we indicated by examples, MOVC instructions read from the Program Memory, MOV instructions access the internal Data Memory and MOVX instructions work with external Data Memory. In addition, bit-oriented instructions are capable of manipulating the bit-addressable area.

The data transfer instructions can use a variety of addressing modes. For example,

        MOV   A,1FH

moves a value from location 1FH (direct address) to the accumulator. The instruction applies register addressing mode for the first operand and direct addressing mode for the second. In fact, both source and destination are placed in the internal Data Memory address space.

Likewise, the instruction

        MOV   A,P1

reads Port 1 and moves the value to the accumulator. As you might expect, P1 is a predefined symbol of the **ASM51** assembler. When the assembler passes through the source file, P1 is replaced by 90H. Eventually, if you rewrite the instruction into

        MOV   A,PORT1

the assembler will look for a directive, such as

        PORT1   EQU   90H

to find the code that corresponds to the name PORT1. All SFR names are predefined symbols and can be used freely in the program (see Figure 2.4, section 2.3 or Appendix B).

The distinction between different addressing modes is useful and it helps organize our program using the most suitable instructions. For example, we want to move a constant (10H) to accumulator. There are two options to do this and they are illustrated by the following instructions:

        74 10   MOV   A,#10H   ; First option
        75 E0 10   MOV   ACC,#10H   ; Second option

The hexadecimal digits in front are the machine code. While in the first instruction the register addressing mode is used for the first operand (A), the second instruction replaces the name ACC with its direct address E0H (direct addressing mode). Consequently, the second instruction becomes one byte longer. Furthermore, the execution time is increased from one to two cycles.

There are cases, however, when we do not have alternative addressing modes. For instance, the access to the external Data Memory always requires indirect addressing. Indeed, we could use different registers to introduce the address indirectly. Glance through the following sequences of instructions and compare the efficiency:

```
74 66      MOV    A,#66H         ; Load accumulator
90 11 22   MOV    DPTR,#1122H    ; Select the address
F0         MOVX   @DPTR,A        ; Move accumulator
```

against

```
74 66      MOV    A,#66H     ; Load accumulator
75 A0 11   MOV    P2,#11H    ; Select the high-order address byte
78 22      MOV    R0,#22H    ; Select the low-order address byte
F2         MOVX   @R0,A      ; Move accumulator
```

Both groups of instructions move a constant (66H) to an address (1122H) in the external Data Memory. The first approach (through register DPTR) is more natural. It consumes less memory and moves the data faster. However, if the program repeats the access several times within the current page (only the low-order byte is changed), the second choice is more attractive. The high-order byte is established by a move to Port 2 and from then on the low-order address byte is what makes the difference. Related examples, which indicate exactly how beneficial it could be to migrate from the first to the second option, are presented in sections 5.3 and 5.4.

There is another method for data transfer which deserves closer attention. Forget for a while about electronics, and think instead about mechanics. Figure 3.9 shows a container used to store different things which, for example, could be coins. When we add a new one, we press the spring to align the new-comer to the edge. Also, when we take a coin out, the spring shifts all coins. In principle, the last coin in will be the first one out (LIFO buffer). The microcontroller's stack memory is analogous to the mechanical model. The stack memory is an area of the internal Data Memory used for temporary storage of data or return addresses. In fact, there is a difference in the implementation. Rather than shifting all bytes, the microcontroller uses the register Stack Pointer (SP) for indirect addressing of the memory.

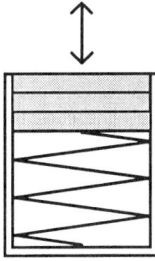

Figure 3.9 The stack memory mechanical analogy.

Figure 3.10 shows how the stack grows up in memory after execution of three instructions PUSH. The SP is incremented by one before the data is stored in the RAM. As a result, the register SP points to the last byte written. When an instruction POP is executed, the SP is decremented by one after the data is moved to the destination. Both instructions combine direct addressing mode with indirect addressing through register SP. For example,

    PUSH 30H

copies a byte from direct address 30H onto the stack.

  Similarly,

    PUSH ACC

saves the accumulator in the stack. When we want to restore the accumulator, we can use the instruction

    POP ACC

Now, the address E0H in the internal Data Memory is the destination address.

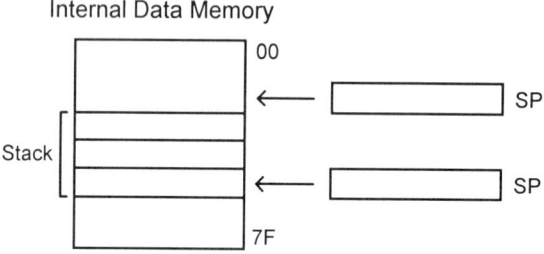

Figure 3.10 The 8051 microcontroller stack memory.

After reset the SP is set to 07H. Therefore, the first PUSH will affect register R0 from bank 1. It is very common for the SP to be initialized to 2FH. In this way, the stack is placed just immediately after the bit addressable area.

One caveat: the SP can be set to any code, however pushes in the address range 80H through FFH (outside the internal RAM) have no effect.

3.6 Arithmetic instructions

Even though the 8051 single chip microcomputer is a control dominated machine, it possesses a rich set of arithmetic instructions. The arithmetic instructions can be further subdivided into

- Addition and subtraction instructions
- Increment and decrement instructions
- Multiplication and division instructions.

Addition and subtraction instructions

Being an 8-bit machine, the 8051 microcontroller is capable of adding two 8-bit numbers together using one of the various ADD instructions. In all cases, the accumulator holds an operand and the result, also, goes to the accumulator. Furthermore, the register PSW, by means of the flags carry and overflow, indicates if the operation has been carried out normally or the result should be tackled in a special way.

The carry flag is set if there is a carry from bit 7. The overflow flag is set if there is a carry from bit 7, but not from bit 6, or there is a carry from bit 6, but not from bit 7.

Figure 3.11 illustrates five typical examples. First, we distinguish between unsigned and signed binary numbers. Second, we present addition examples and discuss the results. Finally, we include assembly language instructions which implement the examples.

The byte patterns can be viewed from different angles. Unsigned numbers swing between 0 and 255, as shown in Figure 3.11. Signed numbers use seven bits for magnitude and a bit for the sign. A zero in the most significant bit indicates a positive number. A one in the same bit is used for negative numbers. As a result, the signed numbers can vary between 0 and 127 and between -1 and -128 (two's complement code). You can follow a simple rule to convert the signed numbers into two's complement code. The positive numbers do not need any change. The negative numbers are converted by complementing each bit and adding 1 to the result. For example,

```
-5      0000 0101
        1111 1010
             +1
        ─────────
        1111 1011
```

• Example U1. The microcontroller adds 10 and 132. As the binary form is essential (the attached decimal numbers are just to make the example more readable), we get 1000 1110. This number converted to decimal is 142. The carry flag is reset. This example can be tested by a MOV and an ADD instruction, as shown in Figure 3.11.

• Example U2. This example is more demanding. The result is bigger than 255 and the carry flag is set. Thus, the actual sum is nine bits long. The example instruction ADD demonstrates register addressing mode for the second operand.

• Example S1. In this case, the microcontroller adds +20 and -10. The result is 0000 1010, which converted to decimal is +10. Since the overflow flag (OV) is reset, we ignore the carry. The example instruction ADD is another option for the second operand, the direct addressing mode.

• Example S2. We reached the point where the situation seems to be out of control. The microcontroller adds 96 and 96. We expect the result to be 192. Unfortunately, the microcontroller suggests -64. Since, we have carry from bit 6 and no carry from bit 7, the overflow flag is set. The carry flag is reset. We can overcome the problem by complementing the sign bit (ACC.7). If we view the carry flag as an extension of the magnitude part of the number, we can, also, complement the carry. In this way, we get a perfect result. For simplicity, the sequence of instructions does not include a test for consistency between the signs.

• Example S3. Again, the result is out of range and we will try to crack the problem. The microcontroller adds -96 and -96. The result, interpreted as a signed number is +64. Both overflow and carry flags are set. The microcontroller added two negative numbers and the result was a positive number. Thus, the only logical thing to do is to complement the sign. In addition, we complement the carry and view the result as a signed number (8-bits plus a sign bit). The most significant bit of the result is the carry flag. In particular, 1 0100 0000 converted into decimal is -192, which is the correct result. Again, the example instructions are only the core of the program.

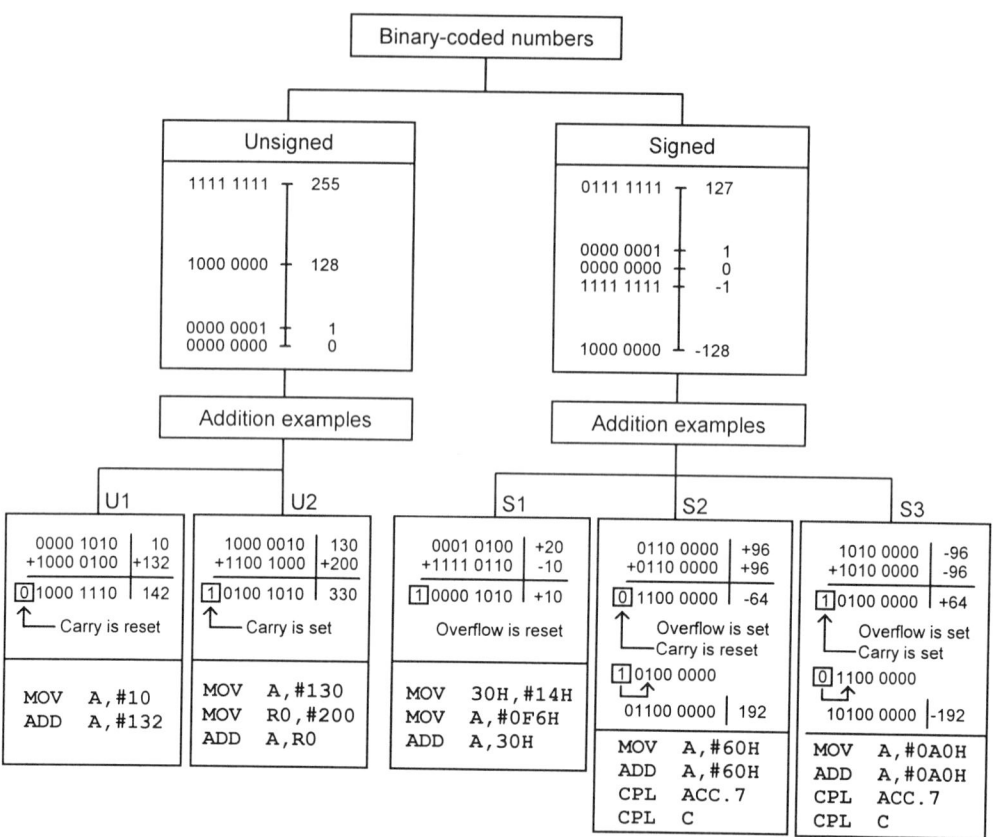

Figure 3.11 Addition examples and the flags carry and overflow.

The overall conclusion for signed numbers is that when the overflow flag alarms for sign inconsistency (OV is set), both the sign and the carry flags must be complemented. The carry is used as an extension of the magnitude field.

The 8051 microcontroller has an instruction add with carry (ADDC) as well. The instruction makes possible for the microcontroller to add numbers which are longer than a byte. For example, the following sequence of instructions performs addition of two unsigned numbers which are two bytes long.

```
; R0 holds the low-order byte of the first number N1_L
; R1 holds the high-order byte of the first number N1_H
; R2 holds the low-order byte of the second number N2_L
; R3 holds the high-order byte of the second number N2_H
; R4 receives the low-order byte of the result SUM_L
; R5 receives the high-order byte of the result SUM_H
        MOV     A,R0    ; Take N1_L
        ADD     A,R2    ; Add the low-order bytes
        MOV     R4,A    ; Save SUM_L
        MOV     A,R1    ; Take N1_H
        ADDC    A,R3    ; Add high-order bytes plus carry
```

```
MOV     R5,A    ; Save SUM_H
```

Along with the addition instruction, the microcontroller possesses a subtraction instruction (SUBB). As discussed earlier, we can avoid the subtraction operation by adding negative numbers. Even though, the SUBB instruction can subtract a number as big as 255 against 128 if the two's complement is used. In fact, the SUBB is a mirror image copy of the ADDC. Thus, the carry flag, when set, indicates a borrow. Consequently, multiple-byte subtraction can be performed. However, an instruction clear carry bit (CLR C) must precede the first subtraction.

For example,

```
MOV     A,#15   ; Load accumulator
CLR     C       ; Clear carry
SUBB    A,#5    ; Subtract
```

So far, we have interpreted the numbers either as unsigned or signed values. There is, however, another option which is termed BCD numbers (Binary-Coded Decimal). Each nibble of such a number represents a decimal digit. Figure 3.12 lays out the range of the one byte long BCD numbers which can swing between 0 and 99. Obviously, that is the most limited range we have discussed. The motivation which underlies the BCD application is the human oriented interface. In many cases, we prefer decimal numbers on the displays rather than hexadecimal. The measurement equipment is a typical example. Naturally, the interface features could tip the balance in favor of BCD numbers.

Moreover, Figure 3.12 includes three useful examples:

• Example BCD1. The microcontroller adds 23 and 45. The result is 68 and the life goes on normally. A short sequence of instructions demonstrates how the example could be tried out. The purpose of the last instruction (DA A) will be explained a little later.

• Example BCD2. Again, the microcontroller follows steadily the addition routine and the programmer must be aware of the side-effects. In particular, when the microcontroller adds 23 and 49 the result is 6CH, which is not a BCD number. Luckily, the situation changes significantly when we add 6 to the result. Finally, we obtain 72 that is the correct number. Traditionally, we suggest instructions to test the example. Now you might be getting closer to the role of the DA A instruction.

• Example BCD3. When we order the microcontroller to add 29 and 39, we get 62. This is also a challenge. Generating a carry from bit 3 to bit 4 (Auxiliary Carry) the microcontroller, in fact, adds 10, not 16. We rectify this omission by adding 6. After that, we get the correct result.

It is not difficult to predict that the same corrections should be applied to the high-order nibble. Consequently, the addition of BCD numbers will require a correction of 00H, 06H, 60H or 66H. The instruction decimal adjust of accumulator (DA A) performs exactly that job.

Increment and decrement instructions

Increment and decrement instructions are simple as action, but useful. There are four increment instructions which are based on different addressing modes. Furthermore, the same addressing modes are used in four decrement instructions. In addition, an instruction increments the 16-bit register DPTR.

Let's consider the following example. There are four bytes in the internal Data Memory which should be moved to the external RAM. The starting address in the internal Data Memory is 10H. The destination starting address is 1000H. There are two basic approaches to build the program. Here is one solution:

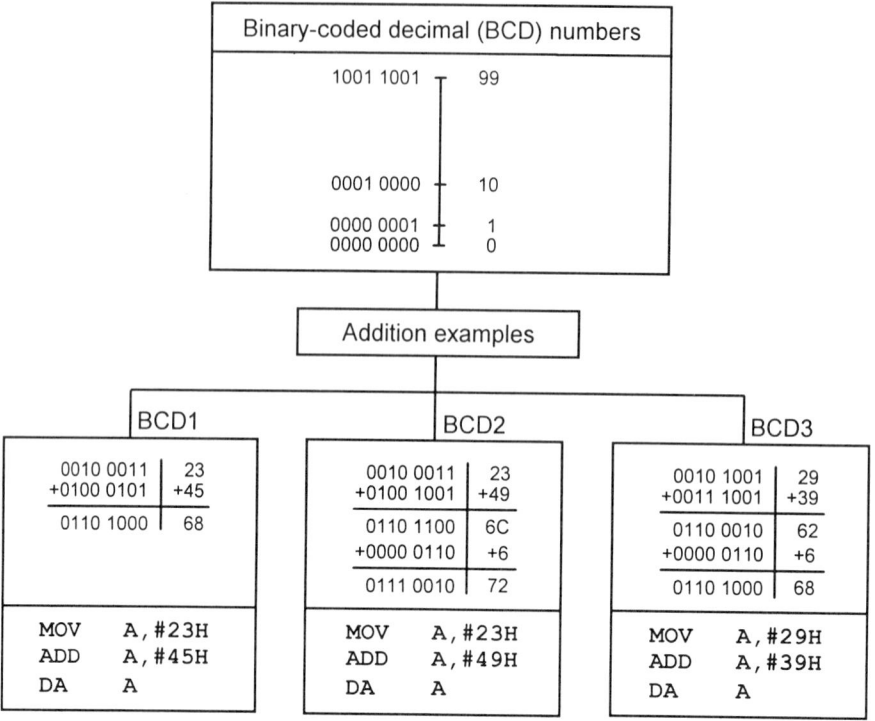

Figure 3.12 Addition examples with BCD numbers.

```
            MOV     R0,#10H      ; Load the first direct address
            MOV     DPTR,#1000H  ; Load the first external RAM
                                 ; address
MOVE        MACRO
            MOV     A,@R0        ; Load accumulator
            MOVX    @DPTR,A      ; Move the byte to the external RAM
            INC     R0           ; Increment the direct address
            INC     DPTR         ; Increment the external RAM
                                 ; address
ENDM
            MOVE
            MOVE
            MOVE
            MOVE
```

We use a macro, called MOVE (see section 3.4). As an alternative solution, we could organize a loop using a program control instruction.

Multiplication and division instructions

Two unsigned numbers, placed in the accumulator and register B, can be multiplied by an instruction

```
            MUL     AB
```

The low-order byte of the 16-bit product goes to the accumulator and the high-order byte to register B. The overflow flag will be set if the result is greater than FFH. However, no law dictates that a set overflow flag necessarily means an error. In this case, the overflow flag says that the result is larger than FFH and the high-order byte is available in register B.

Indeed, the register pair A and B has ample room for the result and there is no need for the microcontroller to indicate overflow. This is proved by the following calculation.

$$FFH \times FFH = 255 \times 255 = 65025 = FE01H < FFFFH$$

The carry flag is always cleared. Here is a multiplication example.

```
MOV     A,#5
MOV     B,#10
MUL     AB   ; Result 50 = 0032H, B = 00H, A = 32H
             ; and OV is cleared
```

As yet another example:

```
MOV     A,#140
MOV     B,#200
MUL     AB   ; Result 28000 = 6D60H, B = 6DH, A = 60H
             ; and OV is set
```

Similarly, division operations are based on the accumulator and register B. For example,

```
DIV     AB
```

divides the value in the accumulator by the number in register B. The accumulator receives the integer part of the result. Register B receives the integer part of the remainder. The carry flag is cleared. The overflow flag indicates a divide-by-zero condition. Any attempt for division by 0 leaves the accumulator and register B undefined and sets the overflow flag.

A division example appears here:

```
MOV     A,#120
MOV     B,#6
DIV     AB   ; Result 20 = 14H, B = 00H, A = 14H and
             ; OV is cleared
```

As an extra division example:

```
MOV     A,#20
MOV     B,#6
DIV     AB   ; B = 02H, A = 03H and OV is cleared
```

The multiplication and division instructions are the longest as far as the execution time is concerned. It takes four machine cycles for a multiplication or division.

3.7 Logical instructions

In this section we discuss byte-wise logical instructions. There are, also, bit-oriented logical instructions which are part of the Boolean processor and will be our focus in section 3.8. Now we distinguish between AND/OR/exclusive-OR instructions, complement/clear instructions and rotate instructions.

AND/OR/exclusive-OR instructions

In essence, the operations are performed on corresponding bits. The accumulator is always one of the operands. However, the destination is either the accumulator or direct address (see Appendix D and E).

We often want to modify one or more bits in a register. If the register is not bit addressable we have to use a byte-wise instruction. Think, for instance, of an example which requires the Timer/Counter 0 to operate in mode 2 (see Figure 2.14). The bit 0 in register TMOD must be reset and bit 1 must be set. The other bits in the same register must not be changed. The following instructions will select mode 2:

```
ANL     TMOD,#0FEH      ; Reset bit M0
ORL     TMOD,#02H       ; Set bit M1
```

Furthermore, the logical instruction exclusive-OR can be used to complement one or more bits in a memory location or in a register. If the target is bit-addressable, we could employ a bit-wise instruction. However, if the target is not bit-addressable or we want to complement more than one bit simultaneously, we can use an exclusive-OR instruction. For example, let's complement bits 0 and 7 which belong to direct address 10H. The instructions appear here:

```
MOV     A,#81H          ; Set ACC.7 and ACC.0
XRL     10H,A           ; Complement
```

Complement and clear instructions

As we have just discussed, we can complement each bit in the accumulator by an exclusive-OR instruction

```
XRL     A,#0FFH
```

However, a better way to do this is to use the instruction CPL A.
Likewise, we can clear the accumulator by

```
MOV     A,#0
```

but we can save one byte using the instruction CLR A.

Rotate and swap instructions

The 8051 microcontroller is capable of performing four rotate and one swap instructions. All of them affect accumulator. Figure 3.13 illustrates the rotate instructions. There are two rotate left and two rotate right instructions. Furthermore, we distinguish between rotate accumulator instructions and rotate accumulator through carry instructions.

We can use the rotate instructions, for example, to organize a ring counter. It is certainly implied that the rotate instructions can be applied to divide or multiply by powers of two.

The SWAP A instruction, shown in Figure 3.14, interchanges the accumulator nibbles. It might be useful, for example, when the program operates with hexadecimal digits (one nibble) and outputs them to seven-segment displays.

3.8 Boolean instructions

A hallmark of microcontrollers is the ability to process individual bits. The 8051 single chip microcomputer possesses a subset of instructions, called a Boolean processor. The Boolean instructions are classified in Figure 3.15. We distinguish between instructions which affect the carry flag, a specific bit and both the carry and a bit.

Observe that two **ANL** instructions are included in the instruction set. The slash in the second one indicates that the selected bit takes part in the operation complemented, but without being altered.

Consider, for example, that the microcontroller inputs three signals through pins P1.0, P1.1 and P1.2. The input logical values define the output P1.7 by means of the following equation:

$$P1.7 = P1.0 \vee \left(P1.1 \wedge \overline{P1.2} \right)$$

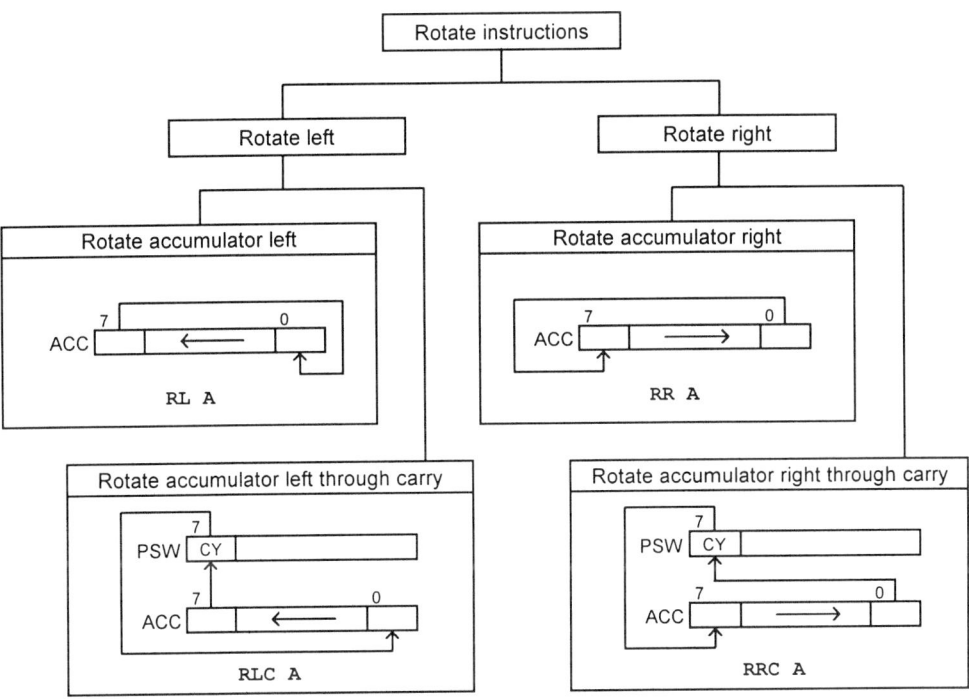

Figure 3.13 Rotate instructions.

In order to update the output, the microcontroller has to execute a sequence of instructions:

```
MOV    C,P1.1    ; Read pin P1.1
ANL    C,/P1.2   ; Read pin P1.2, complement the value and
                 ; calculate AND
ORL    C,P1.0    ; Read pin P1.0 and compute OR
MOV    P1.7,C    ; Update the output
```

It is certainly implied that actual applications would require bigger computational efforts from the microcontroller, however the example above lacks just two details. First, the output must be updated non-stop. This feature can be achieved by an endless loop. Second, we normally organize input and output tables in the memory. In this way, we introduce a kind of clock discipline and maintain consistency between input and output values.

The Boolean processor is the backbone of the PLC (Programmable Logic Controller) software and we will discuss more examples in section 7.11.

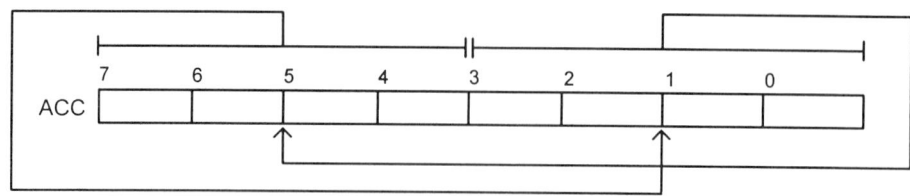

Figure 3.14 The SWAP A instruction.

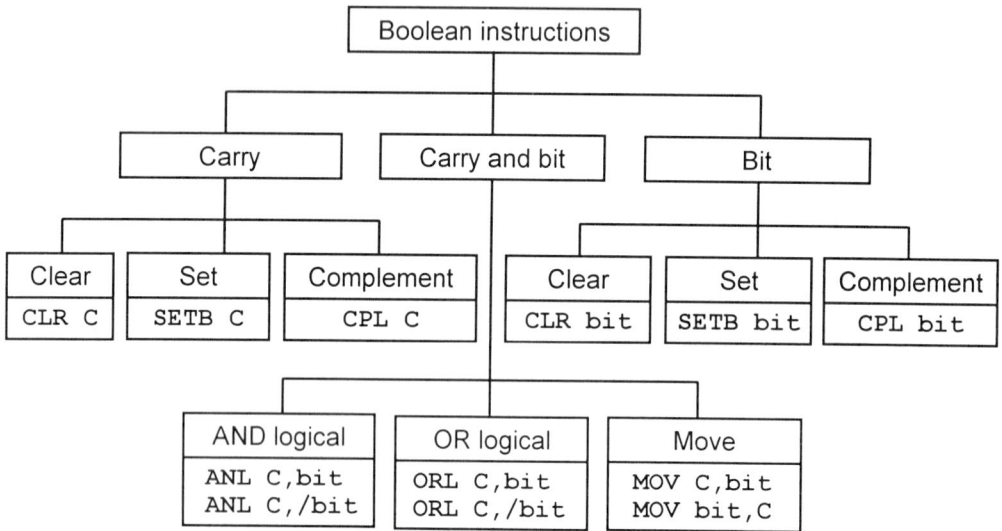

Figure 3.15 Boolean instructions.

3.9 Program control instructions

The microcontroller's program has to react to a large variety of events both internal to the microcontroller and external. Thus, a timer/counter overflow or a specific input signal will require the program to continue with different instructions. Furthermore, certain results from arithmetic operations may lead to a branch in the program. We can cluster the program control instructions into conditional jumps, unconditional jumps and call/return instructions.

Conditional jump instructions

Figure 3.16 specifies the conditional jump instructions. As the name implies, the instructions will implement a jump if a certain condition is met. If the condition is not met, the execution will go on with the following instruction. In all cases, the jump is based on relative addressing mode.

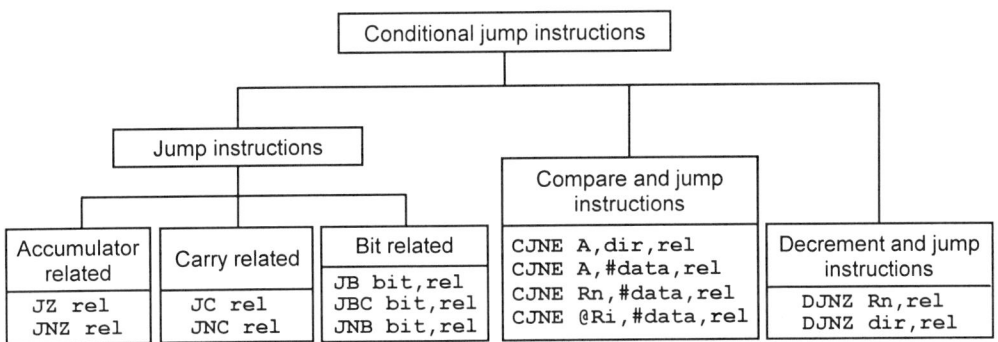

Figure 3.16 Taxonomy of conditional jump instructions.

The advantage of relative addressing is portability. If we move the code in the memory, the offset byte of these instructions should not be changed and the program runs correctly. At the same time, we should be aware of the limitations. How far forward or backward can we jump? To answer this question, let us calculate the maximum and minimum number which could be written in the offset byte.

Offset byte	Binary	Decimal
Maximum	0111 1111	127
Minimum	1000 0000	- 128

Therefore,

$$PC + 2 - 128 \leq \text{Destination address} \leq PC + 2 + 127$$

Usually, we take as a base the first byte of the next instruction. From this point the program can jump 127 bytes forward and 128 bytes backward.

Furthermore, Figure 3.16 shows that the conditional jump instructions can be broken down into three groups. First, we have pure jump instructions. For example, the instruction

        JZ   L1

will jump to a label L1 if the accumulator is zero. In this case, we indicate the jump address by a label. Likewise, the instruction

        JC   CARRY

will jump to the label **CARRY** if the carry flag is set. The same style applies for the carry and bit related instructions. In fact, there is an instruction that is somewhat different. The instruction JBC bit,rel not only jumps if the bit is set, but clears the bit as well.

Second, the compare and jump instructions have a better resolution when the comparison is organized. For instance, the instruction

        CJNE   A,#80H,TEST

introduces a jump if the accumulator is not equal to 80H. Again, **TEST** is a label and its place in the program must be consistent with the calculations above.

Finally, decrement and jump instructions implement a sequence of two actions which is very convenient for organizing loops.

Unconditional jump instructions

We have already discussed an example of unconditional jump instruction called short jump (see Figure 3.6). Along with the relative addressing mode of the **SJMP** instruction, there is a long jump instruction based on extended addressing. The **LJMP** instruction was illustrated in Figure 3.7. The key advantage of the **LJMP** instruction is that it can redirect the execution flow to an address anywhere in the 64K byte memory. On the other hand, the **SJMP** instruction provides portability of the code and is one byte shorter.

There is, also, another instruction called absolute jump (**AJMP**) which lays somewhere in the middle between **SJMP** and **LJMP** in terms of range and portability. The encoding of the **AJMP** is as follows:

a10 a9 a8 0	0 0 0 1		a7 a6 a5 a4	a3 a2 a1 a0

As you can see, the instruction defines the address bits A0 through A10. The address bits A11 through A15 are the current values in the Program Counter. In other words, on the background of one 2K byte page, the instruction **AJMP** is capable of selecting any jump address within the page. The five most significant address bits are out of control and limit the range to the current 2K bytes. On the other hand, they introduce portability. If the chunk of code is moved to another 2K bytes page in the memory, the **AJMP** instructions will require no change.

Problem 3.1

The pin P3.5 of an 8051 microcontroller is controlled by a pushbutton. Write a sequence of instructions which turns Port 1 into a counter. The Port 1 value has to be incremented when the pushbutton is pressed and released.

Solution 3.1

Due to the internal pullup, there is no need of an external resistor. If the pushbutton is not pressed, the P3.5 input level is high. When we toggle the pushbutton, the code in Port 1 must be incremented. Figure 3.17 illustrates the microcontroller's interface and operation.

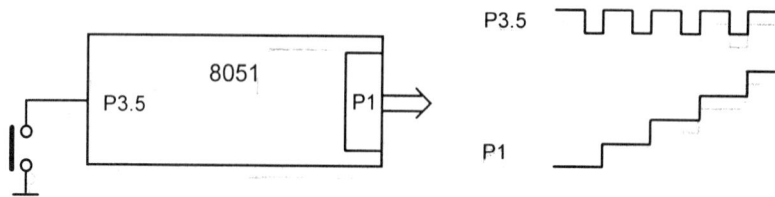

Figure 3.17 Port 1 acts as a counter.

The following sequence of instructions waits for the pushbutton and increments the Port 1 code.

```
        MOV     P1,#0       ; Clear Port 1
L1:     JB      P3.5,$      ; Wait for P3.5 low
        JNB     P3.5,$      ; Wait for P3.5 high
        INC     P1          ; Increment Port 1
        SJMP    L1          ; Loop
```

The dollar sign ($) used as a label indicates that the jump is to the current line. The idea to insert two instructions (**JB** and **JNB**) and to wait for a manual response from the user, may prove useful in many cases, when you debug your program. On the other hand, special care should be taken when mechanical switches are used. They might need a capacitor or latch for debouncing. Also, the mechanical switches could be debounced by software.

Call and return instructions

The call instructions are similar to the jump instructions we have just discussed. Figure 3.18 shows an example of a long call instruction (**LCALL**). By means of this instruction the programmer can call a subroutine.

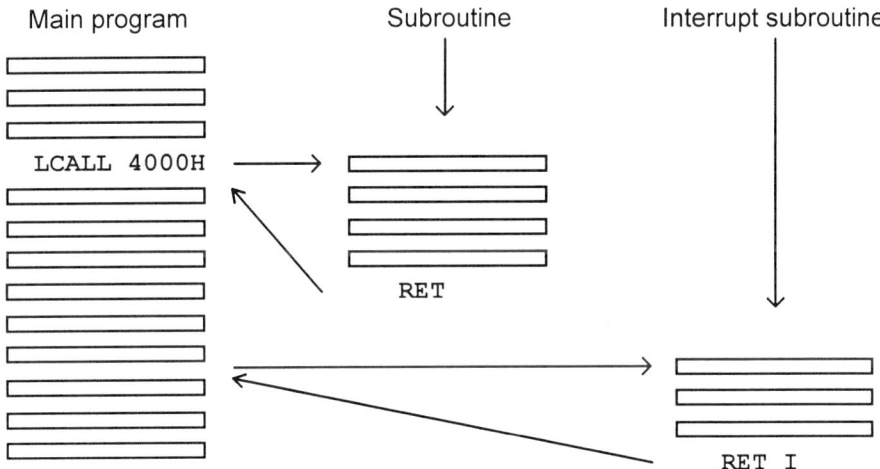

Figure 3.18 Subroutines and interrupt subroutines.

The subroutine starting address in this example is 4000H. Essentially, before branching the **LCALL** instruction pushes the current Program Counter onto the stack. Normally, the last instruction in the subroutine is return (**RET**). The instruction **RET** pops the return address from the stack and reloads the Program Counter. The microcontroller is back on the instruction following immediately the **LCALL 4000H**.

As you might expect, the instruction **ACALL** operates as a **LCALL** except for the range which is limited to the current page of 2K bytes.

Furthermore, Figure 3.18 deals with an interrupt subroutine. As a rule, the last instruction in an interrupt subroutine is **RET I**. The **RET I** instruction pops the Program Counter from the stack. In addition, the instruction resets the flip-flop of the current interrupt. Consequently, the interrupt logic can accept further requests at the same priority level as the one which has been processed. Remember that at least one more instruction will be executed if an interrupt is waiting while the instruction RET I is in progress (see section 2.9).

If you replace the instruction RET I with RET, the microcontroller will leave the interrupt subroutine successfully, however any further request from lower or the same priority level will be ignored.

3.10 Application examples and problems

Programmable pulse source

The first application example we will discuss is a programmable pulse source. The 8051 microcontroller will perform a simple job. It will set and reset a certain pin repeatedly, in order to control a peripheral device. When the pin P1.7 is chosen as an output of the pulse generator, it will cause the following instructions to be executed.

```
L1:         CPL     P1.7            ; Complement bit
            SJMP    L1              ; Loop
```

We use the Boolean instruction CPL bit to toggle the output. In addition, a short jump instruction forms the loop.

If the oscillator frequency f_{OSC} = 12 MHz, the output P1.7 will change its state every 3 µs. Longer on and off times for the square pulses can be achieved by NOP instructions, inserted in the program. The interrupt system could be used for terminating the loop.

If the goal is an output to be pulsed a fixed number of times, the output frequency can be increased (250 KHz) for the LSB of the ports. The instructions appear here:

```
            MOV     P1,#20          ; For 10 pulses
L1:         DJNZ    P1,L1           ; 10 pulses on P1.0
```

An alternative way to write the second instruction is DJNZ P1,$.

The microcontroller generates 10 pulses on the output P1.0 and continues the program. As discussed earlier in section 2.6, the instruction DJNZ P1,L1 is a "read-modify-write" instruction which reads the latch rather than the pin. In this way, the influence of the load, attached to the port, is eliminated.

Software time delay

When microcontrollers have to respond to peripheral devices after a certain delay either timers or software can be employed. The second approach is based on NOP or another instruction which is used just to insert execution time. Depending on the delay, one or more loops might be the best approach.

Adjustable software time delay can be organized by means of the following subroutine
DEL.ASM:

```
;**************************************************
;*    DEL.ASM                                    *
;*    This program provides a delay of           *
;*    103 through 24741 mcs                       *
;*    depending on the code in                    *
;*    register R0 (1 to 255)                       *
;*    Execution time : 97*R0 + 6, mcs            *
;**************************************************
           ORG    2000H
           MOV    R0,#***      ; 1 adds roughly 100 mcs
           LCALL  DEL
           NOP
;
           ORG    2100H
DEL:       PUSH   ACC          ; Execution time: 2 mcs, save ACC
DEL1:      MOV    A,#2FH       ;                 1 mcs
DEL2:      DJNZ   ACC,DEL2     ;                 2 mcs
           DJNZ   R0,DEL1      ;                 2 mcs
           POP    ACC          ;                 2 mcs, restore ACC
           RET                 ;                 2 mcs
           END
```

The required time delay is selected by moving a parameter to the register R0. Naturally,
another option to start the subroutine DEL is the instruction ACALL DEL.

When you debug the delay program you may need to look at the list file, extension .LST.
Here is a printout of the list file, produced from the assembler for the delay example.

```
 3                    ;**********************************************
 4                    ;*    DEL.ASM                                 *
 5                    ;*    This program provides a delay of        *
 6                    ;*    103 through 24741 mcs                    *
 7                    ;*    depending on the code in                *
 8                    ;*    register R0 (1 to 255)                    *
 9                    ;*    Execution time : 97*R0 + 6, mcs         *
10                    ;**********************************************
11  2000  20 00          ORG    2000H
12  2000  78 64          MOV    R0,#100  ; 1 adds roughly 100 mcs
13  2002  12 21 00       LCALL  DEL
14  2005  00             NOP
15  2005             ;
16  2100  21 00          ORG    2100H
17  2100  C0 E0     DEL:  PUSH  ACC      ; Execution time: 2 mcs
18  2102  74 2F     DEL1: MOV   A,#2FH   ;                 1 mcs
19  2104  D5 E0 FD  DEL2: DJNZ  ACC,DEL2 ;                 2 mcs
20  2107  D8 F9           DJNZ  R0,DEL1  ;                 2 mcs
```

21	2109	D0 E0	POP	ACC	;	2 mcs
22	210B	22	RET		;	2 mcs
23	210B		END			

The basic rule is that one instruction is one line. In this example, the only one-byte instructions are NOP (code 00H) and RET (code 22H). There are plenty of two-byte instructions and finally two three-byte instructions, LCALL (line #13) and DJNZ (line #19).

Furthermore, if you need to download the code through a serial link to a single board computer you may bank on a standard called Intel HEX format. Here is the corresponding file DEL.HEX:

```
:0620000078641222100000CB
:0C210000C0E0742FD5E0FDD8F9D0E0223B
:00000001FF
```

The colon sign (:) indicates the start of a line. The first byte (06H) displays the number of the bytes included in this line. The address of the first byte (2000H) can be seen immediately after the number of the bytes. The following byte (00H) defines the line as a piece of code. In contrast, the byte on the same position in the last line is changed to 01H (end of the file line). The next field contains the machine code starting with the first byte 78H and completing the field with a byte 00H. If you calculate the code bytes occupying the first line, you will realize that the number is seven, not six. No blame should be attached to the linker. The last byte (CBH) is a two's complement checksum byte. Let's calculate it manually. Figure 3.19 illustrates the method using the first line as an example.

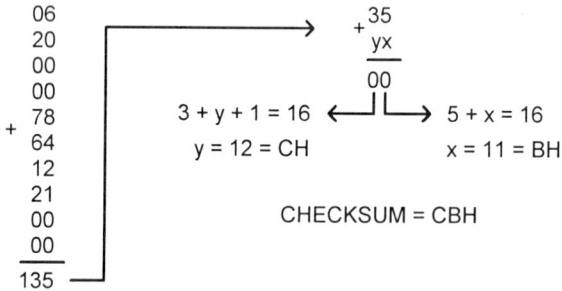

Figure 3.19 Calculation of the two's complement checksum byte.

We add all bytes and get 135H. Furthermore, we define x to be the low nibble of the checksum and y the high nibble. Two simple equations emerge. Observe that the equation for the variable y is somewhat conditional. We must take into account the carry from low to high nibble. In terms of the 8051 microcomputer system it is called auxiliary carry.

As you might predict, the second line cover the code bytes from address 2100H through 210BH. The last line says this is the end of the file. An extra line would have appeared if the program had contained code in another segment in the memory or simply if the bytes had been too many.

Synchronizing timer interrupts

Now, we discuss the case of generating long pulses. For some applications the timer interrupt interval might be so long that the mode 1 must be used. When an interrupt is generated it is not predictable exactly how long it will take to complete the instruction in progress - 1, 2 or 4 cycles. It would cause an offset in the timing and for a lot of applications this could be a problem. The embedded systems designer should find an appropriate method for synchronization.

Figure 3.20 shows the code in a timer and some key points between two consecutive overflows. The aim is all interrupts to come one after the other with a constant delay which we denote by N.

$$-\left[N-(N_1+N_2)\right] = -N + N_1 + N_2$$

The program can simply stop the timer, add to its code $-N + N_2$ and start the timer again.

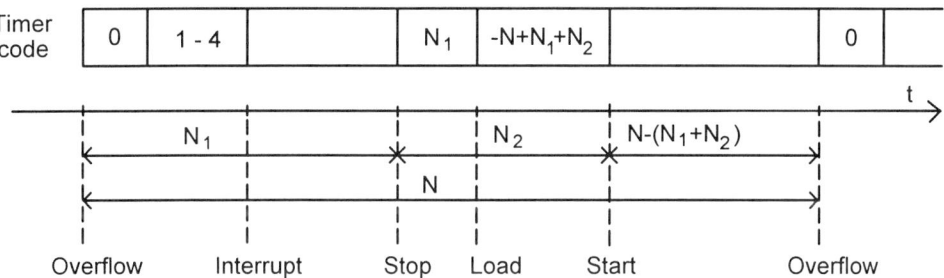

Figure 3.20 Synchronizing timer interrupts.

For example, assume the oscillator frequency f_{OSC} = 12 MHz, the desired interval N = 1 ms and the time between stop and start points N_2 = 7 μs. The value for addition appears here:

$$-N + N_2 = -1000 + 7 = -993 = FC1FH$$

The program might then be as follows:

```
CLR    EA          ; Disable all interrupts
CLR    TR1         ; Stop Timer 1
MOV    A,#1FH       ; Low-order byte        1 mcs
ADD    A,TL1        ;                       1 mcs
MOV    TL1,A        ;                       1 mcs
MOV    A,#0FCH      ; High-order byte       1 mcs
ADDC   A,TH1        ;                       1 mcs
```

```
        MOV     TH1,A       ;                               1 mcs
        SETB    TR1         ; Restart Timer 1               1 mcs
```

The execution time of the instructions above shows that the period N_2 must be 7 µs.

Another approach might be mode 2 plus a software counter. If that is the case, the auto-reload mode does not need synchronization, but the drawbacks are worse resolution and more time used for timer functions.

Problem 3.2

Write a program for the 8051 microcontroller which sets and resets the output P1.7 in a loop. The required times are: the output must be ON for 2 ms and then OFF for 4 ms.

Use oscillator frequency 12 MHz and Timer 0 to interrupt the main program.

Solution 3.2

The program must include initialization instructions and a subroutine which is activated by Timer 0 interrupts. In order to define the current delay, the subroutine checks latch P1.7. It is achieved by a "read-modify-write" instruction JBC.

```
;****************************************
;*    P_2_4.ASM                       *
;*    This program sets and resets    *
;*    the output P1.7                 *
;*    ON time 2 ms                    *
;*    OFF time 4 ms                   *
;****************************************
        ORG     0000H
        LJMP    RESET
;
        ORG     000BH
        LJMP    PULSE
;
        ORG     0030H
RESET:  MOV     TMOD,#01H       ; Timer 0, mode 1
        MOV     TL0,#30H        ; T(ON) = 2 ms = -2000 = F830H
        MOV     TH0,#0F8H
        SETB    PT0             ; High priority
        SETB    EA              ; Enable all interrupts
        SETB    ET0             ; Enable interrupts from Timer
        SETB    P1.7            ; ON
        SETB    TR0             ; Start Timer 0
;
MAIN:   NOP                     ; The main program
;
PULSE:  CLR     EA              ; Disable all interrupts
        PUSH    A               ; Save accumulator
        PUSH    PSW             ; Save PSW
        CLR     TR0             ; Stop Timer 0
```

```
;
            JBC         P1.7,OFF
            SETB        P1.7            ; ON
            MOV         A,#3AH          ; -N + N2 = -1990 = F83AH
            ADD         A,TL0
            MOV         TL0,A
            MOV         A,#0F8H
            ADDC        A,TH0
            MOV         TH0,A
            SETB        TR0             ; Start Timer 0
            SETB        EA              ; Enable all interrupts
            POP         PSW             ; Restore PSW
            POP         A               ; Restore accumulator
            RETI
;
OFF:        MOV         A,#69H          ; -N + N2 = -3991 = F069H
            ADD         A,TL0
            MOV         TL0,A
            MOV         A,#0F0H         ; Load the high-order byte
            ADDC        A,TH0
            MOV         TH0,A
            SETB        TR0             ; Start Timer 0
            SETB        EA              ; Enable all interrupts
            POP         PSW             ; Restore PSW
            POP         A               ; Restore accumulator
            RETI
```

The parameter N_2 is 10 μs for the ON part and 9 μs for the OFF segment.

Two typical cases of interaction between the microcontroller's program and embedded peripherals have been discussed so far. First, we used flags to communicate between software and hardware. Second, the serial port receive and transmit operations were organized as a byte-wise interaction. Now, we face the demand of reading two bytes from the hardware which runs concurrently with the program. This is the case with reading a timer/counter "on-the-fly".

In particular, we often want to read the current value from a timer/counter without interfering with the counting process. Unfortunately, a two byte timer/counter must be read in two steps. It is always possible for the first byte read to be changed while the microcontroller is accessing the second one. Let's get down to a practical example.

Assume the subroutine RDTIME should return in the registers R1 and R0 a 16-bit value indicating the count in Timer 0. The low-order byte runs faster and therefore it must be the last read. The problem is that between reading the two halves, a low-order register overflow could increment the high-order register, and the two data bytes returned would be "out of phase". The solution is to read the high-order byte first, then the low-order byte and to check if the high-order byte has not changed. If it has, we repeat the procedure.

```
RDTIME: MOV     A,TH0           ; Sample Timer 0 high-order byte
        MOV     R0,TL0          ; Sample Timer 0 low-order byte
        CJNE    A,TH0,RDTIME    ; Repeat if necessary
```

```
MOV     R1,A                    ; Store Timer 0 high-order byte
RET
```

The subroutine RDTIME supplies the background program with the value of the low-order byte (TL0) in the register R0 and the value of the high-order byte (TH0) in the register R1.

Single-step operation

Typically, we develop and test programs by breaking them down into smaller parts. Once we have made sure the isolated segment is correct, we can move to the next one. Both breakpoints and execution one instruction at a time could be organized by replacing the original code in the memory with jump to the Monitor program. Unfortunately, this approach would be feasible, only if the program resides in RAM. Luckily, we could overcome the problem if we harness the interrupt system to the task of freezing the normal execution. In this case, either a timer or an interrupt input may be used. In many circumstances single-step operation might prove useful. Similarly to Problem 3.1, we will demonstrate how single-step operation can be organized. Now, in light of the new strategy, the input must be an interrupt input.

Figure 3.21 reveals the basic idea in two steps. We assume that the address of the target instruction (INSTR1) has been pushed in the stack. Hence, if a level activated interrupt input (P3.2 / $\overline{\text{INT0}}$) is asserted and the polling take place while the instruction **RET I** is in progress, the microcontroller will execute one more instruction (INSTR1). Next, the microcontroller will enter the interrupt subroutine again. If the mentioned interrupt subroutine is embedded in the Monitor, the user would be able to inspect the system and to interpret the results. The trick here is to fit two processes which are not correlated (the transition on the input P3.2 / $\overline{\text{INT0}}$ and the instruction **RET I**).

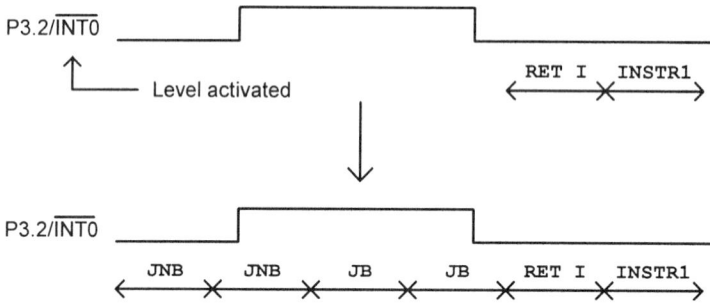

Figure 3.21 Single-step operation achieved by the interrupt system.

The second waveform and the corresponding instructions in Figure 3.21 outline the synchronization. Two extra instructions are used - **JNB** and **JB**. They help the microcontroller to execute the instruction **RET I** at the right moment. The **JNB** and **JB** instructions are used as conditional jumps to the current lines (one instruction loops) which test the pin P3.2 / $\overline{\text{INT0}}$. Even though the interrupt input is asserted before the instruction **RET I**, the request has a priority equal to the one currently in progress and therefore is blocked. The instruction **RET I** clears the priority flip-flop and makes further interrupts possible. As a result, there is always an overlap between the successful polling and the instruction **RET I**. Consequently, the instruction following immediately **RET** **I** is executed before the interrupt subroutine is entered again.

Let us practice this approach by the following example. There is a sequence of three instructions, starting from address 4000H.

```
4000    75 90 11        MOV    P1,#11H    ; Example
4003    75 90 22        MOV    P1,#22H    ; instructions
4006    75 90 33        MOV    P1,#33H
```

If we single-step through these instructions, it will be very easy to observe the effect.

When we pack the instructions from Figure 3.21, the following subroutine emerges:

```
2003    30 B2 FD        JNB    P3.2,$     ; Wait here until *INT0=0
2006    20 B2 FD        JB     P3.2,$     ; Wait here until *INT0=1
2009    32              RETI
```

The symbol * indicates inversion. For example, we write *INT0 instead of $\overline{INT0}$.

Obviously, the subroutine was written with the assumption that the pin P3.2/$\overline{INT0}$ will be the control input. Unlike what you may expect, the interrupt subroutine starts from address 2003H (the interrupt vector for this input is 0003H). The explanation is simple. Usually Monitor programs, such as the one we use, redirect the execution flow from unused interrupt vectors to higher addresses. In particular, the Monitor contains the instruction

```
0003    02 20 03        LJMP   2003H      ; This jump moves the
                                          ; vector address to 2003H
```

Furthermore, the single board computer has a RAM with start address 2000H and we are in position to store the interrupt subroutine from address 2003H for our experiment.

Before activating the interrupt subroutine by a jump (only the first time), two things must be done. First, the interrupt system must be initialized. Second, the address of the first instruction to be single-stepped must be stored in the stack. Thus, the procedure follows the steps below:

• The high-order byte (40H) of the address of the first instruction (MOV P1,#11H) is stored in the internal memory (at address 70H). The low-order byte (00H) of the address of the same instruction is stored at direct address 71H. This is done using the Monitor utilities.

• The interrupt input P3.2/$\overline{INT0}$ is pulled high and then the following handler is executed from address 2020H.

```
2020    53 88 FC        ANL    TCON,#FCH  ; Clear IE0, low level
                                          ; activated
2023    43 B8 01        ORL    IP,#01H    ; High-priority for *INT0
2026    43 A8 81        ORL    IE,#81H    ; Enable interrupt
2029    C0 71           PUSH   71H        ; Store the address
202B    C0 70           PUSH   70H        ; in the stack
202D    02 20 03        LJMP   2003H      ; Enter the subroutine
                                          ; by a jump
```

• The interrupt input P3.2/$\overline{INT0}$ is pulled down. From now on each pulse (transitions 0-1 and 1-0) will cause the microcontroller to execute a single instruction and stop again. Figure 3.22 indicates the state of Port 1 as a response of the input pattern.

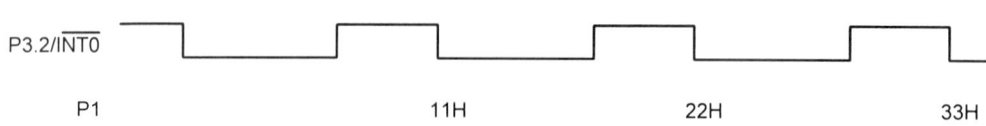

P3.2/INT0

P1   11H   22H   33H

Figure 3.22 Timing diagram for the Port 1 response.

Since we do not replace any code, this approach is consistent with both RAM and EPROM components.

3.11 Supplementary problems

Problem S3.1
The instruction in process of completion is

`MOV 17H,#25H`

State the addressing mode of this instruction. Fill in the correct hexadecimal numbers in the marked areas of Figure 3.23, which apply to the new value of the Program Counter, the address and the code of the affected memory location.

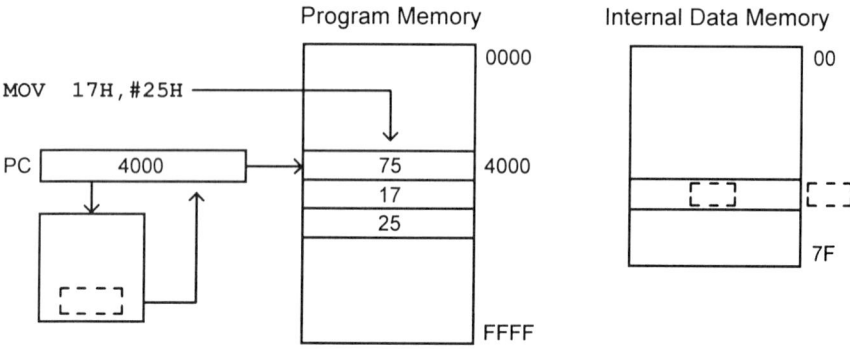

Figure 3.23 Fill in the marked areas.

Problem S3.2
Find the mnemonic of a two-byte instruction which alters a direct byte from 0FH to 10H. Fill in the code in the marked areas in the Program Memory (Figure 3.24).

Problem S3.3
Write the appropriate assembly language instructions to program the pins P1.0, P1.1 and P3.5 as inputs. In addition, program the lines P1.2 and P3.4 as outputs. The other pins of Port 1 and Port 3 should keep their status of either inputs or outputs.

Problem S3.4
Write a subroutine which provides a delay of 1 through 25 seconds, step 0.1. The desired delay time is selected by loading the register R0. Assume that the oscillator frequency is 12 MHz.

Problem S3.5

Create a source file for the program in Problem S3.4. Assemble and link the program. Generate an Intel HEX file. Type the file on the screen.

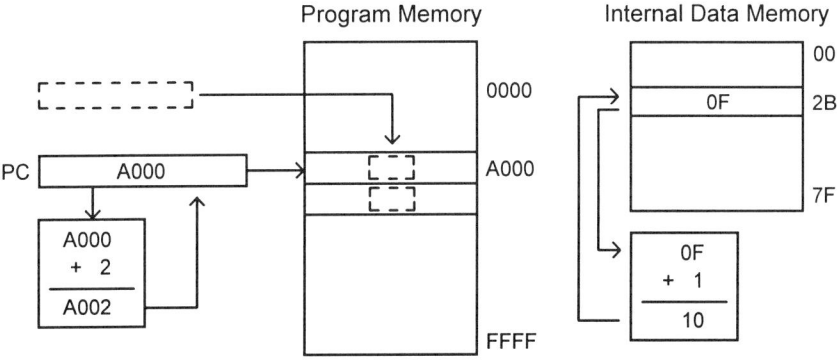

Figure 3.24 Fill in the instruction mnemonic and the marked areas in the Program Memory.

Problem S3.6

Write an interrupt subroutine which toggles the output P1.7. The subroutine is activated after one hundred pulses applied to the input P3.4/T0. Write the initialization code as well.

Problem S3.7

Assume that you lost your source and other related files except the Intel HEX file. Recover the program using the following Intel HEX file:

```
:0D20000075900020B5FD30B5FD059080F60F
:00000001FF
```

Problem S3.8

Write an assembly language program which multiplies two 16-bit numbers.

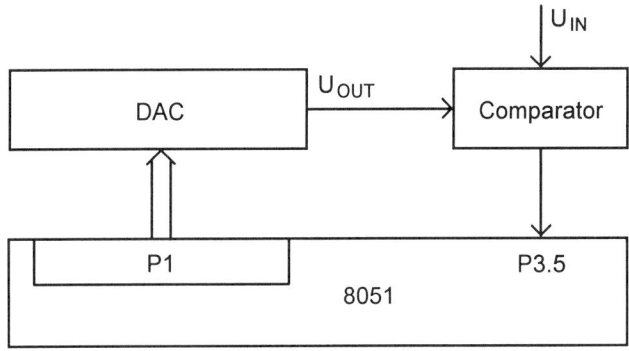

Figure 3.25 The microcontroller converts analog voltage into digital code.

Problem S3.9

Find a method to measure the input voltage U_{IN} (0 to 5 V) by generating a sequence of output voltages U_{OUT} and comparison (Figure 3.25). The Digital-to-Analog Converter (DAC) is capable of emitting voltages within the range 0 to 5 V. The comparator's digital output indicates which input voltage is higher. Write the correspondent assembly language program for the method you suggest. Assume that the DAC and the comparator are fast enough and only the microcontroller's execution time defines the rate.

3.12 References

Atmel, *Microcontroller Data Book*, 1997.

Kenneth J. Ayala, *The 8051 Microcontroller*, West Publishing Company, 1991.

Knud Smed Christensen, *User Manual for The System51*, KSC Software Systems, 1995.

Rudolf Graf, *Simula51 : Program, Usage, Application, Examples*, Siemens Aktiengesellschaft, 1992.

Intel, MCS 51 *Microcontroller Family User's Manual*, 1994.

Intel, *Embedded Microcontrollers*, 1996.

Zdravko Karakehayov and Stanislav Grigorov, *Single Chip Microcomputers*, Technica, Sofia, 1992.

Zdravko Karakehayov and Emil Saramov, *Applied Microcomputer Systems*, Technical University of Sofia, 1995.

Philips Semiconductors, *80C51-Based 8-Bit Microcontrollers, Data Handbook IC20*, 1997a.

James W. Stewart, *The 8051 Microcontroller, Hardware, Software and Interfacing*, Prentice Hall, 1993.

Sencer Yeralan and Ashutosh Ahluwalia, *Programming and Interfacing the 8051 Microcontroller*, Addison-Wesley Publishing Company, 1995.

A limited assembler / linker system can be downloaded from
    KSC Software Systems   http://www.ksc-softsys.com/c.htm

Chapter 4

DIGITAL INTERFACING

4.1 Introduction

In this chapter we examine the microcontroller's interface to digital components and peripherals. This theme is addressed at two different hierarchy levels. First, we discuss interfacing to a couple of typical embedded systems components, such as EPROMs and RAMs. We need them in many cases, when the on-chip resources are insufficient. Second, at a higher level, we investigate the digital interface between embedded systems and the outside world. In this situation, we explain how to increase the number of available I/O lines. Next, we introduce two serial interfaces for interaction between embedded computers and other systems or peripherals. In section 4.5, we combine the covered topics in order to design a single board computer. The single board computer has its own importance as both a control device and a development tool. In addition, the small-scale computer design can be viewed as a jumpboard to the more sophisticated case studies in Chapter 11. Finally, we discuss how to organise the interface to typical digital actuators, stepper motors.

The digital interfacing theme is not confined to the examples included in this chapter. Chapter 7 adds extra topics, such as interfacing displays and programmable logic controllers.

4.2 Memory design

The following motivations dictate the use of additional memory components in the embedded systems:

• The minimal size of memory, which is required by the application, is larger than the microcontroller's on-chip memory.

• The user is faced with situations in which the memory must be reorganized in terms of capacity and addresses. Consequently, the embedded computers are designed to accommodate a different number of memory components which could be placed at different addresses.

• There are applications which require vital data to be saved in the event of power failure. As a result, the design involves specific techniques, such as battery-backed RAMs or nonvolatile memory solutions.

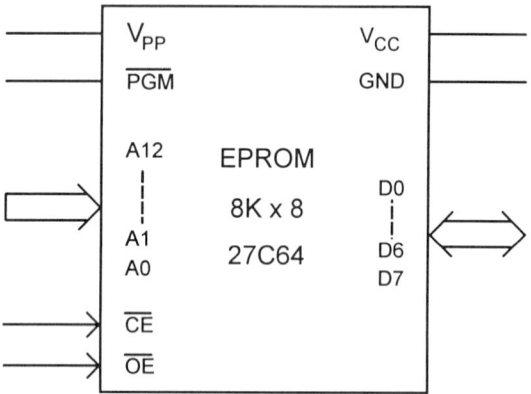

Figure 4.1 An EPROM memory component.

We should look at some memory components first. Figure 4.1 shows an 8K x 8 EPROM. The power supply pins are V_{CC} and GND. Figure 4.2 lays out the three major operation modes.

Initially, the EPROM must be programmed. The EPROM is selected by an input named chip enable $\overline{CE}$. The chip enable input is asserted low. An input termed output enable $\overline{OE}$ is driven high to disable the output buffers. The programming input $\overline{PGM}$ is pulled down several times. In parallel, the input V_{PP} receives 12.5 V. The code which must be written is applied to the bidirectional lines D0 - D7. More details about the programming procedure and EPROM programmers can be found in section 11.2.

		Pins				
		$\overline{CE}$	$\overline{OE}$	$\overline{PGM}$	V_{PP}	D0 - D7
Mode	Program	0	1	⊓⊔	12.5 V	D_{IN}
	Read	0	0	1	5 V	D_{OUT}
	Standby	1	x	x	5 V	OFF

Figure 4.2 EPROM operation modes.

The read mode (the mode that is used from the microcontroller to fetch instructions) is activated by applying low levels to the inputs $\overline{CE}$ and $\overline{OE}$. The program enable input $\overline{PGM}$

must be kept high. The programming voltage pin V_{PP} is driven by 5 V. The byte to be read out appears on the lines D0 - D7.

Furthermore, the standby mode saves power and is desirable when the device is not selected. The standby mode is asserted by pulling high the input $\overline{CE}$. In this mode, the supply current will drop, for example, to 100 µA against 20 mA for the read mode. In standby mode and also when the input $\overline{OE}$ is pulled up, the EPROM's output buffers are in the OFF state. Thus, other memory components can use the bus.

Besides the requirement for memory capacity, there could be a need to adjust the speed of the memory to the microcontroller timing by allocating the right component. The question arises with the modern microcontrollers which run at oscillator frequencies higher than 30 MHz.

But how fast should the EPROM be? The microcontroller timing, discussed in section 2.5 and the EPROM timing diagram in Figure 4.3 form the basis of this part of the design. The main objective of the timing diagram (as outlined in Figure 4.3) is to introduce the essential information without being overwhelming. The EPROM will output correct data if the address bits have been established for a period of time called access time t_{ACC}. In parallel, the EPROM must have been selected for another period of time indicated by t_{CE}. Finally, the last delay, which runs concurrently, is a period of time labelled output enable t_{OE}. When all inputs are valid or active for the required duration, the addressed byte comes out.

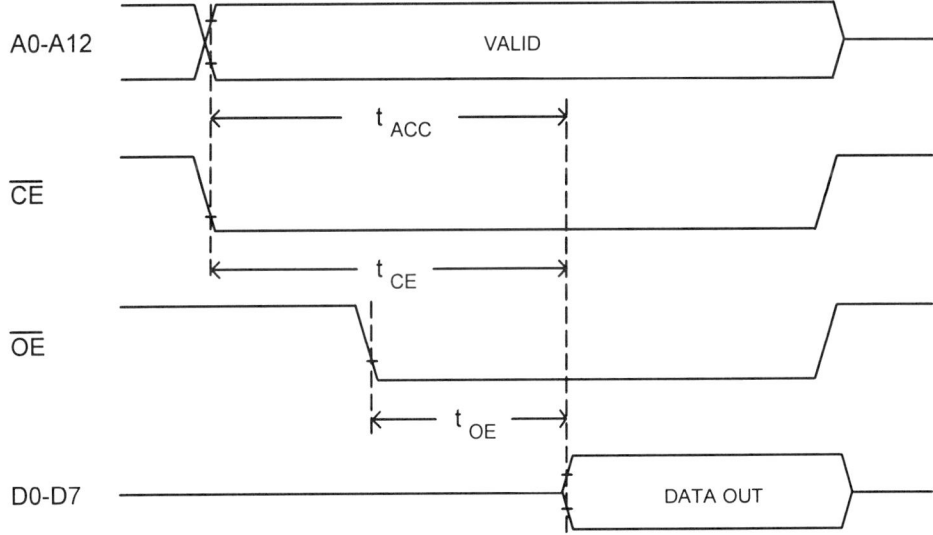

Figure 4.3 The EPROM read mode timing.

In an 8051 based embedded system, both the EPROM address inputs and the input $\overline{CE}$ are driven by the microcontroller's address pattern. The input $\overline{OE}$ is normally derived from the strobe $\overline{PSEN}$. Inevitably, if decoding logic is used, the constraints on the memory component will be somewhat higher.

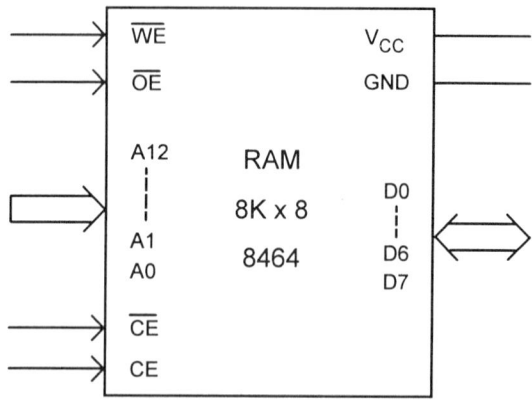

Figure 4.4 A RAM memory component.

Along with external Program Memory, a typical embedded system will need a certain amount of external Data Memory. The most widespread ICs for this purpose are static RAMs. Figure 4.4 presents a RAM equivalent of the 8K x 8 EPROM. A new, active-low input is write enable $\overline{WE}$. We need this input to determine the direction of data transfer during a memory access. Logically, the standard 28-pin package gives an ample number of pins for 8K x 8 RAMs and an extra chip enable input (CE) is implemented. The availability of one active-low and one active-high chip enable inputs helps the designer to simplify or even to avoid the decoding logic.

		Pins				
		$\overline{CE}$	CE	$\overline{OE}$	$\overline{WE}$	D0 - D7
Mode	Read	0	1	0	1	D_{OUT}
	Write	0	1	1	0	D_{IN}
	Standby	1	x	x	x	OFF

Figure 4.5 RAM operation modes.

The correspondence I/O pins - operation mode is described in Figure 4.5. The RAM read mode timing is identical to the EPROM diagram in Figure 4.3. The RAM write mode timing is presented in Figure 4.6. A principle requirement is valid data to be applied to the RAM for at least a certain duration, denoted as t_{DW} (data valid time). In addition, there must be an overlap between the data, valid address bits and active inputs $\overline{CE}$ and $\overline{WE}$. Finally, a certain period of data hold time t_{DH} must be provided. Thus, we have five periods of time (t_{ACC}, t_{CE}, t_{WP},

t_{DW} and t_{DH}), which minimal values must be met. Of course, reasonable margins are also beneficial.

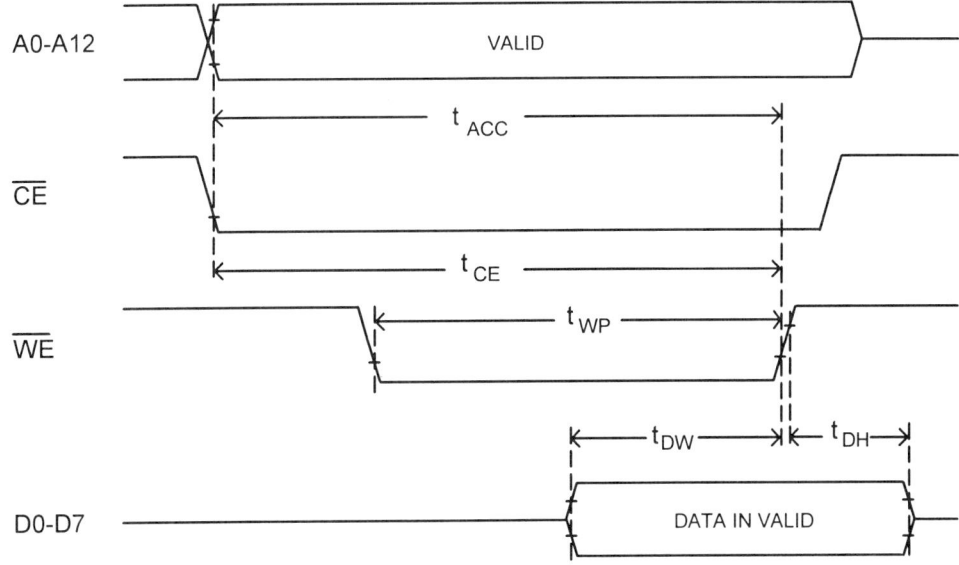

Figure 4.6 The RAM write mode timing.

Our examples so far have dealt with typical memory components. Now we move on to the memory organization phase. While the application dictates the size of the memory, the microcomputer system, in particular the 8051 microcontroller, determines the memory organization. The most important point to note is that the first address of the Program Memory must be covered. If the internal Program Memory is used (the 8051's pin $\overline{EA}$ is pulled high), it comes automatically. However, if the Program Memory is entirely external, an EPROM component must start from address 0000H.

	Address bits			
	A15 A14 A13 A12	A11 A10 A9 A8	A7 A6 A5 A4	A3 A2 A1 A0
EPROM	0 0 0 x	x x x x	x x x x	x x x x
RAM	1 0 0 x	x x x x	x x x x	x x x x

Figure 4.7 An example memory map.

Since we have to define all lower and upper addresses for each memory component, we often find it easiest to use a scheme, such as the one presented in Figure 4.7. For each memory component we distinguish between fixed address values and address bits which can be either 1 or 0. The first group address bits (A15, A14 and A13) are used to select the component. The second group address bits indicate a certain location within the memory component. They can be seen as "x" in Figure 4.7.

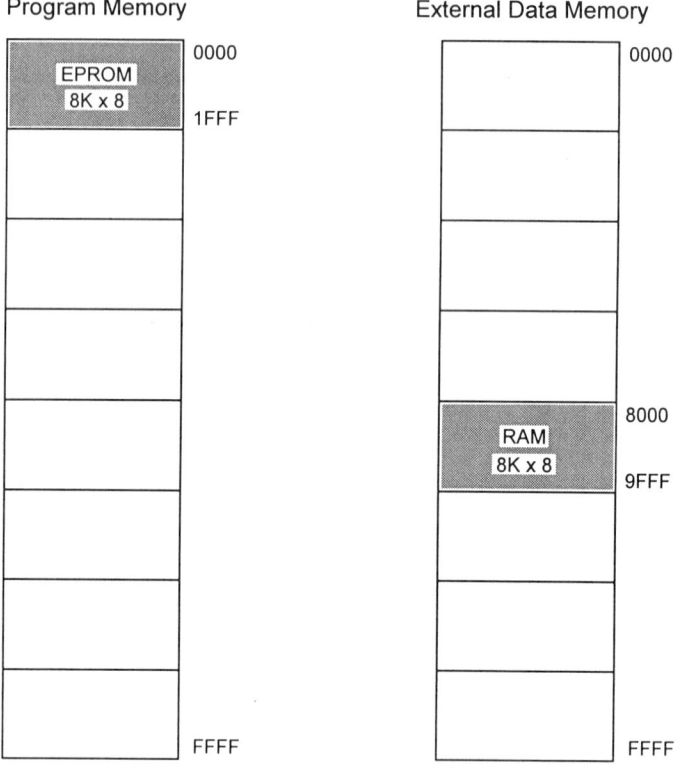

Figure 4.8 An alternative presentation of the memory map.

As discussed earlier, the start address of the EPROM must be 0000H and therefore we fill in 0s for A15, A14 and A13. Replacing all "x" on the same line with 1s, we obtain the EPROM upper address (1FFFH).

Similarly, we mark the address bits which scan the RAM's locations by "x". While this is done automatically, we have to think it over where to place the RAM. There are eight options. In this example, we choose the first RAM location to be addressed at 8000H. It is a simple matter to calculate the upper RAM address (9FFFH). Recapitulating briefly, the EPROM will be selected when the address bits A15, A14 and A13 are low. The correspondence between address values and selected ICs and bytes is termed memory map (Figure 4.7). An alternative presentation of the memory map can be seen in Figure 4.8. The approach used in Figure 4.8 seems to be more attractive, if for example, we want to add an extra component in the empty space.

Once we have completed the memory map, we can approach a schematic diagram. Figure 4.9 shows the connections between the microcontroller and the memory components. A 3-line to 8-line decoder has been allocated to select the components according to the memory map (see section 1.2). Although the eight outputs are not used to the full, the 74HCT138 decoder makes the link to the following schematics more gradual.

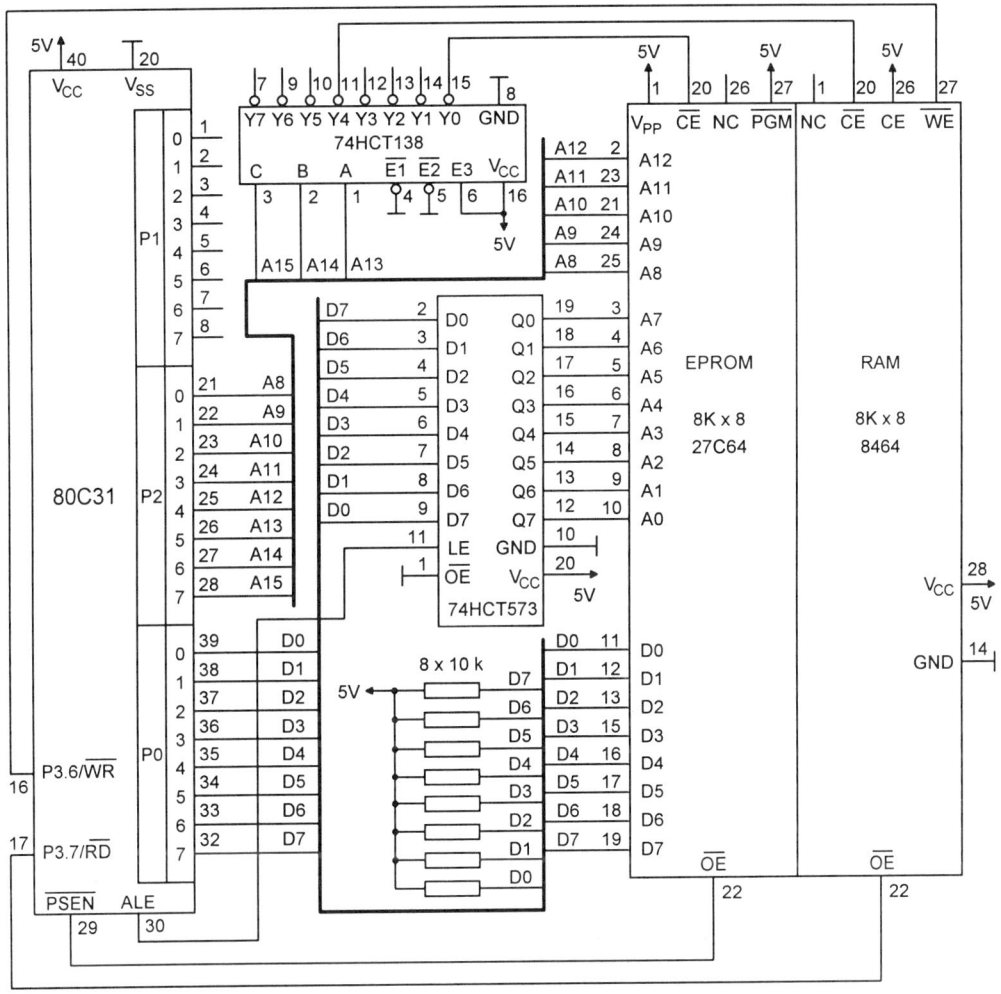

Figure 4.9 The interface between the microcontroller and memory components.

In addition, a register 74HCT573 is used for temporary storage of the low-order address byte. The register contains eight latches. The latch inputs are denoted as D0 through D7. Similarly, the latch outputs are indicated as Q0 through Q7. All latches have a common active high clock input LE (Latch Enable). The register 74HCT573 possesses three-state output buffers. The buffers must be permanently open. As a result, the input $\overline{OE}$ is connected to the ground.

The EPROM and RAM, used in our design, share many common signals which the manufacturers place to the same pins for both integrated circuits. This approach not only alleviates the PCB (Printed Circuit Board) layout but give us the opportunity to draw simple and readable schematic diagrams. Thus, the common lines (addresses, data and power supply) are shown horizontally and only once. Contrary to that, the other signals (control inputs) can be seen to start their individual connections vertically (Figure 4.9).

It is, perhaps, worth mentioning at this point that 8K byte Program Memory is enough amount for a Monitor program plus a Basic interpreter. Moreover, 8K byte of external Data Memory capacity is ample room for storing temporary data.

To this point, we have demonstrated how to organize the interface between the microcontroller and the memory components. Now, we have to prove that the timing constraints are met. This can be done only by relating the microcontroller's timing parameters with those of the memory.

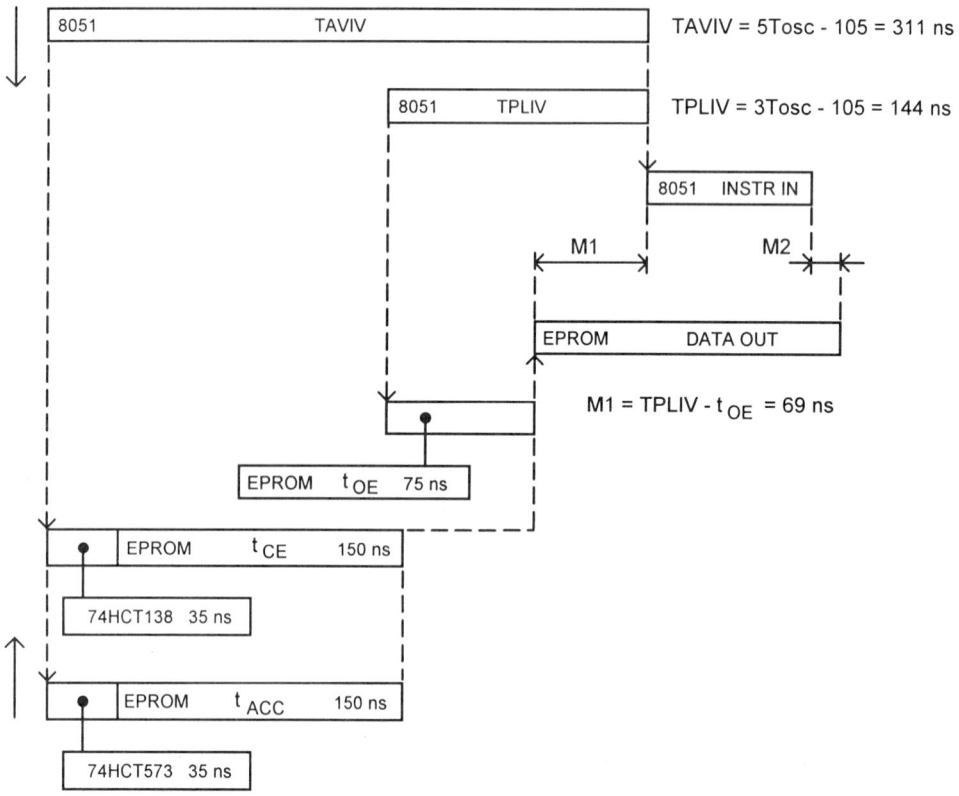

Figure 4.10 The microcontroller and EPROM related timing ($f_{OSC} = 12\,MHz$, $T_{OSC} = 83.3\,ns$).

Figure 4.10 shows a related timing diagram which corresponds to the schematic diagram in Figure 4.9. The drawing is laid out top-down for the microcontroller and bottom-up for the EPROM. Essentially, the period of time used by the microcontroller to read an instruction byte (INSTR IN) is compared with the actual time when the EPROM generates valid data (DATA OUT). There could be no doubt that the proper execution of the program will require the period of time DATA OUT to overlap completely the region INSTR IN.

We obtain the origin of the period INSTR IN on the base of two microcontroller's parameters TAVIV and TPLIV. Once the microcontroller has output the address, it can wait for a certain period of time. This interval is termed TAVIV (address to valid instruction in). The EPROM output data must be available at the microcontroller pins latest at the end of the interval TAVIV. Likewise, the microcontroller's output $\overline{PSEN}$, when asserted, triggers an interval labelled TPLIV ($\overline{PSEN}$ to valid instruction in). We calculated the timing for

$$TAVIV = 5T_{OSC} - 105 \quad \text{(all values in ns)}$$
$$TPLIV = 3T_{OSC} - 105$$

Different versions 8051 microcontrollers may have different values for these parameters.

Against both microcontroller's intervals, which run in parallel, TAVIV and TPLIV, the EPROM has three similar periods: access time t_{ACC}, chip enable time t_{CE} and output enable time t_{OE}. Once these three intervals have elapsed, the time window DATA OUT starts. Note that the intervals t_{ACC} and t_{CE} were shifted right to reflect the delay of the glue logic (see Figure 4.9).

We obtain two margins: M1 and M2. The margins indicate how reliable the microcontroller fetches instructions in case of delay variations. The EPROM, register-buffer and decoder delays are taken from the databooks for the worst case (maximal values). Again, we link the key points by dashed lines. You have probably noticed that the EPROM delay could be minimized if the input $\overline{OE}$ were activated earlier.

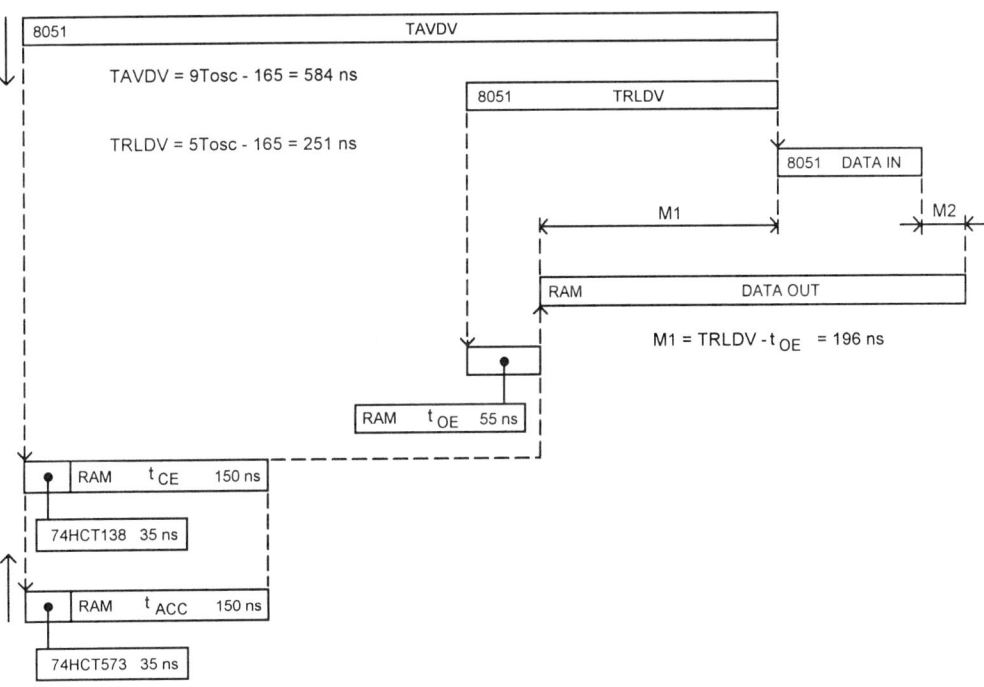

Figure 4.11 The microcontroller and external RAM related timing for read instructions ($f_{OSC} = 12\,MHz$, $T_{OSC} = 83.3\,ns$).

We get M1 = 69 ns. For simplicity, we do not evaluate the margin M2. The 8051 single chip microcomputer does not demand any hold time when the strobe $\overline{PSEN}$ is changed from low to high. Furthermore, the EPROM can not alter its outputs immediately, which is also beneficial in this case.

Similarly, Figure 4.11 shows the microcontroller - external RAM related timing for read instructions. Again, we want to compare the period of time when the microcontroller is sensitive for data (DATA IN) with the actual time domain when the external RAM provides correct data (DATA OUT). The timing parameter TAVDV indicates how long the microcontroller can wait for valid data. The period TAVDV starts when the address is established. In parallel with TAVDV runs another period named TRLDV ($\overline{\text{RD}}$ low to valid DATA IN).

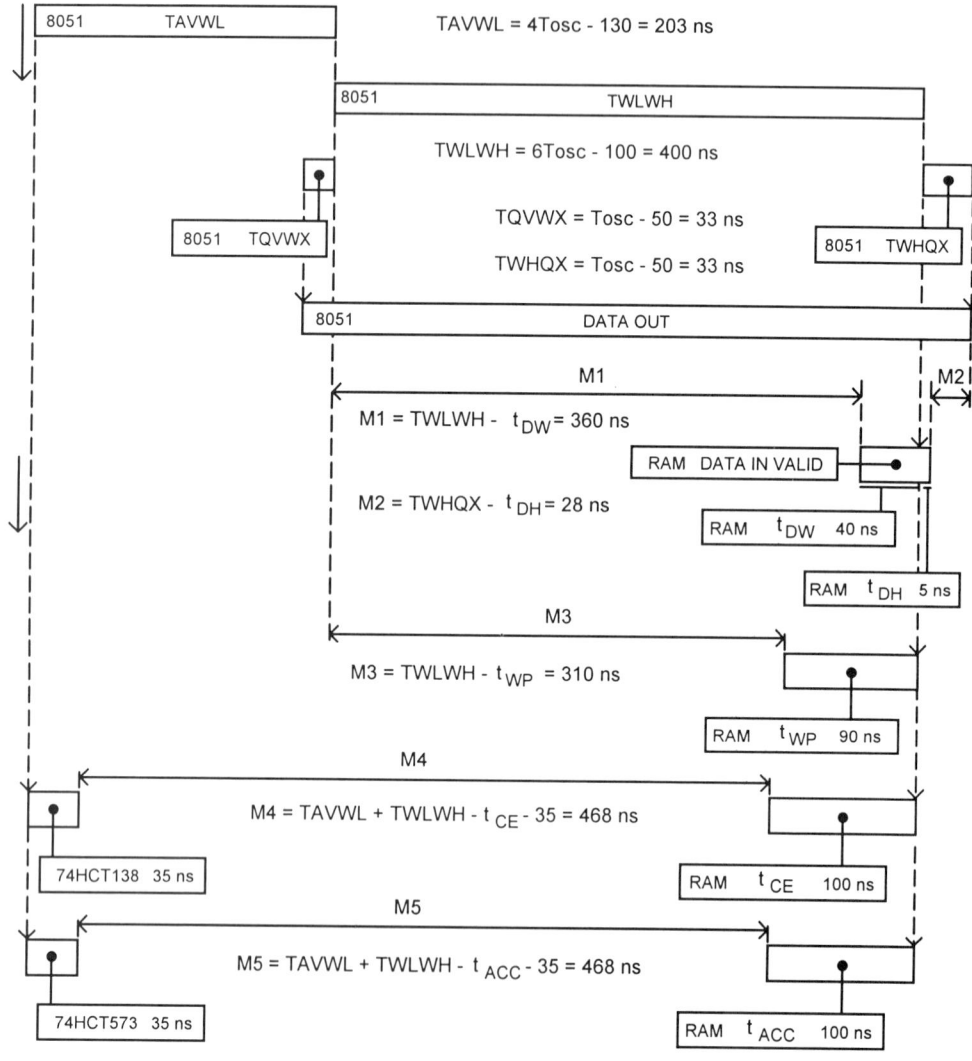

Figure 4.12 The microcontroller and external RAM related timing for write instructions ($f_{OSC} = 12\,\text{MHz}$, $T_{OSC} = 83.3\,\text{ns}$).

Since the related timings maintain one scale for all times involved, it is easy to notice that the designers of the 8051 microcontroller overspecified the problem of interfacing external RAM. Once the RAM delay might have been a problem. Now, the fast modern RAMs (access time in the area of 10 ns) have ample reserve to follow the swift march of the high-performance microcontrollers. However, in the majority of embedded applications the dominant factor is the price. Slower microcontrollers and RAMs are cheaper. On top of that, they consume less power.

Figure 4.12 shows the microcontroller - external RAM related timing for write instructions. This is the last step in evaluating the timing constrains of the microcomputer shown in Figure 4.9. As a principle, the timing in Figure 4.12 resembles to both previous schemes. Following the 8051 timing parameters, we obtain the period of time when the microcontroller outputs correct data (DATA OUT). In this drawing however, we present the RAM timing parameters top-down. The motivation behind is that the RAM is the destination now. Thus, we must check all RAM parameters, starting from the DATA IN VALID requirement and completing the evaluation with the access time.

When writing to external RAM, the microcontroller activates the output $\overline{WR}$. The $\overline{WR}$ pulse width is introduced by TWLWH. The valid output data is centred to the pulse $\overline{WR}$. Furthermore, the data is available some time before the high-to-low transition of the signal $\overline{WR}$ (TQVWX) and some time after the low-to-high transition of $\overline{WR}$ (TWHQX). The parameter TAVWL has the meaning of address to low $\overline{WR}$ time.

Again, we must make sure that the source overlaps completely the region DATA IN VALID by correct data. In fact, this period of time consists of three components.

First, the normal write mode of the RAM requires a certain duration of valid data overlapped by active write input t_{DW}. Since the RAM input $\overline{WE}$ is driven by the microcontroller's output $\overline{WR}$, we align the end of the period of time t_{DW} with the end of the active strobe $\overline{WR}$. Second, there is a short transition time of the output $\overline{WR}$. Finally, the RAM demands a certain amount of hold time t_{DH}.

Once we have outlined the DATA IN VALID region, we can predict the margins M1 and M2. Both equations and the calculated values are included in Figure 4.12.

Next, we evaluate the RAM input signal $\overline{WE}$. The signal is derived directly from the strobe $\overline{WR}$. The RAM manufacturer requires a minimal duration of the write pulse t_{WP}. This duration must be met with some excess. Likewise, the margins for the chip enable time t_{CE} and the access time t_{ACC} must be positive.

The related timing diagrams keep the design pretty transparent. In addition, they help the user to organize consistent replacement of components when the system is upgraded. Even though some of the margins we calculated were quite large, the temptation to skip this phase of the design, especially for higher oscillator frequencies, might lead to unfeasible solutions.

Single board computers, especially if used as development tools, might need more memory for both Program Memory and external Data Memory. Furthermore, it would be useful, if the distinction between both types of external memory was not rigid. Practically, a few sockets on the board can be used either for EPROMs or for RAMs. The user will be able to strike the balance between both types of memory. In addition, there might be need to move the components in the memory space by microswitches. The following Problem/Solution deals with the practical aspects of the memory organization.

Problem 4.1

Design a memory subsystem for a single board computer which has three sockets for 8K bytes memory components. The microcomputer must be capable of implementing the options indicated in Figure 4.13.

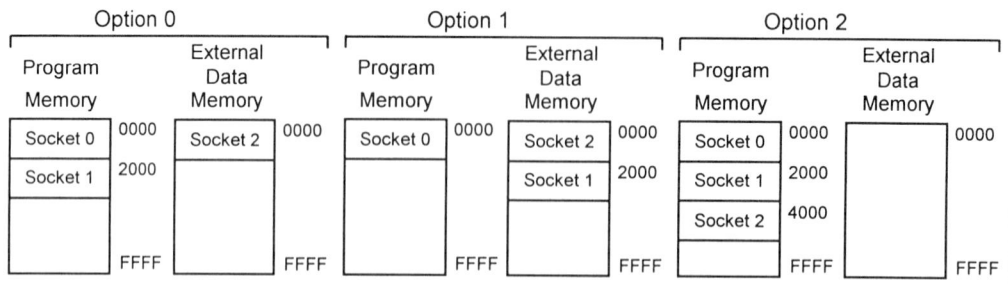

Figure 4.13 Three options for the memory map of the microcomputer.

Solution 4.1

Figure 4.14 shows the solution. We move the memory component in socket 2 by jumpers

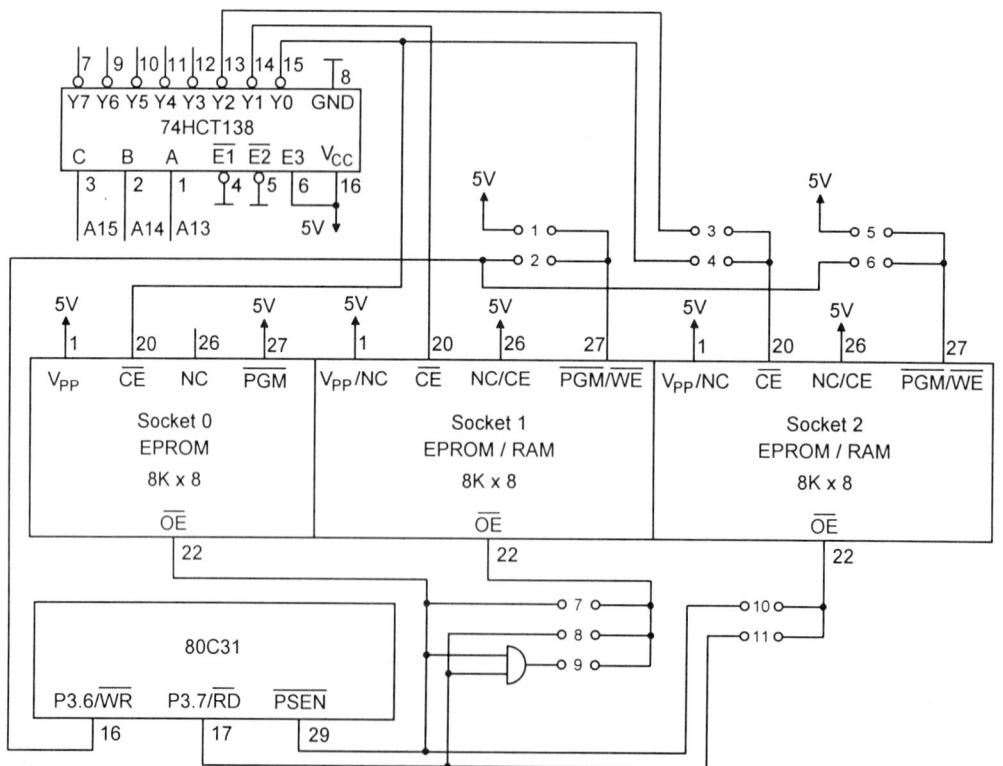

Figure 4.14 The memory subsystem.

(microswitches) #3 and #4. The type of the memory (Program or external Data) is selected by microswitches #7, #8, #10 and #11. In addition, microswitch #9 allows execution of programs from the RAM (socket 1) in option 1. The overall picture of the microswitches can be seen in Figure 4.15.

The address and data interface is identical to the one presented in Figure 4.9. Option 1 allows the code in socket 1 RAM to be treated as Program Memory. We emphasize this possibility by closing switch #9 in Figure 4.15. It is certainly implied that the code must be moved to the RAM in socket 1 before the execution of the program.

	Closed switches
Option 0	1, 4, 6, 7, 11
Option 1	2, 4, 6, 9, 11
Option 2	1, 3, 5, 7, 10

Figure 4.15 The correspondence closed switches - options.

Now we move on to the next key feature of the design - nonvolatile memory. Figure 4.16 shows four basic nonvolatile memory solutions.

Battery-backed CMOS RAM	Nonvolatile storage upon power failure
NOVRAM	Nonvolatile storage upon power failure Configuration data
EEPROM	Upgrade or change of software from a remote location Configuration data Calibration parameters
Flash memory	Upgrade or change of software Nonvolatile storage upon power failure

Figure 4.16 Nonvolatile memory solutions.

The first possibility is battery-backed CMOS RAM. This traditional approach is still popular. The schematic diagram in Figure 4.17 shows a battery-backed CMOS RAM solution controlled by a microprocessor supervisory circuit MAX805L produced by Maxim [Maxi 1994]. The supervisory circuit provides the following four functions:

• Power-fail warning

Unregulated DC through input PFI (Power Fail comparator Input) is compared to an internal 1.25 V reference. In the event of impending power loss, the voltage across resister R2 will be less than 1.25 V and the output power fail $\overline{PFO}$ will be brought down and the microcontroller interrupted. The minimal input voltage of 7805 regulator must be approximately 7 V and it is normally chosen between 8 V and 10 V. Therefore, the regulator will continue functioning for 50 to 100 ms until the large filter capacitor discharges to about 7 V. During that period, the single chip microcomputer can save its vital information in the RAM and prepare the system for the impending power loss.

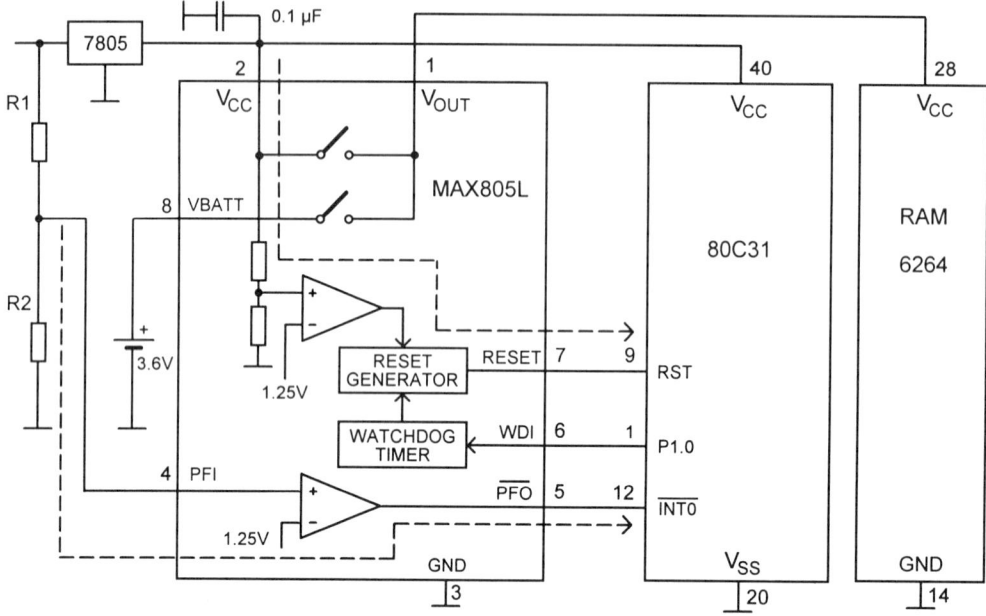

Figure 4.17 Microprocessor supervisory circuit.

• Reset

The second path in Figure 4.17 links another comparator, a reset generator and the 80C31 RST input. The supervisory circuit brings the RESET output high when the supply voltage V_{CC} drops below 4.65 V. When V_{CC} exceeds the threshold again, an internal timer keeps RESET output high for 200 ms. In the event of power failure, the RESET output voltage is equal to supply voltage V_{CC} or battery voltage whichever is higher.

• Battery-backup switchover

When the supply voltage falls below the threshold of 4.65 V, the output V_{OUT} is switched to V_{CC} or VBATT whichever input has a higher voltage level. The circuit is stabilized by hysteresis.

• Watchdog timer

Noise is a major source of problems and normally manifests itself as an altered address or data code and finally the wrong sequence of instructions. For many applications, the problem will be practically eliminated if the wrong sequence is detected within a certain period and the microcontroller is reset. This could be done by adding a Watchdog timer and extra instructions which write to the Watchdog regularly. When an instruction clears the Watchdog it assumes that everything is fine and life goes on normally. If the Watchdog has not been cleared for a certain period it will time out and reset the microcontroller. The Watchdog embedded in MAX805L requires its input WDI (WatchDog Input) to be toggled within 1.6 sec.

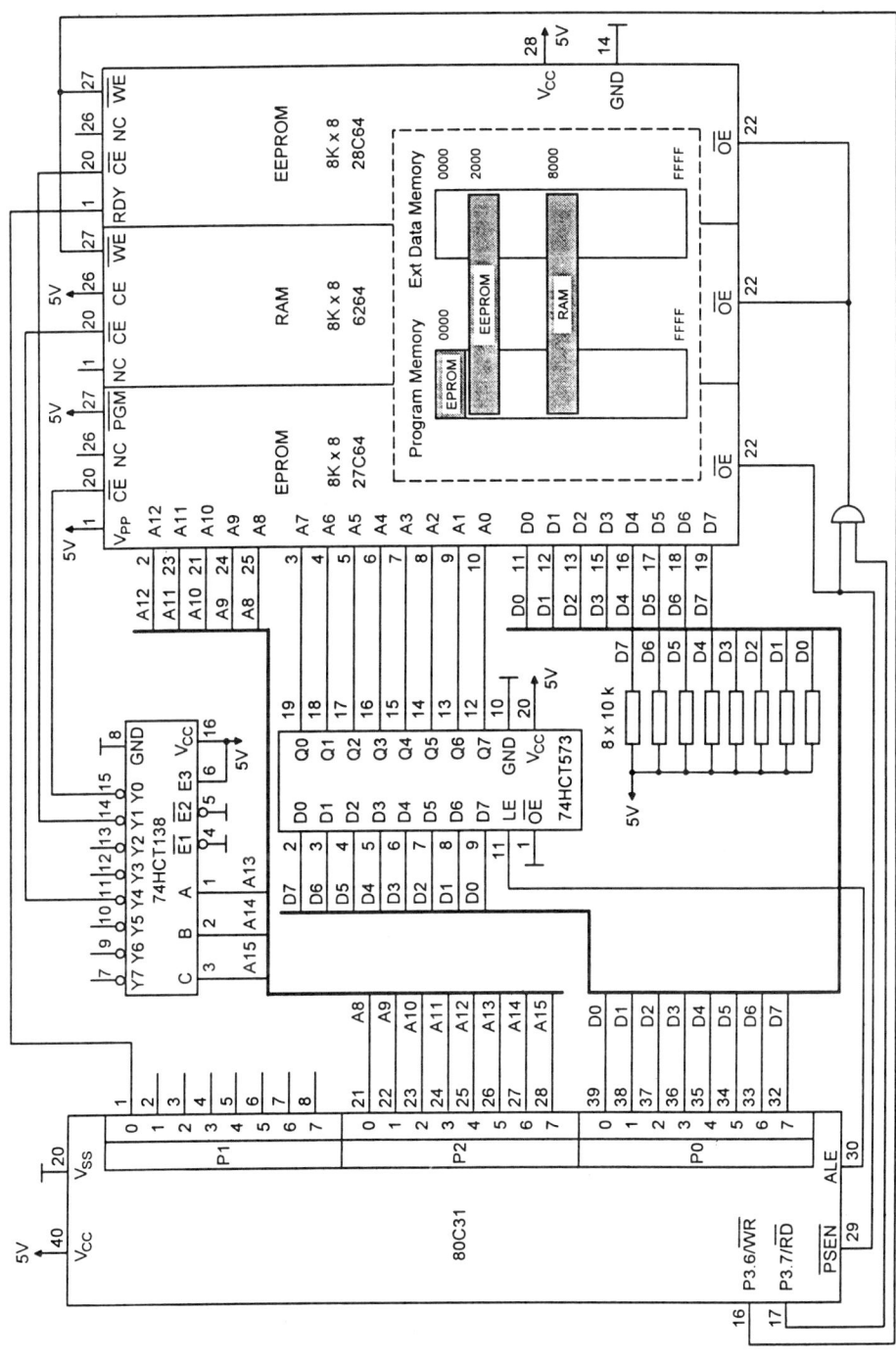

Figure 4.18 EEPROM nonvolatile memory solution.

In many cases the requirements which the applications dictate can be satisfied by either NOVRAM (NOnVolatile RAM) or EEPROM (Electrically Erasable PROM). The EEPROM cell is small and simple. It allows higher density storage and therefore lower price. Also, flash memories are modern fast and low-power devices.

Figure 4.18 shows an example of a system, which is capable of upgrading or changing its software. The EEPROM 28C64 is used as a static RAM for the read or write cycles. During a write instruction, the address and data are latched internally, releasing the address and data bus for the next instruction.

There are two possibilities for the microcontroller to determine if the write cycle has been completed. First, the ready flag RDY switched back to high level is an indication the write cycle is over. Second, a polling can be done by a read instruction accessing the same address. If the MSB D7 comes out complemented, the write mode is still in progress. The write cycle time for 28C64 has a maximum of 1 ms. Some versions have write cycle time of 200 µs.

Both the RAM and the EEPROM are driven by $(\overline{PSEN})(\overline{RD})$. Thus, they can be accessed as Program Memory. This feature is indicated in Figure 4.18 by a built-in memory map.

When you choose your nonvolatile chip watch out for this detail: some more sophisticated memories possess a buffer for the low order address byte, which must be clocked by ALE and low order address/data lines connected to Port 0 directly.

4.3 Parallel interface

The 8051 microcontroller has four parallel ports which can be used to interact with the environment. As is frequently the case, Port 0 and Port 2 are occupied to link the microcontroller to the external memory. Some other microcontrollers might have more I/O lines, however sooner or later the designer of embedded systems will face the problem of insufficient I/O pins. Along with the other considerations which are not trivial, the embedded systems design is a battle for minimization of the number of connections microcontroller - outside world (see the EPROM emulator case study in section 11.3).

There are three basic approaches to expand the microcontroller's I/O interface. First, we can use the external Data Memory address space to organize bidirectional input/output. This implementation is termed memory-mapped I/O. Second, port based expansion schemes will give us multiple-byte input or output. Finally, a serial-to-parallel and parallel-to-serial conversion through the serial port (mode 0) can be considered (see section 2.8). We concentrate on the first two possibilities, which are pure parallel methods for I/O expansion.

Memory-mapped I/O

Reduced to its fundamental principle, the functioning of a memory-mapped I/O is not difficult to understand. What we need is a decoder. However, one decoder can be used, for example, to select a set of outputs. If we have to organize both inputs and outputs, we allocate two decoders. The decoder which controls the outputs redirects the strobe $\overline{WR}$ to clock a certain register. The decoder which takes care of the inputs redirects the microcontroller strobe $\overline{RD}$ to a set of eight three-state buffers. Naturally, the three-state buffers have a common input $\overline{OE}$. On the base of two 3-line to 8-line decoders, such as the 74HCT138, we can implement eight 8-bit input and eight 8-bit output ports. The way the microcontroller's address lines are connected to the decoder inputs determines the locations which are used in the external Data Memory address space.

In general, the advantages of the memory-mapped I/Os are the huge number of I/O ports which could be implemented and the possibility to apply different instructions and addressing modes. However, the 8051 microcontroller possesses only two read and two write instructions which deal with the external Data Memory. The first advantage is also to some extent, a controversial subject. Practical designs compromise between the complexity of the decoding logic and the number of locations occupied by a single I/O port.

Logically, the chip-makers developed LSI devices which contain the required decoders, latches and buffers. They are termed Programmable Peripheral Interface (PPI) or Programmable Input Output (PIO). As the name indicate, the support ICs possess programmability. For example, a certain port can be used either as an input or as an output port, depending on the application.

Figure 4.19 shows the PIO 82C55A interface and programming. The 82C55A links the microcontroller with the outside world. The data bus (D7-D0) is transformed to three 8-bit ports - Port A, Port B and Port C. So, the communication capability, directed to the environment, is increased three times.

Also, Figure 4.19 attempts to show the relationship between the PIO address inputs (A1 and A0) and the port selected. As you might expect, the system address lines A1 and A0 are connected to the corresponding inputs of the PIO. For instance, if the system address bits A1 and A0 are both low, the selected port is Port A.

At the same time, Port C can be broken down into Port C upper nibble and Port C lower nibble. Thus, two groups of I/O lines, labelled A and B, are formed. The group A unites Port A and Port C-upper. The group B includes Port B and Port C-lower.

The 82C55A occupies four locations from the memory. When both inputs A1 and A0 are high a Control Register (CR) is selected. When the microcontroller writes a byte to the CR we distinguish between mode definition and Port C bit set/reset. In contrast to the popular 8255A, the CMOS version 82C55A allows the CR to be both written and read.

The PIO can be programmed by mode definition if the byte sets CR.7. If the byte written to the CR resets the bit CR.7, the current mode is not changed, but a selected bit from Port C is set or reset. All of the output registers are reset when the mode is changed. Providing an active-high pulse for at least 50 μs on the RESET input will leave the three ports programmed as inputs.

Defining a mode, we can choose between mode 0, mode 1 and mode 2. Bits CR.6, CR.5 and CR.2 determine the mode.

• Mode 0 offers three ports without any additional functions.

• Mode 1 specifies Port A and Port B as data ports and Port C is used for handshaking and interrupt-driven interface.

• In mode 2, Port A becomes a bidirectional data port. Five handshaking signals are taken from Port C. Port B can be used in mode 0 or mode 1.

We focus our attention on mode 0. When the PIO is in mode 0, there are sixteen possible configurations. Bits CR.4 and CR.3 define the direction for Port A and Port C-upper. Likewise, bits CR.1 and CR.0 determine the direction for Port B and Port C-lower.

Similarly to the 8051 microcontroller, the 82C55A is designed to set or reset individual bits of Port C. When we want to use this feature, along with the logic 0 for CR.7, we send the address of the bit to CR.3, CR.2 and CR.1. The actual value of the bit is moved to CR.0.

The following Problem/Solution demonstrates initialization in mode 0 and access to individual bits from Port C.

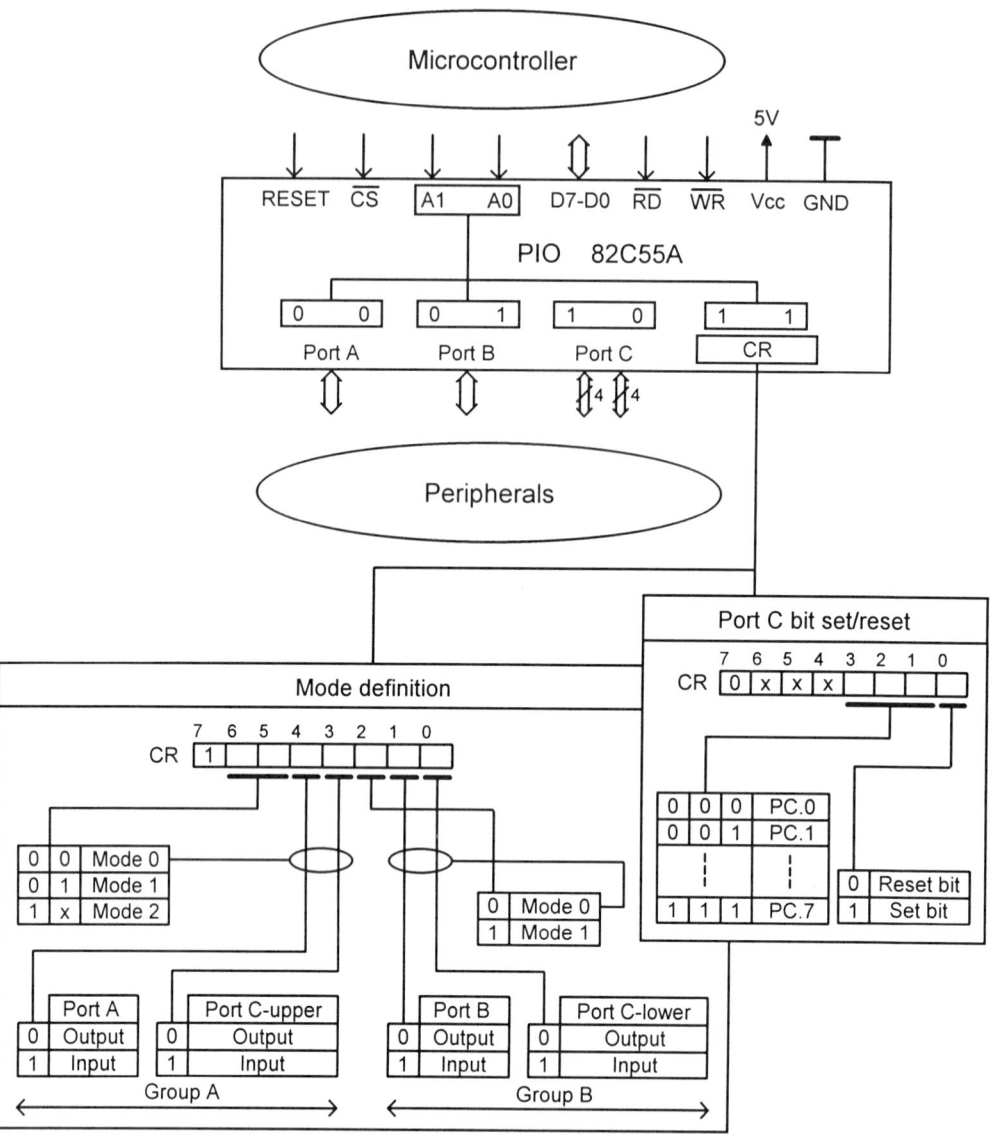

Figure 4.19 The PIO 82C55A interface and programming.

Problem 4.2

An 8051 microcontroller works in conjunction with an 82C55A peripheral I/O. The PIO is assigned at address 8000H in the external Data Memory space. Write a sequence of instructions which initializes the 82C55A and performs an endless loop.

• The microcontroller programs ports A and B as input ports. Port C is programmed as an output port.

• The loop includes the following actions: the microcontroller reads ports A and B, compares both bytes and indicates the relationship by pulsing high a bit in Port C.

Solution 4.2

We are going to pulse high the PIO output PC.7 when the value applied to Port A is lower than the Port B one. The output PC.6 will be switched twice in the same vein when both values are equal. Finally, the output PC.5 will be pulsed high when the Port A byte is higher than the Port B one.

```
            MOV     DPTR,#8003H     ; PIO CR address
            MOV     A,#10010010B    ; Mode definition, mode 0
                                    ; Port A and Port B - input
                                    ; Port C - output
            MOVX    @DPTR,A
;
LOOP:       MOV     DPTR,#8001H     ; Port B address
            MOVX    A,@DPTR         ; Read Port B
            MOV     TEMP,A          ; Save Port B
            MOV     DPTR,#8000H     ; Port A address
            MOVX    A,@DPTR         ; Read Port A
            CJNE    A,TEMP,L1       ; Compare Port A and Port B
            MOV     DPTR,#8003H     ; PIO CR address
            MOV     A,#0DH          ; A=B, set PC.6
                                    ; 0CH + 01H = 0DH
            MOVX    @DPTR,A
            DEC     A               ; Reset PC.6
            MOVX    @DPTR,A
            SJMP    LOOP
L1:         JC      L2              ; Carry flag might have been
                                    ; set from CJNE if A<B
            MOV     DPTR,#8003H     ; PIO CR address
            MOV     A,#0BH          ; A>B, set PC.5
                                    ; 0AH + 01H = 0BH
            MOVX    @DPTR,A
            DEC     A               ; Reset PC.5
            MOVX    @DPTR,A
            SJMP    LOOP
L2:         MOV     DPTR,#8003H     ; PIO CR address
            MOV     A,#0FH          ; A<B, set PC.7
                                    ; 0EH + 01H = 0FH
            MOVX    @DPTR,A
            DEC     A               ; Reset PC.7
            MOVX    @DPTR,A
            SJMP    LOOP
```

Port based expansion schemes

This approach is simple and effective. Figure 4.20 gives an example of a 16-bit output. Two registers (74HCT573) are driven by Port 1. We distinguish between the registers by separate clock signals (P3.3 and P3.5).

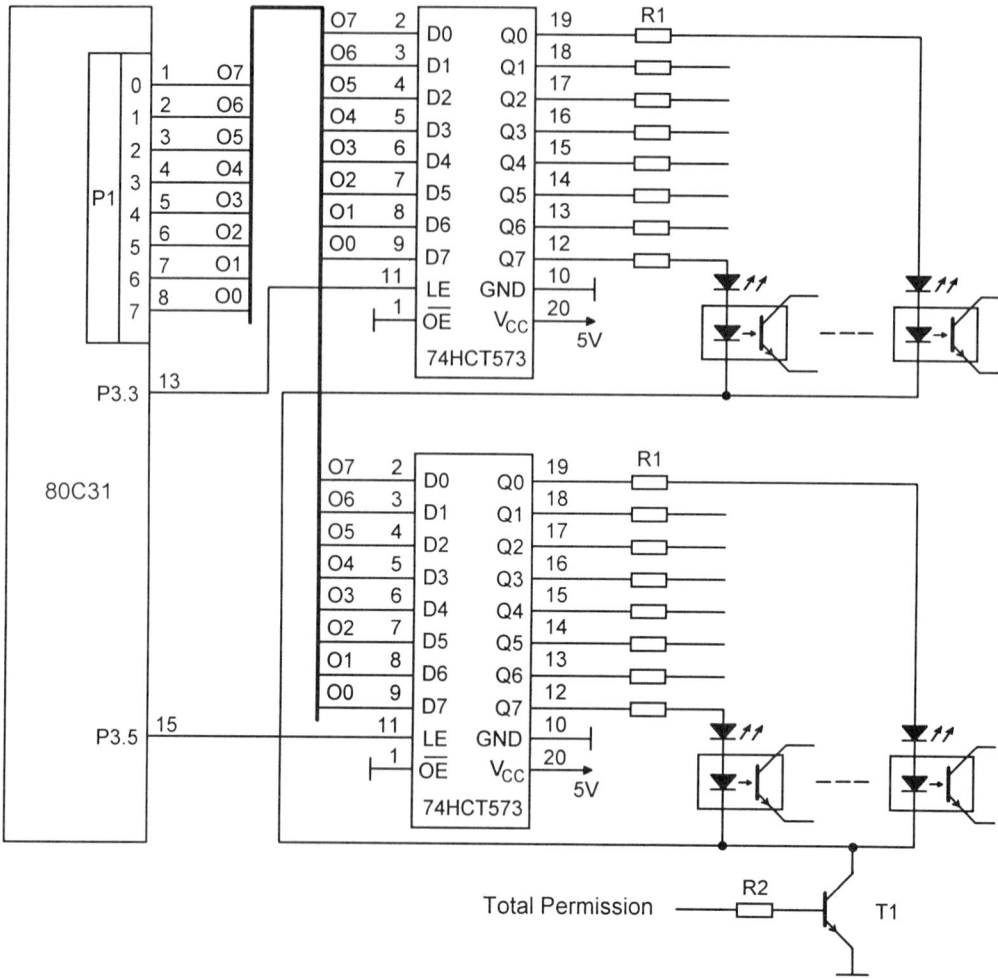

Figure 4.20 Parallel interface showing a port based expansion scheme.

Furthermore, in an attempt to design reliable systems we use optocouplers. The ground isolation may give us results which we would not otherwise be able to obtain. Next, we introduce a total permission for output control. When the transistor T1 is cut off all outputs are disabled. It allows a supervisory circuit to switch off all outputs when the microcontroller fails or during the transient conditions that accompany application of power. Finally, each output has an indicating LED (Light Emitting Diode) lamp.

Problem 4.3

Design a subsystem which expands the 8751 Port 1 into three 8-bit output ports. Use one pin from Port 3 to clock the outputs. Assume that the flip-flops have only a clear ($\overline{CL}$) asynchronous input.

Solution 4.3

This problem is relevant in case of internal Program Memory. All of the microcontroller pins are harnessed to the task of communicating with the environment. However, the designer still needs more I/O lines. The considerations above justify the limited number of pins we can use to build the required output ports.

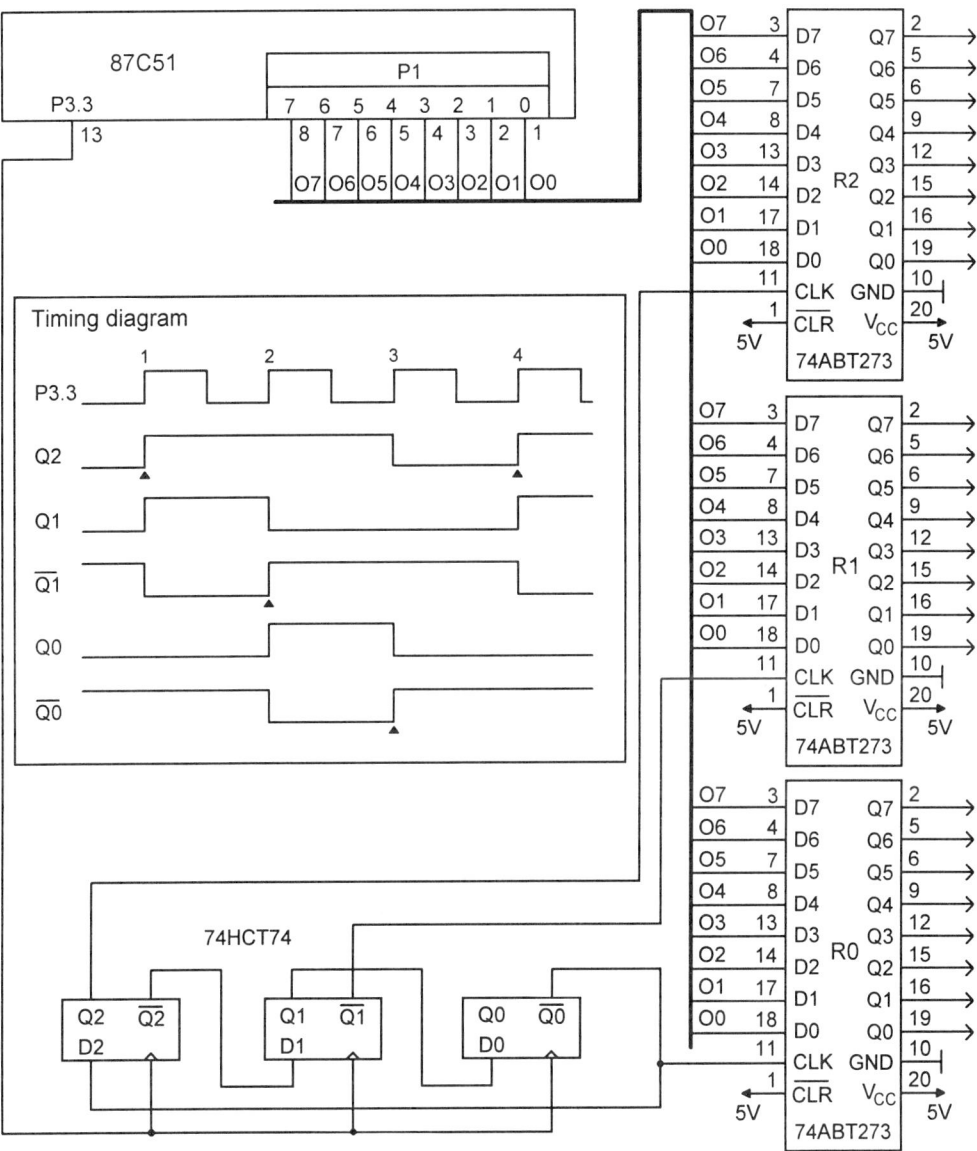

Figure 4.21 Parallel outputs expansion.

Furthermore, if we manage to complete the design using only flip-flops with a clear ($\overline{CL}$) asynchronous input, we will be in a better position to implement the subsystem by modern programmable LSI devices.

Figure 4.21 shows a solution for the problem. We can break down the subsystem into output register buffers and clock circuit. We allocate registers 74ABT273 for the output buffers [Texa 1993]. The registers contain eight edge triggered D-type flip-flops (see also section 1.3). The flip-flops possess a clear input ($\overline{CL}$) which must be pulled down for at least 3.5 ns.

The clock circuit consists of three edge triggered D-type flip-flops which can be taken from two ICs 74HCT74 [Phil 1994a]. The idea is to organize a ring counter. A logic 1 circulates and clocks the register buffers. Banking on a ring counter we face the problem with the flip-flops initial state. All the flip-flops are reset when the power is switched on. We could overcome this problem by closing the feedback from the output $\overline{Q0}$. As a result, we are forced to rotate logic 0s in flip-flops Q1 and Q0, which in turn leads to driving the corresponding clock inputs by the outputs $\overline{Q1}$ and $\overline{Q0}$.

It is certainly implied that the system must maintain a rigid correlation between the code emitted by Port 1 and the activated clock signal. For example, the first low-to-high transition of the output P3.3, which follows immediately the application of power, must be accompanied by code for the buffer R2. The next step is to move the data, which destination is R1, to Port 1 and to toggle the output P3.3 again. Regardless of our intention to change or not change the output pattern of a certain channel, we must emit the new or old code through Port 1.

The following concluding points are in order for the presented expansion scheme. First, the structure is homogeneous and we can easily add more outputs. Second, the number of output channels defines the updating rate when the system overrides the ring counter sequence. As long as the output operations are consistent with the rotation of the clock, practically there is no significant degradation of the performance for extra channels.

4.4 Serial interface

In contrast to the parallel interface, serial communication systems transmit and receive data bits one after the other. The number of wires is reduced to one or two. Consequently, the serial approach is cost-effective, especially for long distances. At the same time, the serial interface brings the rate down.

Figure 4.22 shows two basic types of serial interfaces: synchronous and asynchronous.

In synchronous systems there is a common clock which synchronizes the data transfer. One of the devices generates the clock. The other computer or peripheral receives the clock. The 8051 microcontroller can communicate synchronously using the serial port mode 0.

Next, we can save the clock line and work asynchronously with regard to the bytes. The devices use separate clocks. The clocks must be adjusted to an agreed frequency. Furthermore, the receiver needs an indication that the transmission will start. When the line is not busy it is always high. Logically, the transmitter inserts a low bit before the actual byte to advice the receiver that data bits are to come. Finally, a high stop bit completes the data frame. Different devices may use different number of stop bits. In particular, the 8051 microcontroller works with one stop bit (serial port mode 1, 2 and 3). As discussed earlier in section 2.8, the microcontroller may insert an extra bit (D8) between the MSB bit D7 and the stop bit (serial port mode 2 and 3). The additional bit can be used either for second stop bit or for parity check.

Moreover, the bit D8 is related to the interrupt system. When microcontrollers communicate over a network, the bit D8 makes the interaction more efficient (see section 6.2).

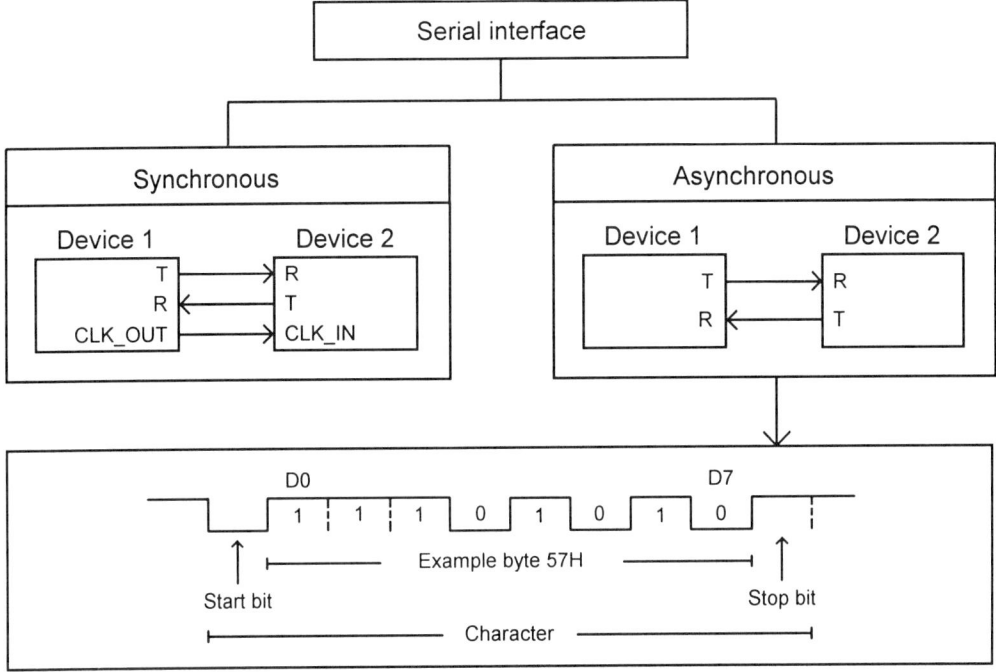

Figure 4.22 Serial interface - timing taxonomy.

Figure 4.23 illustrates a few commonly used serial interface terms. We distinguish between full-duplex, half-duplex and simplex serial interfaces. While the full-duplex system can communicate in both directions simultaneously, half-duplex interface assumes one line to be used either for transmission or reception at a certain moment. Finally, simplex systems communicate only in one direction.

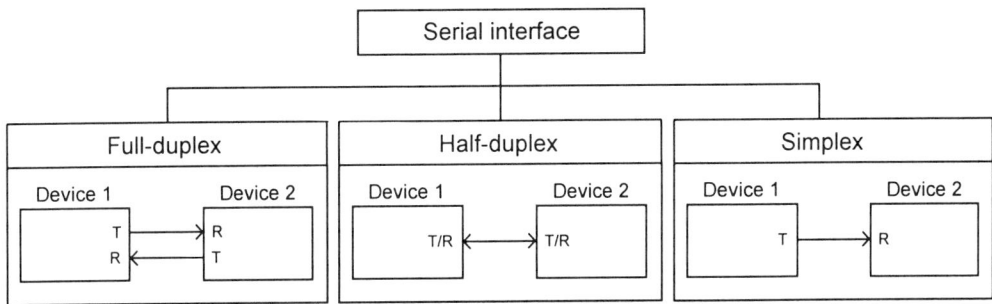

Figure 4.23 Serial interface - data flow terminology.

American manufacturers proposed several standards for serial communication through their EIA (Electronic Industries Association) organization.

Interface RS-232

The majority of computers are equipped with a RS-232-C interface. The standard was updated in 1969. It enables a full-duplex link to be established between two devices. The RS-232 is a single-ended approach. The input and output voltages are referenced to a common ground. The driver output levels for a logic low are between 5 V and 15 V. The receiver recognizes logic low within the range 3 V through 25 V. Thus, the noise margin for logic low is 2 V. Likewise, the driver logic high area is from -5 V through -15 V. The receiver logic high specification is from -3 V through -25 V. Again, the noise margin is 2 V.

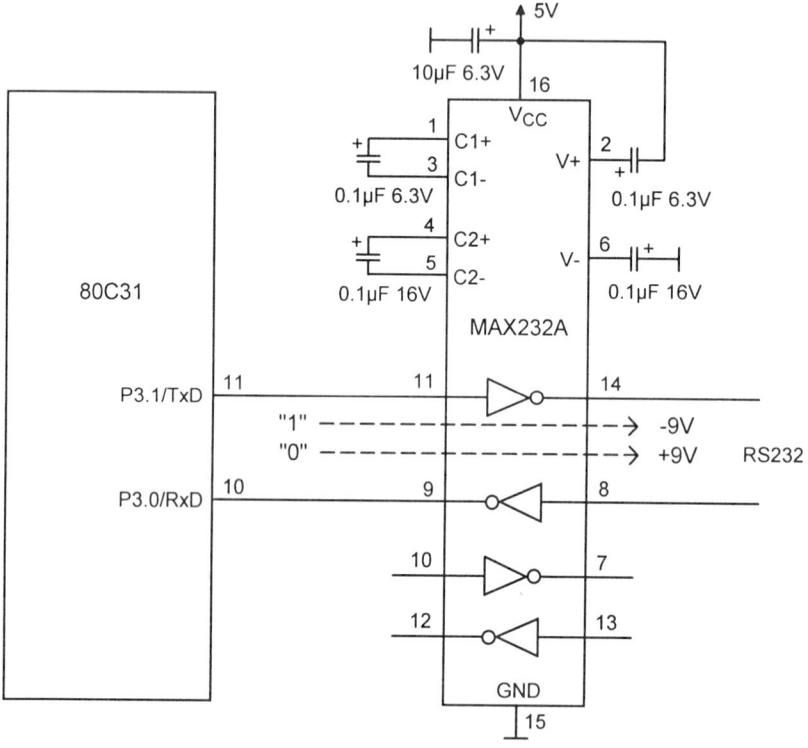

Figure 4.24 Interface RS-232.

The RS-232 standard does not specify the baud rate, however it imposes limitation on the transition time from one logic level to the other. The transition time must not exceed 4% of one bit time. The resulting maximum cable length is restricted to 50 ft for 19200 bps.

The schematic diagram in Figure 4.24 shows a RS-232 buffer which translates the microcontroller TTL signals to the levels specified by the standard. The IC MAX232A is produced by Maxim [Maxi 1994]. The buffers (shown as inverters) are four: two drivers and two receivers. Essentially, the drivers have RS-232 outputs (pins #7 and #14). The receivers possess RS-232 inputs (pins #8 and #13). One IC can be used for two RS-232 interfaces. The typical output voltages 9 V and -9 V are achieved by an internal charge pump.

Note that the MAX232A buffer is suitable for point-to-point communication. If you want to organize a network based on the RS-232 interface, you should surround the microcontroller with some extra logic to expand the serial port to two channels. One buffer MAX232A is sufficient to link a microcontroller to two other devices.

So far, we have covered the introductory part of the RS-232 interface that we need for the single board computer design. In addition to the data signals, the interface also includes control/status signals. Their meaning and placement in a physical interface (a standard connector) is discussed in section 6.3.

Interface RS-485

The standard was established by the EIA in 1983. The application field is inexpensive local area networks. An essential improvement over the RS-232 interface is the use of symmetrical link. Inevitably, the differential approach makes the link suitable only for half-duplex connections. The transmission medium is a 120 Ω twisted pair line.

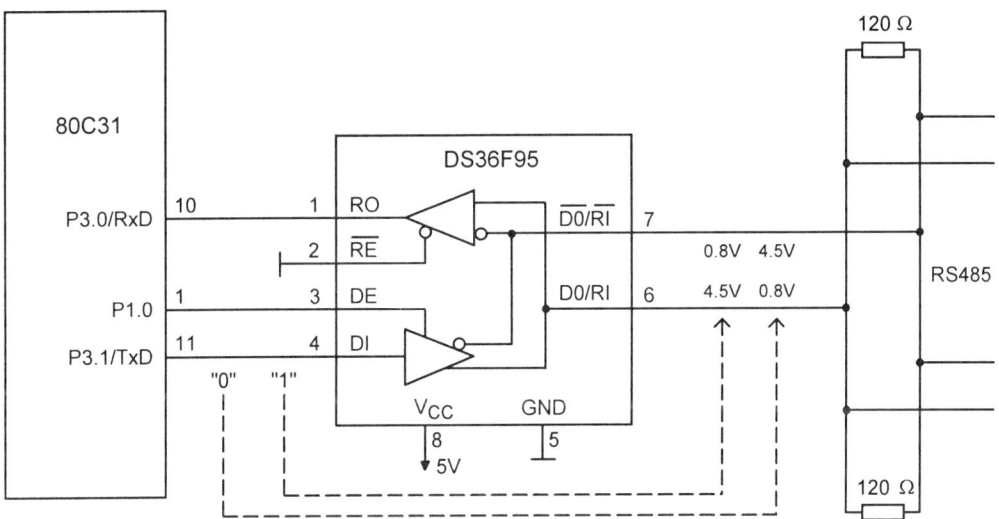

Figure 4.25 Interface RS-485.

An example of a RS-485 buffer can be seen in Figure 4.25. The bus transceiver DS36F95 is produced by National Semiconductor [Nati 1994]. As shown, the input receive enable $\overline{RE}$ can be tied to ground and the microcontroller will listen to the bus all the time. On the contrary, the input driver enable DE must be controlled by a dedicated output in order to avoid bus contention.

When the microcontroller output TxD is low the driver output level is 0.8 V. The driver inverting output emits 4.5 V. Both voltages are with respect to the ground. The voltage between the lines is -3.7 V. Likewise, a high level on the microcontroller output TxD results in 3.7 V between the RS-485 lines.

The RS-485 interface is very flexible as far as the number of currently attached devices is concerned. However, all participants must obey the bus discipline.

4.5 Single board computer

We can now draw a schematic diagram of a real single board computer. We should unite the memory design, parallel interface and serial interface. Figure 4.26 shows the single board computer hardware.

It is possible different versions of 8051 microcontrollers to be plugged in the board. The computer could accommodate any 8051 pin compatible microcontroller. The jumpers #19 and #20 determine if the code is fetched from external Program Memory. In the case of the 80C31 microcontroller the jumper #19 is always installed and #20 not installed.

A switch S1 is used for manual reset of the microcontroller and the PIO 82C55A. We need this feature when the computer is out of control.

The memory components EPROM and RAM, which the sockets can accommodate, are 8K x 8 and 32K x 8. The second option is indicated by dashed lines. It applies to the size, the name of the IC and the name of the pins which are different. For example, the EPROM pin #26 is not used for 8K byte versions (no internal connection). The same pin for 32K byte EPROMs is used to accept the address signal A13. We can adjust the settings for a certain pair of memory components by switches (jumpers). Furthermore, we will see how to place the EPROM and the RAM in the memory map. Before that, however, a few comments about the PIOs addressing are in order.

As discussed in section 4.3, the PIO 82C55A occupies four locations from the memory. There are, also, other programmable I/O peripherals which require a small amount of addresses. The full decoding of such a small set of addresses might outweigh all other aspects of the design. We could work around this problem by applying an approach termed partial decoding. The idea is to ignore a few address signals in order to simplify the decoder. For instance, if we design a decoder for a 82C55A PIO and do not use the lines A2, A3 and A4, the PIO will occupy 32 locations from the memory. Certainly, we have ample reserve to simplify the decoder logic around the 82C55A. Designing embedded systems, we often find it convenient to equalize the number of locations used by a candidate for partial decoding to the most common bigger number. In our particular case this is 8K. Consequently, each one of the PIO's four locations will have more than one address. Practically, we use the lowest address possible. Even though we sacrifice almost 8K locations from the external Data Memory, we still have free room for three other PIOs of the same type.

Figure 4.27 shows a possible memory map for the single board computer. In addition, it indicates the switches pattern. An essential feature of this arrangement is that code could be fetched from the RAM. Typically, a Monitor program resides in the EPROM. The Monitor controls the interaction with the personal computer. Code is downloaded from the PC to the RAM. We delegate the right to run programs to the RAM by installing the jumper #17. As a result, the RAM input $\overline{OE}$ is driven by $\overline{RD}\,\overline{PSEN}$.

Figure 4.28 shows another solution for the memory map. Due to the overlap between the EPROM and RAM addresses the single board computer lacks the possibility to run programs from the RAM. The switch #17 must be off and #18 on.

When you design a new system it is desirable to check the hardware by oscilloscope test loops. For example, download the code below in the RAM and execute it.

```
        MOV  DPTR,#2000H ; Select an address within the range
L1:     MOVX A,@DPTR     ; Read
        SJMP L1
```

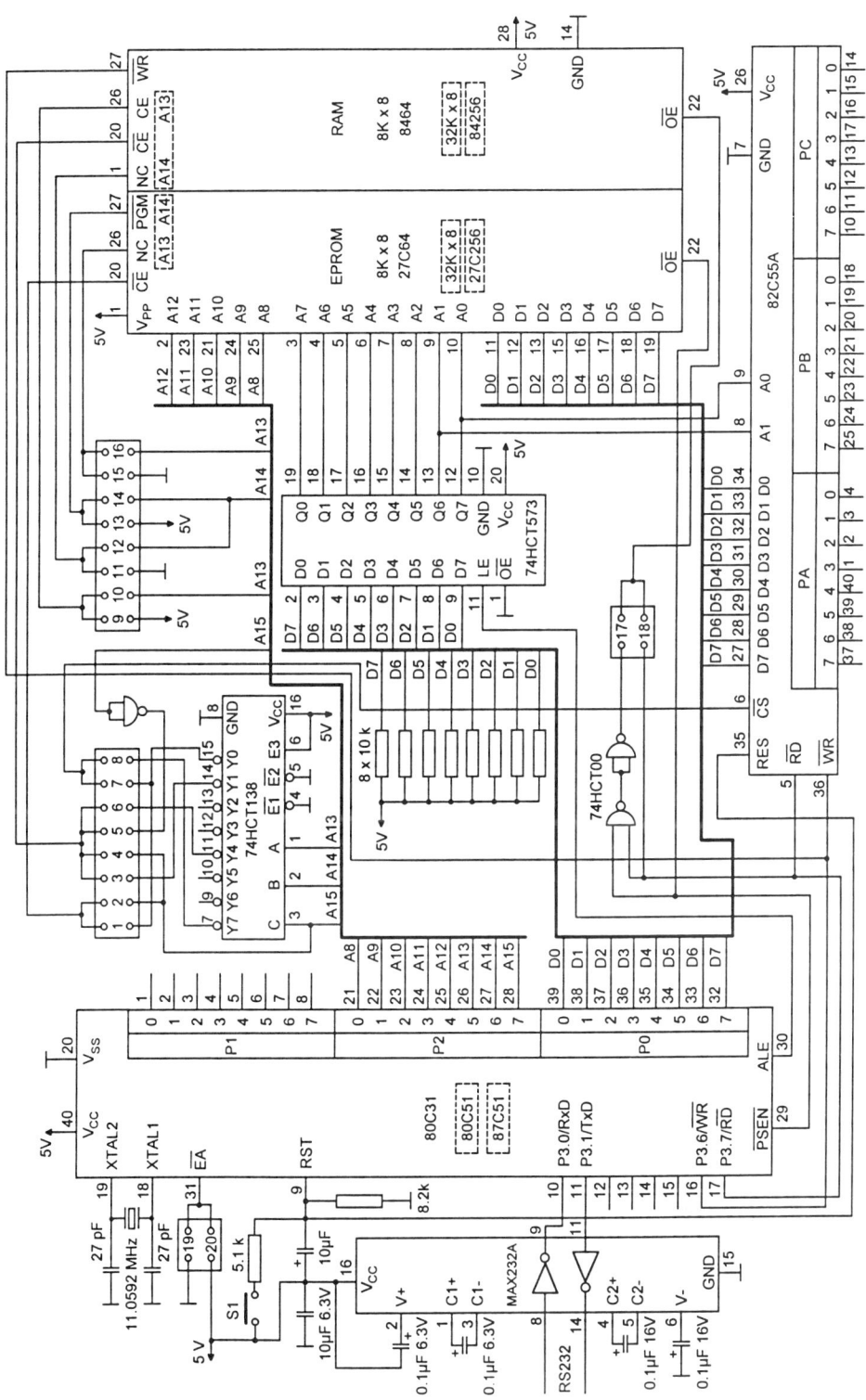

Figure 4.26 The single board computer.

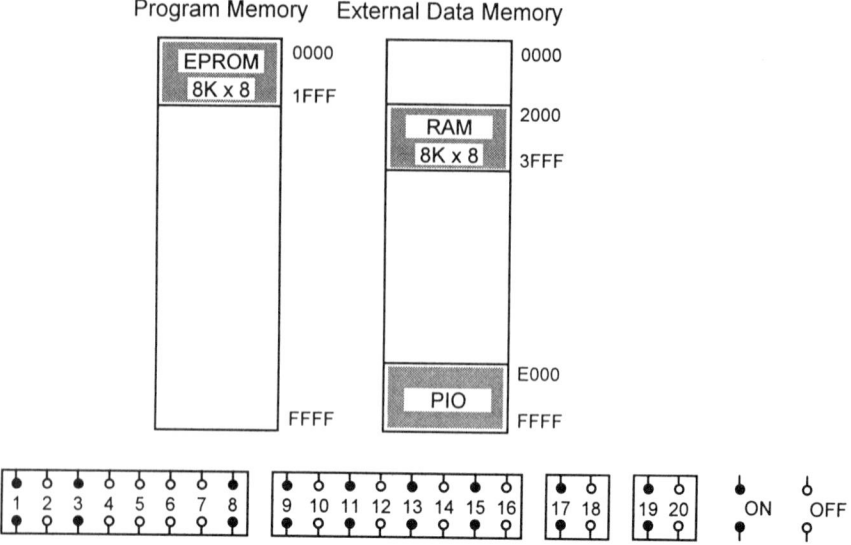

Figure 4.27 Memory map - 8K byte EPROM and RAM.

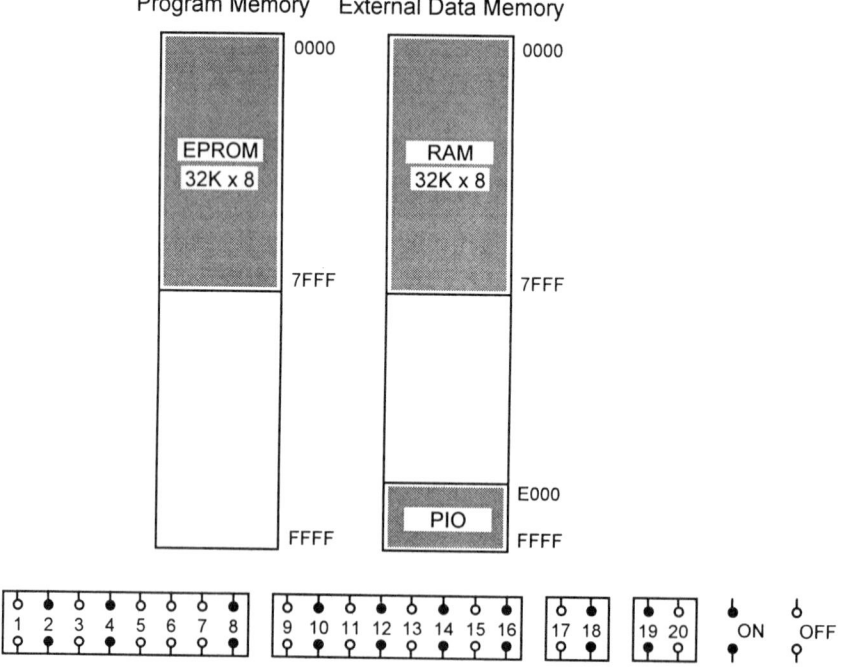

Figure 4.28 Memory map - 32K byte EPROM and RAM.

It does not matter where the code is stored. The microcontroller executes the instructions in a loop until you press the reset switch S1. As the address 2000H falls within the RAM area, the input $\overline{CE}$ (pin #20) must be pulsed down. If the memory map is the one presented in Figure 4.27, the decoder (74HCT138) output Y1 will be asserted repetitively. In case of the memory map from Figure 4.28, the code above must be included in the EPROM. We can watch the waveform of the signal $\overline{CE}$ on the oscilloscope screen. The microcontroller output $\overline{RD}$ is pulsed once per loop and would be used to synchronize the oscilloscope.

Finally, let us consider a situation which manifests how hardware and software are intermixed. The single board computer shown in Figure 4.26 forms the core of an embedded system. We assume that the PIO Port A and Port B will communicate with some other devices as input ports and Port C will control peripherals as an output port. The memory is organized according to Figure 4.27.

Following the most natural way, we plug an EPROM with Monitor program. The user programs are downloaded into the RAM and debugged. Apparently, the first task of the user software is to initialize the system, including the PIO ports. The next step is to program an EPROM with the user code and to plug it in the board. Unlike what you might expect, the result is different now. All PIO ports remain in input mode which is done automatically during reset.

It is incomprehensible that the instructions can do the job when they are executed from the RAM, but they have no effect when they reside in the EPROM. What is superficial is that the codes are moved from one IC to another. What is less obvious is that the initialization instructions are executed from the RAM after a certain number of Monitor instructions. When they are fetched from the EPROM, on the contrary, it is done just after reset. The microcontroller and PIO reset circuits are not identical. Recollect that the microcontroller is reset for two machine cycles, which is always a fraction of the 50 µs required to reset the 82C55A. This comparison signals that the PIO may have a relatively long recovery time after reset. When microcontroller fetches from address 0000H the PIO could still be in reset and the first instructions will have no impact on it. The problem will be easily solved by means of delay in the program and the PIO control register will be accessed when the device is able to listen to the bus.

4.6 Stepper motor interfacing

Stepper motors, as the name prompts, rotate from one fixed position to another. Many embedded applications, such as printers, plotters, floppy disk drives and robots, require moving parts in small increments. Stepper motors satisfy this demand. Stepper motors are alternatively labelled as stepping or step motors.

Figure 4.29 shows a simplified model of a two-phase stepper motor. The motor consists of a permanent magnet rotor and two sets of windings. The movement of the motor from one fixed position to the next is a result of switching the currents in the phases.

An example of rotating sequence can be seen in Figure 4.30. The moral of this example is that the drive circuitry has to provide bidirectional currents in the windings. Figure 4.31 shows a schematic diagram which can be viewed as a first step toward a bidirectional driver. Coming to the explanation of the circuitry, consider the condition A high and B low. There will be a conducting path from V_S to ground via the transistor T_1, the coil and the transistor T_4. The current direction in the other winding is determined by a low input C and a high input D. Changing the input A to low and the input B to high will produce current in the opposite direction.

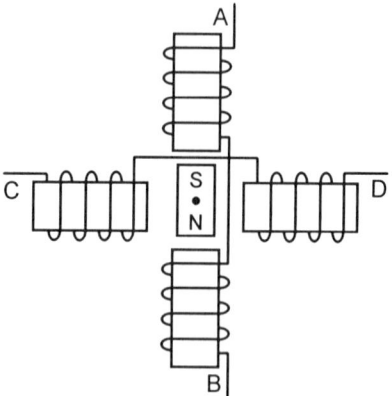

Figure 4.29 A two-phase stepper motor.

Before we follow the way to a practical solution for a stepper motor controller, we can create a timing diagram for the clockwise rotation in Figure 4.30. Figure 4.32 depicts the correspondence between the rotor position and the control signals, which must be applied.

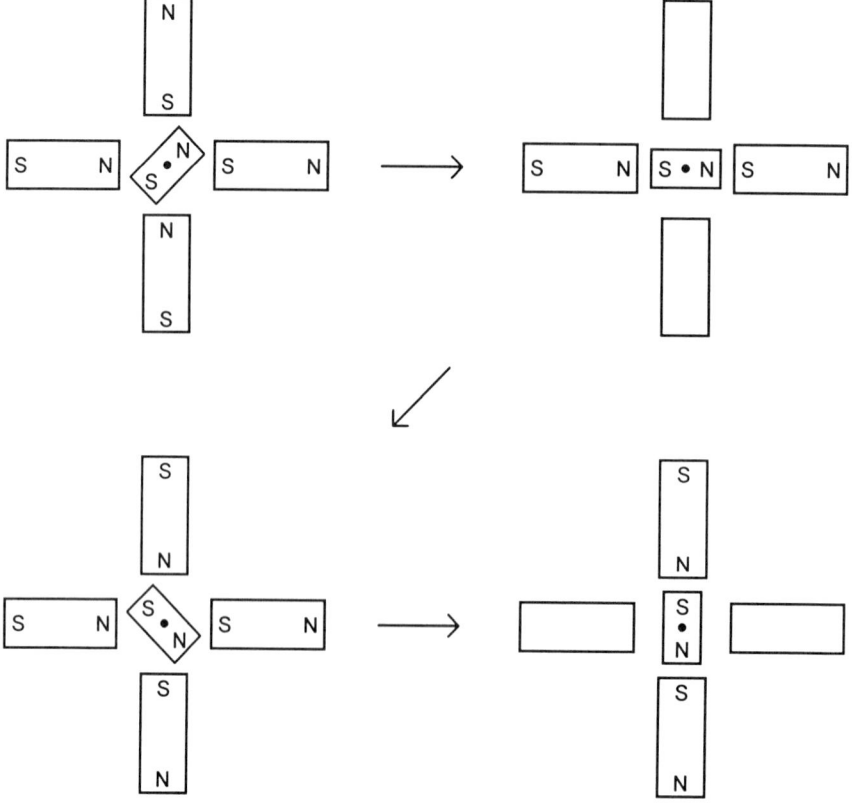

Figure 4.30 Clockwise rotation of a stepper motor.

While the drive circuitry in Figure 4.31 demonstrates the basic idea, the schematic diagram in Figure 4.33 already shows a practical solution. Two improvements have been made in the controller. First, four protection diodes create paths for the currents when the phases are switched off. Second, by introducing sense resistors (R_S), we are in position to switch off the current when it exceeds a specified value.

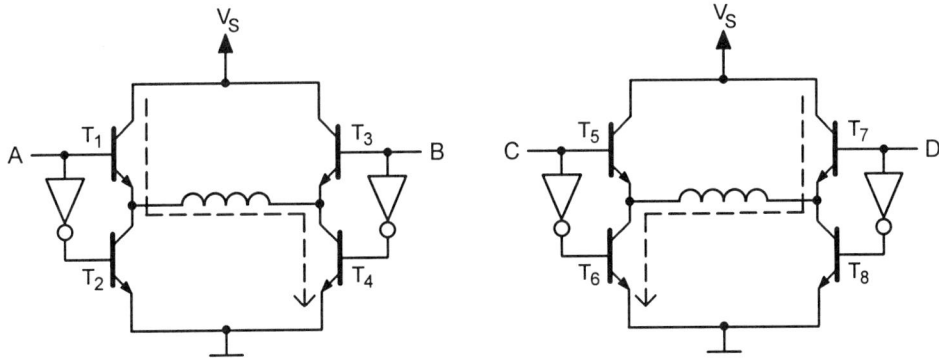

Figure 4.31 A basic drive circuitry.

Using two ICs specially developed for stepper motor control [SGS 1992], we get the schematic diagram shown in Figure 4.34. The interface between the 8051 microcontroller and the stepper motor is based on a L297 controller and a L298N dual full-bridge driver. The L298N driver implements the required electrical parameters of the interface. The L297 controller performs the specific control functions.

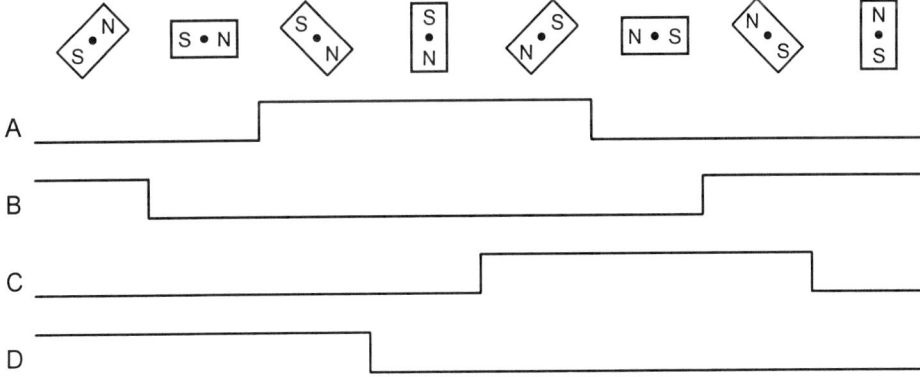

Figure 4.32 A timing diagram for the drive circuitry.

The L298N driver internal structure is almost a direct copy of the basic drive circuitry presented in Figure 4.31. In addition, the designers included two disable (inhibit) inputs, $\overline{\text{INH1}}$ and $\overline{\text{INH2}}$. A phase current will be switched off if the corresponding disable input is pulled low.

The electrical characteristics of the driver and the protection diodes are vital. On the top of this, the diodes must be rather fast. Switching times not higher than 200 ns are required.

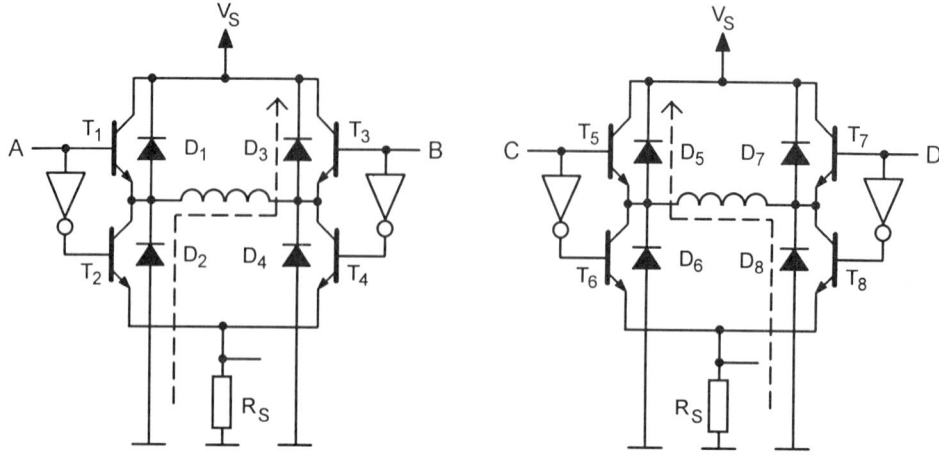

Figure 4.33 A drive circuitry with protective diodes and load current control.

The external components R and C determine the frequency of the built-in oscillator. The oscillator's output, SYNC, can be used for synchronization purposes. Usually, this oscillator is referred to as a chopper oscillator. In fact, what chops the current in the windings is the value of the voltage drops across the sense resistors. Both voltages are compared with the threshold voltage applied to the input V_{ref}. If a sense voltage exceeds the threshold, the L297 controller will switch off the current in the corresponding winding. Automatically, the current is switched on in the beginning of the next oscillator period. Due to the complex load, the current in the windings is changed gradually. However, the protective diodes create a parallel conductive path and the sense resistor current is pulsed down. In essence, each new oscillator cycle allows the load currents to clime up and within each cycle there is a point when the feedback signals for a peak current and the controller intervenes to switch off. As you might expect, the L297 controller can use for this procedure either the phase outputs (A, B, C and D) or the disable outputs ($\overline{INH1}$ and $\overline{INH2}$). An input named CONTROL determines which approach is to be taken. In this example, we tied the input CONTROL low to chop through the outputs $\overline{INH1}$ and $\overline{INH2}$.

The L297 controller possesses a chip enable input (ENABLE). While this input is low, the outputs A, B, C, D, $\overline{INH1}$ and $\overline{INH2}$ are pulled down. Another input, labelled $\overline{RESET}$, initializes the L297 controller and emits the pattern 0101 via the outputs ABCD. This state of the controller is indicated by a high level on the output HOME. More precisely, the transistor of the open collector output HOME is cut off while the controller is in its initial state.

The direction of the motor rotation is controlled by an input named ClockWise / $\overline{CounterClockWise}$. Note that the actual direction of rotation depends on the connection of the windings as well.

Generating different phase sequences the L297 controller can implement different rotation modes. Figure 4.35 shows the available operation modes. The half-step mode has already been illustrated in Figures 4.30 and 4.32. The half-step mode doubles the resolution of the motor.

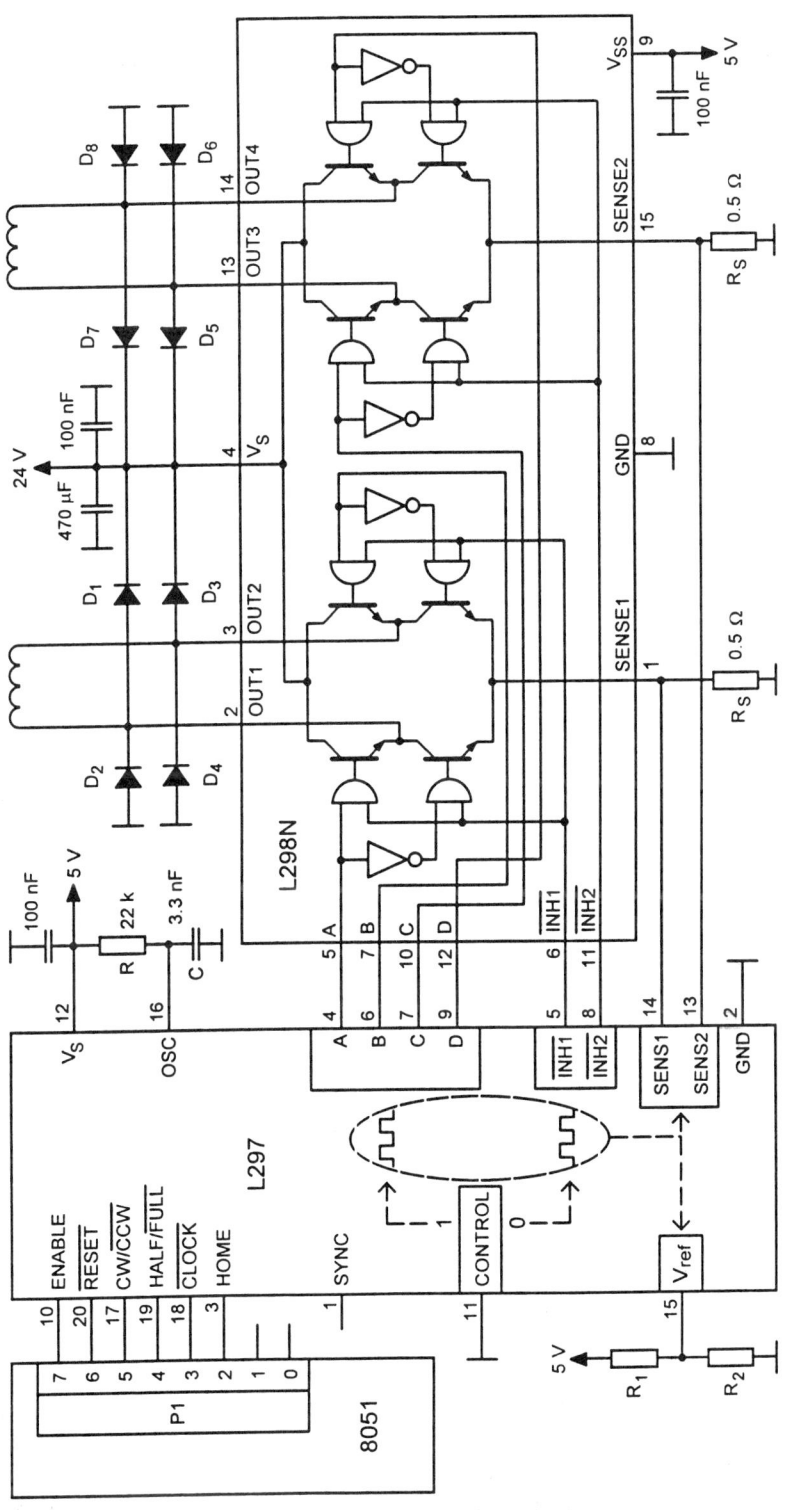

Figure 4.34 A two-phase bipolar stepper motor control circuit.

For example, a 3.6 degree stepper motor will have a 1.8 degree angle in half-step mode. As a result, the total number of steps per revolution becomes 200. We can instruct the controller to use half-step mode by applying a high level to the input HALF / $\overline{\text{FULL}}$. The half-step mode graph in Figure 4.35 shows a loop of 8 states. The states are numbered from 1 through 8. Each state is characterized with a specific combination of output values. The home state is number 1 (ABCD = 0101). One iteration in the loop will result in 8 half-steps of rotation.

The phase outputs are changed synchronously with the rising edge of an input labelled $\overline{\text{CLOCK}}$. Applying a low level to the input HALF / $\overline{\text{FULL}}$ we select one of the two full-step modes. If the controller is in an odd state (1, 3, 5 or 7) when the mode is changed, "Normal drive mode" will be entered. Alternatively, if the controller is in an even state (2, 4, 6 or 8) and the input HALF / $\overline{\text{FULL}}$ is pulled down, from then on the loop will include only even states. The mode is referred to as "Wave drive mode".

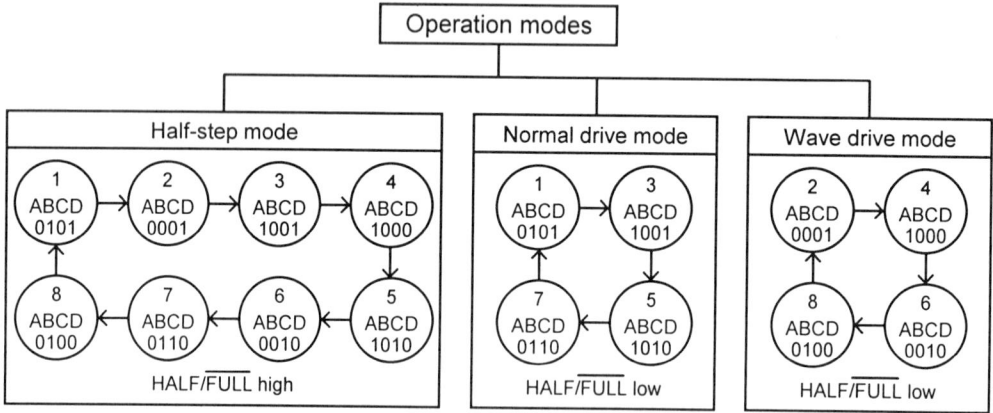

Figure 4.35 Operation modes.

Since the output currents of the L298N driver can be up to 2 A, when bigger currents are required the L297 controller could be combined with discrete power components.

Problem 4.4

Write the initialization part and an interrupt subroutine for the 8051 microcomputer to control an electromechanical clock by a stepper motor. The motor has 1.8 degree angle.

Solution 4.4

The ideal step motor for this application is the one with 6 degree angle. An 1.8 degree angle motor has to work in sequences of 3, 3 and 4 steps. The resulting rotation angles will be 5.4 degrees, 10.8 degrees and 18 degrees. In most clock applications, it will be difficult to narrow down the difference between a 3/6 degree motor and a 1.8 degree one which works in outbursts of 3, 3 and 4 steps. Figure 4.36 shows a flowchart for the electromechanical clock control. In fact, this is the essential core of the algorithm shaped as an interrupt handler. Timer 0 interrupts the program every 250 µs. In order to expand this interval to one second, we organize two software counters (C_200 and C_20). An extra counter named STEPS operates in the range 0 through 4. Every second a flag SECOND is set. In addition, one of the flags ST_3_1, ST_3_2 or ST_4 is set to indicate which sequence (three steps, three steps or four steps) is under way. The flowchart is written with the assumption that the drive circuit is the one shown in Figure 4.34.

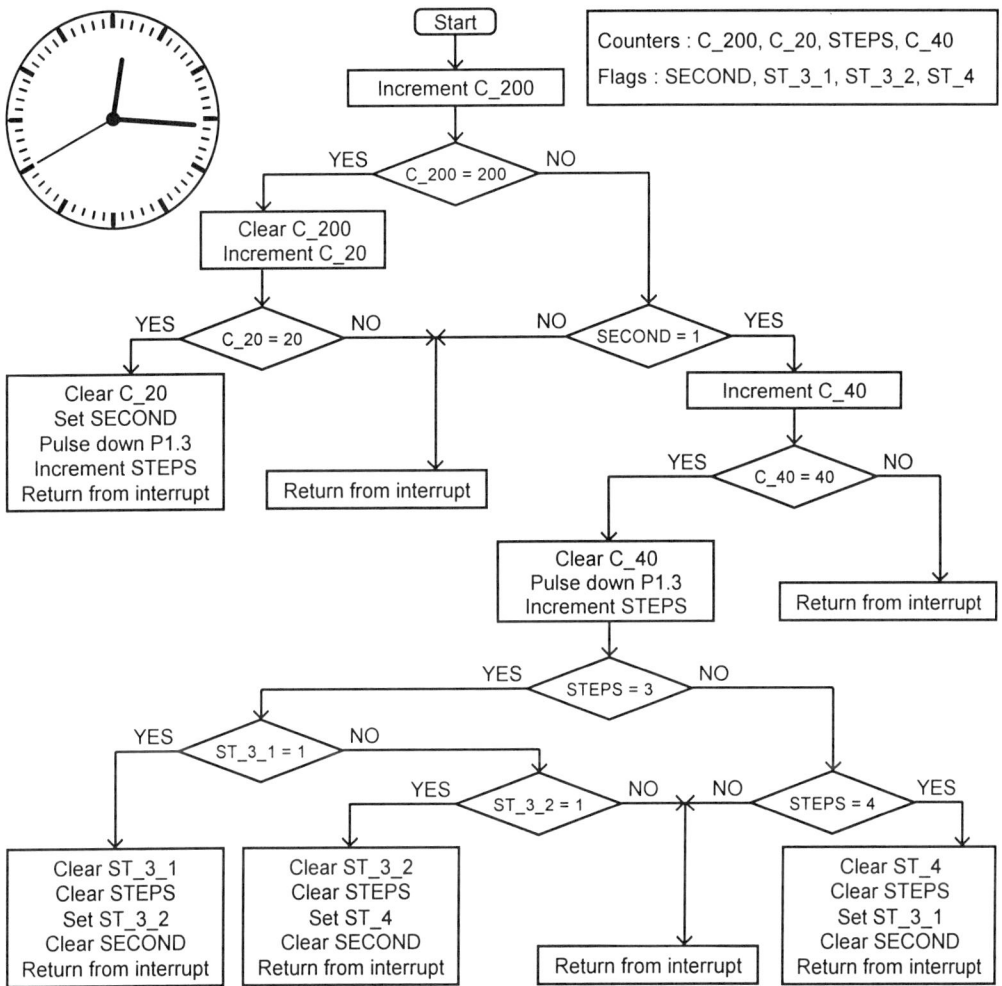

Figure 4.36 The flowchart for the electromechanical clock.

Using the flowchart, the following code emerges almost automatically:

```
;********************************************************
;*    CLOCK.ASM                                        *
;*    This program controls an electromechanical clock *
;*    driven by a stepper motor                        *
;********************************************************
C_200    IDATA   04H
C_20     IDATA   05H
STEPS    IDATA   06H
```

```
C_40        IDATA    07H
SECOND      BIT      00H
ST_3_1      BIT      01H
ST_3_2      BIT      02H
ST_4        BIT      03H
;
            ORG      0000H
            LJMP     RESET
;
            ORG      000BH
            LJMP     INT_T0
;
            ORG      0030H
RESET:      MOV      R4,#0       ; Clear C_200
            MOV      R5,#0       ; Clear C_20
            MOV      R6,#0       ; Clear STEPS
            MOV      R7,#0       ; Clear C_40
            CLR      SECOND
            SETB     ST_3_1
            CLR      ST_3_2
            CLR      ST_4
;
            MOV      SP,#10H
            MOV      TMOD,#02H   ; Timer 0, mode 2
            MOV      TH0,#6      ; Delay 250 mcs, 256-250=6, 12 MHz
                                 ; crystal
            SETB     ET0         ; Enable interrupts Timer 0
            SETB     EA          ; Enable all interrupts
            SETB     TR0         ; Start Timer 0
;
            SETB     P1.7        ; Enable the L297
            CLR      P1.6        ; Reset the L297
            SETB     P1.5        ; Clockwise direction
            CLR      P1.4        ; Full-step mode
            SETB     P1.6        ; Disable the reset input
;
            SJMP     $           ; Wait for interrupt
;
INT_T0:     INC      R4          ; Increment C_200
            CJNE     R4,#200,T_SEC
            MOV      R4,#0       ; Clear C_200
            INC      R5
            CJNE     R5,#20,RETURN
            MOV      R5,#0       ; Clear C_20
            SETB     SECOND      ; Start a sequence of 3 or 4 steps
            CLR      P1.3        ; Pulse down, one step
            SETB     P1.3
            INC      R6          ; Increment STEPS
```

```
RETURN:     RETI
T_SEC:      JNB     SECOND,RETURN
            INC     R7              ; Increment C_40
            CJNE    R7,#40,RETURN
            MOV     R7,#0           ; Clear C_40
            CLR     P1.3            ; Pulse down, one step
            SETB    P1.3
            INC     R6              ; Increment STEPS
            CJNE    R6,#3,T_STEPS
            JNB     ST_3_1,T_ST32
            CLR     ST_3_1
            MOV     R6,#0           ; Clear STEPS
            SETB    ST_3_2
            CLR     SECOND
            RETI
T_ST32:     JNB     ST_3_2,RETURN
            CLR     ST_3_2
            MOV     R6,#0           ; Clear STEPS
            SETB    ST_4
            CLR     SECOND
            RETI
T_STEPS:    CJNE    R6,#4,RETURN
            CLR     ST_4
            MOV     R6,#0           ; Clear STEPS
            SETB    ST_3_1
            CLR     SECOND
            RETI
```

We use registers (R4 through R7) for the counters which are manipulated by software. The counter C_40 helps us to obtain an interval of 10 ms (40 x 250 = 10000). This period of time separates the clock pulses sent to the stepper motor controller L297, while an increment is active.

4.7 Supplementary problems

Problem S4.1
Design a memory subsystem for the 8051 microcontroller which has 64K byte Program Memory and 64K byte external Data Memory.

Problem S4.2
The single board computer presented in Figure 4.26 accommodates an 32K x 8 EPROM and a 80C31 microcontroller which runs at 16 MHz [Phil 1997a]. The microcontroller possesses the following timing parameters :

$$TAVIV = 5T_{OSC} - 55$$
$$TPLIV = 3T_{OSC} - 45$$

Calculate the upper limits (zero margins) of the EPROM timing parameters (see Figure 4.10).

Problem S4.3

The single board computer presented in Figure 4.26 accommodates a 32K x 8 RAM and a 80C31 microcontroller which runs at 24 MHz [Phil 1997a]. The microcontroller possesses the following timing parameters:

$$TAVDV = 9T_{OSC} - 165 \qquad TWLWH = 6T_{OSC} - 100$$
$$TRLDV = 5T_{OSC} - 90 \qquad TQVWX = T_{OSC} - 20$$
$$TAVWL = 4T_{OSC} - 75 \qquad TWHQX = T_{OSC} - 20$$

Calculate the upper limits (zero margins) of the RAM timing parameters (see Figure 4.11 and 4.12).

Problem S4.4

Write an instruction sequence for the single board computer in Figure 4.26 which initializes the PIO 82C55A. Assume that switch 7 is closed and switch 8 is open. The group A lines (Port A and Port C-upper) must be programmed as inputs. The group B lines (Port B and Port C-lower) must be programmed as outputs (see Figure 4.19).

Problem S4.5

Write an instruction sequence for the 87C51 microcontroller. The microcontroller Port 1 is surrounded by the logic shown in Figure 4.21. The microcontroller must perform the following actions:

- Read Port 0
- Wait 200 μs
- Output the data to register R2
- Wait 400 μs
- Output the data to register R1
- Wait 800 μs
- Output the data to register R0.

In summary, one byte is read through Port 0 and is sent out to different peripherals after increasing delays.

Problem S4.6

Design a logic circuit which expands the 8051 microcontroller serial port into two serial ports. Use a dedicated pin to select the ports.

Problem S4.7

Describe a procedure which allows the single board computer from Figure 4.26 to execute code from the RAM. The starting address must be 0000H. The procedure should specify how to use the switches and the reset switch S1. Even though the Monitor program is disactivated and breakpoints are not possible, all interrupt vectors are available and the designer could obtain valuable information through the I/O interface.

Problem S4.8

Design a system based on the 80C31 microcontroller which is capable of accepting interrupt requests from four sources. A high-to-low transition asserts the corresponding interrupt request. All requests have equal priority. When activated, the interrupt subroutines toggle a dedicated pin from Port 1. Draw the schematic diagram interrupt sources - microcontroller. Write the initialization part and the interrupt subroutines.

Problem S4.9
Write a sequence of instructions for an oscilloscope test loop. The single board computer shown in Figure 4.26 must turn the Port A of the PIO 82C55A into a counter. If you do not have an oscilloscope, you could slow down the frequency of the counter and observe the result by LED lamps.

Problem S4.10
Modify the schematic diagram of the single board computer from Figure 4.26 in order to expand the external Data Memory to 128K bytes. The external Data Memory should be split into two banks of 64K bytes. Use a dedicated output to select a memory bank.

Problem S4.11
Write an assembly language program for the 8051 microcontroller which controls a stepper motor according to the requirements in Figure 4.37. Insert a delay of 20 ms between the steps.

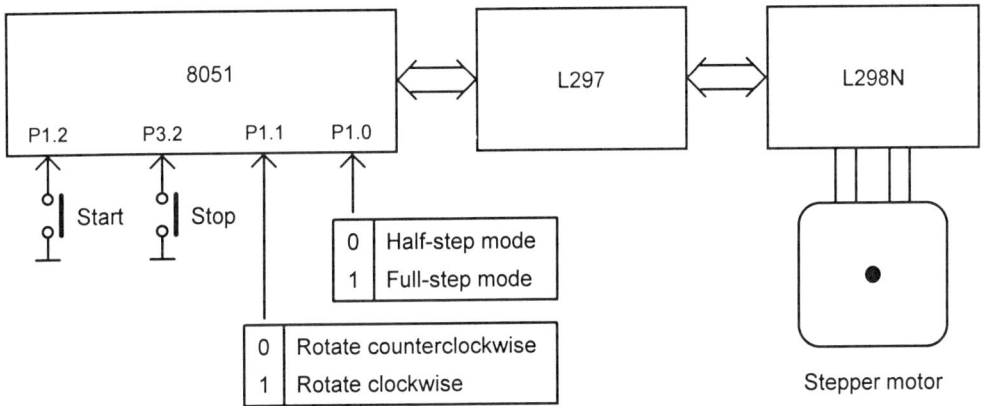

Figure 4.37 Interfacing a stepper motor.

4.8 References

Atmel, *Nonvolatile Memory*, 1996.

Fujitsu, *Static RAM Products*, 1991.

Douglas V. Hall, *Microprocessors and Interfacing*, McGraw-Hill, 1992.

Intel, *Embedded Microcontrollers and Processors*, Volume I, 1992.

Intel, *Peripheral Components*, 1993.

Intel, *Embedded Microcontrollers*, 1996.

Zdravko Karakehayov and Stanislav Grigorov, *Single Chip Microcomputers*, Technica, Sofia, 1992.

Maxim, *New Releases Data Book*, Vol. III, 1994.

Microchip, *Data Book*, 1993.

Mitsubishi, *Semiconductors, Memories, SRAM*, 1997.

Motorola, *Fast Static RAM*, 1995.

National Semiconductor, *INTERFACE : Data Transmission Databook*, 1994.

Philips Semiconductors, *High-speed CMOS Logic family, Data Handbook IC06*, 1994a.

Philips Semiconductors, *80C51-Based 8-Bit Microcontrollers, Data Handbook IC20*, 1997a.

Richard C. Seals, *Microprocessor-Based Systems*, Stanley Thornes (Publishers) Ltd, 1992.

SGS-Thomson, *Industrial and Computer Peripheral ICs*, 1992.

SGS-Thomson Microelectronics, *Memory Products*, 1994.

Clarence W. De Silva, *Control Sensors and Actuators*, Prentice Hall, 1989.

Texas Instruments, *The TTL Data Book*, Volume 1, 1989.

Texas Instruments, *Advanced BiCMOS ABT Bus Interface Logic, Data Book*, 1993.

Information about components for digital interfacing can be found at :

Atmel	http://www.atmel.com
Fujitsu Microelectronics	http://www.fujitsumicro.com
Maxim Integrated Products	http://www.maxim-ic.com
Motorola	http://www.mot.com
National Semiconductor	http://www.national.com
Philips Semiconductor	http://www.philips.com
SGS-Thomson	http://www.st.com
Sipex	http://www.sipex.com
Texas Instruments	http://www.ti.com
WaferScale Integration, Inc.	http://www.wsiinc.com

Information about stepper motors can be obtained from

http://www.eio.com/stepindx.htm

Chapter 5

ANALOG INTERFACING

5.1 Introduction

So far we have discussed the opportunities to interface the microcontroller to digital components and peripherals. Also, embedded systems are required to interface analog peripherals. In this case, the interface deals with signals which can assume any value in some predetermined range. We employ A/D converters and D/A converters for fast and accurate transfer of data between the analog and digital domains. Initially, the analog interface can be viewed as an additional digital interface which links the microcontroller and the converters. On top of that, we use analog components which sense the environment or process the signals.

In this chapter we first consider the basic concepts related to a subset of embedded systems termed data acquisition systems. Next, we discuss digital-to-analog conversion. The chapter includes basic specifications of the D/A converters and an example of interfacing a real device. An analog-to-digital conversion section gives some insight on the methods for A/D conversion. Again, the essential parameters are followed by an example which deals with a popular device. Finally, we demonstrate how a microcomputer system can be harnessed to a specific task: temperature measurement.

5.2 Data acquisition systems

The embedded system architecture outlined in Figure 1.22 includes sensors, actuators and a processor. As discussed in section 1.9, the sensors convert physical variables, such as temperature, pressure or flow rate into proportional voltage or current. Moreover, the embedded system influences the environment by actuators. When the analog interface outweighs all other aspects of the application, we often use the term Data Acquisition System (DAS).

Figure 5.1 shows the architecture of a typical DAS. An input signal conditioning circuit immediately follows the sensor. The sensor output signal might be small and mixed with noise. An operational amplifier and sometimes a filter form the input signal conditioner. In addition, the high-impedance input of the amplifier is the ideal load for the sensor. The low-impedance output of the operational amplifier optimizes the link to the next stage. Using analog sensors, we inevitably need Analog-to-Digital Converters (ADC). The ADC parameters, together with the other analog components, have significant impact on the overall system characteristics. When the ADC input is changed relatively fast there might be a need to sample and hold the voltage. This is done by analog memory called Sample/Hold (S/H) circuit. A S/H circuit contains a capacitor which is switched to the input to follow the voltage. When the switch is opened the voltage across the capacitor remains almost constant for a certain period of time. Using a fast ADC we can eliminate the need for a S/H circuit.

A processor gives the computing power of the DAS. As is frequently the case, the processor is a microcontroller which could, also, incorporate an ADC or DAC. The 83C552 microcontroller, discussed in chapter 7, contains an ADC. Furthermore, the 83C552 microcontroller possesses two pulse-modulated outputs which can be used for digital-to-analog conversion.

The last stage in a DAS are one or more actuators. The actuators are the mirror image of the sensors. They influence the environment by movement, heat or light. Typical actuators are motors, solenoids, heaters, lamps and spark plugs. Clearly, the actuators must be driven by voltages or currents much higher than a microcontroller can provide. Usually, we connect the actuators to the microcontroller through output signal conditioners. For this purpose we normally employ specialized ICs, relays, Darlington transistors or triac drivers. An example of an output signal conditioner is the set of two ICs, L297 and L298N, designed for stepper motors interfacing.

The applications often demand the DAS to monitor several variables and to have an effect on more than one parameter of the control object. Figure 5.2 shows a DAS which is capable of doing this. The new block in the architecture is termed analog multiplexer. The multiplexer connects a selected channel to the S/H circuit. The microcontroller selects the channel by means of the multiplexer address inputs. We could avoid the analog multiplexer by using a separate ADC for each channel. This approach will give us a few substantial advantages. First, it allows higher sample rate. Second, the error introduced by the multiplexer is eliminated. Finally, in the case of sensors placed far from the system core, signals may be converted on the spot. It is much easier to protect digital signals from noise than analog. Using ADC with serial output could be an appropriate solution in this situation.

The 83C552 microcontroller, which integrates not only an ADC but also and an 8-input analog multiplexer, can be used as the backbone of the architecture presented in Figure 5.2.

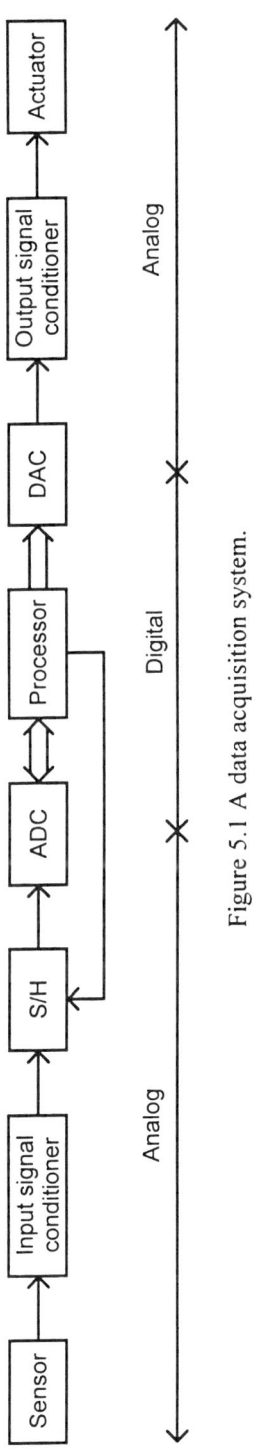

Figure 5.1 A data acquisition system.

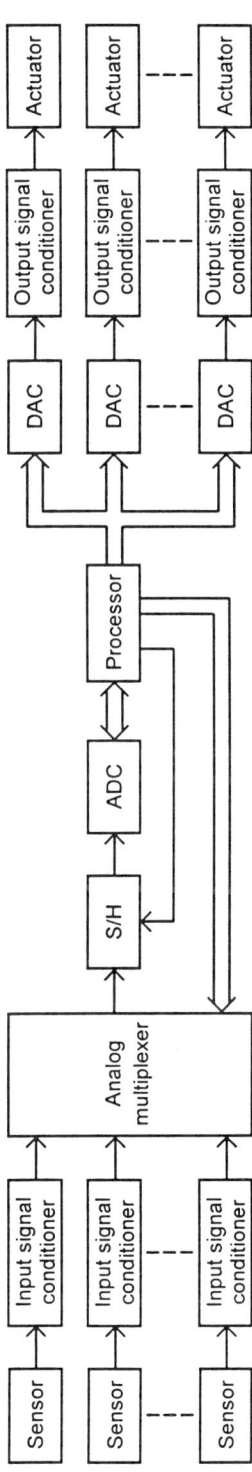

Figure 5.2 A multichannel data acquisition system.

5.3 Digital-to-analog conversion

Although there are many available DACs produced from different manufacturers, the process of selecting a DAC for a certain application is not an easy task. A brief overview of DAC specifications begins with the most widely used parameter, the number of bits (inputs) n. For the vast majority of DACs, this specification varies between 8 and 16. Figure 5.3 includes essential DAC parameters which can be derived from the number of bits.

Bits n	8	10	12	14	16
States 2^n	256	1024	4096	16384	65536
Fractional binary weight 2^{-n}	0.00391	0.000977	0.000244	0.0000610	0.0000153
Quantum (LSB value) $q = \dfrac{FS}{2^n}$ $(FS = 10\,V) \rightarrow$	39 mV	9.77 mV	2.44 mV	610 µV	153 µV
$(FS = 5.12\,V) \rightarrow$	20 mV	5 mV	1.25 mV	312 µV	78 µV

Figure 5.3 Specifications of the DACs.

The fractional binary weight indicates how we can influence the analog output changing the LSB. Furthermore, the quantum (LSB value) express this influence in volts. However, we should introduce two other specifications to be able to calculate the quantum. They are called output range and full scale. Some typical output ranges and the corresponding full scale (FS) are shown in Figure 5.4.

Range	Full scale (FS)
0 V to 5 V	5 V
0 V to 10 V	10 V
-2.5 V to 2.5 V	5 V
-5 V to 5 V	10 V
-10 V to 10 V	20 V

Figure 5.4 Typical output ranges and FSs.

Now, we can calculate the quantum. Figure 5.3 includes two examples. The first one deals with FS = 10 V. The second one is based on FS = 5.12 V, which simplifies the calculations with the LSB value. The quantum is also termed resolution or step size.

So far, we have discussed the DAC parameters which would be sufficient to characterize ideal devices. Unfortunately, there are a great number of sources of error which make the DACs behave differently from what we expect. Logically, we compare the theoretical transfer characteristic, shown in Figure 5.5, to real transfer curves. Figure 5.6 shows a shift of the transfer characteristic termed offset. Practically, the offset error can be adjusted to zero by the operational amplifier that converts the DAC output current to voltage. A trimmer is connected to the amplifier for this purpose.

Figure 5.7 shows another deviation from the ideal transfer function. The gain error, which is the term in this case, is measured for maximum input code. We normally adjust the value of DAC's reference in order to eliminate the gain error.

Both offset and gain error adjustments are done for a certain temperature. Inevitably, the temperature changes will affect the DAC operation. As a result, the offset and gain error will be zero for only one point of the temperature range.

Along with the static characteristics we have just discussed, the DACs have also dynamic parameters. The most important of them is termed settling time. The settling time is an interval which is required the analog output to settle and remain close enough to the desired value. Usually, the settling time is specified for a full scale output change and 0.5 LSB deviation from the required value.

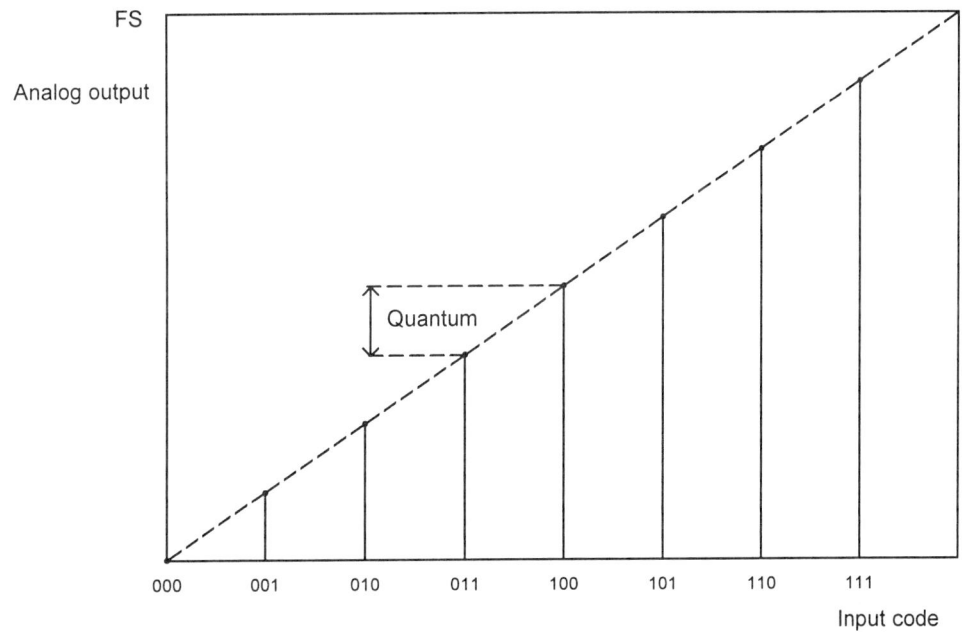

Figure 5.5 The DAC theoretical transfer characteristic.

The Analog Devices bipolar monolithic D/A converter AD565A is illustrated in Figure 5.8 [Anal 1992a]. The device contains a 12-bit DAC plus an internal voltage reference. The DAC includes application resistors which can be used for output voltage scaling and offset implementation.

The 10 V internal reference is used as a source of the 0.5 mA reference current.

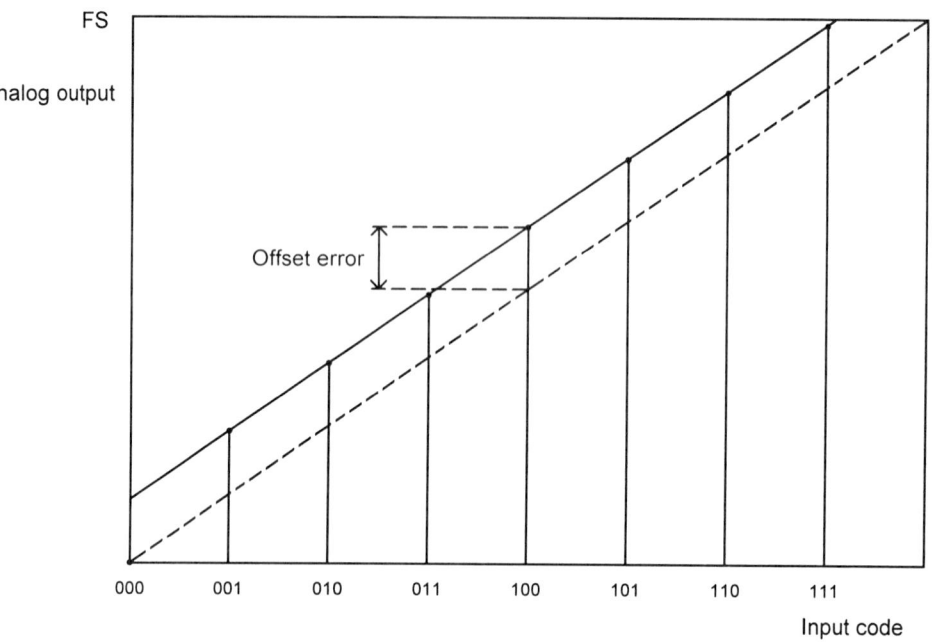

Figure 5.6 The DAC offset error.

The DAC output current I_{OUT} is changed between 0 and 2 mA, as a function of the input code. The exact value which corresponds to input code FFFH is 2 mA minus one quantum. If only the MSB (BIT1 IN) is high, the DAC output current I_{OUT} is 1 mA. However, if we neglect the operational amplifier input current and the current through the resistor 8k (the voltage across the resistor is approximately zero), the actual output current is

$$I_O = I_{OUT} - I_{OFF}$$

It should be pointed out, that we can benefit from the offset current I_{OFF} in two cases. First, there might be need of small current to adjust the offset as shown in Figure 5.8. Second, we would be able to implement bipolar ranges. If the offset current I_{OFF} is 1 mA, the resulting output current I_O will vary between -1 mA and 1 mA. That in turn will produce output range either -5 V to 5V (scaling resister 5k) or -10 V to 10 V (both scaling resistors of 5k connected in series). Moreover, we can connect both scaling resistors in parallel. Thus, the effective output range is -2.5 V to 2.5 V.

It is sound engineering practice to use the built-in scaling resistors. As a rule, they are part of the DAC circuitry and provide optimum temperature tracking characteristics.

The offset error is eliminated by the potentiometer R_1. When all inputs are low we adjust R_1 to obtain output voltage 0.000 V. It is frequently the case that the trimmer is not needed. Then pin 8 must be connected to pin 12 (POWER GND).

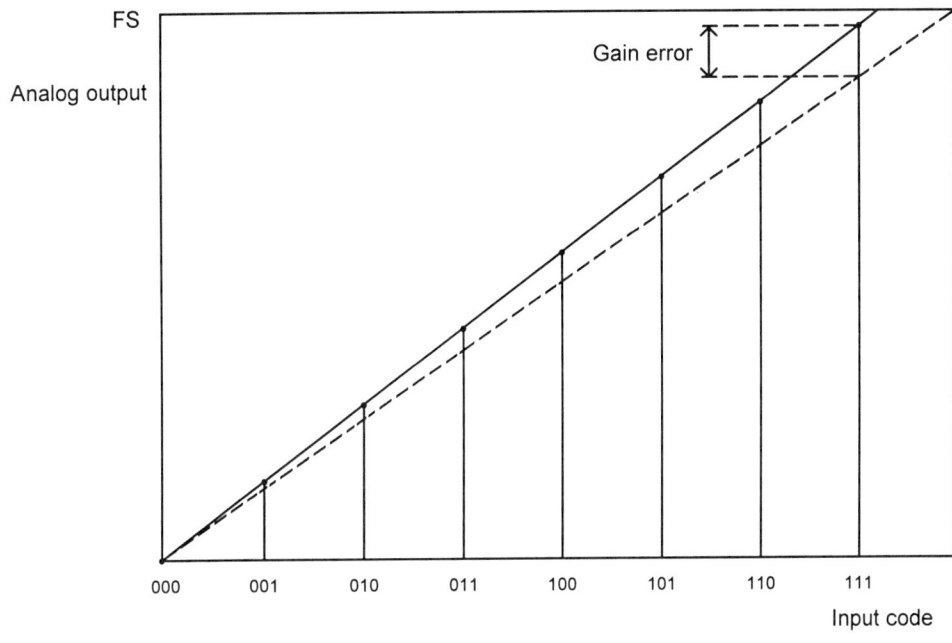

Figure 5.7 The DAC gain error.

The gain error is removed by the potentiometer R_2. When all DAC inputs are high, the output voltage is adjusted to

$$10\,V - 2.44\,mV = 9.99756\,V$$

Now, in light of the description of the example DAC AD565A we can organize the interface with the 8051 microcontroller. Figure 5.9 lays out the essential core of a DAS. The analog output channel can be used in five different ranges as shown in the right table of Figure 5.9. A mark indicates the selected range (-10 V to 10 V).

In addition to the different ranges, we admit that the applications might require different distances between the DAS and the actual load. In case of remote load, the long wires would introduce additional offset error. We work around this problem by opening the jumpers J11 and J12. As a result, we use three wires: feedback (FB), output (U_{OUT}) and ground (GND).

We allocate Port B and a nibble from Port C of a PIO 82C55A to control the digital inputs of the DAC. The PIO's Port A is used for the analog input subsystem, which is discussed in section 5.4.

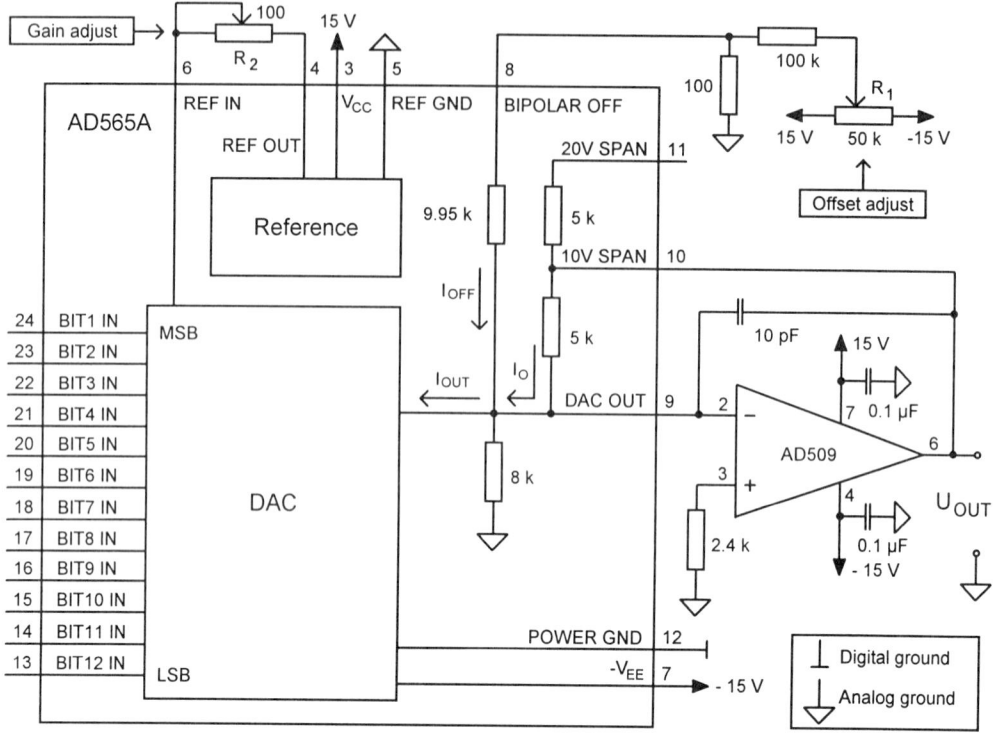

Figure 5.8 The DAC AD565A interface, output range 0 to 10 V.

A typical function of the DAS is to read data from the external RAM and to convert it into analog output. The following handler can be used for this purpose.

```
;************************************************
;*      DAC1.ASM                                *
;*      Analog output 12-bit                    *
;*      XRAM start address                      *
;*              High-order byte    R0           *
;*              Low-order byte     R1           *
;*      Number of conversions      R2           *
;************************************************
DAC1:   MOV     DPH,R0          ; High-order byte XRAM address   2
        MOV     DPL,R1          ;                                2
        MOVX    A,@DPTR         ; Read data high-order byte      2
        INC     DPTR            ; Increment XRAM address         2
        MOV     R0,DPH          ; Save XRAM address              2
        MOV     R1,DPL          ;                                2
        MOV     DPTR,#PBADDR    ; DAC high-order byte address    2
        MOVX    @DPTR,A         ; Output high-order byte         2
;
        MOV     DPH,R0          ; Low-order byte XRAM address    2
        MOV     DPL,R1          ;                                2
        MOVX    A,@DPTR         ; Read data low-order byte       2
```

```
        INC    DPTR           ; Increment XRAM address          2
        MOV    R0,DPH         ; Save XRAM address               2
        MOV    R1,DPL         ;                                 2
        MOV    DPTR,#PCADDR   ; DAC low-order byte address      2
        MOVX   @DPTR,A        ; Output low-order byte           2
;
        DJNZ   R2,DAC1        ; End?                            2
        RET
```

Consider the oscillator frequency f_{OSC} = 12 MHz, where the subroutine DAC1 will update the output voltage every 34 µs.

As far as the DAC settling time is concerned, it could be less than 1 µs (including the amplifier delay). Therefore, the calculated execution time 34 µs indicates significant disproportion. We should investigate how to minimize the execution time, even though we may impose limitations on the system functionality.

Problem 5.1

Rewrite the subroutine DAC1.ASM in order to decrease the execution time. Assume that the data is located in one 256-byte page.

Solution 5.1

If the data is located in one 256-byte page, we can use indirect addressing through R1. Here is the subroutine:

```
;*********************************************************
;*      DAC2.ASM                                        *
;*      Analog output 12-bit                            *
;*      XRAM start address                              *
;*              High-order byte  R0                     *
;*              Low-order byte   R1                     *
;*      Number of conversions    R2                     *
;*      All data codes are located in one 256-byte page *
;*********************************************************
DAC2:   MOV    P2,R0          ; Select the page
DAC21:  MOVX   A,@R1          ; Read data high-order byte       2
        INC    R1             ; Increment address               1
        MOV    DPTR,#PBADDR   ; DAC high-order byte address     2
        MOVX   @DPTR,A        ; Output high-order byte          2
;
        MOVX   A,@R1          ; Read data low-order byte        2
        INC    R1             ; Increment address               1
        INC    DPTR           ; DAC low-order byte address      2
        MOVX   @DPTR,A        ; Output low-order byte           2
;
        DJNZ   R2,DAC21       ; End?                            2
        RET
```

In this case, the DAC is accessed every 16 µs.

Finally, 8-bit digital-to-analog conversion could be performed.

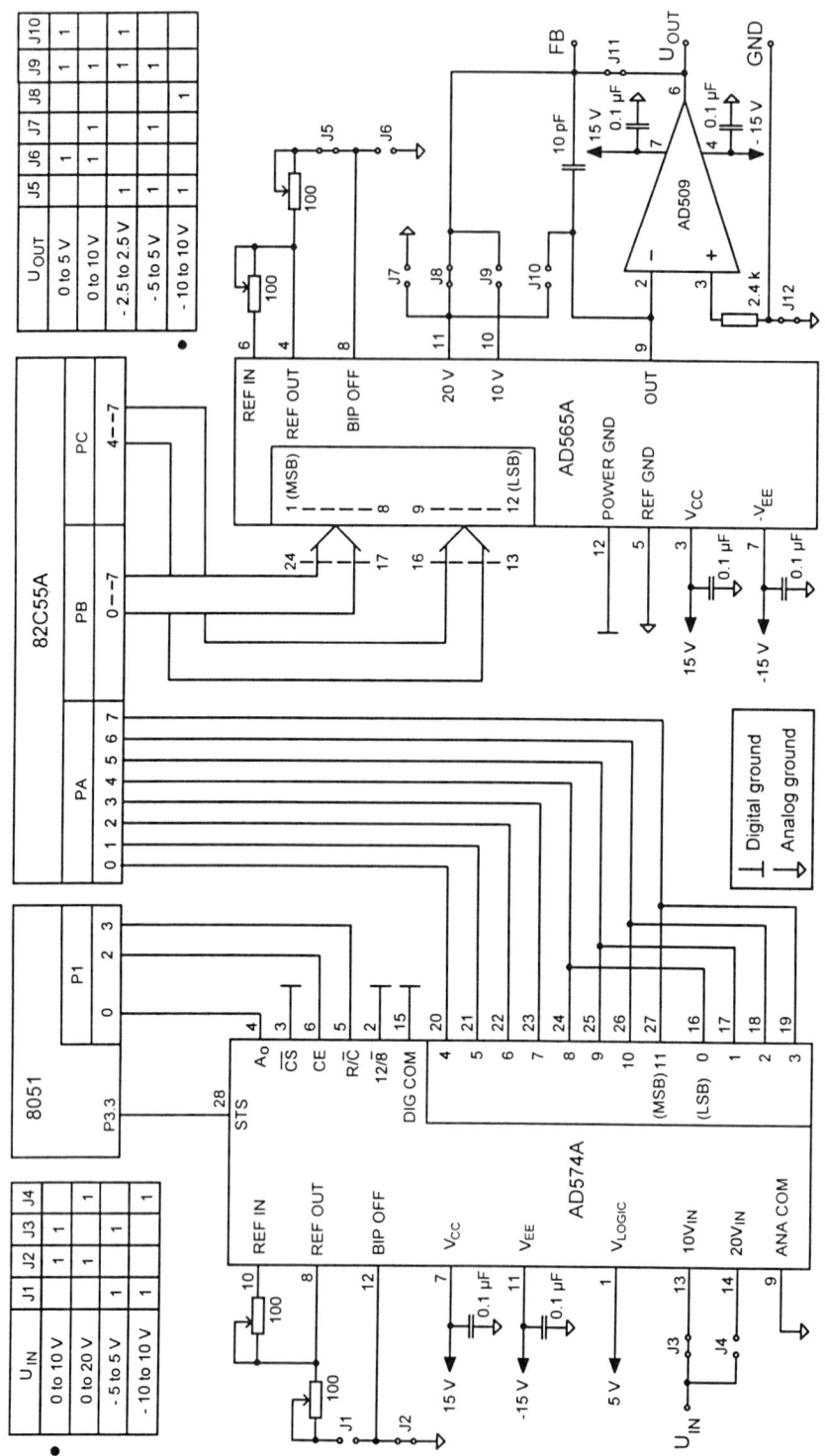

Figure 5.9 Analog interfacing of the 8051 microcontroller.

Problem 5.2

Calculate the execution time for both 8-bit D/A conversion and one 256-byte page.

Solution 5.2

The subroutine is simplified to the following version:

```
;*****************************************************************
;*      DAC3.ASM                                              *
;*      Analog output 8-bit                                   *
;*      XRAM start address                                    *
;*            High-order byte      R0                         *
;*            Low-order byte       R1                         *
;*      Number of conversions      R2                         *
;*      All data codes are located in one 256-byte page       *
;*****************************************************************
DAC3:   MOV     P2,R0         ; Select the page
        MOV     DPTR,#PBADDR  ; DAC address
DAC31:  MOVX    A,@R1         ; Read data              2 cycles
        INC     R1            ; Increment address      1
        MOVX    @DPTR,A       ; Output                 2
;
        DJNZ    R2,DAC31      ; End?                   2 cycles
        RET
```

Thus, the output voltage is altered every 7 µs.

5.4 Analog-to-digital conversion

There are three basic methods for conversion of analog values into digital representation. Figure 5.10 illustrates the methods which are termed counter, successive-approximation and flash. According to the technique used, the converters are named counter ADC, successive-approximation ADC and flash ADC. Discussing the methods, we will use an analogy based on length measurement.

The counter approach is the simplest one. It requires just one standard length. We use the standard to count how many times it must be added to exceed the unknown analog value. In fact, there are two versions of this approach which differ in the final step. We could either stop immediately when the sum of standard lengths becomes bigger than the analog input. Most ADC designs use an internal DAC and comparator. Following this method, the DAC is driven by a counter. The counter is cleared at the beginning of the conversion and then incremented one bit at each clock cycle. The comparator stops the counter as soon as the DAC output exceeds the input. The first version of the counter method accepts the final count for the digital output. The second version, as illustrated in Figure 5.10, assumes that the actual result is one standard lower. The simple implementation of the counter method should be opposed to the variable conversion time. The number of the steps for an n-bit ADC could be up to $2^n - 1$.

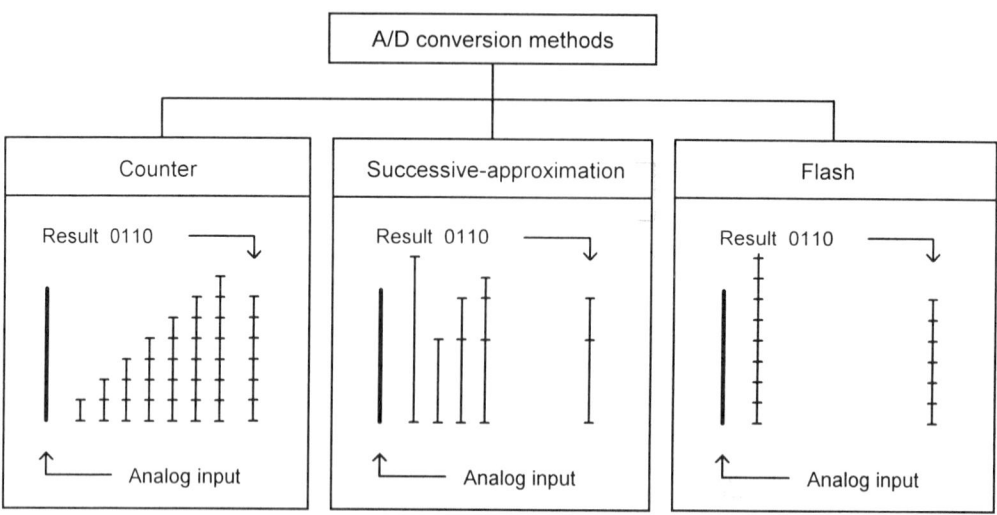

Figure 5.10 A/D conversion methods.

The successive-approximation method is the most popular approach. In contrast to the counter method, the number of standards is increased to n. The first step is to compare the analog input with the biggest standard, which is equal to half of the input range. As can be seen in Figure 5.10, if the standard is bigger, it is removed and the MSB is assigned low. The next standard, which is twice as small, is added. In this particular example, the analog input is bigger now and the corresponding bit in the result is set. The procedure is repeated n times for an n-bit ADC. The successive-approximation ADC employs a DAC and a comparator. The essential core of the control logic is termed Successive-Approximation Register (SAR).

Finally, the flash (parallel) method is based on a complete set of standard lengths (a ruler), which in turn requires only one comparison to give the result. High speed flash ADC have always been available, but at a high price. An n-bit flash ADC includes $2^n - 1$ comparators which not only increase the cost, but form significant capacitive load. A typical application, which relies on flash ADCs, is the digital oscilloscope.

Many of the ADC specifications are identical to the DAC parameters discussed in the previous section. Some others use identical approaches, but of course, have a different graphical appearance. Figure 5.11 shows the transfer characteristic of an 3-bit ideal ADC. Furthermore, Figure 5.12 depicts a real transfer curve, which reflects offset error. Moreover, Figure 5.13 shows a real transfer characteristic which diverges from the ideal shape due to gain error.

A mirror parameter for the DAC's settling time is the ADC's conversion time. The conversion time indicates how long the ADC processes the analog input in order to yield the digital output. The conversion time determines the conversion rate, which is measured in samples per second. The maximum acceptable conversion time will vary from application to application. In addition, the analog input in the end of the conversion time might be different from the value the ADC started with. As a result, there would be inconsistency between the output code and the analog input in the beginning of the conversion. The parameter which is relevant in this case is termed dynamic error. As discussed in section 5.2, we can eliminate the dynamic error by a S/H circuit.

The Analog Devices bipolar monolithic A/D converter AD574A is illustrated in Figure 5.14 [Anal 1992b]. Even though the picture is a simplification of the internal structure, it provides sufficient information for us to understand its operation.

The device AD574A includes a 12-bit DAC, a voltage reference (10 V), ADC control logic, three-state output buffers, a comparator and a few scaling resisters. The control logic has been implemented by I^2L (Integrated Injection Logic) technology.

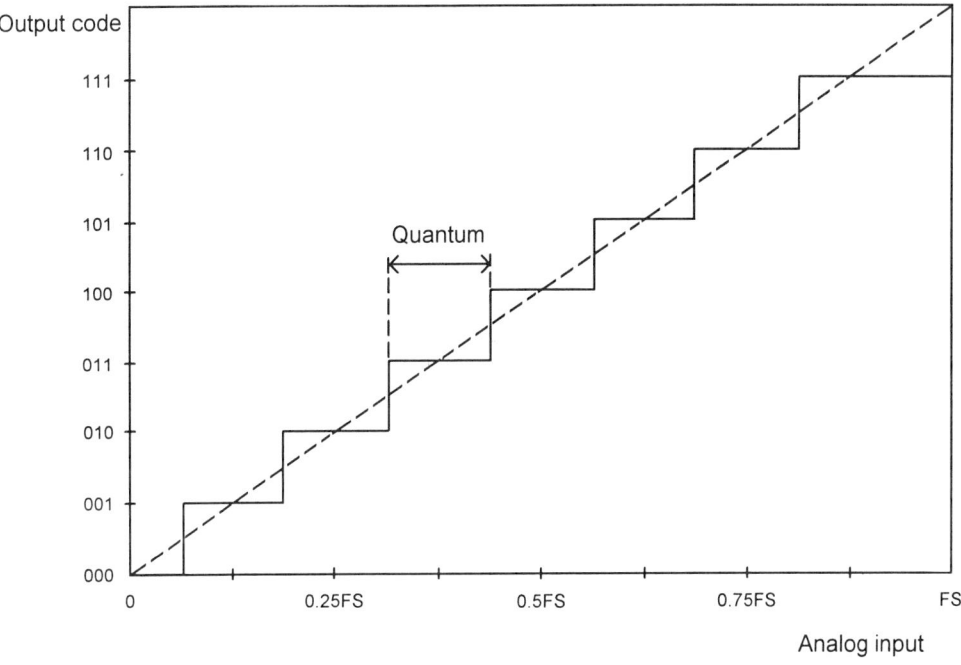

Figure 5.11 The ADC theoretical transfer characteristic.

The digital outputs of the AD574A are numbered from DB0 (LSB) through DB11 (MSB).

The method of conversion is successive-approximation. The analog input is compared with standard values, generated by the internal DAC. In addition, offset current (I_{OFF}) may be involved. This is attractive in double sense. First, the offset current could remove the offset error. Second, bipolar ranges might be implemented.

A key point is the comparator input current

$$I_C = I_{IN} + I_{OFF} - I_{DAC}$$

Figure 5.15 details the possible input ranges and the corresponding currents.

The offset error is eliminated by the trimmer R_1. Since our goal is the transfer characteristic shown in Figure 5.11, we adjust the trimmer R_1 until the transition between the output codes 0000 0000 0000 and 0000 0000 0001 occurs for an input voltage of 0.5LSB. For the input range 0 to 10 V, we have

$$0.5 \frac{10\,V}{2^{12}} = 1.22\,mV$$

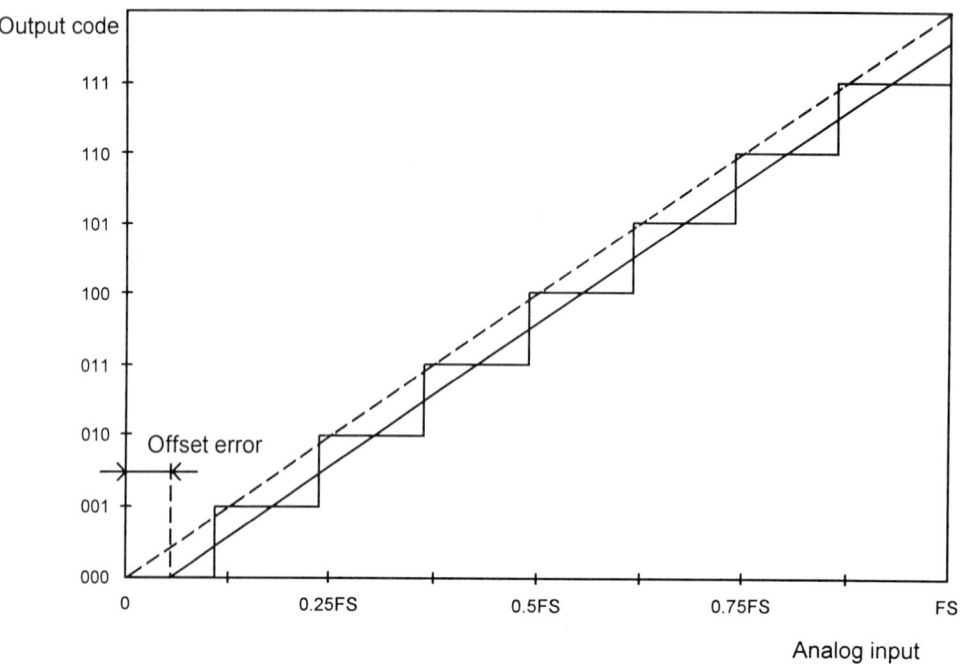

Figure 5.12 The ADC offset error.

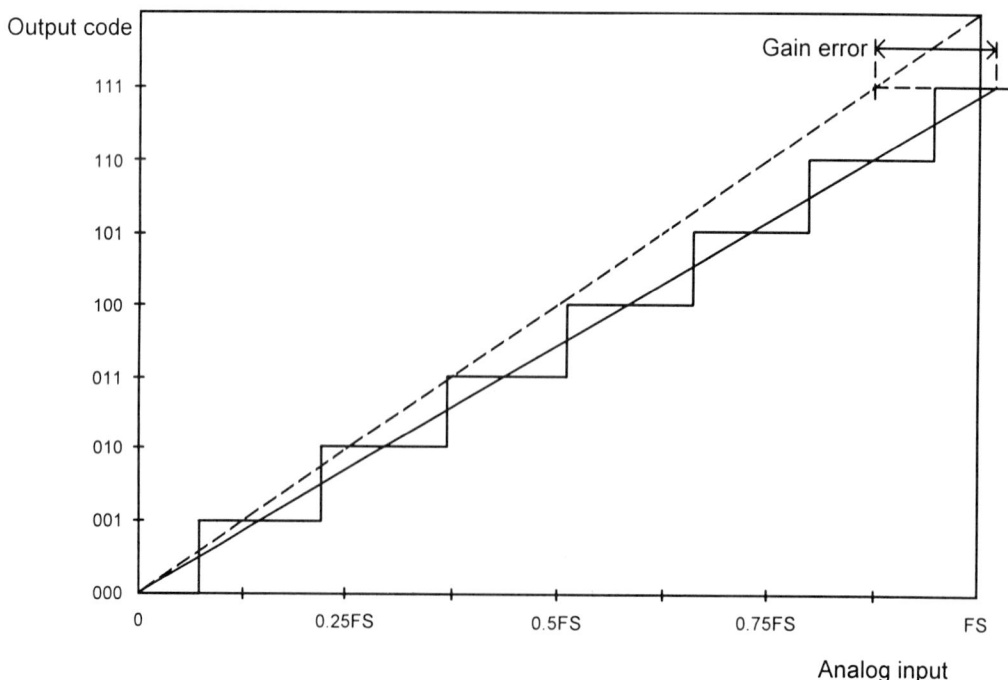

Figure 5.13 The ADC gain error.

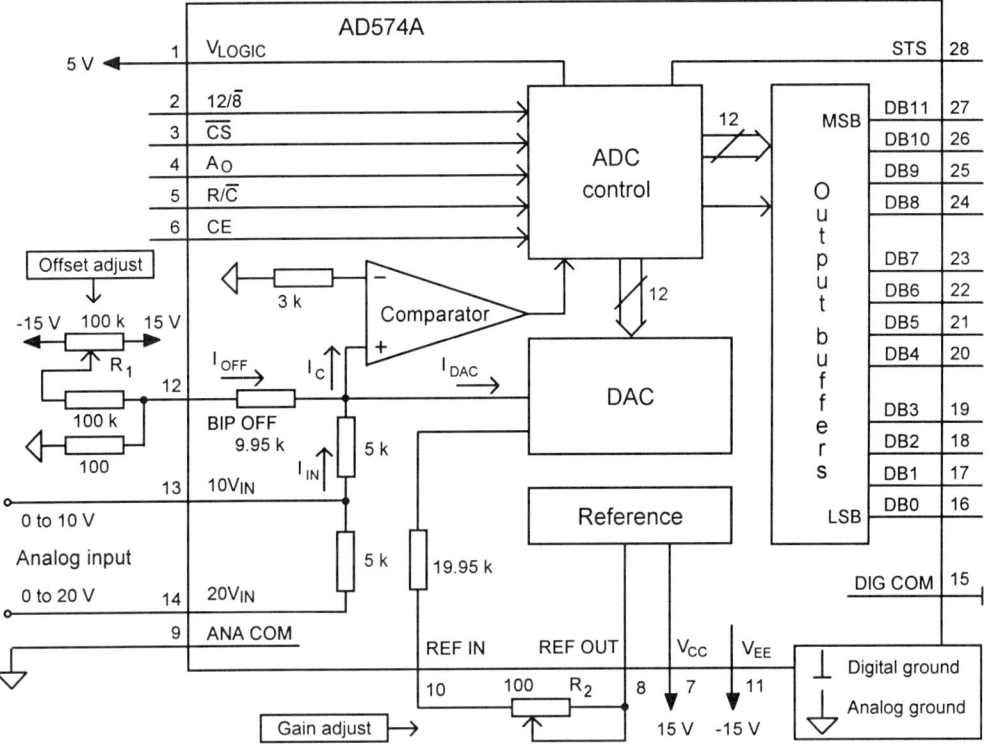

Figure 5.14 The A/D converter AD574A internal structure.

The gain trim is accomplished by applying an input voltage 1.5LSB below the FS (9.9963 V for the range 0 to 10 V). We adjust the trimmer R_2 until the transition between the output codes 1111 1111 1110 and 1111 1111 1111 occurs for the given input voltage.

The ADC has six control/status lines. A conversion is initiated in the following cases:

• The inputs read/convert $R/\overline{C}$ and chip select $\overline{CS}$ are low and the input chip enable CE is switched from low to high.

• The input read/convert $R/\overline{C}$ is low, the input chip enable CE is high and the input $\overline{CS}$ is pulled down.

• Both inputs CE and $\overline{CS}$ have been asserted and the input $R/\overline{C}$ goes low.

When the input CE is switched from low to high if the input A_O is low, a 12-bit conversion will be initialized (conversion time 15 - 35 μs). In the same case, if the input A_O is high, an 8-bit conversion will be implemented (conversion time 10 - 24 μs).

When the AD574A completes its conversion cycle, the output STS goes low.

As the 8051 microcontroller is an 8-bit machine, it is logical to use the 8-bit bus interface mode. Thus, the input $12/\overline{8}$ is pulled down. The ADC indicates the end of conversion by a STS low output level. If we assume that the input $\overline{CS}$ is low and the input $R/\overline{C}$ is high, we just have to pull the input CE high. Then, it will not take more than 200 ns for the outputs to emit the result. However, the input A_O must be established at least 150 ns before the transition of the signal CE. Low level at A_O specifies that the high-order byte is read. High level at A_O enables the low-order byte. The low-order byte contains four bits of the result plus four trailing zeroes.

Range	Input pin	I_{IN}	I_{OFF}	I_{DAC}
0 to 10 V	$10V_{IN}$	0 to 2 mA	0	0 to 2 mA
0 to 20 V	$20V_{IN}$	0 to 2 mA	0	0 to 2 mA
-5 V to 5 V	$10V_{IN}$	-1 mA to 1 mA	1 mA	0 to 2 mA
-10 V to 10 V	$20V_{IN}$	-1 mA to 1 mA	1 mA	0 to 2 mA

Figure 5.15 The AD574A input ranges.

Now we are back on Figure 5.9 to discuss the A/D part of the system. The AD574A interacts with the microcontroller through Port 1, the pin P3.3 and Port A of the PIO 82C55A.

The available input ranges are listed in the left table of Figure 5.9. A mark indicates the selected range (0 to 10 V).

To convert the analog input voltage and store the results in the external RAM, the following subroutine can be used.

```
;*********************************************
;*      ADC1.ASM                          *
;*      Analog input 12-bit               *
;*      XRAM start address                *
;*            High-order byte    R0       *
;*            Low-order byte     R1       *
;*      Number of conversions    R2       *
;*********************************************
ADC1:   MOV    P1,#04H        ; Start, CE high, R/*C low, Ao low
        MOV    P1,#08H        ; Stop, CE low, R/*C high
        MOV    DPTR,#PAADDR
;
        JB     P3.3,$         ; Wait until STS goes low
        MOV    P1,#0CH        ; R/*C high, CE high
        MOVX   A,@DPTR        ; Read high-order byte
        MOV    DPH,R0         ; XRAM address
        MOV    DPL,R1
        MOVX   @DPTR,A        ; Store high-order byte
        INC    DPTR           ; Increment XRAM address
        MOV    R0,DPH         ; Save XRAM address
        MOV    R1,DPL
        MOV    P1,#08H        ; CE low (disable outputs)
                              ; R/*C high
        MOV    P1,#09H        ; Pull Ao up
        MOV    P1,#0DH        ; CE high, R/*C high, Ao high
        MOV    DPTR,#PAADDR
```

```
          MOVX    A,@DPTR        ; Read low-order byte
          MOV     DPH,R0         ; Load XRAM address
          MOV     DPL,R1
          MOVX    @DPTR,A        ; Store low-order byte
          INC     DPTR           ; Increment XRAM address
          MOV     R0,DPH         ; Save XRAM address
          MOV     R1,DPL
          MOV     P1,#09H        ; Disable all outputs
;
          MOV     P1,#0          ; Ready for a new conversion
          DJNZ    R2,ADC1        ; End?
          RET
```

If the oscillator frequency f_{OSC} = 12 MHz and the conversion time is 25 μs, ADC will sample the input voltage every 72 μs.

Problem 5.3

Rewrite the subroutine ADC1.ASM in order to increase the sample rate. Assume that the data is located in one 256-byte page.

Solution 5.3

With the assumption that all codes will be located in one 256-byte page, the following subroutine could be performed:

```
;*******************************************************
;*      ADC2.ASM                                       *
;*      Analog input 12-bit                            *
;*      XRAM start address                             *
;*            High-order byte      R0                  *
;*            Low-order byte       R1                  *
;*      Number of conversions      R2                  *
;*      All codes will be located in one 256-byte page *
;*******************************************************
ADC2:  MOV    P2,R0          ; Select the page
       MOV    DPTR,#PAADDR
ADC21: MOV    P1,#04H        ; Start, CE high, R/*C low, Ao low
       MOV    P1,#08H        ; Stop, CE low, R/*C high
       JB     P3.3,$         ; Wait until STS goes low
       MOV    P1,#0CH        ; CE high, R/*C high
       MOVX   A,@DPTR        ; Read high-order byte
       MOVX   @R1,A          ; Store high-order byte
       INC    R1             ; Increment address
       MOV    P1,#08H        ; CE low, R/*C high
       MOV    P1,#09H        ; Pull A0 up
       MOV    P1,#0DH        ; CE high, R/*C high, Ao high
       MOVX   A,@DPTR        ; Read low-order byte
       MOVX   @R1,A          ; Store low-order byte
       INC    R1             ; Increment address
       MOV    P1,#09H        ; Disable all outputs
       MOV    P1,#0          ; Ready to start a new conversion
```

```
        DJNZ    R2,ADC21        ; End?
        RET
```

In this case, the analog input is sampled every 52 μs.

Problem 5.4

Calculate the speed-up for both 8-bit A/D conversion and one 256-byte page.

Solution 5.4

Finally, if 8-bit resolution is sufficient, the program can be streamlined to the following instruction layout:

```
;************************************************************
;*      ADC3.ASM                                          *
;*      Analog input 8-bit                                *
;*      XRAM start address                                *
;*              High-order byte      R0                   *
;*              Low-order byte       R1                   *
;*      Number of conversions        R2                   *
;*      All codes will be located in one 256-byte page    *
;************************************************************
ADC3:   MOV    P2,R0           ; Select the page
        MOV    DPTR,#PAADDR
        MOV    P1,#01H          ; Ao high > 8-bit conversion
ADC31:  MOV    P1,#05H          ; Start, CE high, R/*C low, Ao high
        MOV    P1,#01H          ; Stop, CE low, R/*C low, Ao high
        JB     P3.3,$           ; Wait until STS goes low
        MOV    P1,#0CH          ; CE high, R/*C high, Ao low
        MOVX   A,@DPTR          ; Read
        MOVX   @R1,A            ; Store
        INC    R1               ; Increment address
        MOV    P1,#01H          ; Ready to start a new conversion
        DJNZ   R2,ADC31         ; End?
        RET
```

In this case, the analog input is sampled every 29 μs.

5.5 Temperature measurement

Measurement of temperature is a typical case of interfacing in the analog world. We need to monitor temperature or to control it in applications such as automotive industry, solar energy conversion, food production processes and heating systems. Of course, the list might go on, but only the fields mentioned are sufficient to demand temperature sensors with different parameters. Figure 5.16 illustrates a taxonomy of temperature sensors, based on various technologies.

The Resistance Temperature Detectors (RTD) are produced by metals, usually platinum, nickel or copper. The resistance versus temperature characteristic is almost linear. Normally, the RTD sensors are used in the temperature range -200°C to 800°C. Compared to the other sensors, the RTDs are most stable and most accurate.

Thermistors are built from different semiconducting materials, mixed and encapsulated. The resistance of the thermistor decreases when the temperature is increased. The advantage of using thermistors is that they have high output and are relatively fast.

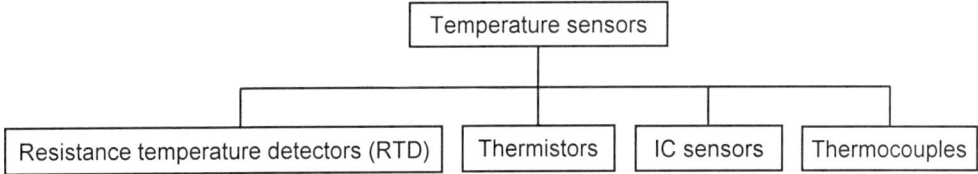

Figure 5.16 Different technologies of temperature sensors.

The IC temperature sensors use the fact that the base-emitter voltage of a transistor is related to temperature at a constant collector current. The distinctive features of the IC sensors are linearity and high output. On the other hand, ICs become rather useless when the temperature exceeds 200°C.

Finally, thermocouples can be used in wide temperature ranges, for example, 0°C to 1700° C. They are produced from dissimilar metals and are very robust. The major shortcoming of thermocouples is that they are not stable. Another drawback is the non-linear transfer characteristic, but the embedded systems are intelligent enough to mitigate this problem.

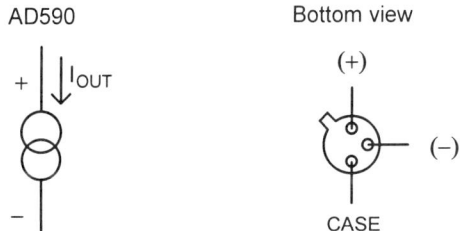

Figure 5.17 The temperature sensor AD590, the circuit symbol and pin designations.

Figure 5.17 shows an example of an IC temperature sensor. The Analog Devices AD590 can be used in the temperature range -55°C to 150°C [Anal 1992c]. The power supply voltage must be between 4 V and 30 V. The temperature sensor equation is:

$$I_{OUT} = \frac{1\,\mu A}{°K} T_K$$

Assuming that the absolute temperature

$$T_K = T_C + 273$$

we rearrange the sensor equation in the following way:

$$I_{OUT} = \frac{1\,\mu A}{°C} T_C + 273\,\mu A$$

For example, if the temperature range is 0°C to 100°C, the sensor output current I_{OUT} will vary between 273 µA and 373 µA.

So far, we have allocated a sensor which can be used in typical embedded systems applications. Also, we have discussed the A/D converter AD574A. Now we need an input signal conditioner to link the sensor and the ADC.

In the majority of cases, the key component of the input signal conditioner is an operational amplifier. In our particular example, the input current must be converted into output voltage. In addition, the signal must be shifted. We need this offset when the input current is 273 µA and the output voltage must be 0 V.

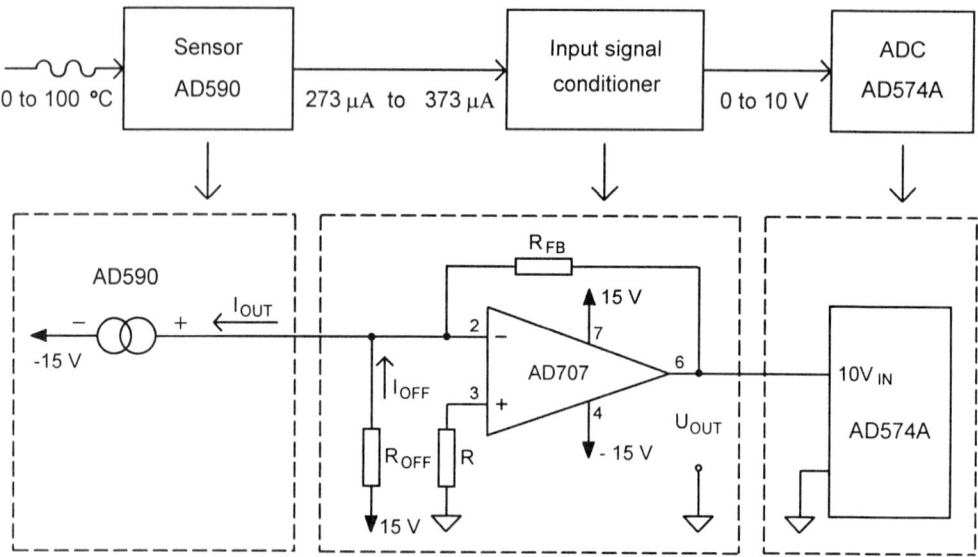

Figure 5.18 The transition from specification to a schematic diagram.

Figure 5.18 depicts the transition from specification to a schematic diagram. First, we must provide power supply for the sensor AD590. As the sensor output will go to the amplifier's inverting input (virtual ground), we tie the other sensor's terminal to negative voltage.

Second, we have to calculate the offset resistor R_{OFF} which is used to bias the amplifier AD707 [Anal 1994].

$$I_{OFF} = 273 \, \mu A$$
$$I_{OFF} \times R_{OFF} = 15 \, V$$

$$R_{OFF} = \frac{15 \, V}{273 \, \mu A} = 54.9 \, k\Omega$$

Finally, we get the value of the feedback resistor R_{FB}

$$R_{FB} = \frac{10 \, V}{100 \, \mu A} = 100 \, k\Omega$$

The resistor R equalizes the resistance of both circuits tied to the amplifier's inputs. We need equal equivalent resistance to make the amplifier's differential voltage independent from temperature variations.

$$R = \frac{R_{OFF} \times R_{FB}}{R_{OFF} + R_{FB}} = 35.7\,k\Omega$$

In many applications the measurement results are involved in calculations. If the present form, in which the temperature is given indirectly, would be a convenient starting point for the calculations depends of the specific case. Furthermore, if the results must be displayed, it is very likely that BCD code will be employed. Following this direction, we should convert the result (0 to 255) to temperature (0°C to 100°C). Then, the temperature code must be converted to BCD numbers. Before that, however, we should decide how to round off the numbers.

As you might expect, general purpose arithmetic subroutines can be used to multiply the results by

$$\frac{100}{255} = 0.392157$$

Again, a subroutine might be used to convert the temperature data to BCD form. We can use a look-up table as an alternative solution. Admittedly, this approach will require more memory space.

Problem 5.5

Write a subroutine which performs the following actions:
- 8-bit A/D conversion with an AD574A converter (Figure 5.9).
- Conversion of the result into temperature sample in consistence with Figure 5.18.
- Final conversion into BCD number with rounded off temperature to the nearest tenth of a degree.

Solution 5.5

The conversion part of the subroutine will be a close version of the ADC3.ASM, discussed in Problem/Solution 5.4. We are going to use a short cut to implement the rest of the specification. We will show how to use a look-up table.

There are 256 possible conversion results and in principle, the table must contain 256 temperature values. Glance through the table in Figure 5.19 to get an idea of the possible indications. The practical implementation of the table, however, should be done by two sections (two tables). We display four decimal digits. Therefore, we must read two bytes each time. We organize the high-order two digits in a table called **TABLE_H**. Likewise, the calculations for the low-order two digits are arranged in a table named **TABLE_L**. We use the assembler directive define byte (DB) to prepare the tables in the Program Memory.

Code	00	01	02	03	04	---	FE	FF
Temperature, °C	000.0	000.4	000.8	001.2	001.6	---	099.6	100.0

Figure 5.19 The look-up table.

The tables and the subroutine appear here:

```
          ORG     TABLE_H          ; High-order digits table
DB 00,00,00,00,00,00,00,00,00,00,00,00,00,00,00,00
DB 00,00,00,00,00,00,00,00,00,00,01,01,01,01,01,01
DB 01,01,01,01,01,01,01,01,01,01,01,01,01,01,01,01
DB 01,01,01,02,02,02,02,02,02,02,02,02,02,02,02,02
DB 02,02,02,02,02,02,02,02,02,02,02,02,02,03,03,03
DB 03,03,03,03,03,03,03,03,03,03,03,03,03,03,03,03
DB 03,03,03,03,03,03,04,04,04,04,04,04,04,04,04,04
DB 04,04,04,04,04,04,04,04,04,04,04,04,04,04,04,04
DB 05,05,05,05,05,05,05,05,05,05,05,05,05,05,05,05
DB 05,05,05,05,05,05,05,05,05,06,06,06,06,06,06,06
DB 06,06,06,06,06,06,06,06,06,06,06,06,06,06,06,06
DB 06,06,06,07,07,07,07,07,07,07,07,07,07,07,07,07
DB 07,07,07,07,07,07,07,07,07,07,07,07,08,08,08,08
DB 08,08,08,08,08,08,08,08,08,08,08,08,08,08,08,08
DB 08,08,08,08,08,08,09,09,09,09,09,09,09,09,09,09
DB 09,09,09,09,09,09,09,09,09,09,09,09,09,09,09,10
;
          ORG     TABLE_L          ; Low-order digits table
DB 00,04,08,12,16,20,24,27,31,35,39,43,47,51,55,59
DB 63,67,71,75,78,82,86,90,94,98,02,06,10,14,18,22
DB 25,29,33,37,41,45,49,53,57,61,65,69,73,76,80,84
DB 88,92,96,00,04,08,12,16,20,24,27,31,35,39,43,47
DB 51,55,59,63,67,71,75,78,82,86,90,94,98,02,06,10
DB 14,18,22,25,29,33,37,41,45,49,53,57,61,65,69,73
DB 76,80,84,88,92,96,00,04,08,12,16,20,24,27,31,35
DB 39,43,47,51,55,59,63,67,71,75,78,82,86,90,94,98
DB 02,06,10,14,18,22,25,29,33,37,41,45,49,53,57,61
DB 65,69,73,76,80,84,88,92,96,00,04,08,12,16,20,24
DB 27,31,35,39,43,47,51,55,59,63,67,71,75,78,82,86
DB 90,94,98,02,06,10,14,18,22,25,29,33,37,41,45,49
DB 53,57,61,65,69,73,76,80,84,88,92,96,00,04,08,12
DB 16,20,24,27,31,35,39,43,47,51,55,59,63,67,71,75
DB 78,82,86,90,94,98,02,06,10,14,18,22,25,29,33,37
DB 41,45,49,53,57,61,65,69,73,76,80,84,88,92,96,00
;
;*************************************************
;*      TEMP.ASM                                 *
;*      Temperature measurement                  *
;*      8-bit A/D conversion                     *
;*      The subroutine stores                    *
;*            High-order BCD numbers in R6       *
;*            Low-order BCD numbers in R7        *
;*************************************************
TEMP: MOV     DPTR,#PAADDR
      MOV     P1,#01H          ; Ao high > 8-bit conversion
                               ; R/*C low, ready for conversion
      MOV     P1,#05H          ; Start, CE high, R/*C low, Ao high
      MOV     P1,#01H          ; Stop, CE low, R/*C low, Ao high
      JB      P3.3,$           ; Wait until STS goes low
```

```
;
        MOV     P1,#0CH          ; CE high, R/*C high, Ao low
        MOVX    A,@DPTR          ; Read the result of conversion
        PUSH    ACC              ; Save
;
        MOV     DPTR,#TABLE_H    ; High-order digits table
        MOVC    A,@A+DPTR        ; Read high-order digits
        MOV     R6,A             ; Store high-order digits
        POP     ACC              ; Restore the conversion result
;
        MOV     DPTR,#TABLE_L    ; Low-order digits table
        MOVC    A,@A+DPTR        ; Read low-order digits
        MOV     R7,A             ; Store low-order digits
        RET
```

The first part of the subroutine **TEMP.ASM** is very similar to **ADC3.ASM**. The new instructions are more or less self-explanatory. We read an element from **TABLE_H** and the corresponding value from **TABLE_L**.

It is certainly implied that the table approach is much faster than using arithmetic and conversion subroutines. The next task would be to display the measurement results. Display interfacing is discussed in Chapter 7.

In terms of calibration, the temperature measurement system (Figure 5.18) includes three analog stages: the sensor, the input signal conditioner and the ADC. We can adjust each one of them. Logically, we should trim in the last stage, the A/D converter.

The tendencies in ADC's new designs are more on-chip functions for the same or even lower price [Bind 1998a]. Improved parameters, reduced supply voltages, on-chip temperature sensors, serial interfaces and smaller packages are some of the hallmarks of the new designs. New techniques were developed for both converter ICs and embedded analog peripherals. For example, the 8051 compatible microcontroller (MicroConverter) ADuC812 from Analog Devices integrates two 12-bit DACs, a 12-bit self-calibrating successive-approximation ADC and a on-chip temperature sensor [Leon 1998]. The ADC has eight input channels and a conversion time of 5 µs. An extra input channel links the ADC to the on-chip temperature sensor.

Note that the more precise the devices are, the more carefully the circuit layout must be designed.

5.6 Supplementary problems

Problem S5.1
Calculate the basic DACs specifications for an 11-bit DAC, output range 0 V to 10 V.

Problem S5.2
Lay out a DAC transfer characteristic which can not be completely corrected neither by the offset nor by the gain trim.

Problem S5.3
Prove that the output voltage U_{OUT} of the R-2R ladder network shown in Figure 5.20 is proportional to the binary value applied to the digital inputs IN1 through IN4. Calculate the value of the resistor R_1.

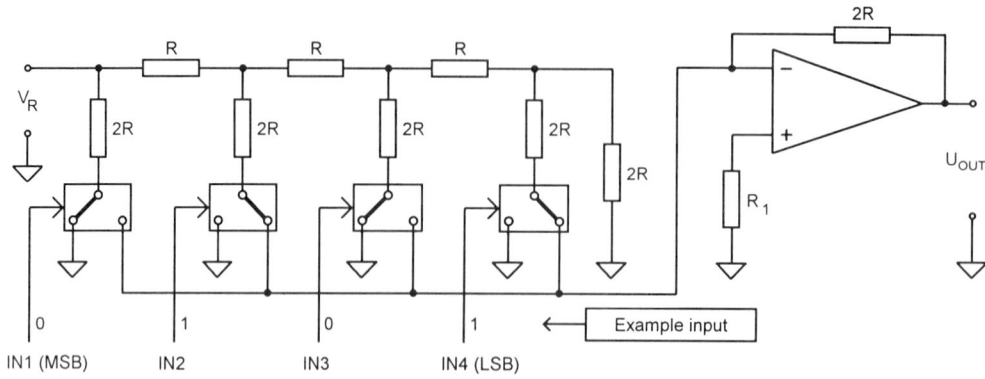

Figure 5.20 An R-2R ladder network.

Problem S5.4

Design an 10-bit DAC on the base of the IC AD565A. The D/A converter has two control inputs (S1 and S0) which select an output range according to the table in Figure 5.21. In fact, you should design the combinational logic circuit CLC that links the actual DACs inputs with the AD565A pins (Figure 5.22).

S1	S0	Range	Code	Transfer curve
0	0	-10 V to 0 V	Offset binary	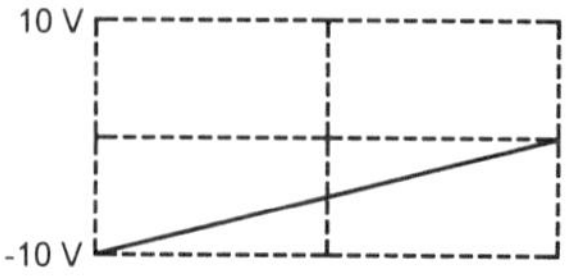
0	1	0 V to 10 V	Ordinary binary	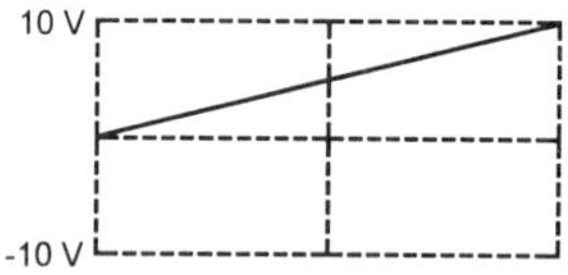
1	0	-10 V to 10V	Two's complement	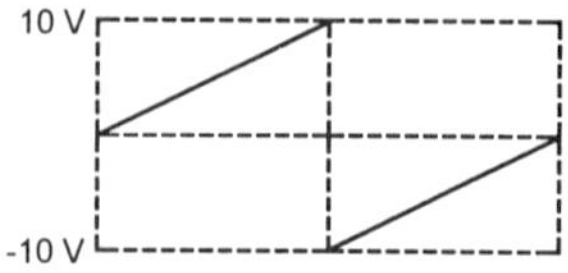
1	1	-5 V to 5 V	Two's complement	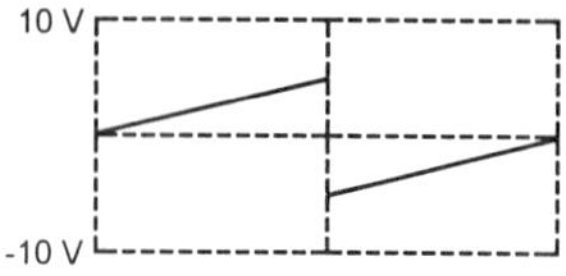

Figure 5.21 The correspondence control inputs and output ranges.

Problem S5.5

Write a sequence of instructions which generate a sawtooth wave for the D/A subsystem shown in Figure 5.9.

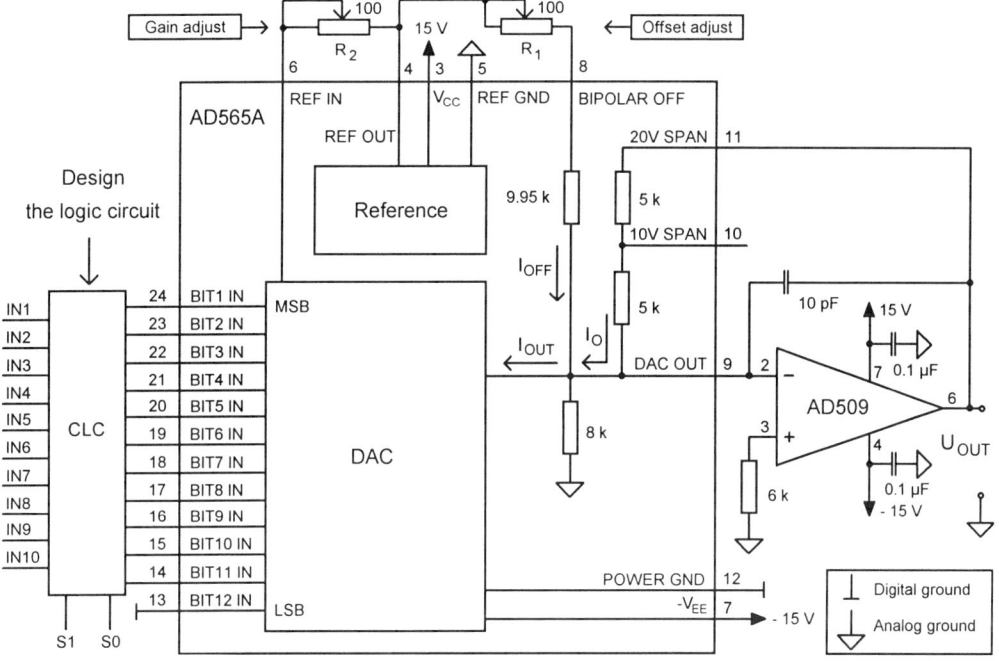

Figure 5.22 The DAC with digitally controlled output ranges.

Problem S5.6

Suggest A/D conversion approaches which lie between the counter and successive-approximation methods shown in Figure 5.10. Also, expand the taxonomy with methods that are faster than the successive-approximation method and slower than the flash method.

Problem S5.7

Assume that you gradually increase the input voltage of an 4-bit ideal ADC. Draw the curve which gives the difference between the analog input and the weight of the digital output (quantization error). Repeat this procedure for a 5-bit ideal ADC.

Problem S5.8

Write an assembly language program to implement software controlled delay of analog signals. The code must be written for the system shown in Figure 5.9. The analog input signal is converted to an 8-bit digital value. After a delay which is accomplished by a call to the subroutine **DEL.ASM** (section 3.10), the digital sample is converted back through the D/A converter AD565A, used as an 8-bit device.

Problem S5.9

Redesign the input signal conditioner shown in Figure 5.18 for a temperature range of -55°C to 125°C.

Problem S5.10
A temperature sensor with frequency output possesses the following temperature-frequency dependence:

$$0°C \rightarrow 2732 \text{ Hz}$$
$$100°C \rightarrow 3732 \text{ Hz}$$

Use a counter input of the 8051 microcontroller to monitor the temperature. Write the code to convert the frequency into temperature sample, rounded off to the nearest tenth of a degree (see also Problem/Solution 5.5).

5.7 References

Analog Devices, *Data Converter Reference Manual, Volume I*, 1992a.

Analog Devices, *Data Converter Reference Manual, Volume II*, 1992b.

Analog Devices, *Special Linear Reference Manual*, 1992c.

Analog Devices, *Design-in Reference Manual*, 1994.

Ashok Bindra, "Commodity ADCs pile more functions on-chip for less", *Electronic Design*, March 9, 1998a, pp. 48-52.

Ashok Bindra, "Commodity DACs wring more performance from smaller packages", *Electronic Design*, April 6, 1998b, pp. 69-80.

Frederick F. Driscoll, Robert F. Coughlin and Robert S. Villanucci, *Data Acquisition and Process Control with the M68HC11 Microcontroller*, Macmillan Publishing, 1994.

K. Euler, "Neue Prinzipen zur Analog-Digital-Umwandlung und deren optimale Auslegung", *Frequenz*, 1963, N10.

Bernard M. Gordon, "Linear electronic Analog/Digital conversion architectures, their origins, parameters, limitations, and applications", *IEEE Transactions on Circuits and Systems*, vol. CAS-25, No. 7, July 1978, pp. 391-418.

Milt Leonard, "Self-programming microcontroller networks sensors and transducers", *Electronic Design*, February 9, 1998, pp. 92-94.

T. P. Morrison, *The Art of Computerized Measurement*, Oxford University Press, 1997.

National Semiconductor, *Linear Applications Handbook*, 1994.

Willis J. Tompkins and John G. Webster, *Interfacing sensors to the IBM PC*, Prentice Hall, 1988.

Information on components and systems for analog interfacing can be found at :

Analog Devices	http://www.analog.com
Burr-Brown Corp.	http://www.burr-brown.com
Linear Technology Corp.	http://www.linear-tech.com
Maxim Integrated Products	http://www.maxim-ic.com
Motorola	http://www.mot.com
National Semiconductor	http://www.national.com
Texas Instruments	http://www.ti.com

Chapter 6

INTERFACING PERSONAL COMPUTERS

6.1 Introduction

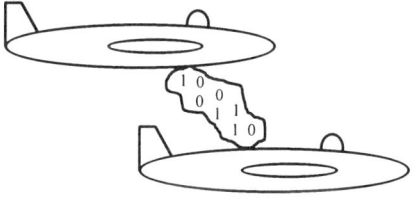

Many embedded systems, such as computer peripherals, development systems and measurement equipment are linked to personal computers. For these applications the interaction with PCs is an integrated part of their operation. There are also embedded computers which are connected to PCs only in the development phase or for test procedures. This distinction, however, lost much of its determination. Currently, there is a trend for designers to put Internet connectivity in a wide range of embedded computers, such as appliances and factory equipment. Many Internet embedded systems are linked to the network through PCs. Some Internet appliances interact with the PC using interfaces discussed in this chapter. Some others use I^2C or CAN buses to communicate with PCs. These interfaces are in focus in Chapter 8.

In this chapter we initially discuss the 8051 microcontroller serial port programming. Since the serial port is closely related to the other subsystems, we'll see how to strike the balance in case of conflicting needs. Next, we deal with the serial ports of the IBM compatible PCs. The beginning of this section can be viewed as an extension of the theme serial interface opened in section 4.4. Following a succession of examples, we finally design a terminal emulator program with interrupt-driven direct read. Since new sophisticated interfaces are taking shape, the last section in this chapter introduces, as an example, the Universal Serial Bus.

6.2 Programming the 8051 microcontroller serial port

The front of the serial port subsystem has two hidden shift registers. The program interact with the shift registers through buffers, which according to the 8051 terminology have a common name, SBUF. The program writes a byte to the transmit serial SBUF (a MOV instruction). At an appropriate time, the byte from the transmit SBUF is moved to the transmit shift register and the flag TI is set. The transmit shift register generates the output stream through the pin TxD. The receive SBUF and receive shift register operation is a mirror-image of the transmit task. Again, a single MOV instruction is the basic core. When the receive shift register becomes full, its contents are moved to the receive SBUF and the flag RI is set. We must be careful not to write too frequently to the transmit buffer. Similarly, in the case of reception the program must not be slow in retrieving characters from the receive SBUF. The term for a lost byte due to a delayed reaction is overrun error.

As with all other microcontroller's subsystems, the serial port must be initialized first. This procedure is somewhat more difficult and we go through an example to explain how it could be done.

We assume that the following parameters are relevant:
• A frame includes a start bit, 8 data bits and a stop bit
• The baud rate is 2400 bps
• The microcontroller runs at oscillator frequency 11.0592 MHz
• All messages should be received.

Figure 6.1 shows revealingly how the register settings are related to the specification. The required frame determines the serial port mode 1 as the only candidate for this example (see section 2.8). As the specification admits, the stop bit can be ignored. We reset the bit SM2 to achieve this feature. The settings for the next three flags in the register SCON are not bound to any specific applications. In all cases, their pattern could be as shown in Figure 6.1. The bit REN is set to enable reception. The flag TI is also set to allow a single subroutine to be used for transmission of the first and the following bytes. The flag RI must be reset to prepare the serial port for reception. Both flags TI and RI will be set by hardware and this will be an indication that the serial port is ready to interact with the CPU.

Furthermore, the desired baud rate of 2400 bps should be accomplished. We are guided by Figure 2.18, which prescribes Timer/Counter 1 to be used as a timer in mode 2. The flag SMOD must be cleared. It is certainly implied that the flag GATE must be reset and the flag TR1 set (see Figure 2.14).

Finally, the auto-reload value F4H must be moved to the register TH1.

The initialization program is shown here:

```
SP_INI: MOV    SCON,#01010010B   ; Serial port, mode 1, set REN
        ORL    TMOD,#00100000B   ; Timer 1, mode 2
        ANL    TMOD,#00101111B   ; Reset GATE, timer, mode 2
        ANL    PCON,#01111111B   ; Reset SMOD
        MOV    TH1,#0F4H         ; Baud rate 2400 bps, oscillator
                                 ; frequency 11.0592 MHz
        SETB   TR1               ; Start Timer 1
```

Of course, we could use a MOV instruction to modify the four bits in register TMOD, however it is a sound practice to leave the other bits unchanged.

As soon as the serial port is initialized, the program can call transmit and receive subroutines. The following subroutine sends the byte in the accumulator over the serial port.

```
SEND:   JNB     TI,$       ; Wait for an empty SBUF
        CLR     TI         ; Reset TI
        MOV     SBUF,A     ; This instruction transmits the byte
        RET
```

Before the MOV instruction we have to wait for a set flag TI. If we did not set the flag TI in the initialization part of the program, we would not be able to send the first byte. As discussed earlier in section 2.8, the flag TI must be reset by software.

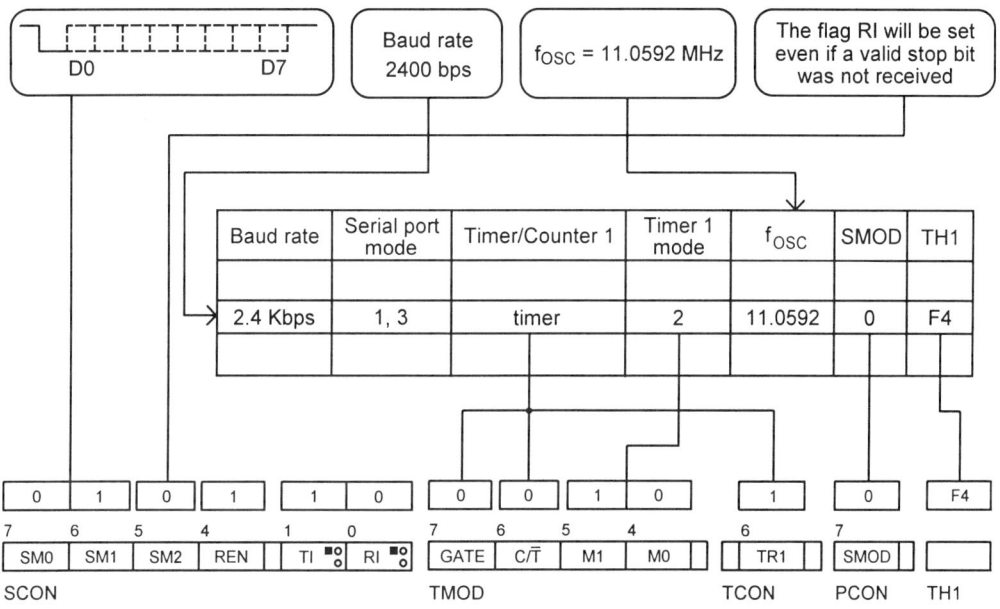

Figure 6.1 The serial port initialization procedure.

Likewise, a subroutine called REC moves the latest received byte from the serial port to the accumulator.

```
REC:    JNB     RI,$       ; Wait for a full SBUF
        CLR     RI         ; Reset RI
        MOV     A,SBUF     ; This instruction moves the byte to ACC
        RET
```

In this case, the loop will be executed until the flag RI is set. Again, the flag RI must be reset by software.

Our serial port examples so far have dealt with oscillator frequency 11.0592 MHz. However, we have never proved that this value is the maximum possible oscillator frequency which allows the microcontroller to communicate at several different baud rates.

Problem 6.1

Define the maximum possible oscillator frequency f_{OSC} and the reload codes if the 8051 microcontroller has to communicate by programmable baud rates in the range from 1200 through 19200 bps.

Solution 6.1

The requirement for programmable baud rates can be met if the serial port works in mode 1 or mode 3 and Timer/Counter 1 is programmed as a timer in mode 2 (see Figure 2.17). If that's the case, the baud rate is given by the following equation

$$\text{BAUD RATE} = \frac{f_{OSC}}{12(256 - TH1)(32 - 16PCON.7)}$$

Let us admit that initially we are aiming for a baud rate of 9600 bps. When we solve this part of the problem, we can easily move to baud rate 19200 bps by setting the bit SMOD (which has been represented also as PCON.7). Baud rates lower than 9600 bps can be achieved by different reload values in register TH1. Thus,

$$9600 = \frac{f_{OSC}}{12(256 - TH1)32} \quad \rightarrow \quad f_{OSC} = 3686400 \, (256 - TH1)$$

Observe that if we start from a minimal value (256 - TH1) and increase by a step of 1, we will hit the ceiling of the oscillator frequency. We arrange the calculations in Figure 6.2.

256 - TH1	f_{OSC} (MHz)		
1	3.68640		
2	7.37280		
3	11.0592	←	$f_{OSC\,MAX} = 12\,MHz$
4	14.7456	←	$f_{OSC\,MAX} = 16\,MHz$
5	18.4320		

Figure 6.2 Maximum possible oscillator frequencies.

Many microcontrollers can run up to 12 MHz. For these, the solution is 11.0592 MHz. For the 16 MHz devices, the maximum possible oscillator frequency is 14.7456 MHz.

The next step is to calculate the reload values. For f_{OSC} = 11.0592 MHz

$$256 - TH1 = 3 \quad \rightarrow \quad TH1 = 253 = FDH$$

If we move one step down in the range (4800 bps) the expression (256 - TH1) must be twice as large, therefore 6. Thus, we get

$$TH1 = 250 = FAH$$

We use the same approach for baud rates of 2400 and 1200 bps. Figure 6.3 shows the serial port and Timer 1 modes plus the reload codes for register TH1. The reload codes were calculated twice per baud rate. The first result is for oscillator frequency 11.0592 MHz. The second value is for oscillator frequency 14.7456 MHz.

Baud rate (bps)	Serial port mode	Timer 1 mode	SMOD	f_{OSC} (MHz)	TH1
19200	1,3	2	1	11.0592	FD
				14.7456	FC
9600	1,3	2	0	11.0592	FD
				14.7456	FC
4800	1,3	2	0	11.0592	FA
				14.7456	F8
2400	1,3	2	0	11.0592	F4
				14.7456	F0
1200	1,3	2	0	11.0592	E8
				14.7456	E0

Figure 6.3 The reload codes for the register TH1.

The following concluding points are in order:
- If we bring down the oscillator frequency from 12 to 11.0592 MHz, the microcontroller will be capable of selecting baud rates in the range 1200 through 19200 bps only by software.
- Extra calculations may be added to the sequence in Figure 6.2, and other oscillator frequencies determined for faster devices.

It is a common practice for terminals or PCs to communicate with single board computers based on microcontrollers. An important issue in such a system is both computers to be initialized at equal baud rates. When we want to change the PC's baud rate, we simply press a key or modify a line of code. The single board computer is decidedly less flexible as far as alterations of the program are concerned. A key point is which computer to dictate the baud rate. Normally, the PC is the master and if it starts to transmit at a certain baud rate, the microcontroller should be able to adapt its serial port to the right pace.

Problem 6.2

Suppose that a terminal works in conjunction with the serial port of the 8051 microcontroller. The oscillator frequency is 11.0592 MHz. Write an assembly language program for the 8051 microcontroller that will recognize the terminal's baud rate in the range from 1200 through 9600 bps and adjust the serial port accordingly. Assume that the first character sent from the terminal is SPACE.

Solution 6.2

As the serial port is initially not functioning, the pin P3.0 can only be used as a general purpose input.

Pressing the SPACE bar at the terminal end will produce the following frame (code 20H):

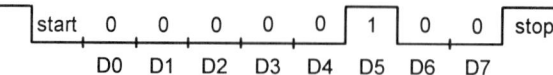

The baud rate could be recognized by measuring the interval between the beginning of the start bit and the end of the last low bit (D4). An efficient approach would be to turn the measurement of the time period into a calculation of the auto-reload value. Lower baud rates will result in longer periods and smaller auto-reload values. Thus, the idea to test the input periodically and decrement a register makes sense. We organize a loop to test the input P3.0/RxD and decrement accumulator. The loop is terminated when a high input is met. Logically, two questions arise: how to calculate the initial code in the accumulator and the delay of the loop.

Figure 6.4 lays out the four waveforms which are relevant. Our goal is to sample the input at intervals as shown in the picture by vertical dashed lines. If a high level is detected, it will occur in the middle of the pulse. This approach brings immunity against possible variations of the transmitter baud rate and excessive transition times.

As shown in Figure 6.4, the auto-reload value for baud rate 9600 bps is FDH. The same code for baud rate 4800 bps is FAH. The difference between both values is FDH - FAH = 3. Taking into consideration that the difference between the initial value in accumulator and FDH is also 3, we get FDH + 3 = 0. Thus, we initialize the accumulator to 00H.

Furthermore, we design the delay loop. It could be done in two steps. First, we outline a sequence of instructions. Second, we calculate a parameter to establish the required delay. The following code will do the job:

```
        CLR   A          ; Initialize accumulator
        MOV   R0,#dd      ; Initialize R0 for the first loop
                         ; dd defines the delay
        JB    P3.0,$      ; Wait for the start bit
DEL: DJNZ   R0,$         ; Execution time 24Tosc for an instruction
        DEC   A          ; Execution time 12Tosc
        MOV   R0,#dd      ; Execution time 12Tosc, dd defines delay
        JNB   P3.0,DEL    ; Execution time 24Tosc
```

The loop's delay T_{DEL} depends on the parameter dd.

$$T_{DEL} = 24T_{OSC} \times (dd) + 48T_{OSC}$$

At the same time,

$$3T_{DEL} = 6.5\,\text{bit}\,\frac{1}{9600\,\text{bps}}$$

Manipulating these equations, we get

$$dd = \frac{T_{DEL} - 48T_{OSC}}{24T_{OSC}} = 102$$

Thus, we replace dd with 102 in the MOV instructions above.

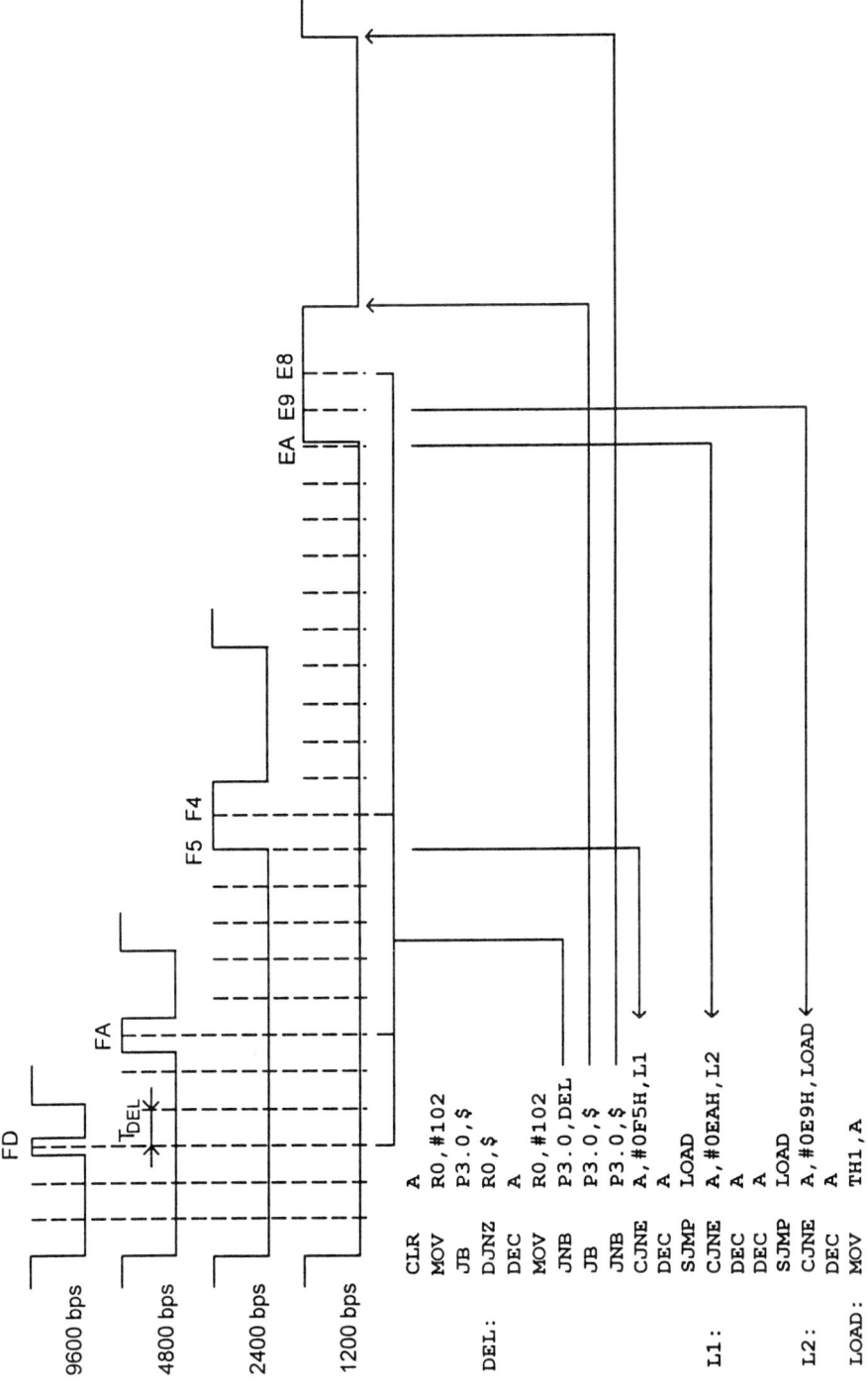

Figure 6.4 Recognizing the baud rate.

You might think we are ready, but we are not. We ran into a problem recognizing both baud rates 2400 and 1200 bps. It stems from the fact that the time quantum T_{DEL} is less than half bit time for baud rates 2400 and 1200 bps. We work around this problem by comparing the code in the accumulator with F5H, EAH and E9H. If a match is found, we correct the value. For example, if the final decrement left E9H in the accumulator, it is replaced by E8H.

Now we are back on the **SEND** and **REC** subroutines. Both pieces of code execute tight loops to test the flags. This approach is termed polled I/O. Let us estimate the performance which is lost, while the microcontroller is waiting for a set flag. For example, if the baud rate is 9600 bps and the serial port runs in mode 1, one frame will last

$$10\,\text{bit} \times \frac{1}{9600\,\text{bps}} = 1042\,\mu s$$

Therefore, we have ample reserve to increase the performance with a more efficient approach. We can use the interrupt system and leave the task of checking the flags to the built-in hardware.

Problem 6.3

Write an initialization part and an interrupt subroutine which has the following features:
• Each received byte is transmitted back twice.
• The device at the other end of the connection waits for the double echo before transmission.
• A frame includes a start bit, 8 data bits and a stop bit.
• The baud rate is 62.5 Kbps and the oscillator frequency is 12 MHz.

Solution 6.3

Figure 6.5 shows the flowchart of the interrupt subroutine. Since the sequence of actions to be done depends of the current status, we often find it easiest to represent an algorithm in this style.

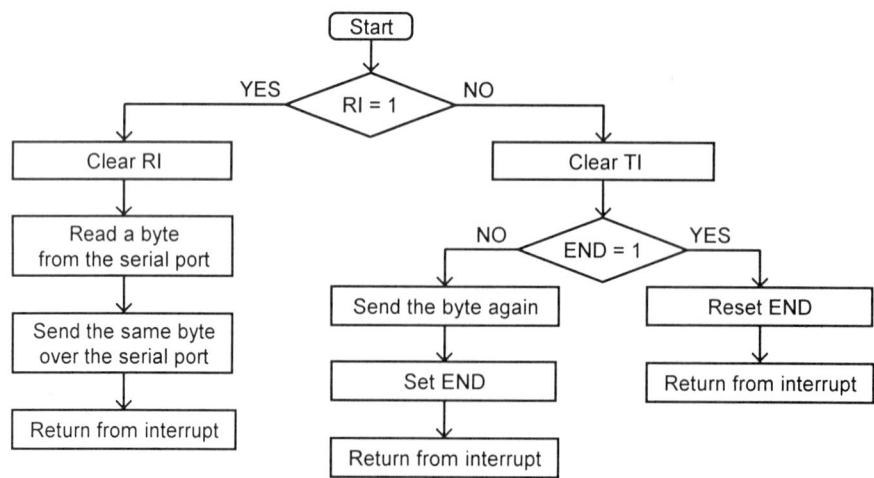

Figure 6.5 The flowchart of the D_ECHO subroutine.

One cycle includes three interrupts: one receive interrupt and two transmit interrupts. We use a flag named **END** to distinguish between the second and third interrupts.

Here is one way to solve the problem:

```
;*******************************************************
;*    D_ECHO.ASM                                      *
;*    Each received byte is transmitted back twice    *
;*    A frame includes a start bit, 8 data bits and a stop bit *
;*    The baud rate is 62.5 Kbps, oscillator frequency 12 MHz  *
;*******************************************************
TEMP      IDATA  07H
END       BIT    78H
;
          ORG    0000H
          LJMP   RESET
;
          ORG    0023H        ; Serial port interrupt
D_ECHO:   JNB    RI,D_ECHO1   ; Jump if the interrupt is due to
                              ; empty transmit buffer
          CLR    RI
          MOV    TEMP,SBUF    ; Read the receive buffer
          MOV    SBUF,TEMP    ; Transmit, first echo
          RETI
D_ECHO1:  CLR    TI
          JB     END,D_ECHO2
          MOV    SBUF,TEMP    ; Transmit, second echo
          SETB   END
          RETI
D_ECHO2:  CLR    END
          RETI
;
; Initialize the serial port
          ORG    0100H
RESET:    MOV    SCON,#01010010B ; Serial port mode 1, set REN
          ORL    TMOD,#00100000B ; Timer 1, mode 2
          ANL    TMOD,#00101111B ; Reset GATE, timer, mode 2
          ORL    PCON,#10000000B ; Set the bit SMOD
          MOV    TH1,#0FFH       ; Baud rate 62.5 Kbps,
                                 ; oscillator frequency 12 MHz
          SETB   TR1             ; Start Timer 1
;
;Initialize the interrupt system
          SETB   PS              ; High priority
          SETB   ES              ; Enable interrupts from the
                                 ; serial port
          SETB   EA              ; Enable all interrupts
;
          SJMP   $
```

The assembler directives **IDATA** and **BIT** are equivalent to **EQU**. In addition, they indicate the exact address space of concern. The instruction **SJMP** symbolizes the main program.

There is another feature of the serial port which is related to the interrupt system as well. Running in mode 2 or 3 of the serial port, the microcontroller is capable of sieving the incoming frames. Some of them will set the receive interrupt flag RI, the other will not (see Figure 2.16). Recollect that both polled and interrupt techniques employ the flag RI.

Occasionally, you want to organize a simple network based on 8051 microcontrollers. Assume that one microcontroller is a master and the rest are slaves. Only the master is entitled to initialize a transfer. In this situation, we consider two key points.

The first point here is how a master microcontroller can select a given slave. As you might expect, a specific address is dedicated to every slave. The second question is how the slave processors could balance between listening to the bus and doing their own business. This is the point where the additional bit D8 can help.

If the bit SM2 of register SCON is set and the receive register is full, the interrupt request will be activated only if the additional bit D8 is high. When the master microcontroller decides to transmit data to a slave, it first sends out an address frame with a set additional bit D8. The address frame will interrupt all slaves and they will be able to know which of them is being addressed. The addressed slave clears its bit SM2 and in this way it does not need D8 high to set the flag RI. The master processor goes on with data frames which contain low bits D8. Only the selected microcontroller is interrupted.

The buffers, which connect microcontrollers to a common bus, may be the RS-485 receivers/transmitters shown in Figure 4.25.

Also, in terms of networks built from microcontrollers, different topologies will require a different number of serial channels. The vast majority of 8051 derivatives have one serial port. However, there are exceptions, such as the DS80C320 from Dallas Semiconductor, which has two full-duplex serial ports [Dall 1994]. If the application requires two or more serial channels based on UARTs, one of the following possibilities can be considered.

• A serial interface can be organized by software. The term software UART indicates that the CPU involvement is vital. Practically, a certain amount of hardware is sacrificed as well. Most likely, one timer/counter should be used for proper timing. Plus, two I/O pins from a parallel port must be allocated for an input (RxD) and an output (TxD). Interrupt capability for the input is highly desirable. As you might expect, even moderate baud rates will require significant percentage of processor time. Also, the code size is increased considerably.

• The second choice is one or more external UARTs to be attached to the microcontroller. The drawbacks of this solution are increased price, board space and power consumption. Often, the embedded applications use a fraction of the capabilities which UARTs offer.

• Finally, the big volumes typical for embedded systems may justify the modular method for design of microcontrollers, which was mentioned in section 1.9. UART's hardware written in a specification language can be implemented in silicon together with the CPU core and other microcontroller peripherals.

6.3 Programming the personal computer serial ports

Along with the term serial port, a number of different names are being used in the PC world. Communication port (COM) or RS-232 port are yet other synonyms. Regardless of the name, the functionality is unchanged. More importantly, each PC's communication port can be programmed to interact with the 8051 serial port.

The PC is capable of running up to four communication ports which are referred to as COM1 through COM4. It is also possible for us to add some more serial ports by means of extra adapter cards. All ports are consistent with the RS-232-C standard (see also section 4.4).

According to the standard, the devices are split into two types - Data Terminal Equipment (DTE) and Data Communication Equipment (DCE). The standard specifies male connectors for DTE and female connectors for DCE.

Figure 6.6 shows a link between two devices using 9-pin connectors. In addition to the indispensable transmit, receive and ground, there are two pairs of signals, which help us to establish communication protocols.

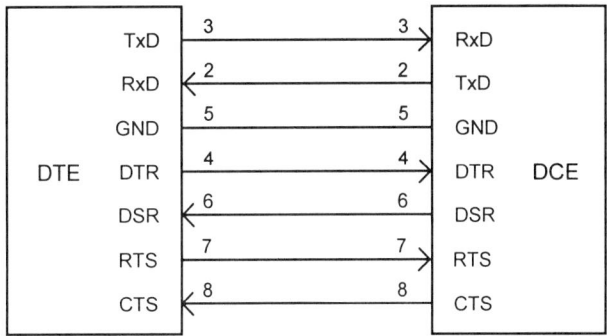

Figure 6.6 Using 9-pin connectors to implement the standard RS-232.

First, the output Data Terminal Ready (DTR), which emits positive voltage, indicates that the DTE is ready for communication. Likewise, positive voltage at the output Data Set Ready (DSR) manifests the DCE's readiness to exchange information. Second, the data flow in a certain direction is controlled by the signals Request To Send (RTS) and Clear To Send (CTS). Positive voltage at the output RTS allows the DCE to send data to the DTE. The same is valid for the output CTS, which in turn, controls the data flow from the DTE to the DCE.

Figure 6.7 shows the correspondence between 9-pin and 25-pin connectors. You can see the names of the signals in the first column as they were defined in the standard. The second column contains the names which gained broader acceptance.

RS-232-C name	Common name	Description	Pin numbers 25-pin connector	Pin numbers 9-pin connector
BA	TxD	Transmitted data	2	3
BB	RxD	Received data	3	2
CA	RTS	Request to send	4	7
CB	CTS	Clear to send	5	8
CC	DSR	Data set ready	6	6
AB	GND	Signal ground	7	5
CD	DTR	Data terminal ready	20	4

Figure 6.7 The RS-232-C major signals with respect to the Data Terminal Equipment.

Normally, the DTE is considered to be a PC and the DCE a modem. However, our goal is to connect a PC and a device, such as a single board computer based on the 8051 microcontroller. Naturally, if we link two DTE devices with a straight-through cable, they will not work as a system. The only way to tiptoe around this problem is to use a set of wires termed a cross-over cable. Figure 6.8 depicts two DTEs connected by a cross-over cable.

Furthermore, we can use a simplification which is shown as a second version in Figure 6.8. Though hardly perfect, it can be adopted in many cases. Nevertheless, if we could compensate simplified hardware by more intelligent software, it will be consistent with the embedded systems design principles.

Moreover, we can exploit the old rule "divide and conquer", which is a sound method for design and test of embedded systems. Thus, in the following chain of examples, our aim will be not only to demonstrate how the serial connection can be debugged, but also to emphasise the separate testing of the devices first.

A lucid example how a PC's communication port can be tried out is to send a character. The simplest way is to use the DOS command COPY. Let's transmit the letters A, B and C over the serial port COM2.

```
COPY CON: COM2: ↵
ABC  F6 ↵
```

When we press the key F6 or Ctrl-Z and the following carriage return, the symbols A, B and C are sent out. The oscilloscope beam can be seen to move up and down. The probe can be attached either to the connector or further in the 8051 system.

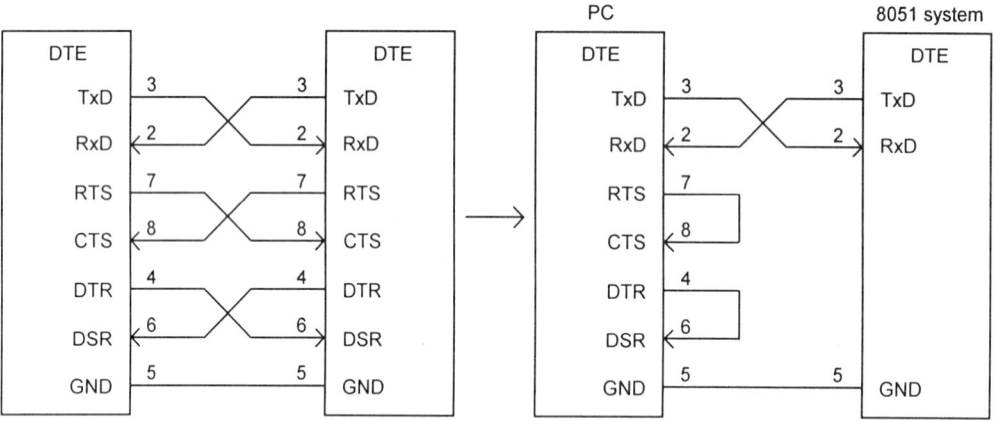

Figure 6.8 Communication between two DTE devices in case of 9-pin connectors.

The next step could be a configuration of the serial port. The DOS MODE command is a simple approach to set the serial port parameters. Occasionally, we want the port COM1 baud rate of 9600, no parity, eight data bits and one stop bit. We type the following line:

```
MODE COM1:9600,N,8,1↵
```

Personal computers contain UARTs to implement serial ports. Also, UARTs are termed Asynchronous Communications Element (ACE). The concept of interfacing UARTs to microprocessors is identical to the one we discussed for the PIO in section 4.3. Again, we are in position to program the device according to our specific requirements. While in section 6.4 we discuss the UART in more detail, and a little peek ahead will help us to deal with the following examples. Essentially, the UART has a receive buffer (or RBR) and a transmit buffer called Transmit Holding Register (THR). As you might expect, both registers occupy one address. The functionality of this register pair is identical to the 8051 registers SBUF. The register RBR must be read to get the byte arrived. The register THR must be written to transmit a byte.

A key point is that the PC's microprocessor has two address spaces: memory and I/O. The UART's registers are assigned to the I/O address space. We can streamline the serial port test using programs called debuggers. Rather than writing code for basic functions and running it, we can instruct a debugger to do the job. Typing at the keyboard is a rather slow process compared to the serial communication. Consequently, we do not have to worry if the register RBR is not full or the register THR is not empty.

For example, a character can be sent out by a DOS debugger named DEBUG. We use the command Output (O), which moves one byte to a specified I/O port.

```
DEBUG↵
-O 02F8 41↵
```

The value 02F8H is an address in the I/O space. The registers RBR and THR from the COM2 UART are assigned to this address. The value 41H is the ASCII code for the letter A. As a result, we transmit the character A through the port COM2.

Also, the debugger can be used to check how many communication ports have been installed in the PC. A piece of configuration data is stored in the RAM at address 00400H (memory address space). In the PC world, however, it is more common to represent addresses as a sum of two partly overlapped digits. In particular, the address 00400H is passed to the PC as 0040:0000. We type DEBUG and choose the function Dump (D).

```
DEBUG↵
-D 0040:0000↵
0040:0000  F8 03 F8 02 00 00 00 00
```

Thus, two serial ports are detected. The COM1 base address is 03F8H. The COM2 base address is 02F8H. You must be aware of the PC microprocessor's trait to store the low-order byte of a word in the lower of the two addresses and the high-order byte in the higher address.

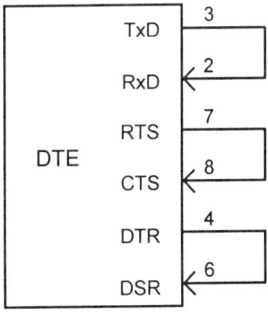

Figure 6.9 A null modem wiring for the 9-pin connector.

As soon as we make sure the PC's serial port transmits properly, we can go on with testing the reception. The most natural thing to do is to tie together the TxD and RxD pins. In this case, the cable is termed a null modem. Figure 6.9 shows the wiring of a null modem. You should be advised, however, that many people do not distinguish between a cross-over cable and a null modem.

Now, with a null modem plugged in the COM2 connector, we are back in the program DEBUG. If a character is written to the port and then the same character is read, it will prove the correctness of both transmission and reception.

```
DEBUG↵
-O 02F8 46↵
-I 02F8↵
46
```

Furthermore, you could use a terminal emulator program under Windows. It can be accessed through the Control Panel. A set of menus will guide you to specify the communication settings [Goft 1994].

At this point, it is probably natural to approach more complex examples. This could be done by using a high level language. The following is a C program for sending out characters through the serial port COM2.

```c
/******************************************************************/
/*    send_ch.c                                               */
/*    This program checks if a key has been pressed           */
/*    Displays the character                                  */
/*    Sends out the character through the serial port COM2    */
/******************************************************************/
#include <stdio.h>
#include <conio.h>
#include <bios.h>
#define COM2         1
#define COM_INIT     0
#define COM_SEND     1
#define ESC          '\x1B'

void main()
{
   char ch, data;

/*** Initialize the serial port ***/
   data = (0x03 | 0x00 | 0x00 | 0x80); /* 8 bits, 1 stop bit */
                                       /*no parity, 1200 baud */
   bioscom(COM_INIT, data, COM2);
   puts("Press any key to send it, ESC to exit");
   do
     {
     if (kbhit())
        {
        ch = getche();
        bioscom(COM_SEND,ch,COM2);        /* Send the character */
        }
     }
   while (ch != ESC);
}
```

```
int bioscom(int cmd, char ch, int port);
```

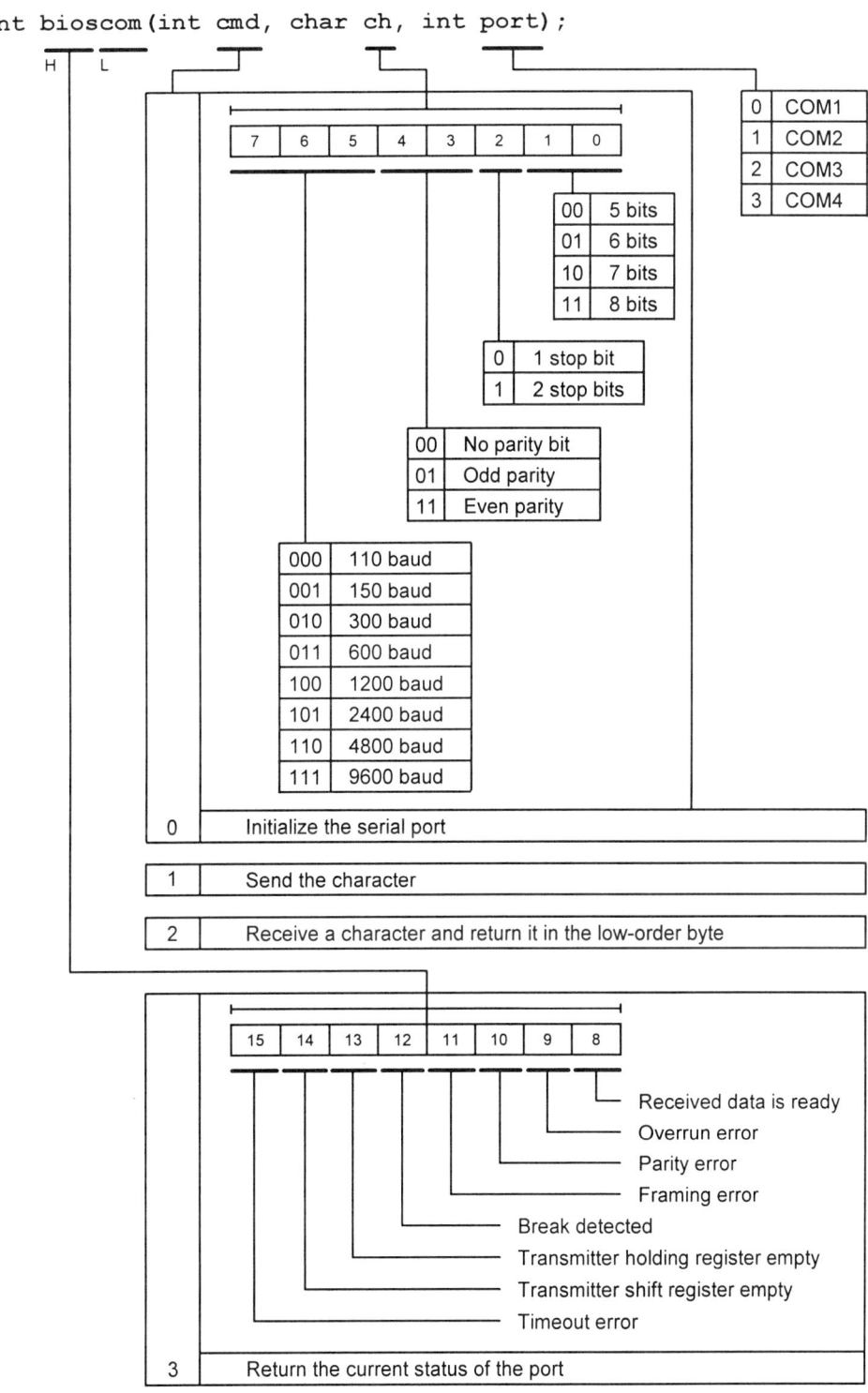

Figure 6.10 The `bioscom` function description.

The serial port is initialized and characters are sent out by the **bioscom** function. The function is based on the PC microprocessor interrupt 14H. The serial port COM2 is used in this example.

Figure 6.10 illustrates the **bioscom** syntax and the most important features. The argument **port** selects the port number. The **bioscom** function acts according to the argument **cmd** and the character **ch** is used if necessary.

• The first possibility, command 0, is used to initialize the serial port. The word length is defined by bits 0 and 1 from the parameter byte **ch**. Bit 2 is employed to choose between one and two stop bits. The parity check completes the frame setting by bits 3 and 4. Finally, bits 5 through 7 define the baud rate in the spectrum from 110 to 9600 bps. When the initializing is completed, the serial port status is reported, just as it is done for service 3.

• The next service, code 1, sends the character **ch** over the serial port **port**. On return, the low-order byte should contain the character just sent out. The high-order byte of the return value indicates error conditions. If the high-order byte is 0, then the service is successful. If the high-order byte is not 0, bit 7 reports an error. In a case of error, it is advisable to use service 3 to check what happened.

• Service 2 reads one character from the serial port. The low-order byte of the return value is the character just received. The high-order byte displays error condition by setting bit 7. Again, it is preferable to use service 3 for more details if an error has been reported.

• The complete serial port status is obtained by service 3. The high-order byte, which can be seen in Figure 6.10, represents the communication port status. The low-order byte contains the modem status and it could be added in Figure 6.10 if there is need for that.

The following example program for the 8051 microcontroller can be used to receive the characters transmitted by **send_ch.c**.

```
;************************************************************
;*    REC_CH.ASM                                          *
;*    This program receives characters from the serial port *
;*    Stores the characters in external RAM from address 2100H *
;*    The program is terminated by reset                   *
;************************************************************
;Serial port initialization
        MOV     SCON,#01010010B   ; Serial port mode 1
                                  ; set REN, set TI
        ORL     TMOD,#00100000B   ; Timer 1, mode 2
        ANL     TMOD,#00101111B   ; Reset GATE, timer, mode 2
        ANL     PCON,#01111111B   ; Reset the bit SMOD
        MOV     TH1,#0E8H         ; Baud rate 1200 bps
                                  ; Crystal 11.0592 MHz
        SETB    TR1               ; Start Timer 1
; Receive and store characters
        MOV     DPTR,#2100H
REC_CH: JNB     RI,$              ; Wait for a full SBUF
        CLR     RI                ; Reset RI
        MOV     A,SBUF            ; Read the character
        MOVX    @DPTR,A           ; Store the character
        INC     DPTR              ; Next address
        SJMP    REC_CH
```

The oscilloscope test loop, illustrated as a technique in section 4.5, can be applied for the serial port as well.

Problem 6.4

Write a C program for the PC which
• Reads the command-line parameters and use them to initialize the serial port by a bioscom function.
• Reads characters from the keyboard and transmits the last one continuously over the serial port.
• Completes its function, when ESC is entered.

Solution 6.4

```
/************************************************************/
/*    send_cyc.c                                         */
/*    This program                                       */
/*        Reads the command-line parameters to initialize */
/*          a serial port by a bioscom function          */
/*        Reads characters from the keyboard             */
/*        Transmits the last character continuously      */
/*          over the serial port                         */
/*        Exits when ESC is entered                      */
/************************************************************/
#include <stdio.h>
#include <conio.h>
#include <bios.h>
#include <process.h>
#define ESC '\x1B'
#define SPC ' '
/***    Communication port parameters    ***/
#define COM1            0
#define COM2            1
#define COM_INIT        0
#define COM_SEND        1
#define COM_BAUD_150    0x20
#define COM_BAUD_300    0x40
#define COM_BAUD_600    0x60
#define COM_BAUD_1200   0x80
#define COM_BAUD_2400   0xA0
#define COM_BAUD_4800   0xC0
#define COM_BAUD_9600   0xE0
#define COM_PAR_NONE    0x00
#define COM_PAR_ODD     0x08
#define COM_PAR_EVEN    0x18
#define COM_STOP_1      0x00
#define COM_STOP_2      0x04
#define COM_WORD_7      0x02
#define COM_WORD_8      0x03
int com=COM2;
/*** Read command-line parameters, set port parameters ***/
void init( int argc, char *argv[] )
```

```
{
  int baud   = COM_BAUD_4800;
  int parity = COM_PAR_NONE;
  int stop   = COM_STOP_2;
  int word   = COM_WORD_8;
  int i;
  for ( i=1; i<argc; i++ )
    switch ( argv[i][0] )
      {
      case '1':
        switch ( argv[i][1] )
          {
          case '2':
            baud = COM_BAUD_1200;
            break;
          case '5':
            baud = COM_BAUD_150;
            break;
          default :
            stop = COM_STOP_1;
          }
        break;
      case '2':
        switch ( argv[i][1] )
          {
          case '4':
            baud = COM_BAUD_2400;
            break;
          default :
            stop = COM_STOP_2;
          }
        break;
      case '3':
        baud = COM_BAUD_300;
        break;
      case '4':
        baud = COM_BAUD_4800;
        break;
      case '6':
        baud = COM_BAUD_600;
        break;
      case '7':
        word = COM_WORD_7;
        break;
      case '8':
        word = COM_WORD_8;
        break;
      case '9':
        baud = COM_BAUD_9600;
        break;
      case 'n':
      case 'N':
```

```
               parity = COM_PAR_NONE;
               break;
            case 'e':
            case 'E':
               parity = COM_PAR_EVEN;
               break;
            case 'o':
            case 'O':
               parity = COM_PAR_ODD;
               break;
            case 'c':
            case 'C':
               switch ( argv[i][3] )
                  {
                  case '1':
                     com = COM1;
                     break;
                  case '2':
                     com = COM2;
                     break;
                  }
               break;
            default:
            printf("Unknown command-line parameter '%s'\n",argv[i]);
            case 'h':
            case 'H':
            case '?':
               puts("Parameters :");
               puts("COM1,COM2 - port");
               puts("150,300,600,1200,2400,4800,9600 - baud rate");
               puts("N,E,O - parity");
               puts("1,2 - stop bits");
               puts("7,8 - word size");
               puts("H,? - help");
               exit(1);
            }
   bioscom(COM_INIT, ( baud | parity | stop | word ), com);
   puts("Press any key to transmit it in a loop, ESC to exit");
}
/*******************************************************************/
void main(int argc, char *argv[])
{
   char ch=SPC;
   init(argc, argv);
   do
      {
      bioscom(COM_SEND, ch, com);
      if (kbhit())
         ch=getche();
      }
   while (ch!=ESC);
}
```

6.4 The 8051 microcontroller - PC serial communications

In this section we are concerned with terminal emulator programs. A typical terminal consists of a keyboard and a screen. The terminal is connected to a computer by a serial link. A character typed at the keyboard may follow different paths to reach the screen. The first possibility is local echo. In this case, the remote computer should not echo the characters. If it does, everything typed at the keyboard will be displayed twice.

The second option is remote echo. The main motivation behind this approach is observability. If we press the key A at the keyboard and the character A is displayed on the screen, it is certain that the code of this character has travelled successfully from the terminal to the host computer and back again. The effect of any failures which influence the echo will be seen immediately.

At the same time, the modes of operation full-duplex and half-duplex, which we discussed in section 4.4, should be taken into account. For systems which operate in half-duplex mode, the option local echo might be better. In such systems, the serial communication could be a bottleneck and the local echo will mitigate the problem. Due to this relation, the term local echo has lost much of its specificity. Selecting local echo in some commercial programs in fact is setting half-duplex mode.

Even though, the full-duplex mode dominates the embedded applications, as a starting point we are going to write a program with local echo. Again, our aim is to test the communication rather than to start with a typical terminal emulator program.

Problem 6.5

> Write a C program for the PC which tests the serial communication. The program
> • Reads the command-line parameters and uses them to initialize the serial port by a bioscom function.
> • Reads characters from the keyboard, displays them on the screen (local echo) and transmits the characters over the serial port.
> • Receives characters from the serial port and prints them on the screen.
> • Completes its function, when ESC is entered.

Solution 6.5

```
/*******************************************************/
/*    test_s_r.c                                      */
/*    This program                                    */
/*       Reads the command-line parameters to initialize   */
/*          a serial port by a bioscom function       */
/*       Reads characters from the keyboard and displays   */
/*          them on the screen (local echo)           */
/*       Transmits the characters over the serial port    */
/*       Receives characters from the serial port and    */
/*          prints them on the screen                 */
/*       Exits when ESC is entered                    */
/*******************************************************/
#include <stdio.h>
#include <conio.h>
#include <bios.h>
```

```
#include <process.h>
#define ESC '\x1B'
#define SPC ' '

/*** Communication port parameters ***/
#define COM1              0
#define COM2              1
#define COM_INIT          0
#define COM_SEND          1
#define COM_RECEIVE       2
#define COM_STAT          3
#define COM_BAUD_150     (1 << 5)
#define COM_BAUD_300     (2 << 5)
#define COM_BAUD_600     (3 << 5)
#define COM_BAUD_1200    (4 << 5)
#define COM_BAUD_2400    (5 << 5)
#define COM_BAUD_4800    (6 << 5)
#define COM_BAUD_9600    (7 << 5)
#define COM_PAR_NONE     (0 << 3)
#define COM_PAR_ODD      (1 << 3)
#define COM_PAR_EVEN     (3 << 3)
#define COM_STOP_1       (0 << 2)
#define COM_STOP_2       (1 << 2)
#define COM_WORD_7        2
#define COM_WORD_8        3
#define COM_TIMEOUT       0x7000
#define COM_DATA_READY    0x0100
int com=COM2;

/*** Read command-line parameters, set port parameters ***/
void init(int argc, char *argv[])
{
   int baud   = COM_BAUD_4800;
   int parity = COM_PAR_NONE;
   int stop   = COM_STOP_1;
   int word   = COM_WORD_8;
   int i;

   for ( i=1; i<argc; i++ )
     switch ( argv[i][0] )
        {
        case '1':
           switch ( argv[i][1] )
              {
              case '2':
                 baud = COM_BAUD_1200;
                 break;
              case '5':
                 baud = COM_BAUD_150;
                 break;
              default :
                 stop = COM_STOP_1;
```

```
                   }
                break;
             case '2':
                switch ( argv[i][1] )
                    {
                    case '4':
                       baud = COM_BAUD_2400;
                       break;
                    default :
                       stop = COM_STOP_2;
                    }
                break;
             case '3':
                baud = COM_BAUD_300;
                break;
             case '4':
                baud = COM_BAUD_4800;
                break;
             case '6':
                baud = COM_BAUD_600;
                break;
             case '7':
                word = COM_WORD_7;
                break;
             case '8':
                word = COM_WORD_8;
                break;
             case '9':
                baud = COM_BAUD_9600;
                break;
             case 'n':
             case 'N':
                parity = COM_PAR_NONE;
                break;
             case 'e':
             case 'E':
                parity = COM_PAR_EVEN;
                break;
             case 'o':
             case 'O':
                parity = COM_PAR_ODD;
                break;
             case 'c':
             case 'C':
                switch ( argv[i][3] )
                    {
                    case '1':
                       com = COM1;
                       break;
                    case '2':
                       com = COM2;
                       break;
```

```
                }
            break;
        default:
        printf("Unknown command-line parameter '%s'\n",argv[i]);
        case 'h':
        case 'H':
        case '?':
            puts("Parameters :");
            puts("COM1,COM2 - port");
            puts("150,300,600,1200,2400,4800,9600 - baud rate");
            puts("N,E,O - parity");
            puts("1,2 - stop bits");
            puts("7,8 - word size");
            puts("H,? - help");
            exit(1);
        }
    bioscom(COM_INIT, ( baud | parity | stop | word ), com);
    puts("Press any key to send it, ESC to exit");
}

/********************************************************/
void main(int argc, char *argv[])
{
    int t, stat;
    char ch=SPC;
    init(argc, argv);
    do
        {
        if (kbhit())
            {
            ch=getch();
            bioscom(COM_SEND, ch, com);
            printf("   Sent      [%c] \n" , ch);
            for (t=0;t < COM_TIMEOUT ; t++);
                {
                stat=bioscom(COM_STAT,0,com);
                if (stat & COM_DATA_READY)
                    {
                    ch=bioscom(COM_RECEIVE,0,com);
                    printf("   Received [%c] \n\n" , ch);
                    break;
                    }
                if (t == COM_TIMEOUT - 1)
                    puts("  Timeout \n ");
                }
            }
        } while (ch!=ESC);
}
```

The program **test_s_r.c** could be initially checked without any connection to the environment. We plug in a null modem cable and run the program. If the result is positive, the next step is to connect the 8051 system with the following code running on it.

```
;****************************************************************
;*     ECHO.ASM                                               *
;*     This program echoes the characters received from       *
;*     the serial port                                        *
;****************************************************************
; Serial port initialization
          MOV     SCON,#01010010B   ; Serial port mode 1, set REN
          ORL     TMOD,#00100000B   ; Timer 1, mode 2
          ANL     TMOD,#00101111B   ; Reset GATE, timer, mode 2
          ANL     PCON,#01111111B   ; Reset the bit SMOD
          MOV     TH1,#0FAH         ; Baud rate 4800 bps
                                    ; Crystal 11.0592 MHz
          SETB    TR1
;
; Receive a character
REC:      JNB     RI,$              ; Wait for a full SBUF
          CLR     RI                ; Reset RI
          MOV     A,SBUF            ; Read the character
;
; Send the character back
SEND:     JNB     TI,$              ; Wait for an empty SBUF
          CLR     TI                ; Reset TI
          MOV     SBUF,A            ; Write the character
          SJMP    REC
```

Now we can easily modify the program **test_s_r.c** into a simple terminal emulator program. When a PC runs a terminal emulator program it acts like a CRT terminal. The computer reads the keyboard and sends the characters out by a serial port and then receives from the serial port and puts the characters on the screen. We remove the local echo from **test_s_r.c** which results in the following C program for terminal emulation.

```
/****************************************************************/
/*     term_1.c                                               */
/*     Terminal emulator program                              */
/*     This program                                           */
/*        Reads the command-line parameters to initialize     */
/*           a serial port by a bioscom function              */
/*        Reads characters from the keyboard and              */
/*           transmits them over the serial port              */
/*        Receives characters from the serial port and        */
/*           prints them on the screen                        */
/*        Exits when ESC is entered                           */
/****************************************************************/
#include <stdio.h>
#include <conio.h>
#include <process.h>
#include <bios.h>

#define ESC '\x1B'
#define SPC ' '
#define CR  '\r'
#define LF  '\n'
```

```c
/*** Communication port parameters ***/
#define COM1             0
#define COM2             1
#define COM_INIT         0
#define COM_SEND         1
#define COM_RECEIVE      2
#define COM_STAT         3
#define COM_BAUD_150    (1 << 5)
#define COM_BAUD_300    (2 << 5)
#define COM_BAUD_600    (3 << 5)
#define COM_BAUD_1200   (4 << 5)
#define COM_BAUD_2400   (5 << 5)
#define COM_BAUD_4800   (6 << 5)
#define COM_BAUD_9600   (7 << 5)
#define COM_PAR_NONE    (0 << 3)
#define COM_PAR_ODD     (1 << 3)
#define COM_PAR_EVEN    (3 << 3)
#define COM_STOP_1      (0 << 2)
#define COM_STOP_2      (1 << 2)
#define COM_WORD_7       2
#define COM_WORD_8       3
#define COM_DATA_READY  0x0100
#define COM_TxHOLD      0x2000
int com=COM2;

/*** Read command-line parameters, set port parameters ***/
void init(int argc, char *argv[])
{
   int baud   = COM_BAUD_4800;
   int parity = COM_PAR_NONE;
   int stop   = COM_STOP_1;
   int word   = COM_WORD_8;
   int i;

   for ( i=1; i<argc; i++ )
     switch ( argv[i][0] )
       {
       case '1':
         switch ( argv[i][1] )
           {
           case '2':
             baud = COM_BAUD_1200;
             break;
           case '5':
             baud = COM_BAUD_150;
             break;
           default :
             stop = COM_STOP_1;
           }
         break;
       case '2':
         switch ( argv[i][1] )
```

```
              {
          case '4':
             baud = COM_BAUD_2400;
             break;
          default :
             stop = COM_STOP_2;
          }
        break;
    case '3':
      baud = COM_BAUD_300;
      break;
    case '4':
      baud = COM_BAUD_4800;
      break;
    case '6':
      baud = COM_BAUD_600;
      break;
    case '7':
      word = COM_WORD_7;
      break;
    case '8':
      word = COM_WORD_8;
      break;
    case '9':
      baud = COM_BAUD_9600;
      break;
    case 'n':
    case 'N':
      parity = COM_PAR_NONE;
      break;
    case 'e':
    case 'E':
      parity = COM_PAR_EVEN;
      break;
    case 'o':
    case 'O':
      parity = COM_PAR_ODD;
      break;
    case 'c':
    case 'C':
      switch ( argv[i][3] )
        {
        case '1':
          com = COM1;
          break;
        case '2':
          com = COM2;
          break;
        }
      break;
    default:
    printf("Unknown command-line parameter '%s'\n",argv[i]);
```

```
        case '?':
        case 'H':
        case 'h':
          puts("Parameters :");
          puts("  COM1,COM2 - port ");
          puts("150,300,600,1200,2400,4800,9600 - baud rate ");
          puts("  N,E,O - parity ");
          puts("  1,2 - stop bits");
          puts("  7,8 - word size");
          puts("  ?,H - help ");
          exit(1);
        }
  bioscom( COM_INIT, ( baud | parity | stop | word ), com );
  puts("Terminal 1 \n");
  puts("Press any key to send it, ESC to exit");
}
/*****************************************************************/
void main(int argc, char *argv[])
{
  char ch=SPC;
  int stat;
  init(argc, argv);

  do
    {
    stat=bioscom(COM_STAT,0,com);
    if (stat & COM_DATA_READY)
      {
      ch=bioscom(COM_RECEIVE,0,com); /* First time it is */
                                     /* a dummy read */
      putch(ch);
      }
    if (kbhit() && (stat & COM_TxHOLD))
      {
      ch=getch();
      bioscom(COM_SEND, ch, com);
      if (ch == CR)
        bioscom( COM_SEND, LF, com );  /* Add a line feed */
      }
    } while (ch != ESC);
}
```

In the program `term_1.c` it does not matter which character is typed at the keyboard. However, there is one exception. In the case of carriage return, a line feed is added.

When we have a functioning program, such as the `term_1.c`, we often want to know how efficient the code is. Does the program exhibit satisfactory utilization of the hardware? To answer this question, we must compare the parameters of the `bioscom` function used in the program with the actual performance built in the hardware. The following points are in order:

• The BIOS serial services based on interrupt 14H can establish baud rates of up to 9600 bps. Some BIOS versions, also, provide a baud rate of 19200 bps. The BIOS services allow only the polling approach to be employed. There is no way to interrupt the PC's microprocessor when the UART receive buffer is full and an overrun is about to occur.

- The UART itself can deliver baud rates much higher than 19200 bps. Furthermore, the UART is capable of generating interrupt requests to the PC microprocessor.

The overall conclusion is that sophisticated serial communications can be organized only by direct control of the UART. The direct control demands a closer look at the UART. The ICs most commonly used in the PCs are 8250, 16450, 16550A and 16C650. The first PCs used the 8250 as an UART. In an attempt to gain the performance, the chip makers released the 16450 UART. The 16450 is an improved specification version of the 8250. Unfortunately, higher operating speeds could not be a remedy for overrun errors.

Finally, the designers understood that the single-character buffer of the UARTs was a major shortcoming. The next UART, the 16550A made a quantum leap in performance providing two 16-byte First In First Out (FIFO) buffers. One buffer acts as a receive buffer which is vital. The other buffer is a transmit buffer. In fact, the 16550A was a successor of a previous version, the 16550 UART, which was proved buggy. The modern PCs are produced by chipsets which usually include two UARTs. This is a matter of technology. The register models of the built-in UARTs remain backwardly compatible.

As we mentioned earlier in this section, two motivations are behind the direct control of the UART: baud rates higher than the ones available in the BIOS services and interrupt capabilities.

The baud rate is derived by dividing down a 1.8432 MHz clock frequency. First, the clock is divided by a divisor which is loaded into a UART register called Divisor Latch. Second, the UART performs a final divide-by-16 of the clock. Thus, the following formula gives the baud rate

$$BAUD\,RATE = \frac{1843200}{16 \times DIVISOR}$$

If we arrange a few calculations in a table (Figure 6.11), two things become obvious. First, the maximum baud rate is 112500 bps. Second, the register Divisor Latch must be a two-byte register. The designers implemented it as a read/write two-byte register.

Baud rate (bps)	110	2400	4800	9600	19200	38400	57600	112500
Divisor	0417H	0030H	0018H	000CH	0006H	0003H	0002H	0001H

Figure 6.11 Some common UART's baud rates and the corresponding divisors.

So, we could make introductory experiments initializing a serial port by the DOS function MODE and reading the actual divisor by the DEBUG service Input. However, in spite of the fact that the test is possible in principle, we must do an adjustment beforehand. The UART contains 12 registers which occupy 8 addresses. A register called Line Control Register (LCR) has a bit named Divisor Latch Access Bit (DLAB). The bit DLAB, which is the LCR.7, in combination with the address selects the UARTs registers. Figure 6.12 shows how the registers are accessed, the type of the registers and 'he addresses related to ports COM1 and COM2. The register FCR is available only in the 16550A.

The divisor value is stored in the register pair DLL and DLM. For both registers the bit DLAB must be set to allow access. We use the DOS MODE function to set baud rate of 4800 bps for the port COM2. Then we employ the program DEBUG to read the register LCR value, modify it by setting the bit DLAB, and store the register code again. We check the divisor value for 4800 bps and make a correction to obtain a baud rate of 9600 bps. We type the following lines:

```
MODE COM2:4800,N,8,1↵
DEBUG↵
-I 2FB↵
03
-O 2FB 83↵
-I 2F8↵
18
-I 2F9↵
00
-O 2F8 0C↵
```

Selecting a particular baud rate in a program can follow the scheme above. Direct control of the hardware does not necessarily mean assembly language programs. Many high level languages are capable of reading a specific port or of writing a byte to a port. For example, Borland's Turbo C/C++ language has supported the statement below since its early versions.

```
outport(0x2f8, 0x2) ;
```

Offset	DLAB	Register		Read/Write	COM1 addresses	COM2 addresses
0	0	Receive Buffer	RBR	Read	3F8H	2F8H
0	0	Transmitter Holding Register	THR	Write	3F8H	2F8H
0	1	Divisor Latch, LSB	DLL	Read/Write	3F8H	2F8H
1	0	Interrupt Enable Register	IER	Read/Write	3F9H	2F9H
1	1	Divisor Latch, MSB	DLM	Read/Write	3F9H	2F9H
2	x	Interrupt Identification Register	IIR	Read	3FAH	2FAH
2	x	FIFO Control Register	FCR	Write	3FAH	2FAH
3	x	Line Control Register	LCR	Read/Write	3FBH	2FBH
4	x	Modem Control Register	MCR	Read/Write	3FCH	2FCH
5	x	Line Status Register	LSR	Read	3FDH	2FDH
6	x	Modem Status Register	MSR	Read	3FEH	2FEH
7	x	Scratch Register	SCR	Read/Write	3FFH	2FFH

Figure 6.12 The UART registers access.

Writing a word of two bytes (0002H), we initialize the serial port COM2 to a baud rate of 57600 bps. The same result can be achieved by two output byte statements.

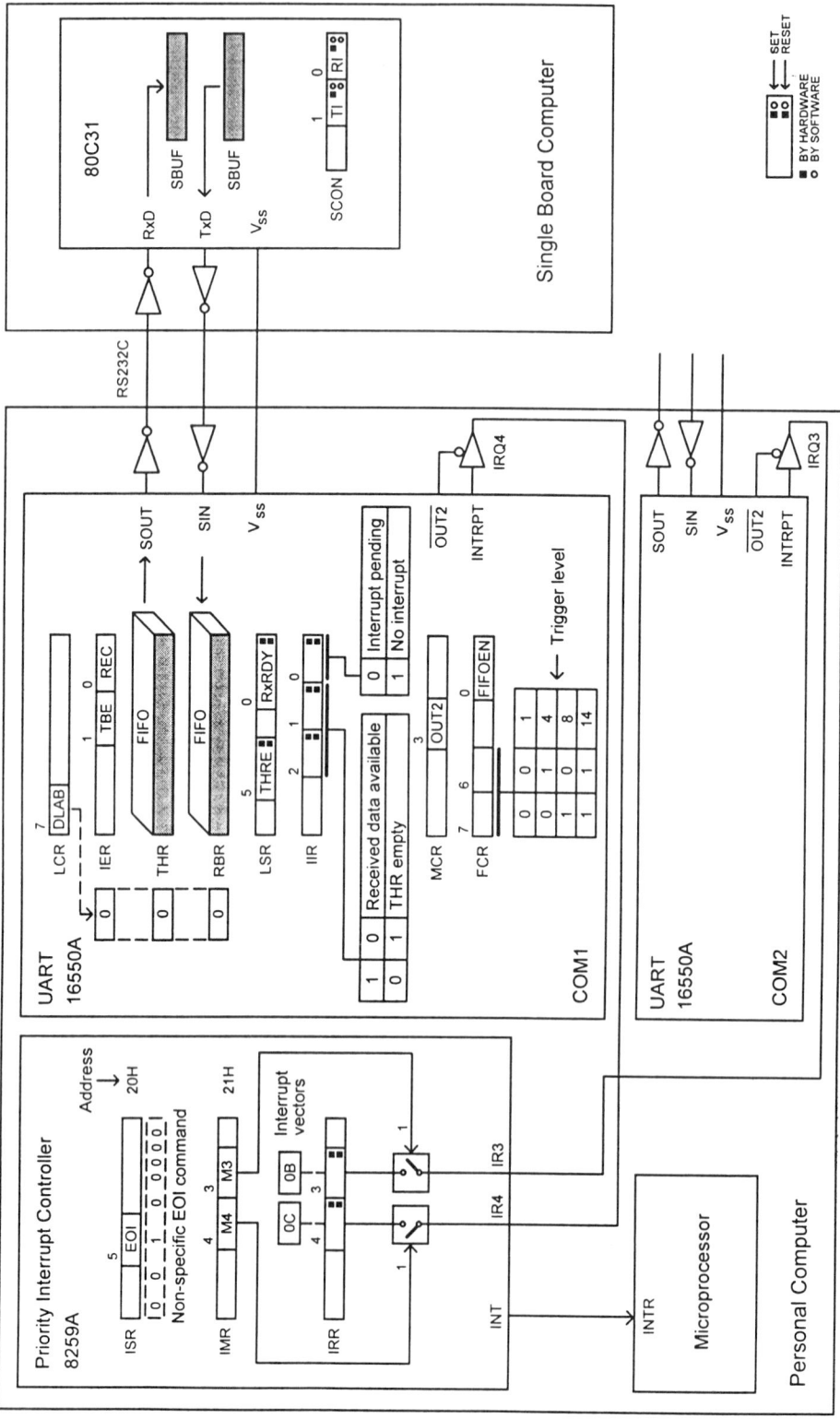

Figure 6.13 The 8051 microcontroller - PC serial communications.

```
outportb(0x2f8, 0x2) ;
outportb(0x2f9, 0x0) ;
```

As yet another example, we read from the port COM2 by executing the following statement in the C program :

```
ch = inportb(0x2f8) ;
```

The interrupts generated by the UART are related to the operation of the FIFO buffers. Figure 6.13 shows the connections between the essential components involved in the communication and a part of the internal structure. The FIFO Control Register (FCR) contains a bit named FIFO enable (FIFOEN). This bit must be set to allow FIFO operations. Furthermore, bits 6 and 7 from the same register define the receive FIFO trigger level. The trigger level is the number of characters received in the FIFO buffer which generate an interrupt request. For example, if we choose a trigger level of 8, the UART will make an attempt to interrupt the PC's microprocessor when 8 received characters are available in the FIFO buffer. Incorporating a FIFO buffer for received data and specifying a trigger level promises to minimize the probability of overrun error. However, it also creates a new threat. If the 8051 system sends 6 characters and waits for response while the trigger level has been programmed to 8, the interaction will be blocked. The designers of the 16550A overcame this problem by introducing a time-out counter. The time-out period is as long as the time taken to receive four bytes. When the microprocessor reads the register RBR within the time-out period, the time-out counter restarts. However, if the time-out period has expired and there is at least one character available in the FIFO buffer, the UART will generate an interrupt request.

When interrupted, the PC's microprocessor must identify the interrupt source. First, more than one UART can share an interrupt vector. Second, the microprocessor should determine if a character has been received or a character can be sent out. The Interrupt Enable Register (IER) possesses two flags to control the interrupts. The bit 0, named in this text as REC, enables the receive interrupt when is set. Likewise, the bit 1, called Transmitter Buffer Empty (TBE), must be set to enable transmit interrupts. For simplicity, we will not discuss interrupts caused by the modem or receiver line status. So, the microprocessor has to distinguish between different types of interrupts and to take an appropriate action.

The Interrupt Identification Register (IIR) indicates if the UART has an interrupt pending. Moreover, the IIR register points out the interrupt source. If we enabled only receive and transmit interrupts, we could go directly to the Line Status Register (LSR) to test the flags RxRDY (received data ready) and THRE (Transmitter Holding Register Empty). Both flags are very similar to the 8051's flags RI and TI. There is, however, an important difference. The 8051's flags RI and TI must be reset by software. Conversely, the UART's flags RxRDY and THRE are cleared automatically by hardware. This approach is very constructive because the flags indicate, also, when there are no more characters available in the receive FIFO buffer (reset RxRDY) and no free room in the transmit FIFO buffer (reset THRE). Consequently, a polling loop is involved in the interrupt handlers as well.

When the UART generates interrupts, the output INTRPT goes high. The last enable which the UART must provide is the $\overline{OUT2}$ output signal. The signal is asserted low by setting the bit OUT2 in the Modem Control Register (MCR). If that is the case, the three-state buffer is turned on and the interrupt request is sent to the next stage. This is a Priority Interrupt Controller (PIC). The PIC 8259A acts as a functional and input expansion buffer between the PC subsystems, including UARTs, and the microprocessor. The PIC collects interrupt requests from different sources, reacts to them according to the initialization procedure and the current situation, asserts the microprocessor interrupt input INTR and if the request is acknowledged by the microprocessor, sends an interrupt type to it. The interrupt type depends of the initialization procedure and the input asserted. The PC is organized so that an interrupt request from serial port COM1 generates interrupt type 0CH and the interrupt vector 0BH corresponds to serial port COM2.

As shown in Figure 6.13, a bit in the PIC Interrupt Request Register (IRR) can be set if only the corresponding mask bit in the Interrupt Mask Register (IMR) is cleared. The PIC priority hardware makes a decision taking into account the code in the IRR register, the way the interrupt inputs (IR3 and IR4) have been prioritized and the current situation which is displayed by the In-Service interrupt Register (ISR).

Interrupt handlers, when they are under way, normally clear the bit in the IRR register which is specific to that interrupt. It makes possible an interrupt request from the same or lower priority to be considered again. The IRR register is not program accessible and the corresponding bit has to be cleared indirectly. For instance, it can be done by a non-specific End Of Interrupt (EOI) command. The command is produced by setting the bit EOI in the ISR register and resetting all other bits in the same register. Figure 6.13, also, shows the PIC addresses in the PC microprocessor I/O address space.

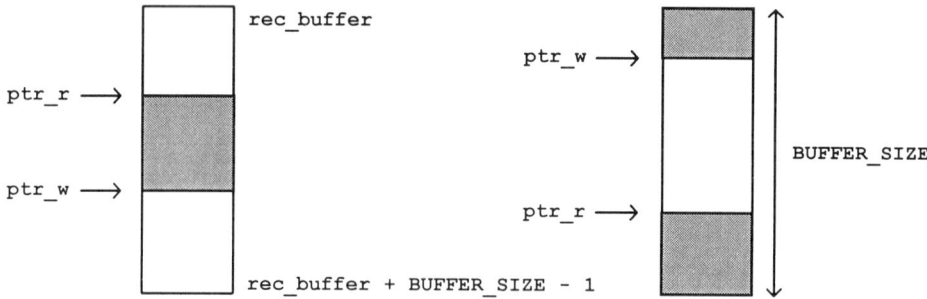

Figure 6.14 Two situations of a ring buffer.

The UART's FIFO buffers can be viewed as a mail box at the lowest hierarchical level in the serial communication. Usually, a bigger buffer is organized in the PC memory. A popular scheme is the ring or circular buffer. Figure 6.14 illustrates the operation of an example ring buffer. In fact, it is a receive ring buffer which we use in the following version of the terminal emulator program. We allocate an area in the memory with a start address **rec_buffer**. The number of the memory locations is **BUFFER_SIZE**. The actual size of the buffer is **BUFFER_SIZE** - 1. The ring buffer is manipulated by a write pointer **ptr_w** and a read pointer **ptr_r**. The used memory from the ring buffer appears shaded in Figure 6.14. Normally, an interrupt handler reads the characters from the UART and writes them to the ring buffer. Tasks from the main program can read characters from the ring buffer. In the current implementation, if the ring buffer becomes full, the incoming characters are not written until free space is available again.

When a variable called **ch** contains a character to be stored in the ring buffer, the following code can be included in the interrupt handler:

```
if (ptr_w == rec_buffer + BUFFER_SIZE - 1)
{
if (ptr_r != rec_buffer)
   {
```

```
      *ptr_w = ch;   /* Store the character in the ring buffer */
      ptr_w = rec_buffer;
      }
  }
else
  {
  if (ptr_r - ptr_w != 1)
      {
      *ptr_w = ch;   /* Store the character in the ring buffer */
      ptr_w++;
      }
  }
```

With the assumption that the main program reads from the ring buffer occasionally and prints the characters on the screen, the following loop can be organized:

```
while (ptr_r != ptr_w)
  {
  putch(*ptr_r);
  if (ptr_r == rec_buffer + BUFFER_SIZE - 1)
    ptr_r = rec_buffer;
  else
  ptr_r++;
  }
```

Now let's discuss how we can gradually move from the first version term_1.c to a more sophisticated program named term_2.c. Figure 6.15 outlines the subtasks which compose the program term_2.c. Since the design and maintenance of the program will benefit if the specification is split into smaller parts, we often find it easiest to think in this style. The subtask T4 will be accomplished by an interrupt handler.

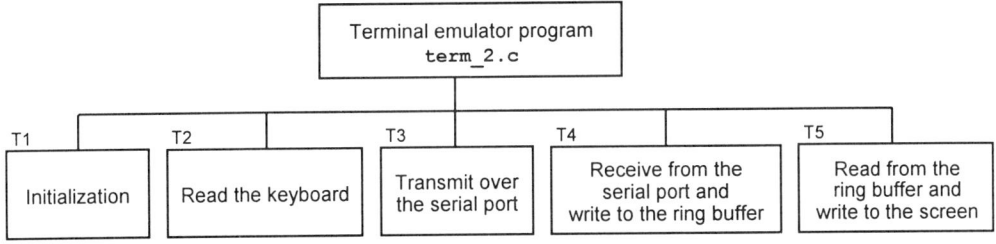

Figure 6.15 Subtasks in the terminal emulator program.

As shown in Figure 6.16, it turned out that the transition from the first to the second version of the terminal emulator program was a two step process. We start with replacing the bioscom receive function with an interrupt handler and direct read. The ring buffer is introduced as well. This version, however, is not feasible. The problem stems from the bioscom transmit function which pulls up the UART's output $\overline{\text{OUT2}}$ and disables the corresponding interrupt source. As a result, in the final version term_2.c the transmit bioscom function is replaced by direct write.

Name	Initialization	Status	Transmit	Receive	Read buffer
term_1.c	bioscom	bioscom	bioscom	bioscom	-
	bioscom	bioscom	bioscom creates the OUT2 problem	FIFO buffer Interrupts Direct read	Ring buffer
term_2.c	bioscom	bioscom	Direct write	FIFO buffer Interrupts Direct read	Ring buffer

Figure 6.16 The transition from the program `term_1.c` to the program `term_2.c`.

The terminal emulator program `term_2.c` replaces the current interrupt handler with a new handler called `rec_handler`. In the end of the program, the old handler is restored. The keyword `interrupt` serves as a qualifier for functions used as interrupt handlers. This approach has been available since early versions of Turbo C/C++.

The program appears here:

```
/*****************************************************************/
/*    term_2.c                                                   */
/*    Terminal emulator program                                  */
/*    This program                                               */
/*        Reads the command-line parameters to initialize        */
/*            a serial port by a bioscom function                */
/*        Reads characters from the keyboard and transmits       */
/*            them over the serial port by direct write          */
/*        Retrieves characters from the receive FIFO buffer      */
/*            by interrupt-driven direct read                     */
/*        Store the characters in the ring buffer                */
/*        Prints the characters on the screen                    */
/*        Exits when ESC is entered                              */
/*****************************************************************/
#include <stdio.h>
#include <conio.h>
#include <process.h>
#include <bios.h>
#include <dos.h>
#define ESC '\x1B'
#define SPC ' '
#define CR  '\r'
#define LF  '\n'

/*  If COM1, then  com = 0  */
/*  If COM2, then  com = 1  */

#define COM_DATA            (0x03F8 - com * 0x0100)
```

```
#define COM_IER              (0x03F9 - com * 0x0100)
#define COM_FCR              (0x03FA - com * 0x0100)
#define COM_LCR              (0x03FB - com * 0x0100)
#define COM_MCR              (0x03FC - com * 0x0100)
#define COM_LSR              (0x03FD - com * 0x0100)

#define PIC_COM_INT          (0x0C - com * 0x01)
#define PIC_IMR_COM_UNMASK   (0xEF + com * 0x08)
#define PIC_IMR_COM_MASK     (0x10 - com * 0x08)

#define COM_LCR_DLAB_0       0x7F
#define COM_FCR_FIFO_EN      0x01
#define COM_FCR_T_LEV        0x80
#define COM_IER_REC_EN       0x01
#define COM_MCR_OUT2_EN      0x08
#define COM_MCR_OUT2_DIS     0xF7
#define COM_LSR_RxRDY        0x01

#define PIC_ISR              0x20
#define PIC_ISR_EOI          0x20
#define PIC_IMR              0x21

#define BUFFER_SIZE          1024   /* Ring buffer size */

/*** Communication port parameters ***/
#define COM1               0
#define COM2               1
#define COM_INIT           0
#define COM_SEND           1
#define COM_RECEIVE        2
#define COM_STAT           3
#define COM_BAUD_300       (2 << 5)
#define COM_BAUD_600       (3 << 5)
#define COM_BAUD_1200      (4 << 5)
#define COM_BAUD_2400      (5 << 5)
#define COM_BAUD_4800      (6 << 5)
#define COM_BAUD_9600      (7 << 5)
#define COM_PAR_NONE       (0 << 3)
#define COM_PAR_ODD        (1 << 3)
#define COM_PAR_EVEN       (3 << 3)
#define COM_STOP_1         (0 << 2)
#define COM_STOP_2         (1 << 2)
#define COM_WORD_7         2
#define COM_WORD_8         3
#define COM_DATA_READY     0x0100
#define COM_TxHOLD         0x2000
int com = COM2;

/*** Read command-line parameters, set port parameters ***/
void init(int argc, char *argv[])
{
   int baud   = COM_BAUD_4800;
```

```
int parity = COM_PAR_NONE;
int stop   = COM_STOP_1;
int word   = COM_WORD_8;
int i;

for ( i=1; i<argc; i++ )
  switch ( argv[i][0] )
    {
    case '1':
      switch ( argv[i][1] )
        {
        case '2':
          baud = COM_BAUD_1200;
          break;
        default :
          stop = COM_STOP_1;
        }
      break;
    case '2':
      switch ( argv[i][1] )
        {
        case '4':
          baud = COM_BAUD_2400;
          break;
        default :
          stop = COM_STOP_2;
        }
      break;
    case '3':
      baud = COM_BAUD_300;
      break;
    case '4':
      baud = COM_BAUD_4800;
      break;
    case '6':
      baud = COM_BAUD_600;
      break;
    case '7':
      word = COM_WORD_7;
      break;
    case '8':
      word = COM_WORD_8;
      break;
    case '9':
      baud = COM_BAUD_9600;
      break;
    case 'n':
    case 'N':
      parity = COM_PAR_NONE;
      break;
    case 'e':
    case 'E':
```

```
                  parity = COM_PAR_EVEN;
                  break;
              case 'o':
              case 'O':
                  parity = COM_PAR_ODD;
                  break;
              case 'c':
              case 'C':
                  switch ( argv[i][3] )
                      {
                      case '1':
                          com = COM1;
                          break;
                      case '2':
                          com = COM2;
                          break;
                      }
                  break;
              default:
              printf("Unknown command-line parameter '%s'\n",argv[i]);
              case '?':
              case 'H':
              case 'h':
                  puts("Parameters :");
                  puts("  COM1,COM2 - port ");
                  puts("300,600,1200,2400,4800,9600 - baud rate ");
                  puts("  N,E,O - parity ");
                  puts("  1,2 - stop bits");
                  puts("  7,8 - word size");
                  puts("  ?,H - help ");
                  exit(1);
              }
    bioscom( COM_INIT, ( baud | parity | stop | word ), com );
    puts("Terminal 2 \n");
    puts("Press any key to send it, ESC to exit");
}

void interrupt rec_handler(void);   /* Our interrupt-driven */
                                    /* read handler */
char rec_buffer[BUFFER_SIZE], *ptr_r, *ptr_w;

/**************************************************************/
void main(int argc, char *argv[])
{
    void interrupt (*old_handler)();
    unsigned char mask;
    int stat;
    char ch;
    init(argc, argv);

/*** UART FIFO initialization ***************************/
    outportb(COM_FCR, COM_FCR_FIFO_EN | COM_FCR_T_LEV);
```

```
                /* Enable the FIFO buffer, set the trigger level */

/*** Interrupt initialization  ****************************/
ptr_r = ptr_w = rec_buffer; /* Initialize the receive buffer */
old_handler = getvect(PIC_COM_INT); /* Get the interrupt */
                                 /* vector current value */
setvect(PIC_COM_INT, rec_handler);  /* Install a new handler */
mask = inportb(PIC_IMR);            /* Read the PIC register IMR */
mask = mask & PIC_IMR_COM_UNMASK;   /* Enable interrupts in */
                                 /* PIC for the corresponding */
outportb(PIC_IMR, mask);            /* serial port */

mask = inportb(COM_LCR);       /* Prepare access to the UART */
mask = mask & COM_LCR_DLAB_0;  /* register IER, by clearing */
outportb(COM_LCR, mask);       /* the bit DLAB */
outportb(COM_IER, COM_IER_REC_EN);    /* Enable the UART */
                                   /* receive interrupt */
outportb(COM_MCR, COM_MCR_OUT2_EN);   /* Assert the UART */
                                   /* output *OUT2 */

/*** Ring buffer initialization  ************************/
ptr_r = ptr_w = rec_buffer;

/*** Read from the ring buffer and print on the screen  ***/
ch = SPC;
do
   {
   while (ptr_r != ptr_w)
     {
     putch(*ptr_r);
     if (ptr_r == rec_buffer + BUFFER_SIZE - 1)
       ptr_r = rec_buffer;
     else
     ptr_r++;
     }

/*** Read the keyboard and transmit over the serial port  ***/
   stat = bioscom(COM_STAT, 0, com);
   if (kbhit() && (stat & COM_TxHOLD))
     {
       ch = getch();
       outportb(COM_DATA,ch);
       if (ch == CR)
         {
         while ((bioscom(COM_STAT,0,com) & COM_TxHOLD) == 0)
           ;
         outportb(COM_DATA, LF);
         }
     }
   } while (ch != ESC);
```

```
/***  Mask the current port interrupt in the PIC and  ***/
                /***  restore the old interrupt handler  ***/
mask = inportb(PIC_IMR);         /* Read the PIC register IMR */
mask = mask | PIC_IMR_COM_MASK;  /* Disable interrupt in PIC */
outportb(PIC_IMR, mask);

mask = inportb(COM_MCR);                 /* Reset the bit OUT2 */
mask = mask & COM_MCR_OUT2_DIS;
outportb(COM_MCR, mask);

setvect(PIC_COM_INT, old_handler); /* Set the old handler */
}

/***  Receive from the port and write to the ring buffer  ***/
void interrupt rec_handler ()
{
   char ch;
   while (inportb(COM_LSR) & COM_LSR_RxRDY)
     {
     ch = inportb(COM_DATA);   /* Read from the UART receive */
                               /* FIFO buffer */
     if (ptr_w == rec_buffer + BUFFER_SIZE - 1)
        {
        if (ptr_r != rec_buffer)
           {
           *ptr_w = ch;   /* Store character in the ring buffer */
           ptr_w = rec_buffer;
           }
        }
     else
        {
        if (ptr_r - ptr_w != 1)
           {
           *ptr_w = ch;   /* Store character in the ring buffer */
           ptr_w++;
           }
        }
     }
   outportb(PIC_ISR, PIC_ISR_EOI);      /* Non-specific EOI   */
                  /* command which makes possible the PIC to */
                  /* respond to further interrupt requests   */
                  /* from the same or lower priority levels  */
}
```

To test the program **term_2.c** and especially its capability to receive a stream of characters, the following program, **F_ECHO.ASM**, is designed. The **F_ECHO.ASM** program receives the characters sent by the program **term_2.c** and accumulates them in the external Data Memory. When a Ctrl Z character is encountered, the bytes stored in the buffer are transmitted over the serial port as an outburst. The intention is to carry out experiments with a different number of characters.

Here is the listing:

```
;*****************************************************************
;*    F_ECHO.ASM                                                 *
;*    This program                                               *
;*       Initializes the serial port for 9600 bps,              *
;*          crystal 11.0592 MHz                                  *
;*       Receives characters from the serial port               *
;*       Stores the characters in a buffer                      *
;*          in the external Data Memory until Ctrl Z is typed   *
;*       Transmits all characters from the buffer               *
;*          over the serial port                                 *
;*****************************************************************
;
BUFFER      EQU     3000H
;
            ORG     2000H
; Serial port initialization
            MOV     SCON,#01010010B  ; Serial port, mode 1, set REN
            ORL     TMOD,#00100000B  ; Timer 1, mode 2
            ANL     TMOD,#00101111B  ; Reset GATE, timer, mode 2
            ANL     PCON,#01111111B  ; Reset the bit SMOD
            MOV     TH1,#0FDH        ; Baud rate 9600 bps
                                     ; crystal 11.0592 MHz
            SETB    TR1
;
START:      MOV     DPTR,#BUFFER     ; Initialize the pointer
WRITE:      JNB     RI,$             ; Wait for a full SBUF
            CLR     RI               ; Reset RI
            MOV     A,SBUF           ; Read the character
            MOVX    @DPTR,A          ; Write to the buffer
            CJNE    A,#1AH,INCR      ; Check for Ctrl Z
            MOV     DPTR,#BUFFER
READ:       MOVX    A,@DPTR          ; Read from the buffer
            CJNE    A,#1AH,TRANSM    ; Check for Ctrl Z
            SJMP    START
TRANSM:     JNB     TI,$             ; Wait for an empty SBUF
            CLR     TI               ; Reset TI
            MOV     SBUF,A           ; Transmit the character
            INC     DPTR
            SJMP    READ
INCR:       INC     DPTR             ; Increment the pointer
            SJMP    WRITE
```

The terminal emulator program `term_2.c` demonstrates the basic principles involved in microcontrollers - PC serial communications. This program can be easily modified or built up to satisfy the requirements of different applications.

6.5 The Universal Serial Bus

In recent years, the PC emerged as a host for a number of peripherals in both business and the home. The PC's communication ports are not able to satisfy the demands for performance and number of links. In order to solve the problem, experts from the well known companies

Compaq, Digital Equipment Corporation, IBM, Intel, Microsoft, NEC and Northern Telecom proposed a solution which allows a wide variety of peripheral devices to communicate with a PC without compromising the performance. This is the history behind the Universal Serial Bus (USB) developed in 1996.

The USB offers two rate modes:

• Low-speed mode which supports 1.5 Mbps data rate.

• Full-speed mode with a rate of 12 Mbps.

The USB specification allows up to 127 external devices to be linked to a PC acting as a host. The list of typical peripherals which were in focus includes keyboards, mice, printers, scanners, and audio and video equipment. Also, monitors are good candidates for USB devices, but in this case, the USB role is confined to transport of signals for software control of picture attributes.

Figure 6.17 outlines the basic features of the USB. The topology of the bus is a tiered star with a hub (buffer/splitter) at the center of each star. The electrical interface is based on a differential approach (see section 4.4). Two extra wires are used for power and ground. As a result, peripherals can draw power directly from the cable. The maximal length of the cable between the hubs is 5 metres. Hub extensions can be done up to seven times.

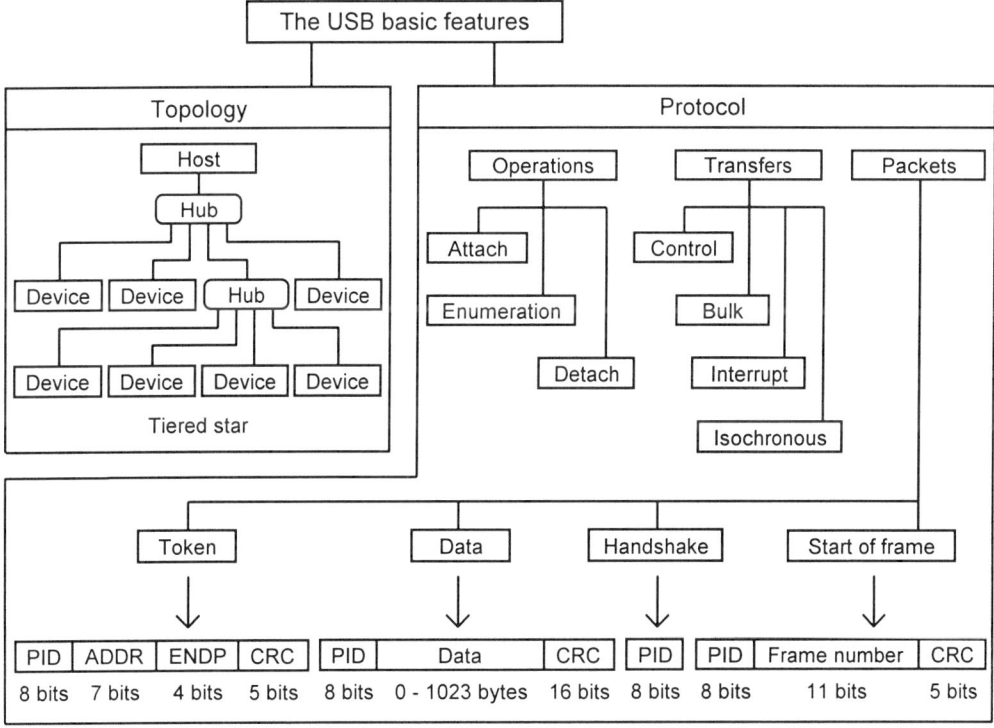

Figure 6.17 The USB profile.

A hallmark of the USB is that the user can plug in peripherals and there will be no need to switch off or reboot the PC. An attach operation starts when a device is connected to the bus. The host recognizes voltage change on the data lines and resets the device.

The second operation is termed enumeration. Once the attach operation has been completed, the host can interrogate the connected device and assign an address and configuration value to it. Finally, the host system software loads an appropriate device driver and the peripheral is ready for use. A detach operation is activated when the host recognizes a voltage change on the data lines caused by a removed device. The host system software frees up the resources related to the detached external device. Thus, a certain amount of bandwidth and power become available to other devices.

The USB standard defines a two level addressing scheme. Each device is viewed as a collection of independently functioning endpoints. Consequently, every effective address is a combination of the device address and the endpoint number. Furthermore, the USB data transfer model is built up by abstractions termed pipes. The main motivation for this approach is that timing parameters are imperative for the intended functionality. Along with the device address and endpoint number, pipes are related to bandwidth and buffer sizes. In addition, the pipes organisation helps to avoid erroneous interactions between devices (functions).

The transfers, which are supported by the USB convention, can be classified as control, bulk, interrupt and isochronous transfers. While the control transfers are based on a bidirectional communication scheme, all other transfers are associated with unidirectional flow. The control transfers are used for configuration purposes. The interrupt transfers, for example, may be an efficient way for a pointing device to deliver updated coordinates to the host. Isochronous transfers are applied for voice related peripherals, such as phones or speakers. This mode must guarantee that the delivery timing is consistent with the received timing. Both the sampling/reproducing rate and delivery delay are important. Bulk transfers are oriented to exchange of large amount of data. Typically, they support peripherals, such as printers and scanners.

The USB packets can be broken down into token, data, handshake and start of frame. A token packet includes four fields. A packet identifier (PID) determines the flow direction, OUT or IN. As you might expect, the transfers are named with regard to the host. A 7-bit address field (ADDR) follows the identifier. Since the address 0 is associated with a reset device, 127 unique addresses can be assigned during the enumeration phase. The endpoint field (ENDP) supports up to 16 endpoints for full speed devices and 2 for low speed devices. The last field in the token packet, Cyclic Redundancy Check (CRC) field protects the packet (except the PID field) against single and double bit errors. The PID field has its own embedded check pattern which includes four bits (one's complements of the actual bits).

Similarly, the data packet can carry up to 1023 bytes supported by a 16-bit CRC field. The status of the transaction is reported by handshake packets. Finally, there is a packet which helps the devices to work synchronously with the host and to keep track of the time. The start of frame (SOF) packet is generated by the host every one millisecond (1 KHz frequency). Thus, the SOF packets underlie the bus clock.

The USB is based on Non-Return-to-Zero (NRZ) data encoding. Each high level is represented by no change in the outgoing stream. Each low level results in a change. However, before the data is NRZI encoded, bit stuffing is implemented. A low is injected after every continuous stream of six high bits. Consequently, a transition is guaranteed at least once every seven bit times. This approach allows the receiver to regenerate the clock directly from the signal.

Typically, microcontrollers are used in the USB hubs and peripherals. A microcontroller capable of communicating in a USB network is termed a USB microcontroller. An example of that type of devices is Intel's 83930AE microcontroller. The main features of this 8051 compatible microcontroller are available in Appendix A.

6.6 Supplementary problems

Problem S6.1
Initialize the serial port of the 8051 microcontroller to meet the following conditions:
- A frame includes a start bit, 8 data bits, an additional bit (D8) and a stop bit.
- The baud rate is 375 Kbps.
- The microcontroller runs at oscillator frequency 12 MHz.
- A byte is ignored if a valid stop bit was not received.

Problem S6.2
Assume that the serial port of the 8051 microcontroller has been initialized for mode 3. Modify the subroutine **SEND** (section 6.2) to make it responsible for creating a parity bit. Employ the additional bit D8 to organize even parity. Also, rewrite the subroutine **REC** (section 6.2) to detect single-bit errors on the base of the additional bit D8. Use a flag to indicate if the received character is considered to be wrong.

Problem S6.3
Calculate the maximum possible oscillator frequency for 24 MHz 8051 microcontrollers which must communicate by programmable baud rates in the range of 1200 through 19200 bps.

Problem S6.4
Use the program **DEBUG** to check how many communication ports have been installed in your PC.

Problem S6.5
Apply the program **DEBUG** to verify if the UART in your PC has a scratch register.

Problem S6.6
Two single board computers, SBC1 and SBC2, are connected to a PC as shown in Figure 6.18.

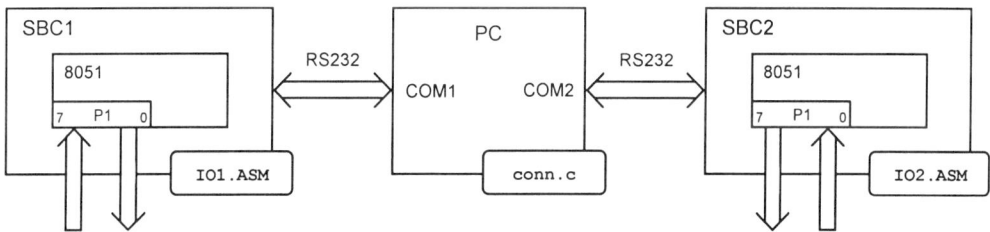

Figure 6.18 Two single board computers exchange information through a PC.

The 8051 microcontroller based in SBC1 reads the high-order nibble of Port 1 and transmits it over the serial port. The PC redirects the four bits to SBC2, where they are emitted through the high-order nibble of Port 1. Similarly, the 8051 microcontroller from SBC2 inputs the bits P1.0 - P1.3 and following the path COM2, COM1 and SBC1, the nibble appears at the lines of Port 1. Write the programs **IO1.ASM, IO2.ASM** and **conn.c** which implement this application. The programs execute endless loops. Assume that the program **conn.c** is based on the **bioscom** function and the baud rate for both communication ports is 9600 bps.

Problem S6.7
Write a C program for terminal emulation which utilizes direct control of the UART 16550A and communicates at baud rate of 112500 bps. Assume that the program controls the port COM2 and both receive and transmit interrupts are enabled.

6.7 References

Jan Axelson, *Serial Port Complete*, Lakeview Research, 1998.

Nabajyoti Barkakati, *The Waite Group's Turbo C Bible*, Howard W. Sams & Company, 1990.

Dallas Semiconductor, *High-Speed Micro User's Guide*, 1994.

Ray Duncan, *Advanced MS DOS Programming*, Microsoft Press, 1988.

Peter Fletcher, "Chip set targets The Universal Serial Bus", *Electronic Design*, November 4, 1996, pp. 69-74.

Peter W. Gofton, *Mastering Serial Communications*, SYBEX Inc., 1994.

Douglas V. Hall, *Microprocessors and Interfacing*, McGraw-Hill, 1992.

Intel, *Embedded Microcontrollers*, 1996.

Marcus Johnson, *PC Programmer's Guide to Low-Level Functions and Interrupts*, SAMS Publishing, 1994.

Robert Lafore, *The Waite Group's Turbo C Programming for the PC*, Howard W.Sams & Company, 1992.

Mark Nelson, *Serial Communications : A C++ Developer's Guide*, M&T Books, 1992.

Winn L. Rosch, *Hardware Bible*, SAMS Publishing, 1997.

Michael Tischer, *PC System Programming*, ABACUS, 1990.

Universal Serial Bus Specification, Compaq, Digital Equipment Corporation, IBM PC Company, Intel, Microsoft, NEC, Northern Telecom, Revision 1.0, 1996.

John Wettroth and Larry Suppan, "The history of the UART and the MAX3100: part 1", *Electronic Engineering*, January, 1998, pp. 20-22.

John Wettroth and Larry Suppan, "The history of the UART and the MAX3100: part 2", *Electronic Engineering*, February, 1998, pp. 24-26.

Sencer Yeralan and Ashutosh Ahluwalia, *Programming and Interfacing the 8051 Microcontroller*, Addison-Wesley Publishing Company, 1995.

Information about UARTs can be found at

    Texas Instruments   http://www.ti.com/sc/docs/msp/datatran/default.htm

    National Semiconductor   http://www.national.com/catalog/sg225.html

Information on USB is located on the Web at

                             http://www.usb.org

                             http://www.intel.com/design/usb/

                             http://www.ti.com/sc/docs/integrat/97apr/usb.htm

Information about another network standard, named FireWire, can be found at

                             http://www.apple.com

Information about a home automation standard, named CEBus, is avaivable at

                             http://www.cebus.org

Chapter 7

THE 83C552 MICROCONTROLLER

7.1 Introduction

The microcontroller manufacturers continued development and created numerous 8051-based versions. The new designs, which adopted the original 8051's instruction set, unlocked the power of sophisticated integral subsystems. At the same time, improvements appeared in the old built-in blocks as well. Figure 7.1 shows a few key tendencies which enabled the microcontrollers to boost up the total performance. In addition, Figure 7.1 indicates sample representative microcontrollers produced by Philips [Phil 1997a].

The microcontroller CPU performance can be improved by different methods. A common approach is to push up the oscillator frequency. For example, the 83C750 microcontroller can run up to 40 MHz. However, using higher oscillator frequencies is both a strong point and weakness. Inevitably, the power consumption and the radio-frequency interference are proportional to the clock speed. The other example microcontroller, the 83C508, has an extra multiply and divide block. As a result, it can perform 24-by-16 multiply and divide operations within 4 machine cycles.

New versions with more on-chip memory are candidates for powering your next embedded system design. For instance, the microcontroller 83C504 offers 16K bytes ROM and 256 bytes internal Data Memory. The internal RAM capacity determines the stack depth which is crucial for running programs written in high level languages, such as C. Furthermore, the version 83C528 expands the internal ROM to 32K bytes and the internal RAM to 512 bytes.

An example of an enhanced digital interface can be seen in the microcontroller 83C451. It has seven 8-bit I/O ports. Of course, the interaction between the microcontroller and the environment is not confined only to simple read and write operations. The timer/counters are also involved in the digital interface. Two aspects are of primary importance in that regard: the number of timer/counters available in the microcontroller and the degree of automation which links the I/O conditions and the current state of the timer/counters. For example, the 83C552 microcontroller possesses an extra timer/counter named Timer/Counter 2. The hardware implementation of this timer/counter allows the current value to be captured automatically upon an input transition. Moreover, the Philips engineers introduced compare logic around the Timer/Counter 2 for direct control of dedicated outputs. The Timer/Counter 2 operation is discussed in more detail in section 7.5.

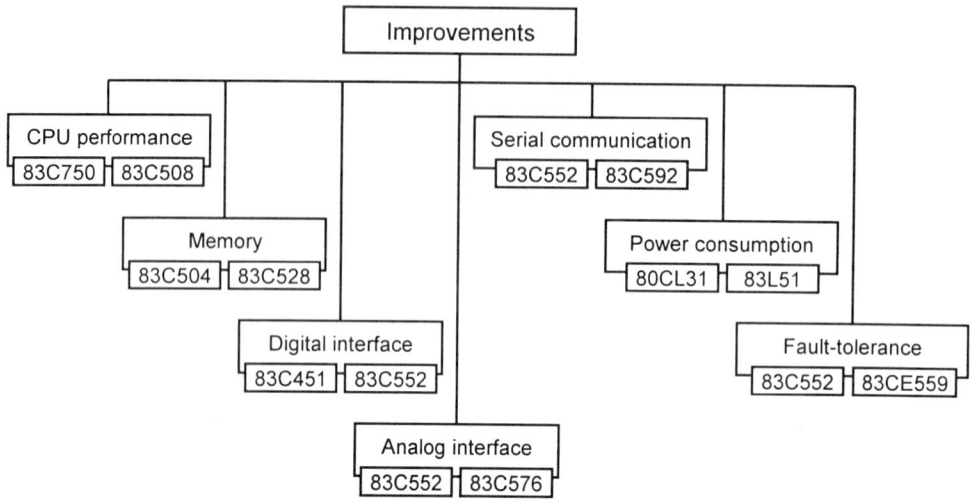

Figure 7.1 Improved 8051-based microcontrollers.

To further facilitate the interaction with the environment, the microcontrollers incorporated A-to-D converters. As is frequently the case, the underlying method of conversion is successive-approximation. The 83C552 microcontroller was equipped with an 8-input, 10-bit ADC. The microcontroller is capable of influencing the outside world by two 8-bit DACs with pulse width modulation. The 83C576 microcontroller also has a 6 channel, 10-bit ADC. In addition, the device integrates 4 analog comparators, which open opportunities for parallel process control.

The microcontroller's serial communications were developed to satisfy the demands for useful and economical embedded systems. An innovation, which was a leap ahead in this direction, is known as Inter-Integrated Circuit or I^2C bus. In fact, any microcontroller could spare two pins from a parallel port and harness software to communicate with other devices. But the radical solution is a built-in I^2C hardware block. This is the case with the 83C552 microcontroller. The fundamental principles of the I^2C bus operation and the 83C552 subsystem are discussed in Chapter 8. The I^2C serial interface in the 83C552 has been replaced with a Controller Area Network (CAN) subsystem in the 83C592 microcontroller [Phil 1997a, Phil 1997b]. The CAN bus allows transfer rates of up to 1 Mbps. Initially, the CAN bus was introduced by Bosh for the domain of vehicle designs. Currently, there is a trend for the CAN interface to link equipment as a field bus.

The power consumption is a crucial parameter for a growing segment of the embedded computers market. The designer's response to this demand was a full static microcontroller. The 80CL31 is characterized by a frequency range of 0 to 16 MHz. Furthermore, the supply voltage can vary between 1.8 and 6.0 V. The 83L51 is a CMOS microcontroller with supply voltage 3.0 to 4.5 V.

The embedded systems involved in safety-critical tasks must stay operational even in the presence of faults. This requirement conflicts with the manufacturers effort to keep the price down to a reasonable level. Often, failures occur due to noisy environment and a recovery scheme would be a sufficient and cost-effective solution. Watchdog timers, once available in the microprocessor supervisory circuits (see section 4.2), are incorporated in microcontrollers. Both microcontrollers, the 83C552 and 83CE559, can be recovered within a certain period of time by watchdog timers. Indeed, it is questionable if a reset could be an acceptable solution for all applications. Not surprisingly, the microcontroller itself is a source of noise emissions. One aspect of the design goal of the Philips engineers was to improve the ElectroMagnetic Compatibility (EMC) of the 83CE559. This was achieved by extra supply and ground pins, internal decoupling capacitance and software control of the output ALE.

In this chapter we discuss the 83C552 microcontroller. Figure 7.2 summarizes the essential features of three versions single chip microcomputers. Actually, the term single chip microcomputer is justified to a higher degree with these devices.

Device	Internal Program Memory	Internal Data Memory	Parallel ports	Analog interface	Serial interfaces
83C552	8K bytes ROM				
87C552	8K bytes EPROM	256 bytes	5	10-bit ADC 8-bit DACs	Serial port I^2C bus
80C552	---				

Figure 7.2 The 83C552 microcontrollers.

First, we deal with the architecture, memory organization and pin designations. Next, we present the 83C552 subsystems which are not available with the 8051 microcontroller. Finally, we demonstrate how to interface the microcontroller to LED and LCD displays and how to approach the design of a digital clock and a simple PLC.

7.2 Architecture

The 83C552 microcontroller internal structure is shown in Figure 7.3. As with the 8051 designation, we use the 83C552 to refer both to the ROM version and the entire family. It is certainly implied that the block 8K ROM in Figure 7.3 is replaced by 8K EPROM for the 87C552 microcontroller. The 80C552 microcontroller lacks internal Program Memory.

Along with the software compatibility, the microcontroller preserves the 8051 method to access the external memory through Port 0 and Port 2. In contrast to the 8051 microcontroller, Port 1 is not only a general purpose port anymore. The interrupt system and the new blocks I^2C interface and Timer/Counter 2 are bound to Port 1. Port 4 and Port 5 are new ports. Port 4 is related to Timer/Counter 2, while Port 5 delivers the analog inputs. Alternatively, Port 5 can be used as a general purpose input port.

While the changes of the interrupt system and the timer/counters were an evolutionary development, the analog interface and the I^2C bus can be considered revolutionary. The combination increased internal memory and analog interface made the 83C552 microcontroller an inspirational choice to people who want to take advantage of a single chip embedded system.

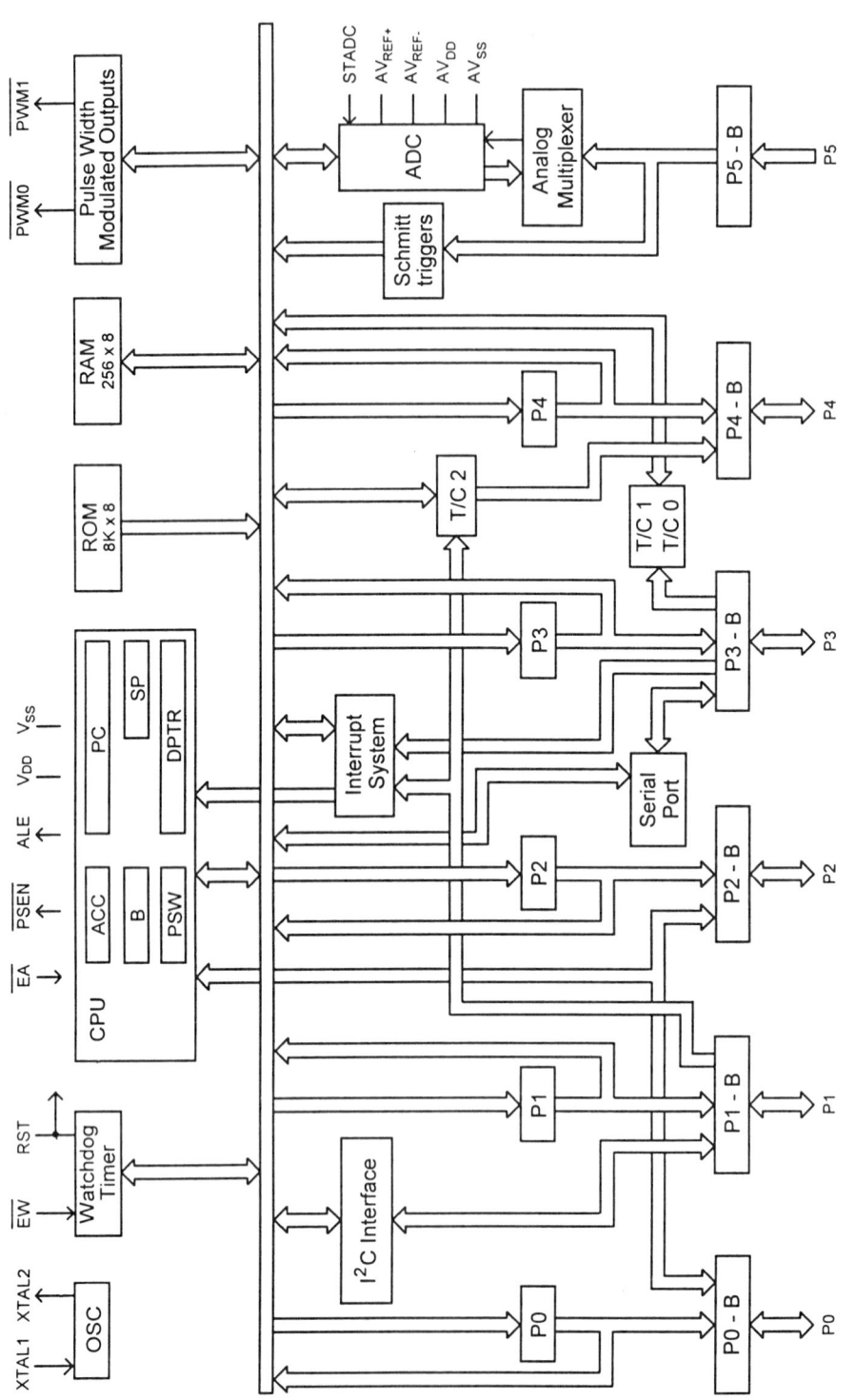

Figure 7.3 The 83C552 microcontroller architecture.

As far as the I^{2}C interface is concerned, it is a compromise between integrating more functions onto the silicon and using complex software against reducing the number of pins and simplifying the connections. So, the serial approach, employed normally over medium and long distances, was harnessed to link components in the close vicinity of the microcontroller. The price and flexibility are strong points and the designers are ready to sacrifice an area on the chip and a piece of code to take full advantage of the improved system parameters.

7.3 Memory organization

There are two differences between the memory structures of the 8051 and the 83C552 microcontrollers. First, the internal Program Memory (ROM or EPROM) is increased from 4K residing in the 8051 microcontrollers to 8K in the 83C552 devices.

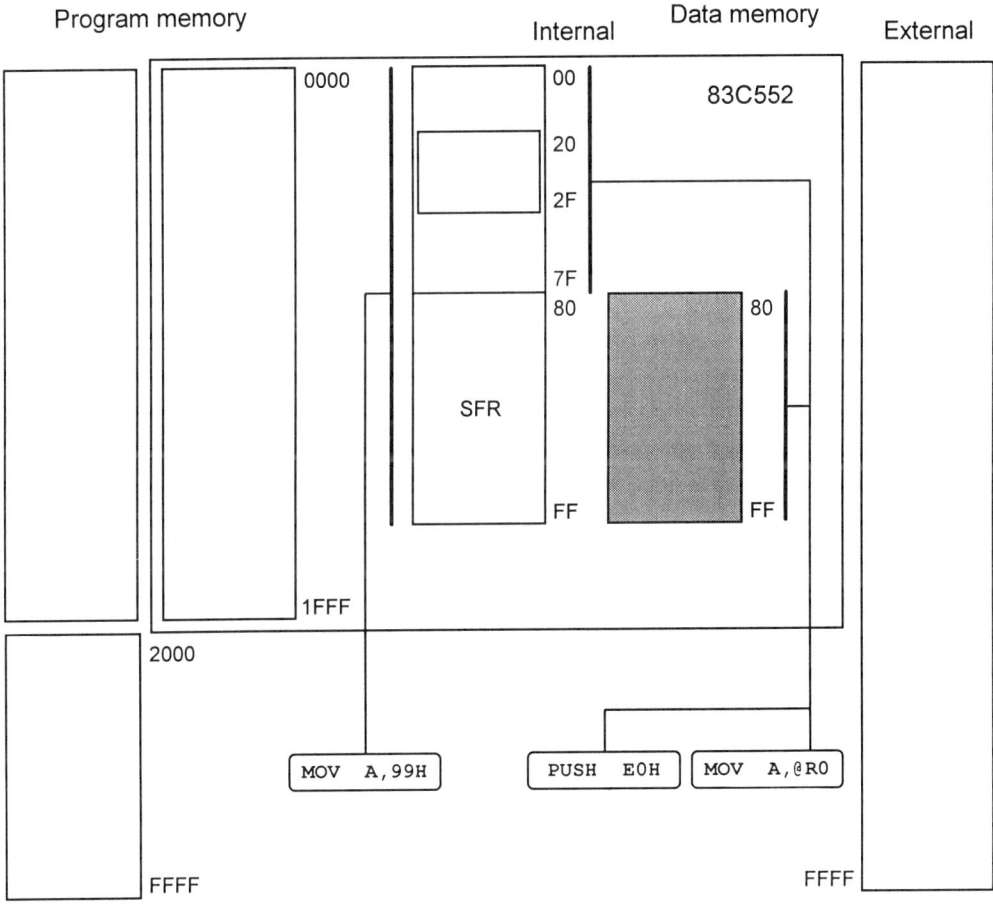

Figure 7.4 The 83C552 memory structure.

Second, the internal Data Memory is also doubled in 83C552 microcontrollers. The additional internal RAM is organized as an extra memory address space, as shown in Figure 7.4. Due to the address overlap between the internal RAM and the SFR area, the microcontroller uses only indirect addressing mode to access the internal RAM from address 80H till FFH. Consequently, each instruction deals with one or more address spaces from the list in Figure 7.5. The stack can be located anywhere in the internal RAM (except the SFR area) and the depth is up to 256 bytes.

Address space	Address range
Program Memory	0000H - FFFFH
Internal Data Memory	00H - FFH
Internal Data Memory ▨	80H - FFH
Bit-addressable area	00H - FFH
External Data Memory	0000H - FFFFH

Figure 7.5 The 83C552 microcontroller address spaces.

Naturally, some gaps in the SFR area visible in Figure 2.4 are now filled with control registers, status registers and output buffers. For example, Port 4 occupies address C0H and Port 5 address C4H.

7.4 Pin definitions and functions

While discussing the 8051 pin definitions and functions earlier in section 2.4, we used an illustration based on a standard Dual In-line Pin (DIP) package. The first Plastic Dual In-line Pin (PDIP) packages were designed in the early 1960s and for many years were the most popular packages. Ceramic Dual In-line Pin (CDIP) packages have also been an industry standard for many years. The CDIP packages are used when the device reliability is critical.

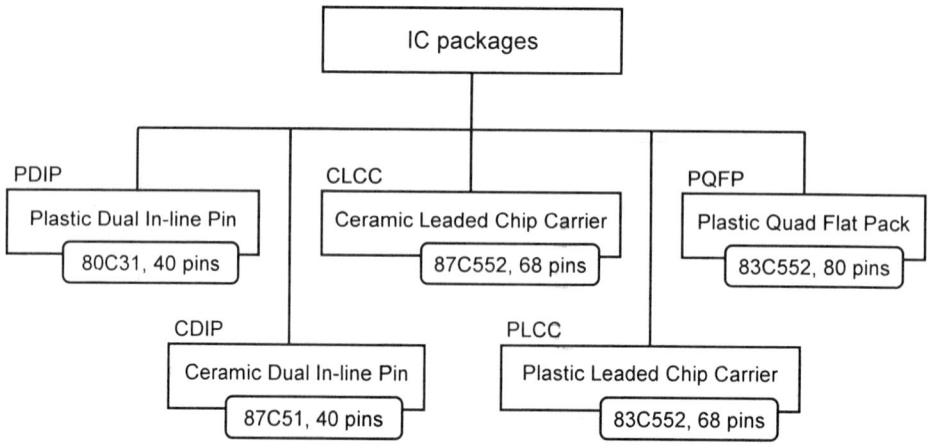

Figure 7.6 A classification for IC packages.

The 8051 microcontrollers have housings with 40 pins. But what if, for example, 68-pin microcontrollers like the 83C552 are manufactured in DIP packages ? The first drawback is that the package would be too long. The second disadvantage is that the difference between the center and the end lead lengths would result in different delays of the signals.

The compact square packages solve these problems. If you intend to use a socket for the 83C552 microcontroller, the solution would be a Ceramic Leaded Chip Carrier (CLCC) or Plastic Leaded Chip Carrier (PLCC) package. Figure 7.6 shows a classification for IC packages. Figure 7.7 depicts a PLCC package. For this package, pin #1 is against the notch. The packages can also be soldered to a printed circuit board by means of surface-mount techniques. This technology is beneficial for both reliability and price.

Another possibility, indicated in Figure 7.6, is the Plastic Quad Flat Package (PQFP) housing, which can be soldered to the printed circuit board. The 83C552 microcontroller is produced in 80-pin PQFP packages.

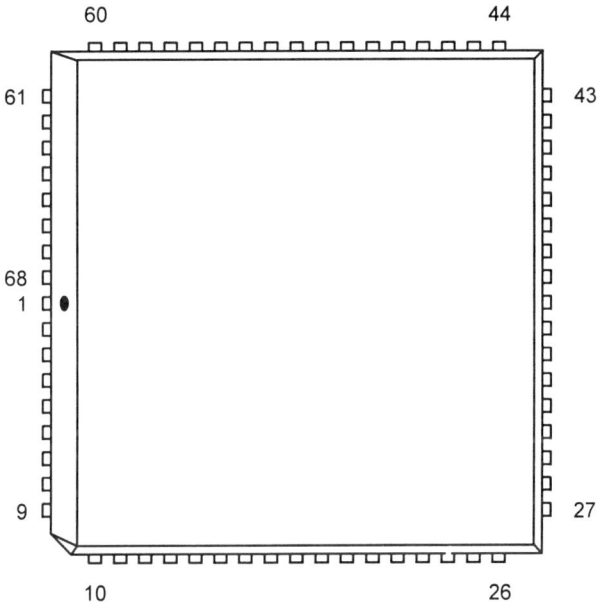

Figure 7.7 The Plastic Leaded Chip Carrier (PLCC) package.

Figures 7.8 and 7.9 show the 83C552 pin numbers for a PLCC package. Furthermore, they lay out the alternative functions and the relation pins - internal blocks. The analog signals, which are related to A-to-D interface, occupy the pins numbered from 58 through 68 and 1. Intentionally, they are grouped in this way to facilitate the Printed Circuit Board (PCB) layout. It is a sound practice to separate the analog pins by a track (a PCB wire) from the other lines. The track must go to the analog ground.

Port 0 output buffers can drive 8 LS TTL loads. As you might expect, Port 0 will require external pull-ups when used as a general purpose output. Port 1.0 to 1.5, Port 2, 3 and 4 are capable of driving 4 LS TTL inputs. The pins P1.6 and P1.7 are the I^2C bus lines and are discussed later in Chapter 8. Port 5 can be used only as an input port.

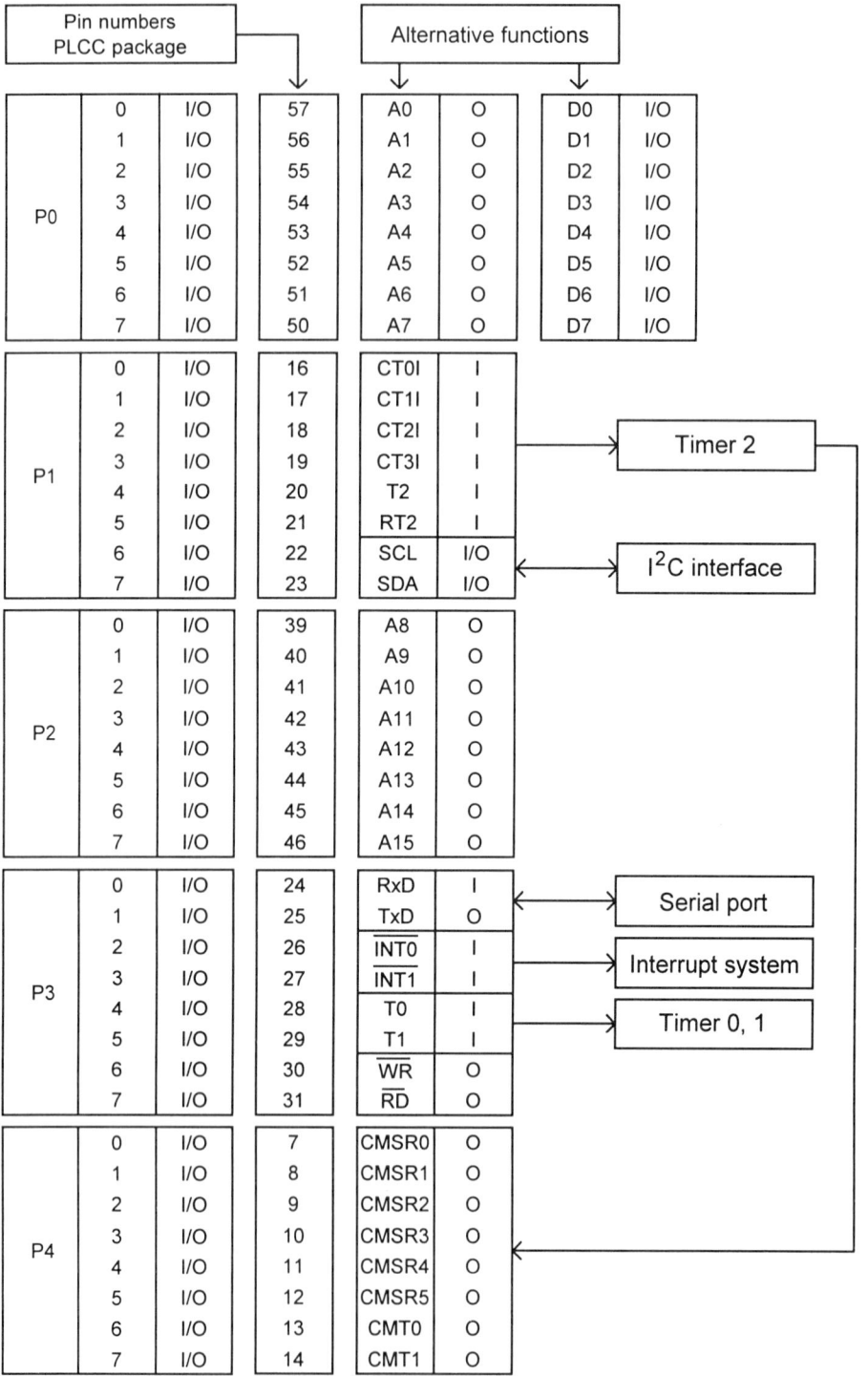

Figure 7.8 The 83C552 pin definitions and functions.

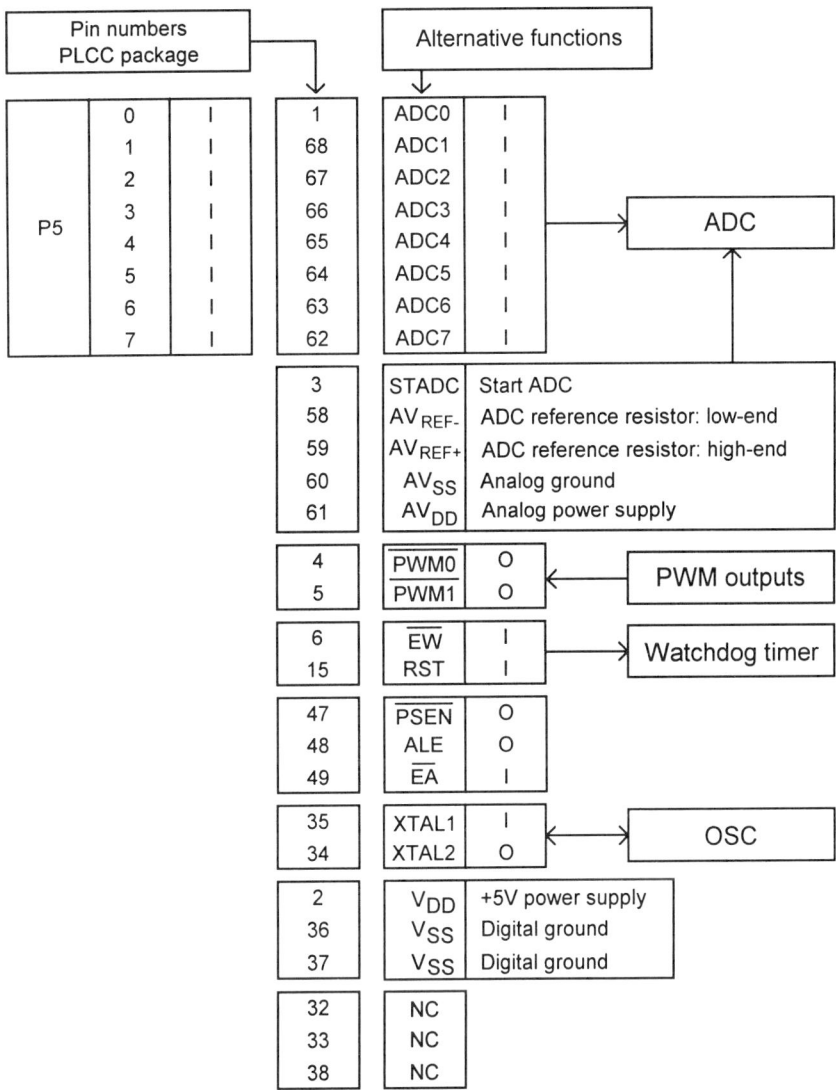

Figure 7.9 The 83C552 pin definitions and functions (continuation).

7.5 Timer/counters

The 83C552 microcontroller has four timer/counters. Timer/Counter 0 and 1 are identical to the 8051 timer/counters. Timer/Counter 3 is a Watchdog timer and it will be discussed later in section 7.6.

Timer/Counter 2 is a matured hardware block and can be viewed as a natural development of the 8051 timer/counters. Embedded computers are reactive systems. They must respond to external events in real time. Generally, the microcontroller's parallel running hardware blocks, such as the interrupt system and the timer/counters, bring the reaction time down.

There are two typical situations, as far as the timer/counters are concerned:

• The current timer/counter code is read when a transition occurs on a certain input pin.

• A particular output is either set or reset when the timer/counter code is equal to a certain value.

The 8051 microcontroller interrupts the program to process both cases. In contrast to that, the 83C552 microcontroller is capable of reacting faster and without interrupting the program.

Figure 7.10 shows the Timer 2 block diagram. Timer 2 is a 16-bit timer/counter. When it is used as a timer, the pulse source is the oscillator. The counter mode implies that Timer/Counter 2 is driven by an external signal applied to the input T2. Timer 2 can generate an interrupt request when an 8-bit or 16-bit overflow occurs. There is one interrupt vector for both cases. Timer 2 code can be moved to one of four 16-bit capture registers when a transition is sensed on inputs CT0I through CT3I. This is done automatically and an interrupt request is an extra option. Likewise, the compare logic includes three 16-bit registers. The compare registers can be used to set or reset Port 4 outputs CMSR0 through CMSR5 and toggle the outputs CMT0 and CMT1.

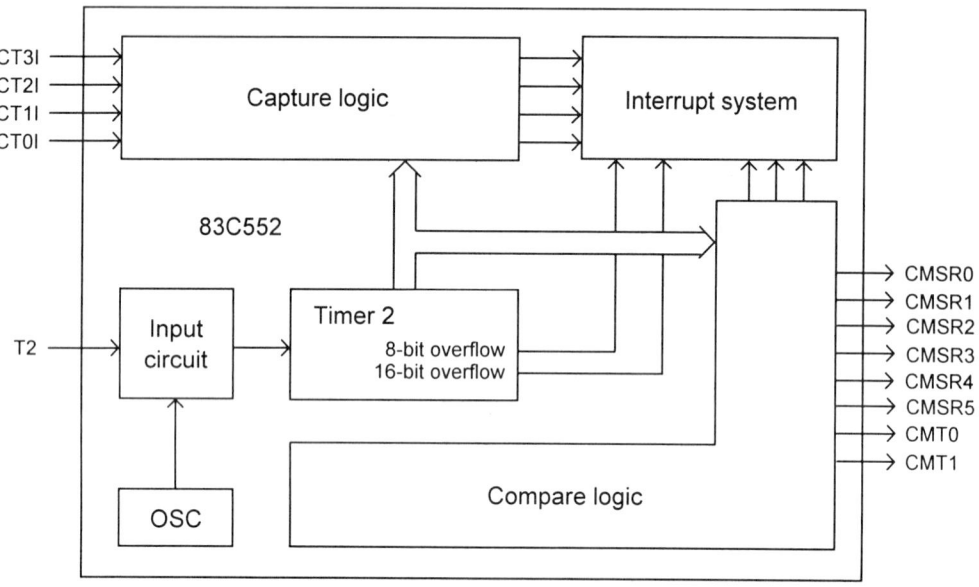

Figure 7.10 The 83C552 Timer 2 block diagram.

Figure 7.11 shows Timer 2 and the connections to the related registers. Figure 7.12 contains the complete register names, addresses and reset values. Timer 2 interrupt flag register is a bit-addressable register (*). Timer 2 consists of two registers, TMH2 (high-order byte) and TML2 (low-order byte). The Input circuit shown in Figure 7.10 includes a switch controlled by the mode selection flags T2MS1 and T2MS0 from the register TM2CON. The three possible modes can be seen in Figure 7.13. The input RT2 resets Timer 2.

When used as a counter, Timer/Counter 2 is twice faster than Timer/Counter 0 and Timer/Counter 1. The counter input T2 is sampled at S2P1 and again at S5P1. If the input T2 is low for at least 1/2 cycle and then high for at least 1/2 cycle, the rising edge will be detected. Consequently, the maximum counter input frequency is 1/12 of the oscillator frequency.

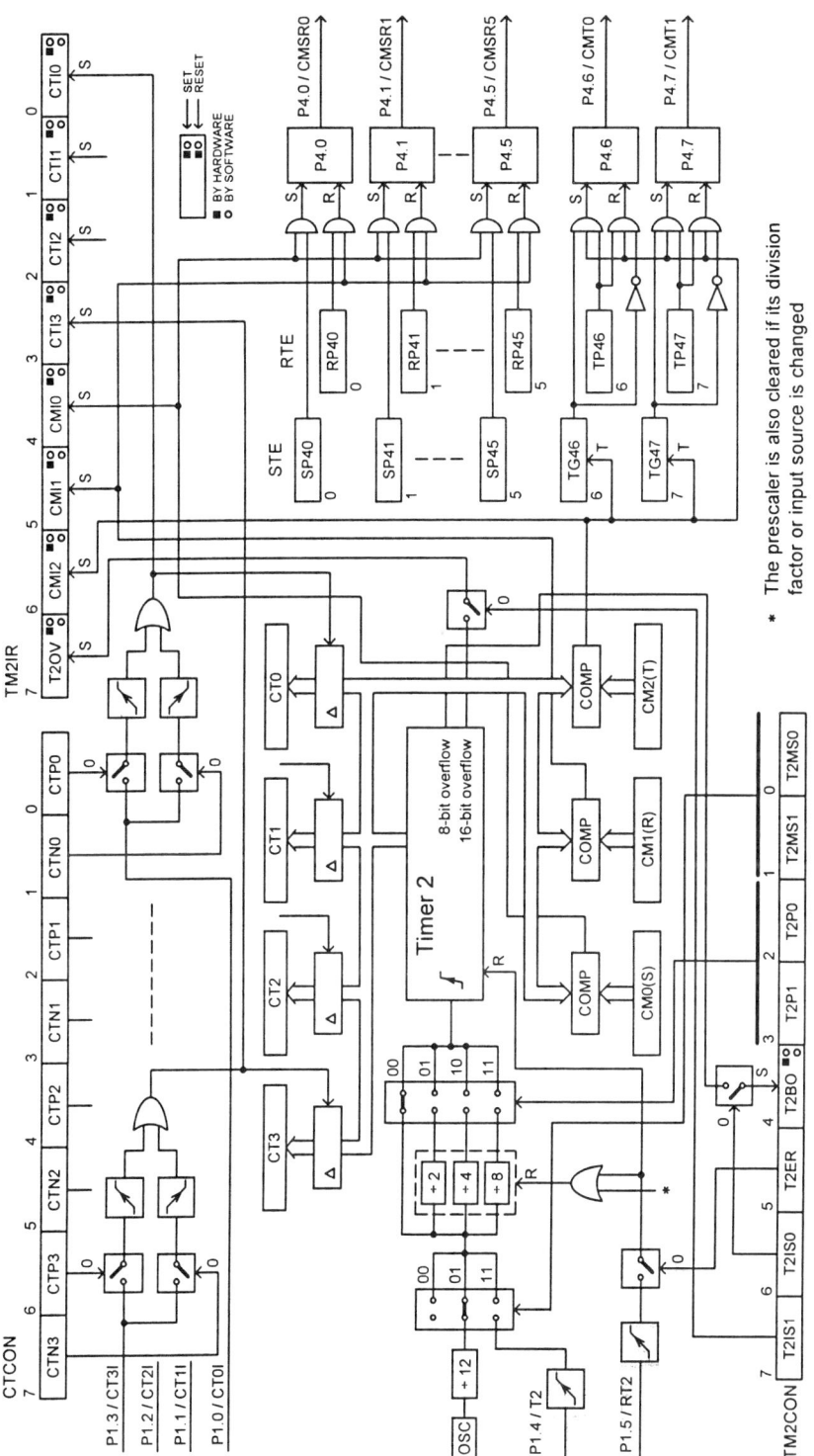

Figure 7.11 Timer 2 and related registers diagram.

The Input circuit also includes a prescaler. The prescaler division factor of 1, 2, 4 or 8 is defined by the flags T2P1 and T2P0. For example, if both flags T2P1 and T2P0 are set, the Timer/Counter 2 clock frequency will be the source frequency divided by 8. The prescaler is cleared when the Timer/Counter 2 is reset. The prescaler is also cleared, when either the source or the division factor is changed.

Name	Address	Explanation	Reset value
TMH2	EDH	Timer high 2	00H
TML2	ECH	Timer low 2	00H
TM2CON	EAH	Timer 2 control register	00H
* TM2IR	C8H	Timer 2 interrupt flag register	00H
CTCON	EBH	Capture control register	00H
CTH3	CFH	Capture high 3	xx
CTL3	AFH	Capture low 3	xx
CTH2	CEH	Capture high 2	xx
CTL2	AEH	Capture low 2	xx
CTH1	CDH	Capture high 1	xx
CTL1	ADH	Capture low 1	xx
CTH0	CCH	Capture high 0	xx
CTL0	ACH	Capture low 0	xx
CMH2	CBH	Compare high 2	00H
CML2	ABH	Compare low 2	00H
CMH1	CAH	Compare high 1	00H
CML1	AAH	Compare low 1	00H
CMH0	C9H	Compare high 0	00H
CML0	A9H	Compare low 0	00H
STE	EEH	Set enable register	C0H
RTE	EFH	Reset/Toggle enable register	00H
* P4	C0H	Port 4	FFH

Figure 7.12 Timer 2 and related registers.

To read Timer/Counter 2 "on the fly", we can use the approach discussed in section 3.10. Timer/Counter 2 is not loadable. As you might expect, Timer 2 is reset when the microcontroller is reset. Timer 2 is also reset when the RT2 input signal is changed from low to high. The RT2 reset function is enabled by setting the bit T2ER from register TM2CON. When Timer 2 is cleared by input RT2, the prescaler is also cleared.

Timer 2 could generate interrupt requests when either the low-order byte TML2 or high-order byte TMH2 overflows. The register TML2 interrupt request output (8-bit overflow) may set the byte overflow flag T2BO in register TM2CON. The flag T2BO will be set if the select bit T2IS0 is set. Likewise, the register TMH2 interrupt output (16-bit overflow) will set the interrupt flag T2OV in the register TM2IR if the select bit T2IS1 is set. Both interrupt flags must be reset by software.

The four 16-bit capture registers are CT0, CT1, CT2 and CT3. The basic idea is for Timer 2 to move its contents to a capture register. Capture functions are practically bound to external events. The four inputs CT0I through CT3I synchronize the functions with whatever happens outside. The capture control register CTCON unites four pairs of flags attached to the capture inputs. This approach allows each input to be individually programmed to trigger the code movement on a rising edge, a falling edge or either a rising or falling edge. The capture inputs are sampled during the state S1P1 of each cycle. If the result is positive, the contents of Timer 2 are captured at the end of the cycle.

As can be seen in Figure 7.11, along with the capture process interrupt flags are set. The four interrupt flags (CTI0 through CTI3) reside in Timer 2 interrupt flag register TM2IR. Thus, if the capture feature is not required, the four pins can be viewed as additional external interrupt inputs.

The second essential microcontroller's feature is the set, reset or toggle outputs control. This option meets the requirement for generating pulses over dedicated outputs. The block Compare logic in Figure 7.10 contains three 16-bit compare registers. They can be seen in Figure 7.11 as register CM0(S), SM1(R) and CM2(T). The register codes are compared with the new Timer 2 value. When the comparators (COMP) detect a match, the corresponding interrupt flag (CMI0, CMI1 or CMI2) in register TM2IR is set at the end of the following cycle.

T2MS1	T2MS0	Timer/Counter 2 mode
0	0	Halted
0	1	Timer
1	1	Counter

Figure 7.13 Timer/Counter 2 modes.

The 83C552 microcontroller provides six set/reset outputs and two toggle outputs. The use of the set/reset outputs is illustrated in Figure 7.14. The example is based on the output CMSR0. With the assumption that the flag SP40 from the set enable register STE and the flag RP40 from the reset/toggle register RTE have been set, the waveform on output CMSR0 will be the one shown in Figure 7.14. If the values in the registers CM1(R) and CM0(S) are changed and the waveforms will be changed. The same approach is used for individual control of all set/reset outputs. If we have to summarize, an output is set/reset when the corresponding flag in register STE/RTE is set and a match occurs.

Similarly, the outputs CMT0 and CMT1 can be toggled when a match with the register CM2(T) happens. The function is allowed if the corresponding bits (TP46 or TP47) in register RTE are set. In fact, the term toggle mode is completely true for the flip-flops TG46 and TG47, but not for the port latches P4.6 and P4.7. The flip-flops TG46 and TG47 are read only and they define the next state of the port latches. A set flip-flop (TG46 or TG47) indicates that the next toggle will set the port latch. A reset flip-flop displays that the following toggle will reset the port latch. Consequently, if the expected action is set, it will not be changed even if the port latch is set by software before the occurrence of the set operation.

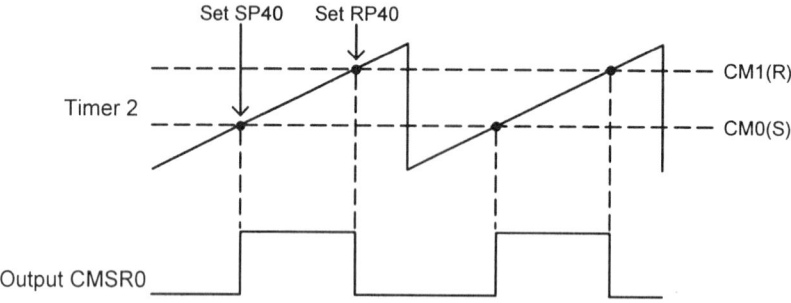

Figure 7.14 The compare set/reset function.

Finally, if equal codes are moved to registers CM0(S) and CM1(R) and the corresponding pair of flags in the registers STE and RTE are set, the port latch will be reset.

7.6 Watchdog timer

The Timer 3, or the 83C552 Watchdog timer provides a recovering opportunity, when the control over the program execution is lost. The approach is identical to the one we have already discussed in section 4.2.

Figure 7.15 attempts to show how the Watchdog timer works. The Timer 3 is an 8-bit timer which occupies the last address in the SFR area - FFH. The timer is cleared after reset. The oscillator frequency clock is divided first by 12 and second by 2048. If the oscillator frequency is 12 MHz, Timer 3 will be incremented approximately every 2 ms.

When Timer 3 overflows, the output pulse will generate internal reset. The reset pin will also be pulsed in a attempt to reset the peripheral devices, attached to the same line. However, if the reset pin is connected to a capacitor, this may not be possible. At any rate, the internal reset will be successful. The T3 output pulse is 3 machine cycles long. A capacitor 2.2 µF forms the power-on reset circuit.

When the Watchdog timer is not used, the enable input $\overline{EW}$ must be pulled high. If the input $\overline{EW}$ is brought low, it will be impossible to disable Timer 3 by software.

Let us describe how the program should avoid reset from the Watchdog timer, in other words, how to avoid overflow in Timer 3, when the program runs normally. The Timer 3 will never overflow if it is reloaded regularly. The procedure is organized in two steps in order to minimize the risk of random reload. First, the enable bit WLE in register PCON must be set. Second, the Timer 3 must be reloaded. The reload instruction will be blocked if the enable bit WLE is not set. When the Timer 3 is loaded, the enable bit WLE is cleared automatically.

The final issue about Timer 3 is its relation with the power reduction modes. Obviously, the Watchdog operation and the Power Down mode are inconsistent. The designers of the 83C552 microcontroller have protected the priority of the Watchdog timer against the Power Down mode. Once the enable input $\overline{EW}$ is pulled down, any attempt to execute the instruction

```
        ORL    PCON,#2    ; Set the bit Power Down
```

will be blocked. The logic gate which sets the flag PD symbolizes the priority of the input $\overline{EW}$ over the instruction ORL.

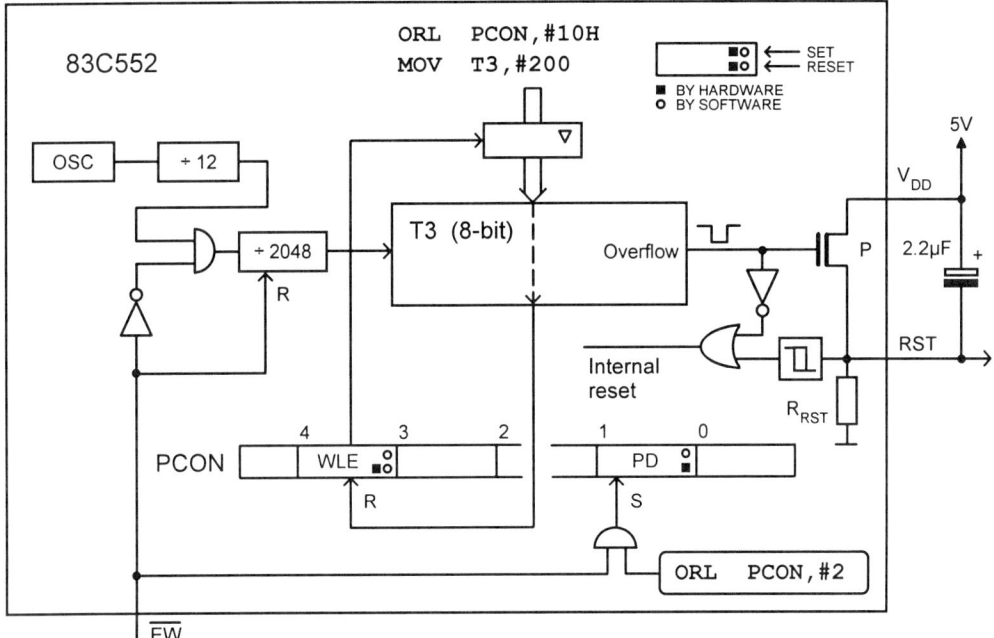

Figure 7.15 The Watchdog timer.

It is certainly implied that the Watchdog timer must be reloaded within shorter periods than the programmed interval. The effect of all possible branches, subroutines and interrupt handlers must be taken into account.

The Watchdog timer, when used, introduces fault security in the system, however some applications might demand higher degree of fault tolerance. In this case, the designers might use replication of microcontrollers and comparison of the program flows.

7.7 Analog-to-digital converter

Now, in light of the basic concepts of the analog interface discussed in Chapter 5, we deal with an actual DAS based on the 83C552 A-to-D subsystem. Admittedly, the integration of a DAS into the microcontroller chip might impose limitations on the completeness and the parameters. Also, rather than solving a specific problem the designers should predict how to support a wide area of applications.

The A-to-D interface of the 83C552 microcontroller consists of an 8-input analog multiplexer and a 10-bit ADC. In addition, two registers are used for interaction with the CPU. Figure 7.16 shows a block diagram of the subsystem.

The ADC is based on the successive-approximation method of conversion, which was discussed in section 5.4. There are two ways to start the ADC. First, this can be done by software. Second, an external event can trigger the conversion process by a dedicated input. Practically, we set or allow to be set a bit named ADCS. The start bit belongs to the register ADCON. As you can see, the flag ADCS can be set by both hardware and software. The hardware approach requires a set bit ADEX and a low to high transition on the start input STADC. As you might expect, the start input STADC must be pulled down for at least one machine cycle and then pulled up for at least one machine cycle. There is, also, another flag which must be taken into account when we start the ADC. This is the ADC interrupt flag ADCI. The flag very much resembles the receive interrupt flag RI. Again, the flag is set by hardware and must be reset by software. However, there is one difference. While the flag RI belongs to a bit-addressable register, the flag ADCI can not be processed by Boolean instructions.

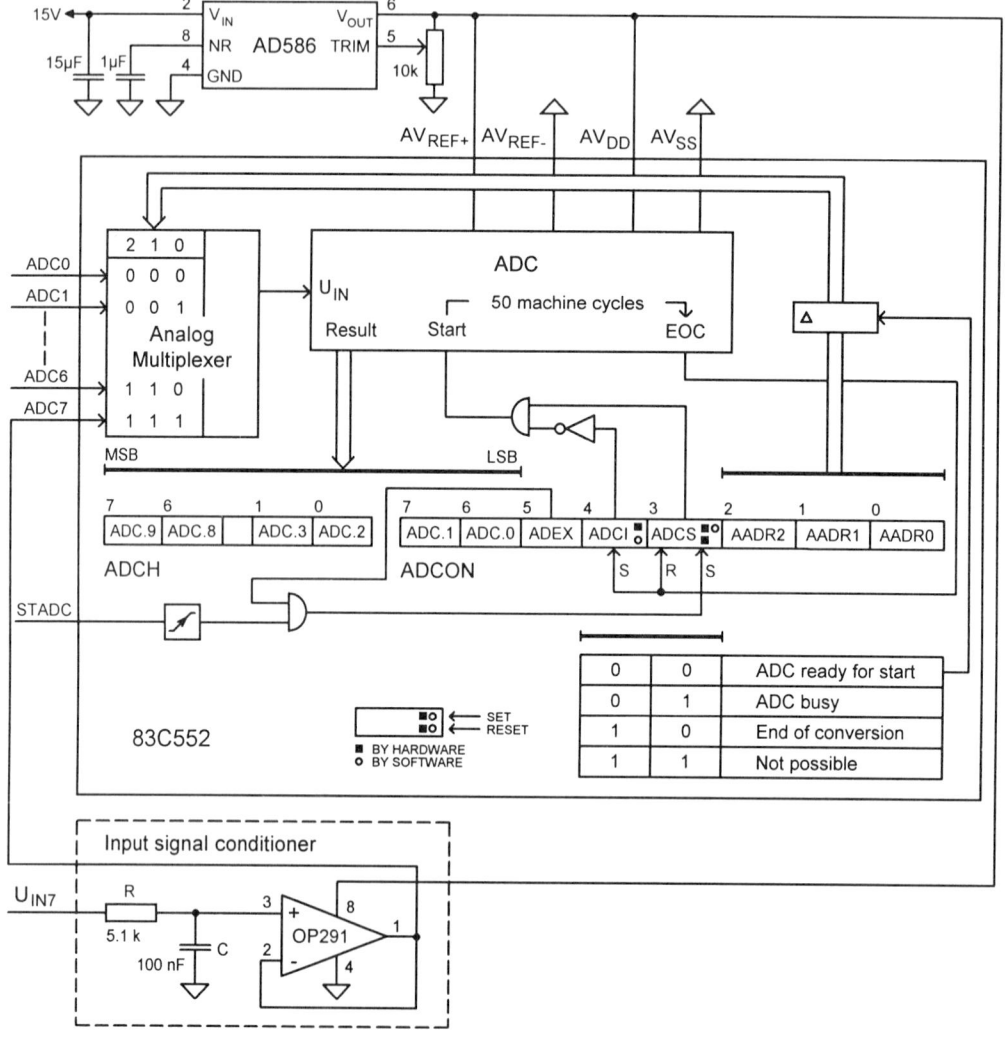

Figure 7.16 A DAS based on the built-in ADC.

The End Of Conversion (EOC) output of the ADC sets the interrupt flag ADCI. Regardless of the technique used (interrupt subroutine or polling), the flag ADCI must be reset by the program to enable a new conversion.

Figure 7.16 includes a table which shows the three possible combinations between the flags ADCI and ADCS. When both flags are low, the bits AADR0, AADR1 and AADR2 instruct the analog multiplexer which channel to select.

The A-to-D conversion takes 50 machine cycles, including 8 cycles to sample the analog input. The slew rate of the input voltage must be less than 10 V/ms. The result goes to the register ADCH (high-order bits) and to the pair of bits ADCON.7 and ADCON.6. The ADC control register ADCON occupies address C5H. The register is not bit-addressable. The register ADCH is addressed at C6H. The reset value of register ADCH and bits 6 and 7 from the register ADCON are not defined. All other bits from the register ADCON are cleared after reset. An A-to-D conversion in progress will be aborted if the Idle or Power Down mode is activated.

Two pins are used to power the ADC, AV_{DD} and AV_{SS}. The reference voltage is applied to the pins AV_{REF+} and AV_{REF-}. The ADC input range is between AV_{REF-} and AV_{REF+}. We employ an IC reference, the Analog Devices AD586 to provide the most commonly used voltage of 5 V [Anal 1992b]. The initial error of this high-precision reference depends on the model, but it can be as small as 2 mV. In addition, an external trim can be applied to adjust the output voltage either to exactly 5.000 V or to 5.12 V (see section 5.3). As a result, the ranges of all inputs will be either 0 - 5 V or 0 - 5.12 V.

Figure 7.16 includes, also, an example input signal conditioner. The conditioner consists of a low-pass filter and a buffer amplifier. The filter removes the high frequency noise introduced by electromagnetic interference. In addition, the filter can be calculated to limit the slew rate of the input voltage to the required value of 10 V/ms. The amplifier, the Analog Devices OP291 [Anal 1994], with its high input impedance guarantees practically no influence toward the signal source. Furthermore, the amplifier's low output impedance releases the error budget from additional items. In particular, the output resistance of the analog input source should be less than 2.5 kΩ. The OP291 is a dual amplifier which possesses rail-to-rail output swing.

A simple test for the 83C552 ADC can be performed by the following instructions:

```
2000 75 C5 05                  MOV  ADCON,#07H   ; Input #7, reset ADCI
2003 43 C5 08      START:  ORL  ADCON,#08H   ; Set the flag ADCS
2006 E5 C5         BUSY:   MOV  A,ADCON       ; End of conversion ?
2008 30 E4 FB              JNB  ACC.4,BUSY
200B 53 C5 E7              ANL  ADCON,#E7H    ; Reset the flag ADCI
200E 85 C6 C0              MOV  P4,ADCH       ; The pattern on Port 4
                                             ; can be observed
                                             ; by LED lamps

2011 80 F0                 SJMP START
```

For simplification, we do not take into account the bits ADC.1 and ADC.0. The instructions were assembled from address 2000H, but the resulting code is movable and can be executed from different addresses. The instructions above and a terminal emulator program can be easily modified to utilize a single board computer and a PC as a digital voltmeter.

Problem 7.1

Write a C program for the PC and an assembly language program for the 83C552 to use the built-in ADC for an 8-bit digital voltmeter.

Solution 7.1

> We adjust the ADC reference to 5.12 V to obtain a quantum of 20 mV.

```
/******************************************************/
/*    d_volt.c                                     */
/*    This program                                 */
/*      Receives the result of  conversion (8 bits)  */
/*      Displays the voltage on the screen         */
/******************************************************/
#include <stdio.h>
#include <conio.h>
#include <bios.h>
#define COM2              1
#define COM_INIT          0
#define COM_RECEIVE       2
#define COM_STAT          3
#define COM_DATA_READY    0x0100
#define ESC               '\x1B'
#define QUANTUM           0.020      /* The ADC quantum is 20 mV */

/************************************************************/
void main()
{
   char data, ch;
   int stat;
   unsigned char result;
   float voltage=0;

   clrscr();

/* Initialize the PC serial port for 8 data bits, 1 stop bit */
                            /* no parity, baud rate 4800 bps */
data = (0x03 | 0x00 | 0x00 | 0xC0);
bioscom(COM_INIT, data, COM2);
puts("    Digital voltmeter");
puts("    Range 0 to 5.12 V");
puts("    Type ESC to exit\n");
do
   {
/* Wait for 1 sec to have a stable image on the screen */
   sleep(1);
   stat=bioscom(COM_STAT,0,COM2);
   if (stat & COM_DATA_READY)
      {
      result=bioscom(COM_RECEIVE,0,COM2);
      voltage=result*QUANTUM;
      printf("    %.2f V \r", voltage);
      }
   if (kbhit())
      ch=getch();
   } while (ch !=ESC);
}
```

The following is the assembly language program for this application.

```
;*****************************************************************
;*    D_VOLT.ASM                                                *
;*    This program                                              *
;*       Converts the analog input ADC7 into digital code       *
;*       Sends the digital result over the serial port          *
;*****************************************************************
;
          ORG     2000H
          MOV     ADCON,#07H      ; Select the input ADC7
START:    ORL     ADCON,#08H      ; Start the ADC
BUSY:     MOV     A,ADCON         ; Read ADCON to test a flag
          JNB     ACC.4,BUSY      ; Test the flag ADCI
          ANL     ADCON,#0E7H     ; Reset the flag ADCI
          JNB     TI,$            ; Wait for an empty SBUF
          CLR     TI              ; Reset the flag TI
          MOV     SOBUF,ADCH      ; Send the result
          SJMP    START           ; Loop
```

Even though the 83C552 serial port is identical to the 8051·subsystem, the register names were slightly changed [Phil 1997a]. The serial buffer becomes S0BUF and the control register new name is S0CON (see Appendix C). In this way, it is possible to distinguish between both serial interfaces.

7.8 Pulse width modulated outputs

In this section the design of digital-to-analog interfaces is in focus again. As can be seen in Chapter 5, a complete DAS influences the outside world by DACs. We demonstrated D-to-A conversion examples in section 5.3. They were based on external DACs. Now, we deal with the 83C552 microcontroller built-in facilities for D-to-A conversion. Incidentally, there is a DAC integrated in the microcontroller. The DAC is constructed by a network of 1025 resistors. However, it is a block from the ADC and can not be used for analog output. The designers of the 83C552 microcontroller did not follow the way which might seem natural, to add one or two extra DACs of the same type. Instead, they provided the microcontroller with two Pulse Width Modulated (PWM) outputs. Thus, the subsystem emerges as a pure digital one. The duty cycle of the PWM outputs is maintained automatically by hardware. Once we have an efficient control over the duty cycle, we can meet the requirements of different applications. For example, if the actuator is a heater or a DC motor we repeatedly switch on and off. When the classical analog outputs are required, we introduce low-pass filters. In short, the PWM approach is a flexible one and it does not burden the microcontroller's chip with additional analog circuitry.

Figure 7.17 shows the 83C552 PWM subsystem. The outputs are named $\overline{PWM0}$ and $\overline{PWM1}$. Moreover, Figure 7.17 depicts a timing diagram for channel 1 and an example schematic diagram for output 0.

The microcontroller contains an 8-bit counter which is compared to the contents of the registers PWM0 and PWM1. The counter counts from 0 to 254 inclusive. If the contents of the counter is less than or equal to the register value, the corresponding output, $\overline{PWM0}$ or $\overline{PWM1}$, will be low. When the contents of the counter becomes greater than the register value, the corresponding output is brought high. Thus, the duty cycle of the output pulses is defined by the values stored in the registers PWM0 and PWM1. The duty cycle can vary between 0 and 1.

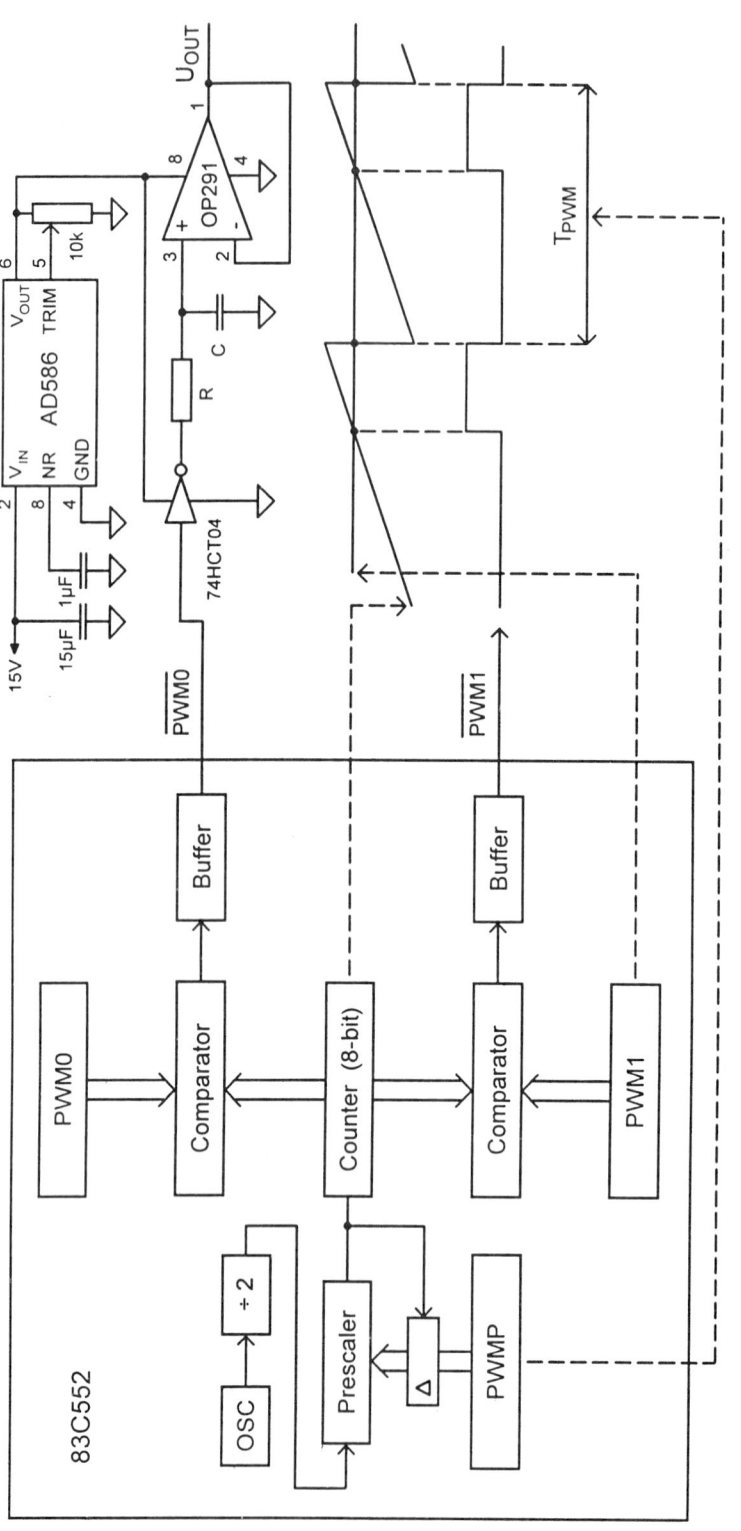

Figure 7.17 The 83C552 pulse width modulated outputs.

The counter frequency depends on the oscillator frequency and the division ratio of the prescaler. The prescaler is an 8-bit down counter, used in an auto-reload mode. The reload code is moved from the register PWMP. The repetition frequency for both outputs naturally emerges

$$f_{PWM} = \frac{f_{OSC}}{2(PWMP+1)255}$$

Let PWMP = 0, than

$$f_{PWM,MAX} = \frac{f_{OSC}}{510} = \left\{ \text{if } f_{OSC} = 12\,MHz \right\} = 23529.4\,Hz$$

Let PWMP = 255, than

$$f_{PWM,MIN} = \frac{f_{OSC}}{130560} = \left\{ \text{if } f_{OSC} = 12\,MHz \right\} = 91.9\,Hz$$

If a new code is moved to a compare register, it will change the corresponding output immediately.

Conditioned pulse width modulated outputs can be used to drive DC motors. The duty cycle of the PWM outputs controls the rotation speed of the motor.

For some other applications a PWM output can be configured as a DAC. This approach is illustrated in Figure 7.17. At this point, it is probably natural to ask why the designers organized active-low outputs. When the code in a PWM register is increased, some intuitive sense prompts that the duty cycle should be brought up too. The reason behind that is an external buffer is required to bind the output high level voltage to a precise and stable reference. If the buffer is an inverter, the expected relationship between the code in a PWM register and the duty cycle is restored.

Figure 7.18 contains the complete register names, addresses and reset values.

Name	Address	Explanation	Reset value
PWMP	FEH	Pulse width modulation prescaler	00H
PWM0	FCH	Pulse width modulation register 0	00H
PWM1	FDH	Pulse width modulation register 1	00H

Figure 7.18 Pulse width modulation registers.

The example circuit attached to channel 0 consists of a CMOS inverter, a low-pass RC filter, an operational amplifier and a voltage reference. The gate supply pins are connected to the reference voltage output V_{OUT} and the analog ground. The reference voltage can be adjusted by a potentiometer. An optional capacitor, connected to the noise reduction input NR, brings the output noise down. In most applications, the resister R and the capacitor C are in the areas of 10 to 30 kΩ and 0.47 to 10 μF.

Problem 7.2

Write a C program for the PC which
• Reads numbers from the keyboard
• Interprets them as voltages in the range 0 to 5 V, quantum 100 mV
• Sends the corresponding codes to the 83C552 microcontroller through COM2.
Write an assembly language program for the microcontroller which
• Receives the characters from the serial port
• Updates the register PWM0.

Solution 7.2

Naturally, we could solve the problem by exploiting the circuit in Figure 7.17. Then the main issue becomes the relation between the quantum and the reference voltage V_{OUT} (see section 5.3). The starting point is the equation

$$q = \frac{FS}{2^n}$$

Taking into consideration that the maximum analog output for our circuit is

$$U_{OUT,MAX} = V_{OUT} = FS - q$$

and the number of bits $n = 8$, we get

$$q = \frac{V_{OUT} + q}{2^8}$$

Therefore, the equation for the reference voltage becomes

$$V_{OUT} = 255q$$

Our aim is a quantum of 100 mV. However, if we face problems with this value, we could try smaller quanta consistent with a step of 100 mV. The first line in the table shown in Figure 7.19 reflects the resulting V_{OUT} for quantum of 100 mV. We go on with better resolution (smaller quanta) until we hit the ceiling of 255 which characterizes every 8-bit register. Consequently, we discard the 10 mV option. The other four cases are evaluated on the base of the corresponding reference voltages.

q	PWM0 code for 5 V	V_{OUT}
100 mV	50	25.500 V
50 mV	100	17.750 V
25 mV	200	6.375 V
20 mV	250	5.100 V
10 mV	500	2.550 V

Figure 7.19 The relation between the quantum and the voltage reference.

Normally, the voltage reference ICs are produced for 2.5 V, 5 V or 10 V. The option $V_{OUT} = 5.1$ V is a feasible and convenient solution. Consequently, we fix the quantum to 20 mV.

Here is the C program:

```c
/************************************************************/
/*    send_pw0.c                                         */
/*    This program                                       */
/*       Reads numbers from the keyboard                 */
/*       Calculates the corresponding PWM codes          */
/*       Transmits the codes over the serial port COM2   */
/************************************************************/
#include <stdio.h>
#include <conio.h>
#include <bios.h>

#define COM2        1
#define COM_INIT    0
#define COM_SEND    1

/**************************************************************/
void main()
{
   float voltage=0;
   char data,ch;
   data = (0x03 | 0x00 | 0x00 | 0x80);
                /* 8 bits, 1 stop bit, no parity, 1200 bps */
   bioscom(COM_INIT, data, COM2);
   puts("Please type PWM0 voltage, range 0 to 5V, quantum 100
mV");
   puts("Exit by input greater than 5");
   while(voltage<=5)
     {
     voltage=50*voltage;       /* The actual quantum is 20 mV */
                    /* The weight for 1 V is 1V/20mV = 50 */
     ch=(char)voltage;   /* Take the least significant byte */
                    /* The maximum code is 5*50=250 */
     bioscom(COM_SEND,ch,COM2);           /* Send the code */
     scanf("%f",&voltage);                /* Read the voltage */
     }
}
```

The 83C552 microcontroller executes the following program:

```
;******************************************************************
;*    REC_PW0.ASM                                              *
;*    This program                                            *
;*       Receives the characters from the serial port        *
;*       Updates the register PWM0                            *
;******************************************************************
          ORG     0000H
          LJMP    RESET
          ORG     0023H
```

```
        LJMP    SP_INT
;
        ORG     0100H
; Serial port initialization
RESET:  MOV     S0CON,#01010010B ; Serial port mode 1
                                 ; set flags REN and TI
        MOV     TMOD,#00100000B  ; Timer 1 mode 2
        ANL     PCON,#7FH        ; Reset the bit SMOD
        MOV     TH1,#0E8H        ; Choose 1200 bps
        SETB    TR1              ; Start Timer 1
;
; Interrupts initialization
        SETB    PS0              ; High priority
        SETB    ES0              ; Enable interrupts from
                                 ; the serial port
        SETB    EA               ; Enable all interrupts
;
; Pulse width modulation set-up
        MOV     PWMP,#00H        ; Repetition frequency 23.5 KHz
;
; Receive voltage code and update the register PWM0
SP_INT: JNB     RI,SP_T          ; Test for a full SBUF
        CLR     RI               ; Reset the flag RI
        MOV     PWM0,S0BUF       ; Update the register PWM0
SP_T:   JNB     TI,SP_END        ; Test for an empty SBUF
        CLR     TI               ; Reset the flag TI
        NOP                      ; You may want to transmit
                                 ; and include additional
                                 ; functions here
SP_END: RETI
```

The 83C552 flags PS0 and ES0 are functional equivalents of the 8051's flags PS and ES. Again, the change in the names was motivated by the presence of two serial interfaces.

Furthermore, we can use the parallel running Timer 2 subsystem and the Port 4 lines to create an extra PWM output.

7.9 Interrupt system

To help us understand the 83C552 interrupt system, we can use the approach, introduced in section 2.9. We divide the system into three blocks - interrupt sources, enable circuit and priority circuit.

Interrupt sources

Even though, the first block has 15 outputs, the microcontroller can react to 17 external or internal events. They are indicated in Figure 7.21 by thick lines.

The interrupt inputs P3.2/$\overline{INT0}$ and P3.3/$\overline{INT1}$, the flags TF0 and TF1, and finally the serial port flags TI and RI, are identical to the 8051 interrupt sources. In addition, the 83C552 microcontroller possesses four extra interrupt inputs - the capture timer interrupt inputs P1.0/CT0I through P1.3/CT3I (see also Figure 7.11). The inputs can be individually programmed, as illustrated in Figure 7.20 by an example input. When the programmed transition occurs on a certain input, the corresponding flag is set. The flags CTI0 through CTI3 must be cleared by software.

CTN3	CTP3	Graphical representation	Input P1.3/CT3I
0	0	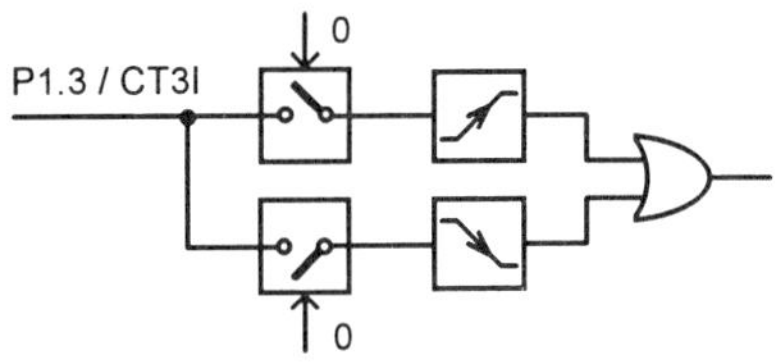 P1.3 / CT3I	Disabled
0	1	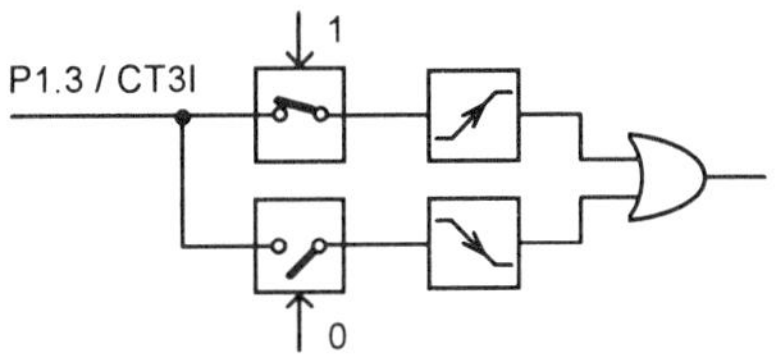 P1.3 / CT3I	Rising edge
1	0	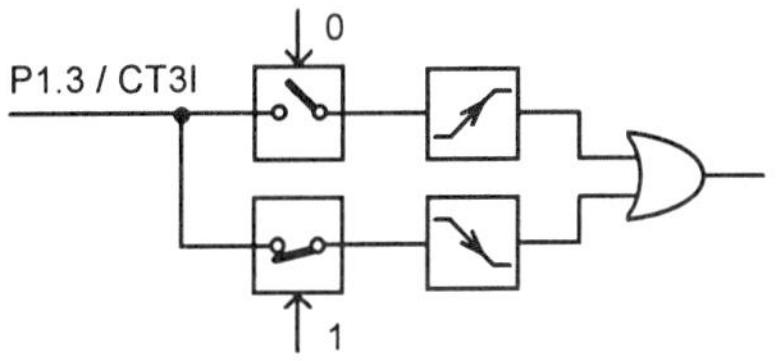 P1.3 / CT3I	Falling edge
1	1	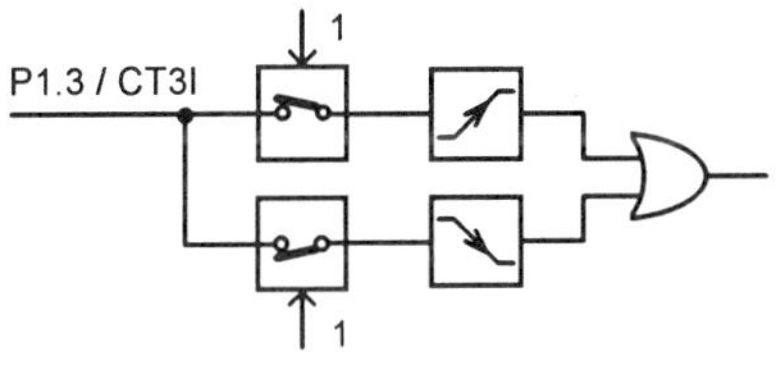 P1.3 / CT3I	Either a rising or falling edge

Figure 7.20 Programming of the input P1.3/CT3I.

The list of interrupt sources goes on with both Timer 2 overflow flags T2OV and T2BO. Furthermore, the register TM2IR also contains the compare interrupt flags CMI0 through CMI2. All flags residing in register TM2IR must be reset by software. Finally, the last two flags are the ADC interrupt flag ADCI and a flag called SI, which belongs to the I^2C bus control register S1CON. Typically, the interrupt system is activated by hardware, but the same result can be achieved by setting a flag by software. The only exception is the flag ADCI which can not be set by software.

Enable circuit

Figure 7.22 attempts to show how the enable circuit works. Each source needs an individual and a total permission. A two register control scheme and one bit occupied for the total permission naturally leads to a 15 inputs enable circuit. The enable registers are IEN0 and IEN1. As you might expect, each source is enabled by setting the corresponding bit.

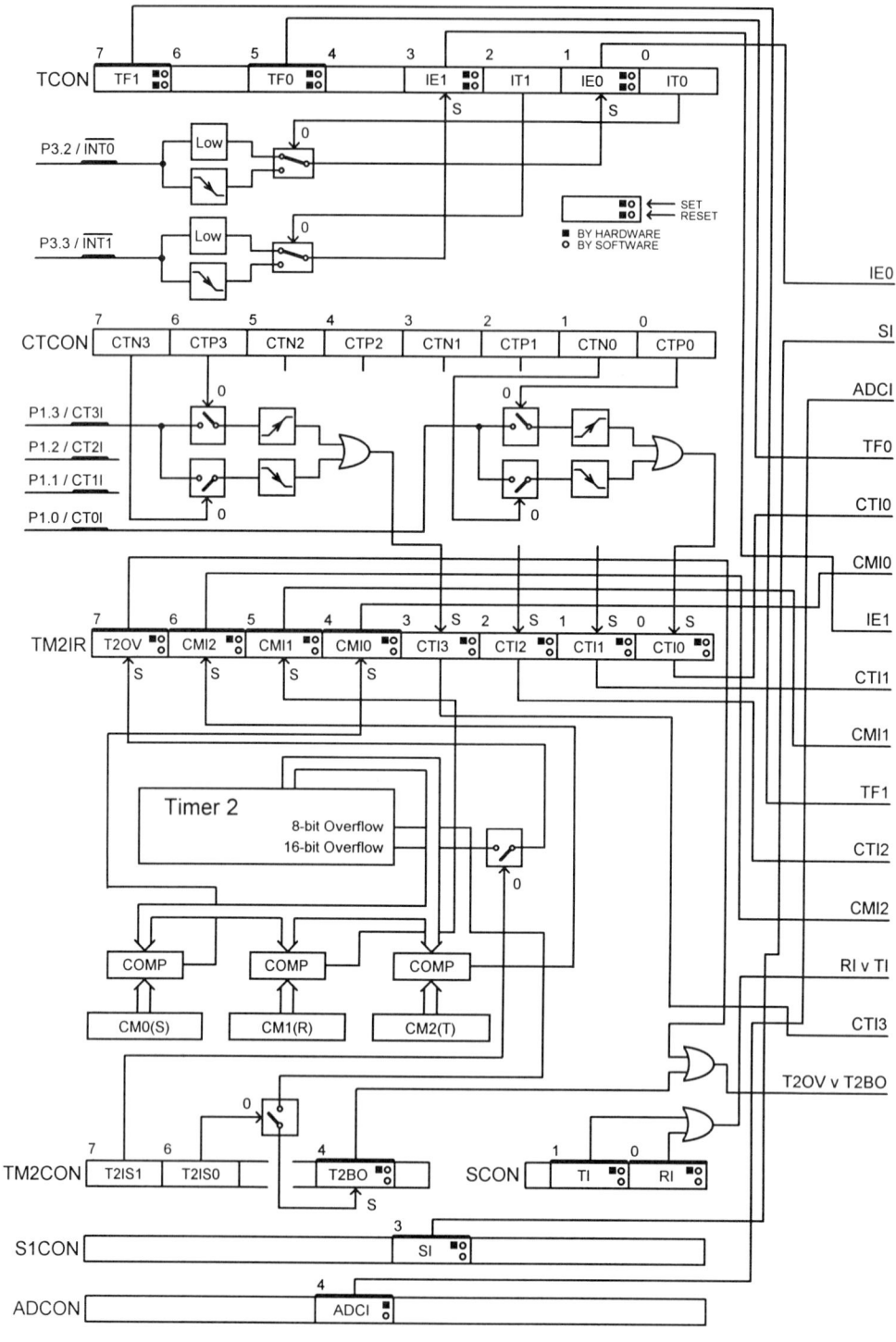

Figure 7.21 The 83C552 interrupt sources.

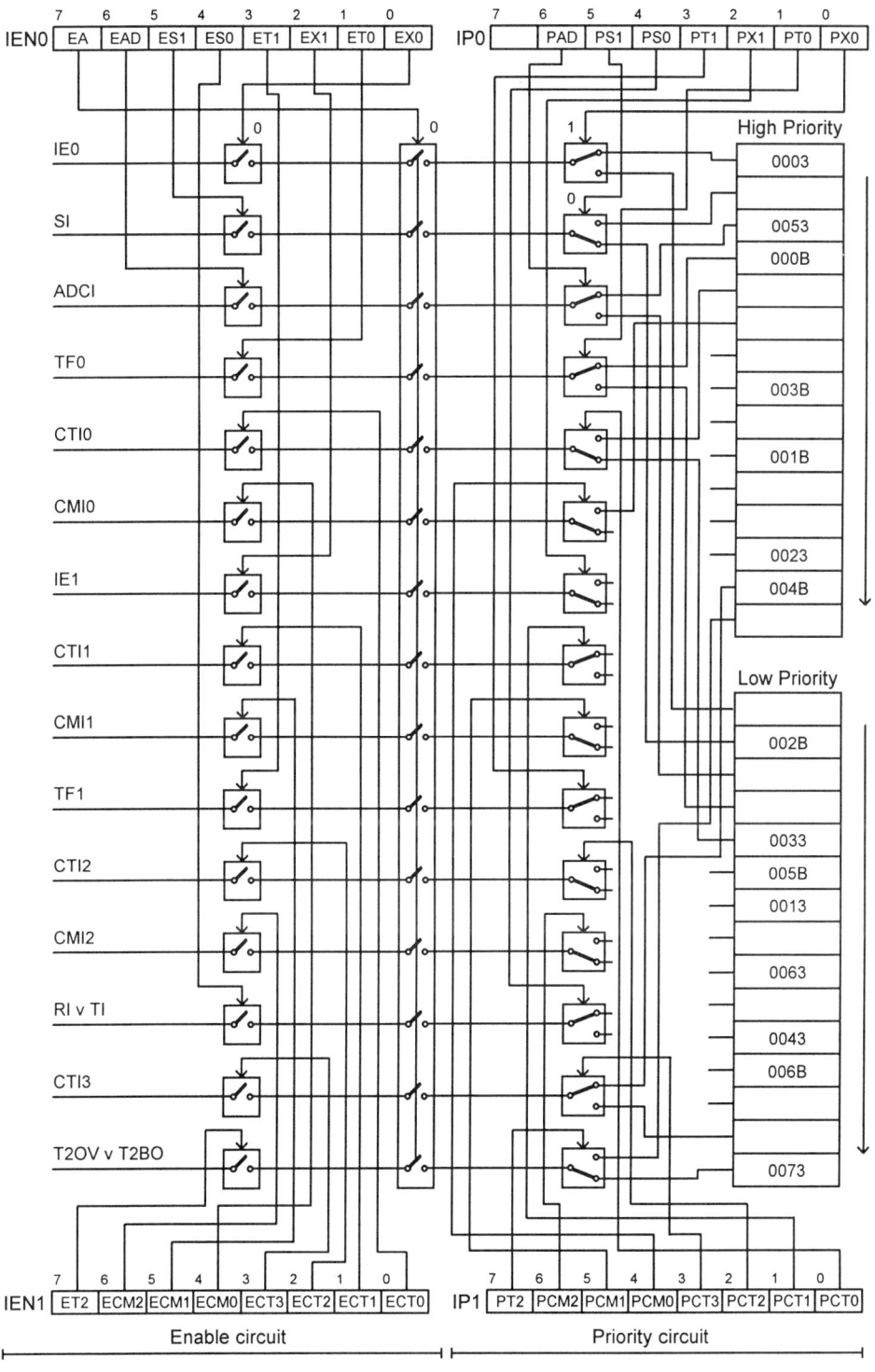

Figure 7.22 The 83C552 interrupt enable and priority circuits.

Priority circuit

Each interrupt source is attached to one of the two priority levels. This process is controlled by the interrupt priority registers IP0 and IP1. All features of the priority circuit are identical to the 8051 microcontroller. Likewise, the interrupt handling and the response time are the same. Figure 7.22 also displays the interrupt vector addresses which can be seen in one of the priority groups. The interrupt vector addresses are listed in Figure 7.23 as well.

Interrupt source	Flag	Vector address
External interrupt 0	IE0	0003H
Timer 0	TF0	000BH
External interrupt 1	IE1	0013H
Timer 1	TF1	001BH
Serial port	TI RI	0023H
I^2C interface	SI	002BH
Timer 2 capture 0	CTI0	0033H
Timer 2 capture 1	CTI1	003BH
Timer 2 capture 2	CTI2	0043H
Timer 2 capture 3	CTI3	004BH
ADC	ADCI	0053H
Timer 2 compare 0	CMI0	005BH
Timer 2 compare 1	CMI1	0063H
Timer 2 compare 2	CMI2	006BH
Timer 2	T2OV T2BO	0073H

Figure 7.23 The 83C552 microcontroller vector addresses.

7.10 Power reduction modes

In principle, the power management of the 83C552 microcontroller is identical to the power saving modes implemented in the 8051 CMOS microcomputers. Again, the power management relies on two modes: Idle and Power Down.

Idle mode

When an instruction sets the bit IDL in the register PCON, it will be the last instruction executed before the Idle mode gates off the clock signal from the CPU, Timer 2, PWM subsystem and the ADC (Figure 7.24). The interrupt system, Timer 0, Timer 1, Timer 3, the serial port and I^2C interface continue to run. The maximum supply current I_{DD} is declined from 45 to 10 mA (f_{OSC} =16 MHz). The Idle mode can be terminated by either interrupt or reset.

Power Down mode

Power Down mode is activated by setting the bit PD in register PCON. The on-chip oscillator is frozen and the supply current I_{DD} is less than 50 μA. The supply voltage can be decreased till 2 V. The only way to terminate the Power Down mode is a reset. Figure 7.25 indicates the microcontroller pin status when power saving modes are used.

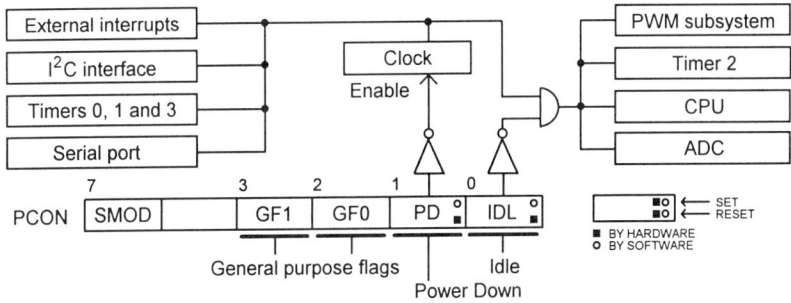

Figure 7.24 The 83C552 microcontroller's clock tree.

Pin	Idle mode		Power Down mode	
	Internal execution	External execution	Internal execution	External execution
ALE	1	1	0	0
$\overline{PSEN}$	1	1	0	0
$\overline{PWM0}, \overline{PWM1}$	High	High	High	High
P0	SFR data	OFF	SFR data	OFF
P1	SFR data	SFR data	SFR data	SFR data
P2	SFR data	PCH	SFR data	SFR data
P3	SFR data	SFR data	SFR data	SFR data
P4	SFR data	SFR data	SFR data	SFR data

Figure 7.25 The 83C552 output signals in power saving modes.

7.11 Application examples and problems

Interfacing LED displays

The specification of many embedded systems includes display output. The majority of designs are based on Light Emitting Diodes (LEDs) and Liquid Crystal Displays (LCDs). Both technologies can be used to provide people with alphanumeric data.

The LED devices are also termed seven-segment displays. The synonym reflects the way the symbols are shaped. The LED displays are cheap and they can provide an indicator panel with high brightness. Alternatively, LCDs have the advantage of consuming practically no power. In addition, they can easily be designed to indicate different symbols.

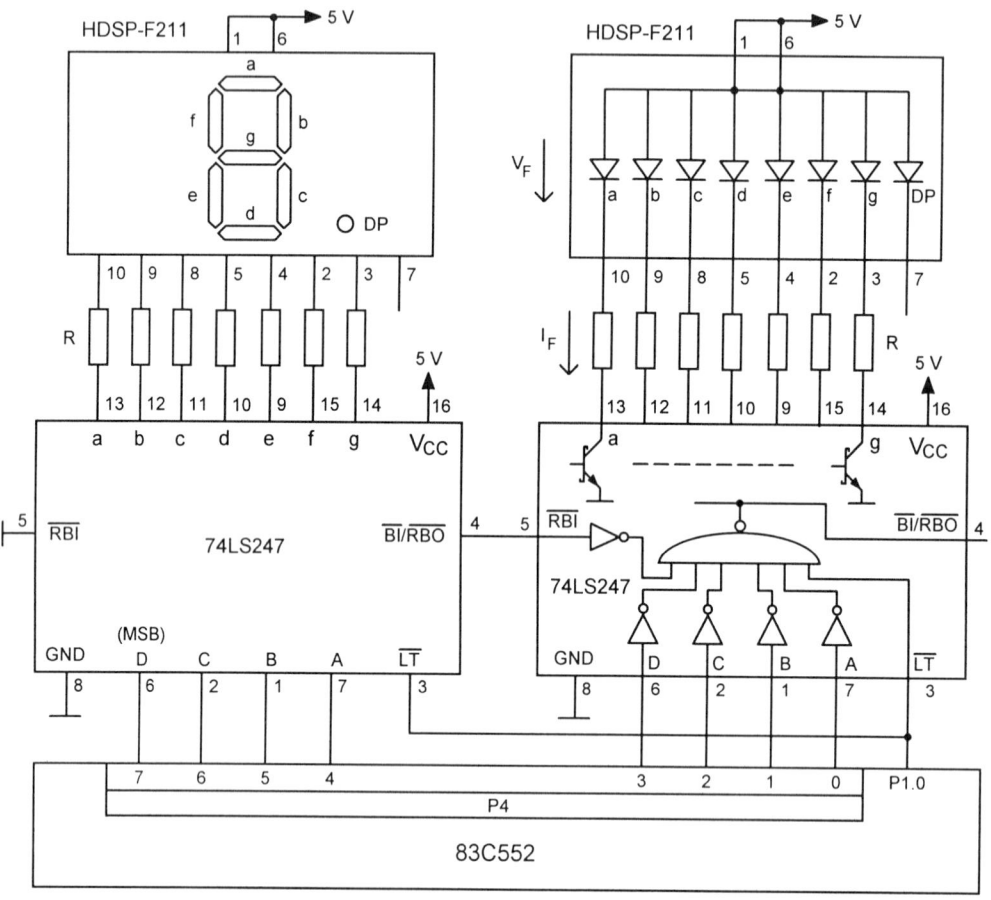

Figure 7.26 Interfacing LED displays by the static method.

Figure 7.26 demonstrates how a microcontroller can interface LED displays. The Hewlett Packard HDSP-F211 is one example of LED devices. The display is capable of illuminating seven segments with the layout shown in the first IC symbol. Moreover, the second display presents a simplification of the internal circuit. To save pins, the displays are produced either common anode or common cathode type. As Figure 7.26 indicates, the HDSP-F211 is a common anode display. If you compare both presentations of the display, you will see which pin must be tied to ground in order to illuminate a certain segment. In any case, current-limiting resistors select the right current I_F (the LED is forward biased). According to the data sheet for the HDSP-F211, a forward current of 5 mA will establish reasonable luminous intensity. From the same source we can obtain the corresponding forward voltage (V_F) which is 1.7 V. Once we have had both values, we can calculate the resistors R.

$$R = \frac{5V - V_F}{I_F} = \frac{5V - 1.7V}{5mA} = 660\,\Omega$$

Practically, we use transistors to control the segments. When turned on, the transistor can be viewed as an additional resistor. Logically, we select a standard resistance value of 620 Ω. Taking into consideration the environmental conditions, the embedded system designer should compromise between sufficient illumination and acceptable power consumption.

In fact, the internal structure of the display exhibits eight LEDs. We need to indicate decimal points in many systems such as calculators and measurement devices.

Traditionally, BCD-to-seven-segment decoders/drivers are placed between the displays and the source of information. As the name prompts, the ICs perform two basic functions. They convert the BCD code to seven-segment pattern and implement appropriate electrical interface. Figure 7.26 includes the Texas Instruments 74LS247 decoder/driver [Texa 1989]. The essential I/Os of this IC are four inputs and seven outputs. The input value (A, B, C and D) is decoded and a set of output transistors (shown in the second IC) are opened. The corresponding segments of the display are forward biased and lighted. For example, if all inputs are low except the input A (0001), two output transistors (b and c) will be switched on to illuminate the segments b and c.

Input codes greater than nine display unique characters which can be found in the data book [Texa 1989]. The 74LS247 decoder is capable of sinking up to 24 mA per output.

Furthermore, an input called Lamp Test ($\overline{LT}$) when asserted low, forces all segments to light. This is an opportunity to test the display. In contrast to the input $\overline{LT}$, another pin named $\overline{BI}\,/\,\overline{RBO}$, when used as Blanking Input ($\overline{BI}$) can blank the display. Again, the input $\overline{BI}$ is an active-low input. In addition, the pin $\overline{BI}\,/\,\overline{RBO}$ can be used as a Ripple Blanking Output. The idea is to organize leading/trailing zero suppression. This improves the readability of the display. To explain this feature, we suggest a simplification of the internal logic network. Look at the IC which is driven by the microcontroller's P4.0 through P4.3. The fact is that the display is blanked when the output of the gate NAND is low. This in turn is related to the following requirements:

• First, the Ripple Blanking Input ($\overline{RBI}$) must be low. This situation will occur if all numbers in front of this display are zeros.

• Second, all four inputs must be low.

• Finally, the lamp test input $\overline{LT}$ must not be asserted.

If the conditions above are met, the display is blanked and the output $\overline{RBO}$ is brought low.

As the main goal of the example in Figure 7.26 is a lucid introduction, we interface only two displays. They will be both blanked, if Port 4 outputs 00H. However, actual designs might keep at least one digit lighted in all circumstances.

The interface example shown in Figure 7.26 is categorized as a static method. For embedded systems with more than four displays, the dynamic method proves useful. The number of interface lines is an important issue. Increasing the number of displays results in expansion of the interface. In many cases it becomes a significant drawback. To work around this problem, the designers employ a dynamic scheme. Glance through Figure 7.27 to get an idea of this approach. Rather than individual connections to the segments of each display, a common bus is used. At the same time, each display is controlled by an individual enable line. The operation of the four-character display shown in Figure 7.27 can be broken down into four steps.

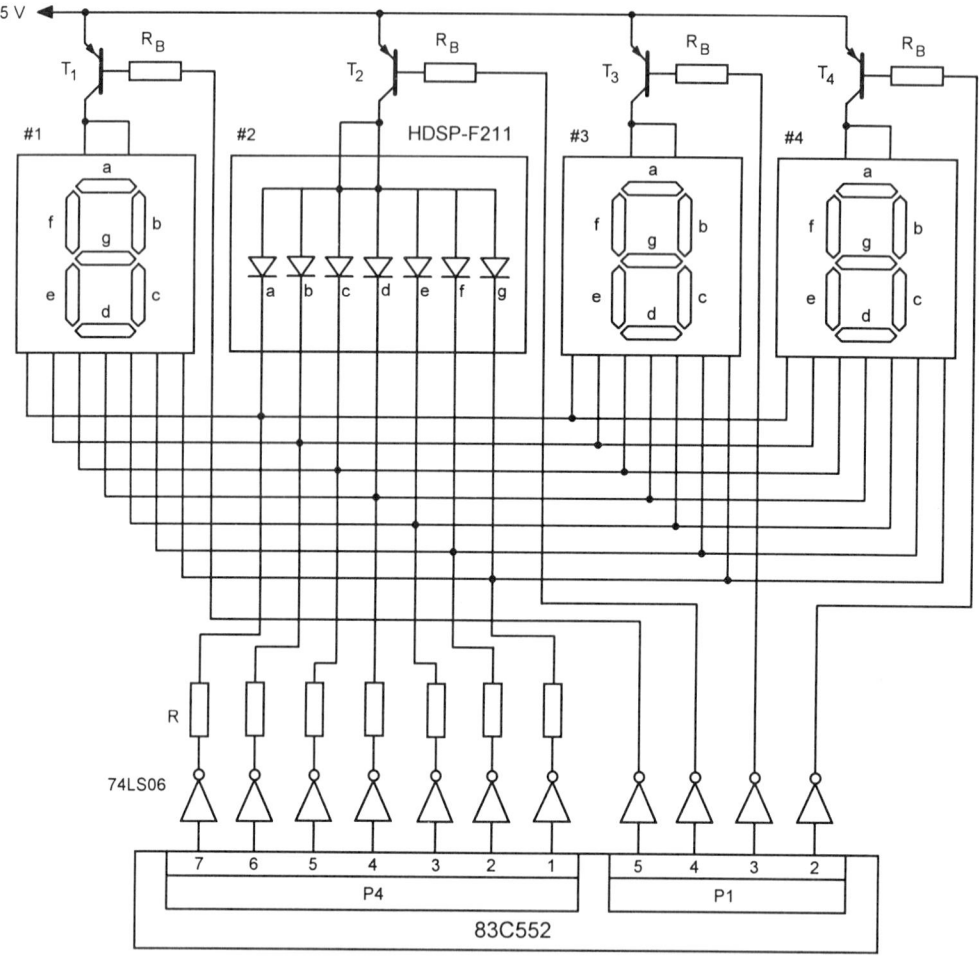

Figure 7.27 Interfacing LED displays by the dynamic method.

First, Port 4 outputs the pattern for display #1. Then the output P1.5 is set. The output drives the IC 74LS06, which possesses open-collector output stage. If the input of the 74LS06 is high, the output transistor is on. As a result, the transistor T_1 is turned on. The selected segments of display #1 are illuminated. Next, the pattern for display #2 is emitted through Port 4. The output P1.5 is reset and the output P1.4 is set. The transistor T_2 is on.

Likewise, display #3 is activated by the transistor T_3. Finally, the transistor T_4 enables the display #4.

Since each display is active one quarter of the time, the current is

$$I_F = 4 \times 5\,mA = 20\,mA$$

This is a compensation for the display multiplexing.

Following the dynamic method we can design cost-effective display interfaces. However, there is a limit to the number of displays. The parameter peak forward current defines a ceiling for the total number of displays. Practically, six or eight displays can be multiplexed with ample margin.

Interfacing LCD displays

Now let's look at an intelligent LCD display of two lines, 16 characters per line. Figure 7.28 shows the LCD LTN211 and the interface to the 83C552 microcontroller [Phil 1991]. Moreover, the picture contains a list of the available instructions and the dot matrix format. Figure 7.28 also gives you a flavor of the internal structure, which includes a Character Generator (CG) and a Display Data RAM (DDRAM).

As a principle, the LCD works very simply. There are two basic functions:
- A certain character is displayed.
- The character is placed on a certain position.

For example, we want to indicate a capital 'A' on the first line, first position. What we do is to move the ASCII code of 'A' to the Display Data RAM, address 00H. Automatically, the character generator will convert the ASCII code into the 'A' dot pattern. Since we deal with the Display Data RAM, address 00H, the letter A will appear on the first line, first position.

The user can either rely completely on the symbols embedded in the Character Generator ROM (CGROM) or define new characters by the Character Generator RAM (CGRAM). We discuss examples that cover both options.

The table in Figure 7.28 contains all possible instructions. The LCD instructions can be viewed, for example, as the procedures to configure the 82C55 PIO's hardware. We distinguish between control operations and data operations.

The control operations are performed, when the Register Select (RS) input is brought low. The data operations are achieved by Register Select input pulled high. Once the data D0-D7 is placed on the bus and the desired levels of inputs RS and R / $\overline{W}$ selected, it is time to pulse high the Enable input (E). Then the operation is under way.

The most important action is to write data to the DDRAM. Since the address in the DDRAM is incremented/decremented automatically, we could apply the same operation again and again. However, if we want to jump to a different address (to select a different position for the current character), we must establish the address first. This is done by a control operation (DDRAM address). Both steps, set-up the address and move the data, form the skeleton of the interface program.

The LCD is a relatively slow device. Efficient dialogue is organized by status inquiry. If the input RS is low and input R / $\overline{W}$ high, the microcontroller can read a status byte. The most significant bit in the status byte is a Busy Flag (BF). The seven least significant bits represent the Address Counter (AC), which is used for both DDRAM and CGRAM. The question arises when the LCD is initialized. The initialization procedure requires the instruction "Function set" to be executed three times. The instruction must be followed by delays. However, the LCD is still not capable of indicating its status by the Busy flag.

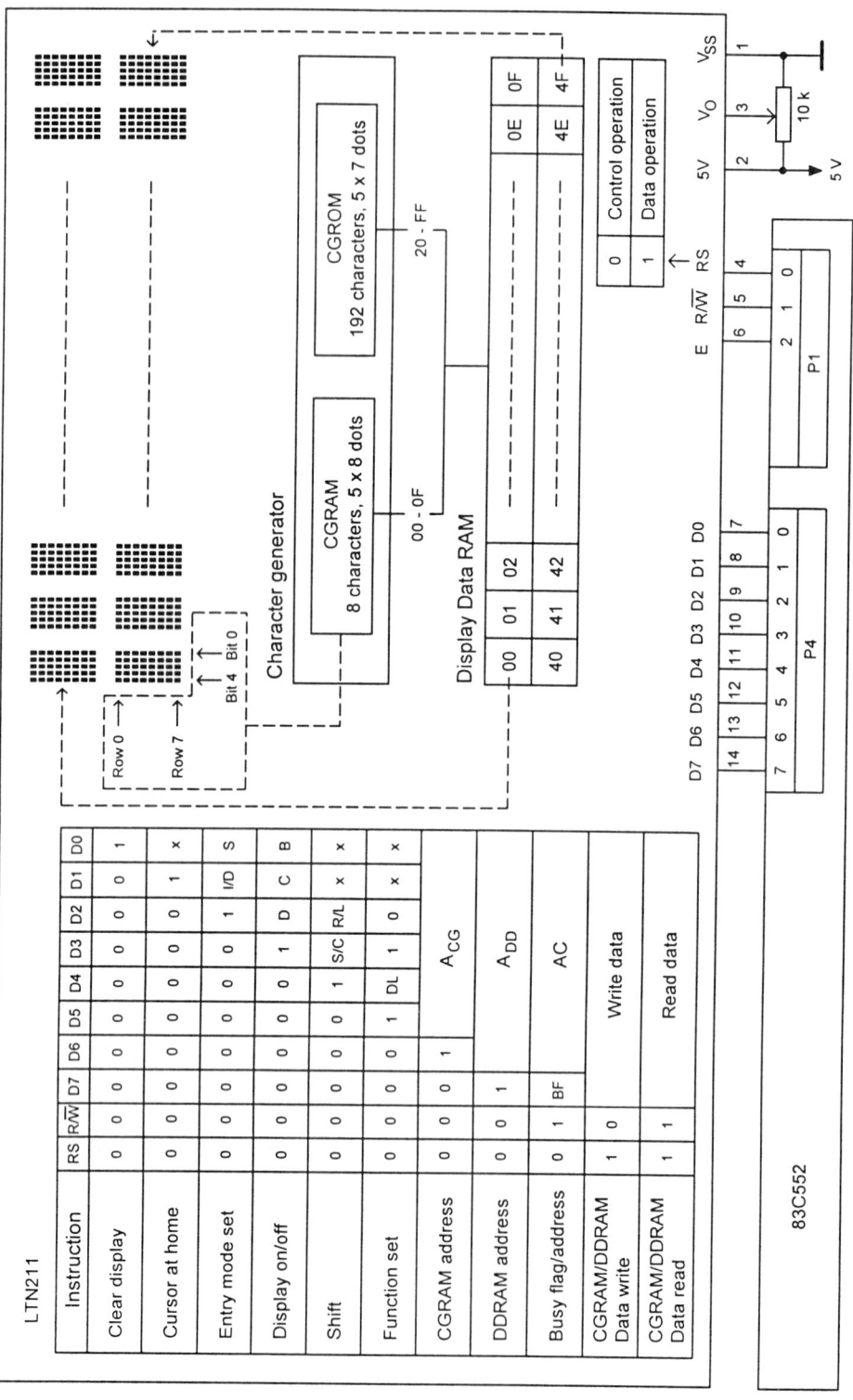

Figure 7.28 Interfacing a LCD display.

The example program LCD1.ASM displays the letters ABC starting from the most-left digit on the first line. With the assumption that the instructions will be single-stepped, there is no need to wait for the LCD to complete the operation. Naturally, executing one instruction at a time we introduce ample delay. By this technique we avoid the time delay problem and are ready for the first touch.

```
;*****************************************************************
;*    LCD1.ASM                                                  *
;*    These instructions if single-step executed display ABC    *
;*****************************************************************
            ORG     2000H
            CLR     P1.0        ; RS=0 > Control operation
            CLR     P1.1        ; R/W=0 > Write
            MOV     P4,#38H     ; DL=1 > 8-bit interface
            SETB    P1.2        ; E=1
            CLR     P1.2        ; E=0
;
            SETB    P1.2        ; E=1
            CLR     P1.2        ; E=0
;
            SETB    P1.2        ; E=1
            CLR     P1.2        ; E=0
;
            MOV     P4,#0FH     ; D=1 > Display ON
                                ; C=1 > Cursor ON
                                ; B=1 > Blink the character at
                                ; the cursor position
            SETB    P1.2        ; E=1
            CLR     P1.2        ; E=0
;
            MOV     P4,#06H     ; I/D=1 >  Increment the cursor
                                ; (move to the right)
                                ; S=0 > The display is not shifted
            SETB    P1.2        ; E=1
            CLR     P1.2        ; E=0
;
            MOV     P4,#01H     ; Clear display
            SETB    P1.2        ; E=1
            CLR     P1.2        ; E=0
;
            SETB    P1.0        ; RS=1 > Data operation
;
            MOV     P4,#'A'     ; Display 'A'
            SETB    P1.2        ; E=1
            CLR     P1.2        ; E=0
;
            MOV     P4,#'B'     ; Display 'B'
            SETB    P1.2        ; E=1
            CLR     P1.2        ; E=0
;
            MOV     P4,#'C'     ; Display 'C'
            SETB    P1.2        ; E=1
```

```
           CLR     P1.2              ; E=0
;
           NOP
```

If you do not see any characters on the display, try to adjust the contrast voltage V_O.

It is certainly implied that the next step will be to reinforce the instructions grouped in the example handler LCD1.ASM by adding two subroutines. The subroutine DEL was introduced in section 3.10 and will perform an adjustable software delay. The subroutine called T_BF will test the Busy Flag. Consequently, the program LCD2.ASM forms a complete software example, that when run by the microcontroller, displays the letters ABC.

```
;************************************************
;*   LCD2.ASM                                   *
;*   This program displays ABC                  *
;************************************************
           ORG     2000H
           CLR     P1.0              ; RS=0 > Control operation
           CLR     P1.1              ; R/W=0 > Write
;
           MOV     P4,#38H           ; DL=1 > 8-bit interface
           SETB    P1.2              ; E=1
           CLR     P1.2              ; E=0
           MOV     R0,#50            ; Delay 5 ms
           LCALL   DEL
;
           SETB    P1.2              ; E=1
           CLR     P1.2              ; E=0
           MOV     R0,#20            ; Delay 2 ms
           LCALL   DEL
;
           SETB    P1.2              ; E=1
           CLR     P1.2              ; E=0
           MOV     R0,#20            ; Delay 2 ms
           LCALL   DEL
;
           MOV     P4,#0FH           ; D=1 > Display ON
                                     ; C=1 > Cursor ON
                                     ; B=1 > Blink the character at the
                                     ; cursor position
           SETB    P1.2              ; E=1
           CLR     P1.2              ; E=0
           MOV     R0,#20            ; Delay 2 ms
           LCALL   DEL
;
           MOV     P4,#06H           ; I/D=1 > Increment the cursor
                                     ; (move to the right)
                                     ; S=0 > The display is not shifted
           SETB    P1.2              ; E=1
           CLR     P1.2              ; E=0
           MOV     R0,#20            ; Delay 2 ms
           LCALL   DEL
;
```

```
        MOV     P4,#01H      ; Clear display
        SETB    P1.2         ; E=1
        CLR     P1.2         ; E=0
        MOV     R0,#20       ; Delay 2 ms
        LCALL   DEL
;
        SETB    P1.0         ; RS=1 > Data operation
;
        MOV     P4,#'A'      ; Display 'A'
        SETB    P1.2         ; E=1
        CLR     P1.2         ; E=0
        LCALL   T_BF
;
        MOV     P4,#'B'      ; Display 'B'
        SETB    P1.2         ; E=1
        CLR     P1.2         ; E=0
        LCALL   T_BF
;
        MOV     P4,#'C'      ; Display 'C'
        SETB    P1.2         ; E=1
        CLR     P1.2         ; E=0
        LCALL   T_BF
;
        SJMP    $
;
;*****************************************************************
;    DEL: This subroutine provides a delay between        *
;         103 and 24741 mcs depending on the code         *
;         in register R0 (1 to 255)                       *
;         Execution time : 97*R0 + 6, mcs                 *
;*****************************************************************
DEL:    PUSH    ACC          ; Save ACC
DEL1:   MOV     A,#2FH
DEL2:   DJNZ    ACC,DEL2
        DJNZ    R0,DEL1
        POP     ACC          ; Restore ACC
        RET
;
;*****************************************************************
;    T_BF:  This subroutine tests the Busy flag           *
;           Prepares the LCD for CGRAM/DDRAM              *
;           Data write operation                          *
;*****************************************************************
T_BF:   MOV     P4,#0FFH     ; Input port
        CLR     P1.0         ; RS=0 > Control operation
        SETB    P1.1         ; R/W=1 > Read
T_BF1:  SETB    P1.2         ; E=1
        MOV     ACC,P4       ; Read the Busy Flag
        CLR     P1.2         ; E=0
        JB      ACC.7,T_BF1
        SETB    P1.0         ; RS=1 > Data operation
        CLR     P1.1         ; R/W=0 > Write
```

```
          RET
```

Up to now, we have used the LCD feature, that the address is incremented or decremented automatically, when data is moved to the DDRAM. It is time to focus on the second basic goal - how to position the symbols over the display, especially how to jump with the cursor.

Problem 7.3

Write an assembly language program for the 83C552 microcontroller and a LCD display according to Figure 7.28. The program displays the letters :

'A' as a first character on the top line   'B' as a last character on the top line
'C' as a first character on the bottom line   'D' as a last character on the bottom line.

Solution 7.3

To this point, we have demonstrated how to display consecutive characters. In this case, we have to jump to the last position on the top row. Hence, the DDRAM address has to be changed to 0FH. In a similar vein, the cursor is moved to DDRAM addresses 40H and 4FH.

```
;*************************************************************
;*    LCD3.ASM                                              *
;*    This program displays                                 *
;*    A                     B                               *
;*    C                     D                               *
;*************************************************************
          ORG     2000H
          CLR     P1.0           ; RS=0 > Control operation
          CLR     P1.1           ; R/W=0 > Write
;
          MOV     P4,#38H        ; DL=1 > 8-bit interface
          SETB    P1.2           ; E=1
          CLR     P1.2           ; E=0
          MOV     R0,#50         ; Delay 5 ms
          LCALL   DEL
;
          SETB    P1.2           ; E=1
          CLR     P1.2           ; E=0
          MOV     R0,#20         ; Delay 2 ms
          LCALL   DEL
;
          SETB    P1.2           ; E=1
          CLR     P1.2           ; E=0
          MOV     R0,#20         ; Delay 2 ms
          LCALL   DEL
;
          MOV     P4,#0FH        ; D=1 > Display ON
                                 ; C=1 > Cursor ON
                                 ; B=1 > Blink the character
                                 ; at the cursor position
          SETB    P1.2           ; E=1
          CLR     P1.2           ; E=0
          MOV     R0,#20         ; Delay 2 ms
```

```
          LCALL   DEL
;
          MOV     P4,#06H        ; I/D=1 > Increment the cursor
                                 ; (move to the right)
                                 ; S=0 > The display is not shifted
          SETB    P1.2           ; E=1
          CLR     P1.2           ; E=0
          MOV     R0,#20         ; Delay 2 ms
          LCALL   DEL
;
          MOV     P4,#01H        ; Clear display
          SETB    P1.2           ; E=1
          CLR     P1.2           ; E=0
          MOV     R0,#20         ; Delay 2 ms
          LCALL   DEL
;
          SETB    P1.0           ; RS=1 > Data operation
;
          MOV     P4,#'A'        ; Display 'A'
          SETB    P1.2           ; E=1
          CLR     P1.2           ; E=0
          LCALL   T_BF
;
          CLR     P1.0           ; RS=0 > Control operation
;
          MOV     P4,#8FH        ; First line, last character
                                 ; address (0FH)
          SETB    P1.2           ; E=1
          CLR     P1.2           ; E=0
          LCALL   T_BF
;
          MOV     P4,#'B'        ; Display 'B'
          SETB    P1.2           ; E=1
          CLR     P1.2           ; E=0
          LCALL   T_BF
;
          CLR     P1.0           ; RS=0 > Control operation
;
          MOV     P4,#0C0H       ; Second line, first character
                                 ; address (40H)
          SETB    P1.2           ; E=1
          CLR     P1.2           ; E=0
          LCALL   T_BF
;
          MOV     P4,#'C'        ; Display 'C'
          SETB    P1.2           ; E=1
          CLR     P1.2           ; E=0
          LCALL   T_BF
;
          CLR     P1.0           ; RS=0 > Control operation
;
          MOV     P4,#0CFH       ; Second line, last character
```

```
                                   ;  address (4FH)
                 SETB    P1.2      ;  E=1
                 CLR     P1.2      ;  E=0
                 LCALL   T_BF
        ;

                 MOV     P4,#'D'   ;  Display 'D'
                 SETB    P1.2      ;  E=1
                 CLR     P1.2      ;  E=0
                 LCALL   T_BF
        ;

                 SJMP    $
;*****************************************************************
;     DEL: This subroutine provides a delay between         *
;          103 and 24741 mcs depending on the code in       *
;          register R0 (1 to 255)                           *
;          Execution time : 97*R0 + 6, mcs                  *
;*****************************************************************
DEL:             PUSH    ACC       ; Save ACC
DEL1:            MOV     A,#2FH
DEL2:            DJNZ    ACC,DEL2
                 DJNZ    R0,DEL1
                 POP     ACC       ; Restore ACC
                 RET
;*****************************************************************
;     T_BF:   This subroutine tests the Busy flag           *
;             Prepares the LCD for CGRAM/DDRAM              *
;             Data write operation                          *
;*****************************************************************
T_BF:            MOV     P4,#0FFH      ; Input port
                 CLR     P1.0          ; RS=0 > Control operation
                 SETB    P1.1          ; R/W=1 > Read
T_BF1:           SETB    P1.2          ; E=1
                 MOV     ACC,P4        ; Read BF
                 CLR     P1.2          ; E=0
                 JB      ACC.7,T_BF1
                 SETB    P1.0          ; RS=1 > Data operation
                 CLR     P1.1          ; R/W=0 > Write
                 RET
```

The designers of the LTN211 broke down the internal character generator into two parts: CGROM and CGRAM. The CGROM covers the most frequently used symbols. The CGRAM gives additional opportunities. The designer of embedded systems can create up to eight extra symbols, usually termed user defined symbols. In principle, the data written in the DDRAM is used as an address pattern for the character generator. However, separating the character generator into two parts, demands the address range to be split into two segments. In particular, codes between 00H and 0FH select symbols from the CGRAM. Codes between 20H and FFH deal with the CGROM. Both the CGRAM and the CGROM are not used completely as address spaces. Consequently, the numbers of available characters, indicated in Figure 7.28, are lower than the ones you might expect. It is certainly implied that the CGRAM contents must be programmed beforehand. This point will be discussed in more detail.

The information for the user defined symbols is stored row by row. When we move a byte to the CGRAM address 00H (following a control operation with code 40H), we program the row 0 of the RAM character 0. The three MSB bits are don't care (they have no influence). A dot is turned on by setting the corresponding bit. The row 1 of the RAM character 0 is programmed by a write to the CGRAM address 01H. Finally, row 7 of the last user defined symbol is written at the CGRAM address 63 (3FH). Once the CGRAM has been programmed, we can use the symbols by writing their CGRAM addresses into the DDRAM.

The following example program LCD4.ASM implements the described opportunity and displays a square pulse shape.

```
;******************************************************
;*    LCD4.ASM                                        *
;*    This program displays a square pulse            *
;******************************************************
            ORG     2000H
            CLR     P1.0        ; RS=0 > Control operation
            CLR     P1.1        ; R/W=0 > Write
;
            MOV     P4,#38H     ; DL=1 > 8-bit interface
            SETB    P1.2        ; E=1
            CLR     P1.2        ; E=0
            LCALL   DEL2        ; DEL2 is a version of DEL for 2 ms
            LCALL   DEL2
            LCALL   DEL2
;
            SETB    P1.2        ; E=1
            CLR     P1.2        ; E=0
            LCALL   DEL2
;
            SETB    P1.2        ; E=1
            CLR     P1.2        ; E=0
            LCALL   DEL2
;
            MOV     P4,#0FH     ; D=1 > Display ON
                                ; C=1 > Cursor ON
                                ; B=1 >  Blink the character
                                ; at the cursor position
            SETB    P1.2        ; E=1
            CLR     P1.2        ; E=0
            LCALL   DEL2
;
            MOV     P4,#06H     ; I/D=1 >   Increment the cursor
                                ; (move to the right)
                                ; S=0 > The display is not shifted
            SETB    P1.2        ; E=1
            CLR     P1.2        ; E=0
            LCALL   DEL2
;
            MOV     P4,#01H     ; Clear display
            SETB    P1.2        ; E=1
```

```
        CLR     P1.2            ; E=0
        LCALL   DEL2
;
        MOV     P4,#40H         ; CGRAM address 00H   (range 0-63)
        SETB    P1.2            ; E=1
        CLR     P1.2            ; E=0
        LCALL   DEL2
;
        SETB    P1.0            ; RS=1 > Data operation
;
        MOV     P4,#00H         ;  CGRAM address 00H
                                ;  Font row 0 of character 0
                                ;  is xxx0 0000
        SETB    P1.2            ;  E=1
        CLR     P1.2            ;  E=0
        LCALL   T_BF
;
        MOV     P4,#00H         ;  CGRAM address 01H
                                ;  Font row 1 of character 0
                                ;  is xxx0 0000
        SETB    P1.2            ;  E=1
        CLR     P1.2            ;  E=0
        LCALL   T_BF
;
        MOV     P4,#0EH         ;  CGRAM address 02H
                                ;  Font row 2 of character 0
                                ;  is xxx0 1110
        SETB    P1.2            ;  E=1
        CLR     P1.2            ;  E=0
        LCALL   T_BF
;
        MOV     P4,#0AH         ;  CGRAM address 03H
                                ;  Font row 3 of character 0
                                ;  is xxx0 1010
        SETB    P1.2            ;  E=1
        CLR     P1.2            ;  E=0
        LCALL   T_BF
;
        MOV     P4,#0AH         ;  CGRAM address 04H
                                ;  Font row 4 of character 0
                                ;  is xxx0 1010
        SETB    P1.2            ;  E=1
        CLR     P1.2            ;  E=0
        LCALL   T_BF
;
        MOV     P4,#1BH         ;  CGRAM address 05H
                                ;  Font row 5 of character 0
                                ;  is xxx1 1011
        SETB    P1.2            ;  E=1
```

```
            CLR     P1.2          ; E=0
            LCALL   T_BF
;
            MOV     P4,#00H        ; CGRAM address 06H
                                   ; Font row 6 of character 0
                                   ; is xxx0 0000
            SETB    P1.2          ; E=1
            CLR     P1.2          ; E=0
            LCALL   T_BF
;
            MOV     P4,#00H        ; CGRAM address 07H
                                   ; Font row 7 of character 0
                                   ; is xxx0 0000
            SETB    P1.2          ; E=1
            CLR     P1.2          ; E=0
            LCALL   T_BF
;
            CLR     P1.0          ; RS=0 > Control operation
;
            MOV     P4,#80H        ; First line
                                   ; first character address
            SETB    P1.2          ; E=1
            CLR     P1.2          ; E=0
            LCALL   T_BF
;
            MOV     P4,#00H        ; Move 00H to the DDRAM
                                   ; This is the CGRAM address 0
                                   ; (range 0-7)
            SETB    P1.2          ; E=1
            CLR     P1.2          ; E=0
            LCALL   T_BF
;
            SJMP    $
;*************************************************************
;    DEL2:  This subroutine provides a delay of 2 ms      *
;*************************************************************
DEL2:       PUSH    ACC
            MOV     R0,#20         ; Delay 2 ms
DEL21:      MOV     A,#2FH
DEL22:      DJNZ    ACC,DEL22
            DJNZ    R0,DEL21
            POP     ACC
            RET
;*************************************************************
;    T_BF:  This subroutine tests the Busy flag           *
;           Prepares the LCD for CGRAM/DDRAM              *
;              Data write operation                       *
;*************************************************************
T_BF:       MOV     P4,#0FFH        ; Input port
            CLR     P1.0            ; RS=0 > Control operation
```

```
           SETB    P1.1            ; R/W=1 > Read
T_BF1:     SETB    P1.2            ; E=1
           MOV     ACC,P4          ; Read BF
           CLR     P1.2            ; E=0
           JB      ACC.7,T_BF1
           SETB    P1.0            ; RS=1 > Data operation
           CLR     P1.1            ; R/W=0 > Write
           RET
```

We streamlined the example program LCD4.ASM by using a delay subroutine named DEL2. The subroutine DEL2 has a fixed delay of 2 ms.

Digital clock

In section 4.6 we discussed the design of an electromechanical clock. The LED and LCD displays provide an alternative which brings advantages and disadvantages. Of course, the precision is better, however if you are interested if the value belongs to a certain range, the digital display is not very convenient. In some cases, both analog and digital displays are combined to form an indication panel.

The following design of a digital clock, compared with the electromechanical clock, shows how the hardware and software are modified to meet the new requirements.

Problem 7.4

Design a digital clock based on an 83C552 microcontroller and a LCD display. The microcontroller runs at 12 MHz.

Solution 7.4

Figure 7.28 is also a good starting point for the digital clock design. We add three pushbuttons to obtain the schematic diagram shown in Figure 7.29. We need the pushbuttons to adjust the clock. Pressing the pushbutton connected to the input P3.5 increments the hours. Likewise, the pushbutton which pulls down the input P3.4 increments the minutes. Our intention is to freeze the counting process while hours or minutes are adjusted. Once we have corrected hours and minutes, we can put the clock to the exact time by resetting the seconds.

In order to make the design process more efficient, we partition the clock functionality into a set of subtasks. Figure 7.30 shows a logical solution. We partition the functionality into four tasks. Furthermore, the task "Count" is divided into three subtasks. However, this decomposition scheme should not be considered as a constant. Similar to Problem/Solution 4.4, we are going to use Timer 0 to clock the software counters. Again, Timer 0 will interrupt the microcontroller every 250 µs. Since this period of time exceeds the LCD delay with sufficient margin, it can be used to synchronize the interaction with the display. Consequently, we merge the tasks "Count" and "Display" into a single task and indicate its relation to Timer 0. The program emerges with three tasks: "Initialize", "Count and display" and "Adjust". We organize an endless loop for the task "Adjust". If we press a pushbutton, the clock will be adjusted. Periodically, the loop is interrupted and the task "Count and display" activated.

Figure 7.31 shows the flowchart for the interrupt handler. Each time an interrupt occurs, the microcontroller increments the software counter C_200. The maximum value for this counter is 199. So, each increment is followed by a test. If the test shows that the counter C_200 contains 200, the next instruction clears the counter. Since the next stage is another software counter named C_20, it is incremented every time C_200 is cleared. The clock's seconds are incremented when the counter C_20 completes one cycle. The hours-tens digit is updated once per second. Also, the corresponding flag F_H_T is set. The next interrupt activates the handler and the hours digit is updated. This sequence continues until the seconds digit is refreshed. The following invocations of the interrupt handler only increment the software counters.

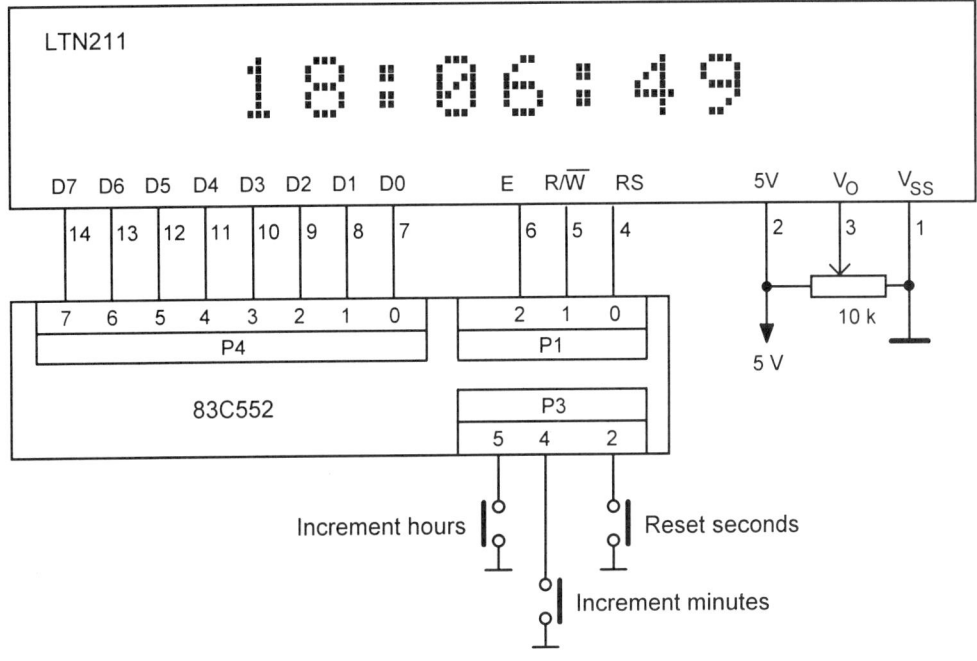

Figure 7.29 The digital clock.

Using a timer and the interrupt system we achieve two advantages:

• The fault-tolerance is improved. Every 250 µs the program is forced to execute the interrupt handler.

• There is no need to check if the LCD display is busy. Consequently, the code is simplified.

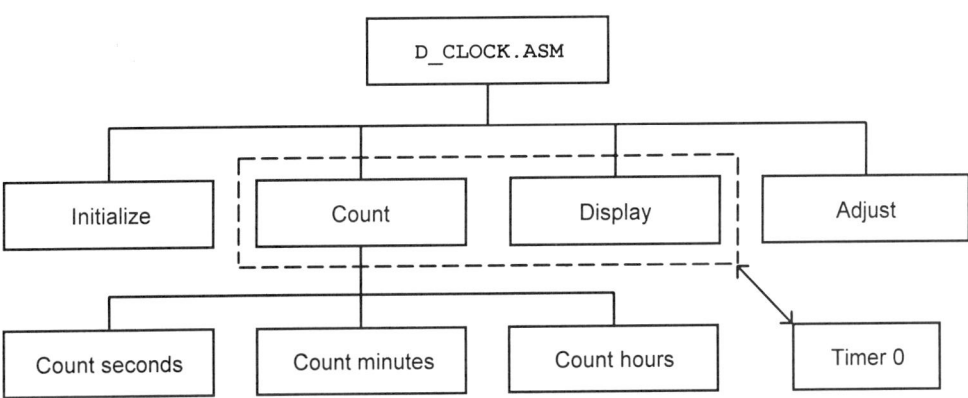

Figure 7.30 The digital clock functionality viewed as a set of subtasks.

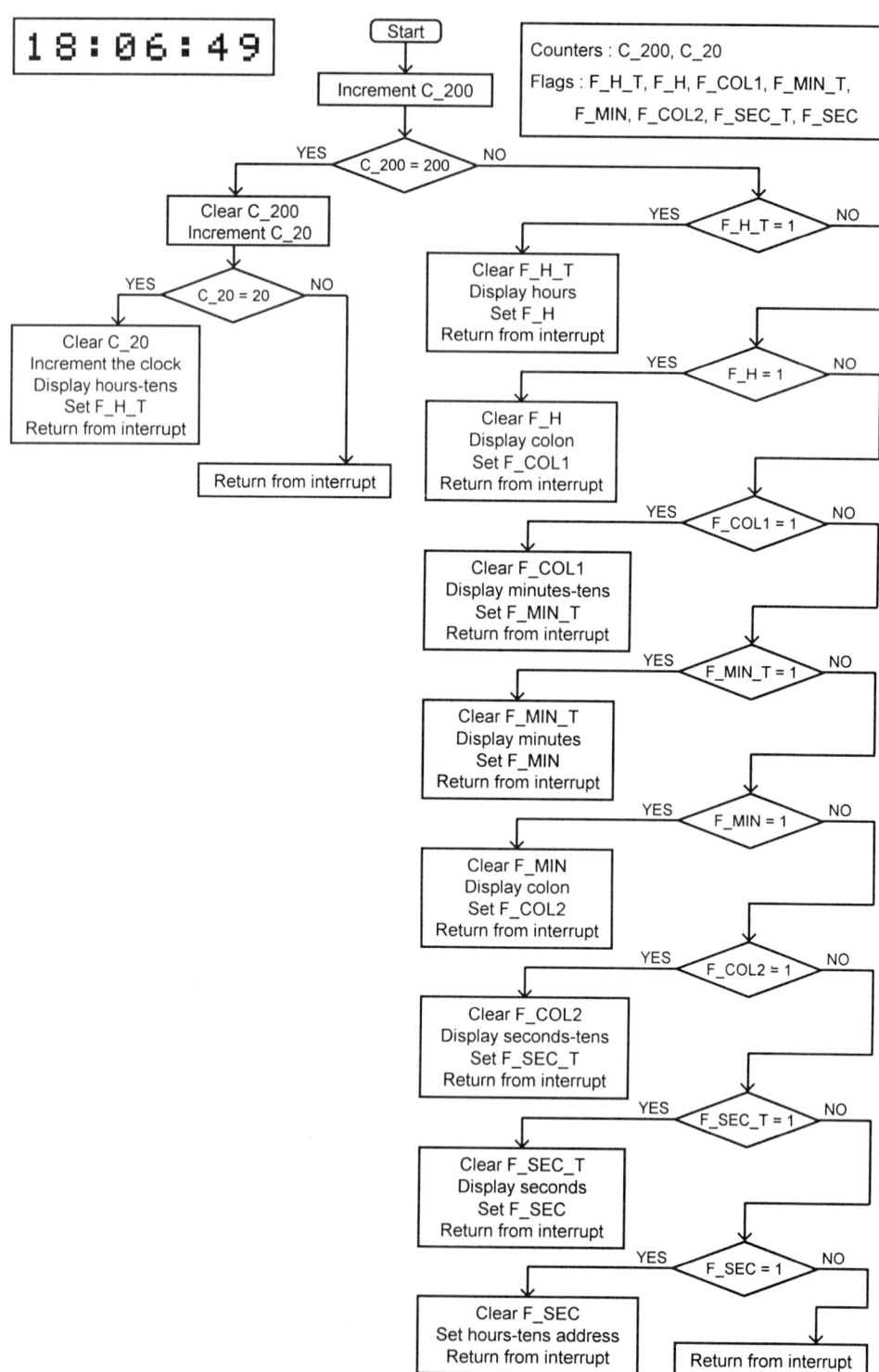

Figure 7.31 The flowchart for the Timer 0 interrupt handler.

The code below is a possible implementation of the digital clock software:

```
;*****************************************************************
;*    D_CLOCK.ASM                                              *
;*    This program runs on an 83C552 microcontroller           *
;*    to implement a digital clock with a LCD display          *
;*****************************************************************
C_200     IDATA   00H       ; Counter, 200 states
C_20      IDATA   01H       ; Counter, 20 states
SEC       IDATA   02H       ; Counter for seconds
SEC_T     IDATA   03H       ; Counter for seconds-tens
MIN       IDATA   04H       ; Counter for minutes
MIN_T     IDATA   05H       ; Counter for minutes-tens
HOUR      IDATA   06H       ; Counter for hours
HOUR_T    IDATA   07H       ; Counter for hours-tens
;
F_H_T     BIT     00H
F_H       BIT     01H
F_COL1    BIT     02H
F_MIN_T   BIT     03H
F_MIN     BIT     04H
F_COL2    BIT     05H
F_SEC_T   BIT     06H
F_SEC     BIT     07H
;
          ORG     0000H
          AJMP    INIT      ; Initialize
          ORG     000BH
          AJMP    COUNT     ; Timer 0 interrupt handler
;
;*** Initialize the LCD display   *********************
          ORG     0100H
INIT:     CLR     P1.0      ; RS=0 > Control operation
          CLR     P1.1      ; R/W=0 > Write
;
          MOV     P4,#38H   ; DL=1 > 8-bit interface
          SETB    P1.2      ; E=1
          CLR     P1.2      ; E=0
          MOV     R0,#50    ; Delay 5 ms
          ACALL   DEL
;
          SETB    P1.2      ; E=1
          CLR     P1.2      ; E=0
          MOV     R0,#20    ; Delay 2 ms
          ACALL   DEL
;
          SETB    P1.2      ; E=1
          CLR     P1.2      ; E=0
          MOV     R0,#20    ; Delay 2 ms
          ACALL   DEL
;
          MOV     P4,#0CH   ; D=1 > Display ON
```

```
                              ; C=0 > Cursor OFF
                              ; B=0 > Character at the cursor
                              ; position does not blink
        SETB    P1.2          ; E=1
        CLR     P1.2          ; E=0
        MOV     R0,#20        ; Delay 2 ms
        ACALL   DEL
;
        MOV     P4,#06H       ; I/D=1 > Increment the cursor
                              ; (move to the right)
                              ; S=0 > The display is not shifted
        SETB    P1.2          ; E=1
        CLR     P1.2          ; E=0
        MOV     R0,#20        ; Delay 2 ms
        ACALL   DEL
;
        MOV     P4,#01H       ; Clear display
        SETB    P1.2          ; E=1
        CLR     P1.2          ; E=0
        MOV     R0,#20        ; Delay 2 ms
        ACALL   DEL
;
        CLR     P1.0          ; RS=0 > Control operation
;
        MOV     P4,#85H       ; Hours digit, address 05H
        SETB    P1.2          ; E=1
        CLR     P1.2          ; E=0
        ACALL   T_BF
;
        MOV     P4,#'0'       ; Display '0'
        SETB    P1.2          ; E=1
        CLR     P1.2          ; E=0
        ACALL   T_BF
;
        MOV     P4,#':'       ; Display ':'
        SETB    P1.2          ; E=1
        CLR     P1.2          ; E=0
        ACALL   T_BF
;
        MOV     P4,#'0'       ; Display '0'
        SETB    P1.2          ; E=1
        CLR     P1.2          ; E=0
        ACALL   T_BF
;
        MOV     P4,#'0'       ; Display '0'
        SETB    P1.2          ; E=1
        CLR     P1.2          ; E=0
        ACALL   T_BF
;
        MOV     P4,#':'       ; Display ':'
        SETB    P1.2          ; E=1
        CLR     P1.2          ; E=0
```

```
        ACALL   T_BF
;
        MOV     P4,#'0'   ; Display '0'
        SETB    P1.2      ; E=1
        CLR     P1.2      ; E=0
        ACALL   T_BF
;
        MOV     P4,#'0'   ; Display '0'
        SETB    P1.2      ; E=1
        CLR     P1.2      ; E=0
        ACALL   T_BF
;
        CLR     P1.0      ; RS=0 > Control operation
        MOV     P4,#84H   ; Hours-tens digit, address 04H
        SETB    P1.2      ; E=1
        CLR     P1.2      ; E=0
        ACALL   T_BF
;***    Initialize the software counters    ****************
        MOV     R0,#0     ; Clear C_200
        MOV     R1,#0     ; Clear C_20
        MOV     R2,#0     ; Clear SEC
        MOV     R3,#0     ; Clear SEC_T
        MOV     R4,#0     ; Clear MIN
        MOV     R5,#0     ; Clear MIN_T
        MOV     R6,#0     ; Clear HOUR
        MOV     R7,#0     ; Clear HOUR_T
;***    Initialize Timer/Counter 0   **********************
        MOV     TMOD,#02H ; Timer 0, mode 2
        MOV     TH0,#6    ; Delay 250 mcs, 256-250=6,
                          ; 12 MHz crystal
        SETB    ET0       ; Enable interrupts Timer 0
        SETB    EA        ; Enable all interrupts
        SETB    TR0       ; Start Timer 0
;**********************************************************
ADJUST: JB      P3.5,TEST_M
        CLR     TR0       ; Stop Timer 0
        JNB     P3.5,$    ; Wait
        SETB    TR0       ; Start Timer 0
        ACALL   COUNT_H
;
TEST_M: JB      P3.4,TEST_S
        CLR     TR0       ; Stop Timer 0
        JNB     P3.4,$    ; Wait
        SETB    TR0       ; Start Timer 0
        INC     R4        ; Increment minutes
        CJNE    R4,#10,ADJUST
        MOV     R4,#0
        INC     R5        ; Increment minutes-tens
        CJNE    R5,#6,ADJUST
        MOV     R5,#0
;
TEST_S: JB      P3.2,ADJUST
```

```
          JNB      P3.2,$
          MOV      R2,#0          ; Clear seconds
          MOV      R3,#0          ; Clear seconds-tens
          SJMP     ADJUST         ; Loop
;*****************************************************************
;     DEL: This subroutine provides a delay between           *
;          103 and 24741 mcs depending on the code in         *
;          register R0 (1 to 255)                             *
;          Execution time : 97*R0 + 6, mcs                    *
;*****************************************************************
DEL:      PUSH     ACC            ; Save ACC
DEL1:     MOV      A,#2FH
DEL2:     DJNZ     ACC,DEL2
          DJNZ     R0,DEL1
          POP      ACC            ; Restore ACC
          RET
;*****************************************************************
;     T_BF:  This subroutine tests the Busy flag              *
;            Prepares the LCD for CGRAM/DDRAM                  *
;            Data write operation                             *
;*****************************************************************
T_BF:     MOV      P4,#0FFH       ; Input port
          CLR      P1.0           ; RS=0 > Control operation
          SETB     P1.1           ; R/W=1 > Read
T_BF1:    SETB     P1.2           ; E=1
          MOV      ACC,P4         ; Read the Busy Flag
          CLR      P1.2           ; E=0
          JB       ACC.7,T_BF1
          SETB     P1.0           ; RS=1 > Data operation
          CLR      P1.1           ; R/W=0 > Write
          RET
;*****************************************************************
;     COUNT: This interrupt handler for Timer 0               *
;     Activates the subroutine COUNT_S once per second        *
;     Updates the display                                     *
;*****************************************************************
COUNT:    INC      R0             ; Increment C_200
          CJNE     R0,#200,TEST_F
          MOV      R0,#0          ; Clear C_200
          INC      R1             ; Increment C_20
          CJNE     R1,#20,RETURN
          MOV      R1,#0          ; Clear C_20
          ACALL    COUNT_S        ; Increment seconds
;
          SETB     P1.0           ; RS=1 > Data operation
          MOV      A,R7           ; Hours-tens
          ADD      A,#30H         ; Convert to ASCII
          CJNE     A,#30H,NO_ZERO
          MOV      A,#20H         ; Blank the digit
NO_ZERO:  MOV      P4,A
          SETB     P1.2           ; E=1
          CLR      P1.2           ; E=0
```

```
            SETB    F_H_T
RETURN:     RETI
;
TEST_F:     JNB     F_H_T,TEST1
            CLR     F_H_T
            MOV     A,R6            ; Hours
            ADD     A,#30H          ; Convert to ASCII
            MOV     P4,A
            SETB    P1.2            ; E=1
            CLR     P1.2            ; E=0
            SETB    F_H
            RETI
;
TEST1:      JNB     F_H,TEST2
            CLR     F_H
            MOV     P4,#':'
            SETB    P1.2            ; E=1
            CLR     P1.2            ; E=0
            SETB    F_COL1
            RETI
TEST2:      JNB     F_COL1,TEST3
            CLR     F_COL1
            MOV     A,R5            ; Minutes-tens
            ADD     A,#30H          ; Convert to ASCII
            MOV     P4,A
            SETB    P1.2            ; E=1
            CLR     P1.2            ; E=0
            SETB    F_MIN_T
            RETI
;
TEST3:      JNB     F_MIN_T,TEST4
            CLR     F_MIN_T
            MOV     A,R4            ; Minutes
            ADD     A,#30H          ; Convert to ASCII
            MOV     P4,A
            SETB    P1.2            ; E=1
            CLR     P1.2            ; E=0
            SETB    F_MIN
            RETI
;
TEST4:      JNB     F_MIN,TEST5
            CLR     F_MIN
            MOV     P4,#':'
            SETB    P1.2            ; E=1
            CLR     P1.2            ; E=0
            SETB    F_COL2
            RETI
;
TEST5:      JNB     F_COL2,TEST6
            CLR     F_COL2
            MOV     A,R3            ; Seconds-tens
            ADD     A,#30H          ; Convert to ASCII
```

```
            MOV     P4,A
            SETB    P1.2            ; E=1
            CLR     P1.2            ; E=0
            SETB    F_SEC_T
            RETI
;
TEST6:      JNB     F_SEC_T,TEST7
            CLR     F_SEC_T
            MOV     A,R2            ; Seconds
            ADD     A,#30H          ; Convert to ASCII
            MOV     P4,A
            SETB    P1.2            ; E=1
            CLR     P1.2            ; E=0
            SETB    F_SEC
            RETI
;
TEST7:      JNB     F_SEC,RETURN
            CLR     F_SEC
            CLR     P1.0            ; RS=0 > Control operation
            MOV     P4,#84H         ; Hours-tens digit, address 04H
            SETB    P1.2            ; E=1
            CLR     P1.2            ; E=0
            RETI
;*****************************************************
;    COUNT_S:   This subroutine counts seconds    *
;*****************************************************
COUNT_S:    INC     R2              ; Increment seconds
            CJNE    R2,#10,RET_S
            MOV     R2,#0           ; Clear seconds
            INC     R3              ; Increment seconds-tens
            CJNE    R3,#6,RET_S
            MOV     R3,#0           ; Clear seconds-tens
            ACALL   COUNT_M         ; Increment minutes
RET_S:      RET
;*****************************************************
;    COUNT_M: This subroutine counts minutes      *
;*****************************************************
COUNT_M:    INC     R4              ; Increment minutes
            CJNE    R4,#10,RET_M
            MOV     R4,#0           ; Clear minutes
            INC     R5              ; Increment minutes-tens
            CJNE    R5,#6,RET_M
            MOV     R5,#0           ; Clear minutes-tens
            ACALL   COUNT_H         ; Increment hours
RET_M:      RET
;*****************************************************
;    COUNT_H: This subroutine counts hours   *
;*****************************************************
COUNT_H:    INC     R6              ; Increment hours
            CJNE    R6,#4,TEST10
            CJNE    R7,#2,TEST10
            MOV     R6,#0           ; Clear hours
```

```
          MOV    R7,#0           ; Clear hours-tens
          RET
TEST10:   CJNE   R6,#10,RET_H
          MOV    R6,#0           ; Clear hours
          INC    R7              ; Increment hours-tens
RET_H:    RET
```

In summary, the endless loop of the task "Adjust" is interrupted every 250 µs. Eight consecutive invocations of the interrupt handler update the LCD display every second.

Programmable Logic Controllers

In the field of process control the Programmable Logic Controllers (PLC) are the most widespread computers. The fundamental features are convenient programming language and ease of programming and reprogramming in-plant. The noise immunity is also a crucial factor.

The PLC manufacturers usually support their products with a variety of programming languages. As a rule, the software systems are capable of converting the user specification from one language to another. Figure 7.32 shows an example specification in three different languages. The system calculates one output value (Q01) on the base of five inputs (I01, I02, I04, I05 and I07). We can employ a small-scale PLC to implement this application. Hence, we will use a PLC programming language as shown in Figure 7.32. If we want to optimize the hardware for our application, we can design our own embedded system. However, in case of software reuse (we have programs previously written for a PLC), we might be forced to also design a compiler for the language used.

For simplicity, we have chosen an example which results in a pure combinational logic solution. This is valid, if we design our own embedded system. In case of commercial PLC, it will run a program which is obtained from the specification. A graphical language, termed Control System Flowchart (CSF), would be a convenient notation for people who are used to working with logic diagrams. Alternatively, the Ladder Diagram (LAD) language could be a better choice for factory stuff. Traditionally, the people who maintain the machines use relay diagrams. Furthermore, with ladder diagram language we are getting closer to the physical reality. Using the LAD notation we distinguish between normally open contacts (switches, sensors) and normally closed contacts. In our example, input I04 is specified as a normally closed contact. Moreover, the LAD notation can be viewed as an electrical circuit. For example, the output Q01 will be activated (energized) if at least one of the parallel branches conducts "current". For the upper brunch, the contact I01 must be closed (active) and the contact I04 must be opened (active). When this condition is met, the PLC will sense a signal at the input I01. No signal will be sensed at the input I04. Likewise, if both contacts I02 and I05 are closed (active), the output Q01 will emit signal (active). Regardless of the other inputs, the input I07 will energize the output Q01 when closed.

The language Statement List (STL) is a specification entry which utilizes text. Again, the abstraction level corresponds to logic functions. The letter A stands for the AND operation. Similarly, the letter O is the abbreviation from OR. By convention, the AND instructions are processed first. Then the results are involved in the OR operation. The final value is assigned to the output Q01.

If the example specification from Figure 7.32 is compiled to the 8051 assembly language instructions, the following sequence will emerge:

```
MOV  C,I01  ; Move the value of input I01 from the
            ; bit-addressable area to the carry flag
```

```
ANL  C,/I04  ; Use a slash for normally closed contacts to
            ; complement their values
MOV  TEMP,C  ; Store the result in a temporary bit-addressable
            ; location
MOV  C,I02  ; Move the value of input I02 from
            ; the bit-addressable area to the carry flag
ANL  C,I05  ; AND operation
ORL  C,TEMP  ; OR operation
ORL  C,I07  ; OR operation
MOV  Q01,C  ; Move the result to a bit-addressable location
            ; labelled Q01
```

The instructions above perform an essential task - calculation of the new output values. In fact, a PLC executes an endless loop which goes through the following steps [Parr 1993]:

• The programmable controller reads all input signals and stores them in the memory. The memory area which contains the current input values is called Process Image Input table (PII).

• The PLC calculates the new output values and stores them in a memory area termed Process Image output table (PIQ).

• The PLC updates the outputs and the program jumps to execute the first task again.

At the same time, it is possible to interrupt the loop and process aperiodic tasks. Either internal or external events can trigger the interrupt subroutines.

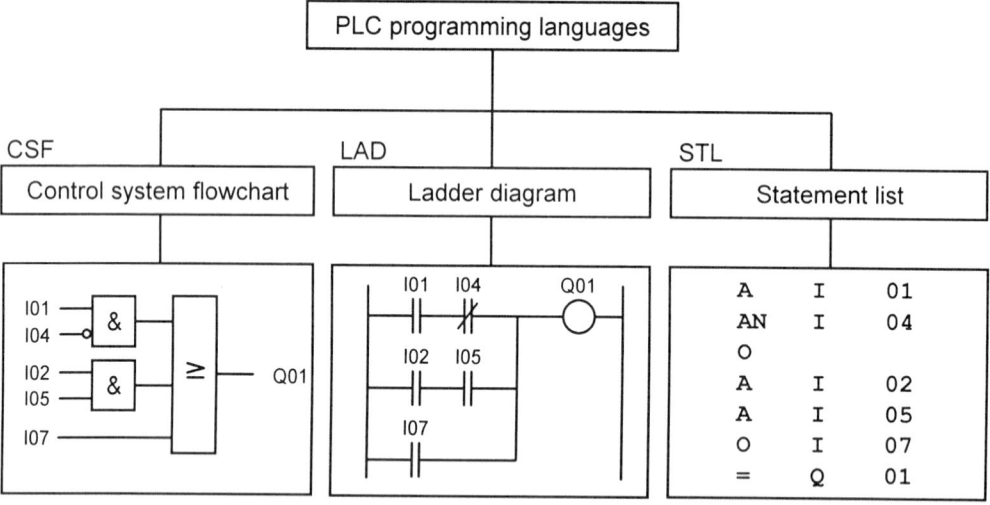

Figure 7.32 One example presented by three different languages.

Obviously, the Boolean (single-bit) processor makes the 8051 compatible microcontrollers suitable candidates for PLC systems. Along with the microcontroller's CPU, the built-in memory and peripherals also outline the 83C552 as a good choice for PLC designs. Figure 7.33 shows a typical PLC structure. In an attempt to evaluate the situation we shaded the blocks which can be partly or completely covered by the 8051 or 83C552 microcontrollers. It gives you some insight into how the microcontroller meets the hardware requirements of this particular application.

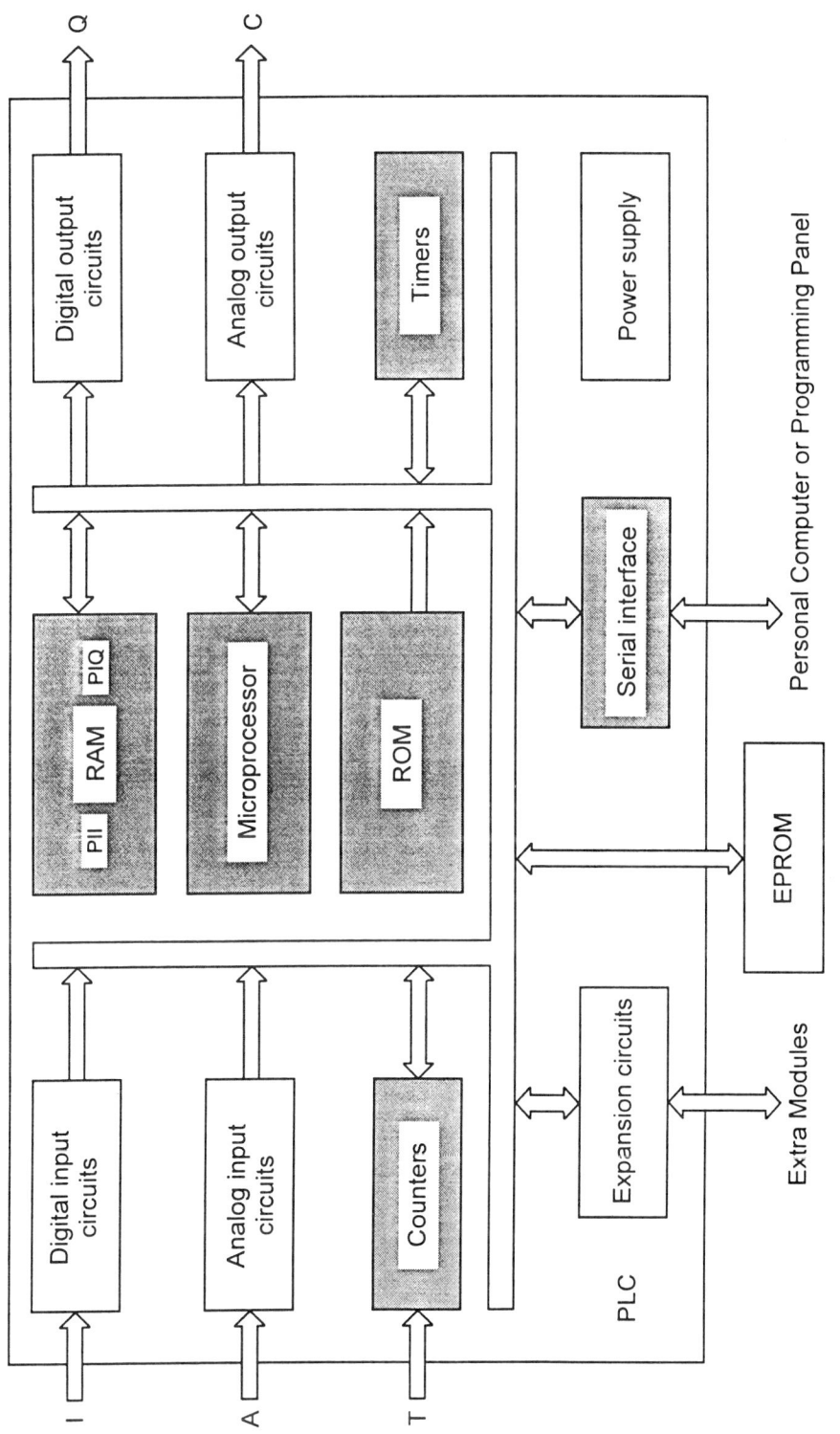

Figure 7.33 A typical PLC structure.

An example of a digital input circuit can be seen in Figure 7.34. The field devices, such as sensors and relays are represented as mechanical switches. The input voltage power supply of 24V complies with the industrial standards. The group R_1 - C forms a low-pass filter which, together with the Schmitt trigger, increases the noise immunity. Normally, each input has an individual LED lamp. Alternatively, optocouplers can be involved as shown in Figure 7.35. An example of a digital output circuit can be found in Figure 4.20, where the parallel interface was discussed.

As far as the analog input and output circuits are in focus, the PLC could use the 83C552 embedded 10-bit ADC and the pulse width modulated outputs $\overline{PWM0}$ and $\overline{PWM1}$ (see sections 7.7 and 7.8). In any case, some extra components, especially filters and amplifiers, should be added.

In the majority of cases, the PLC design will require more memory than the microcontroller can provide. Frequently, a memory block is organized as a removable box which allows fast reprogramming.

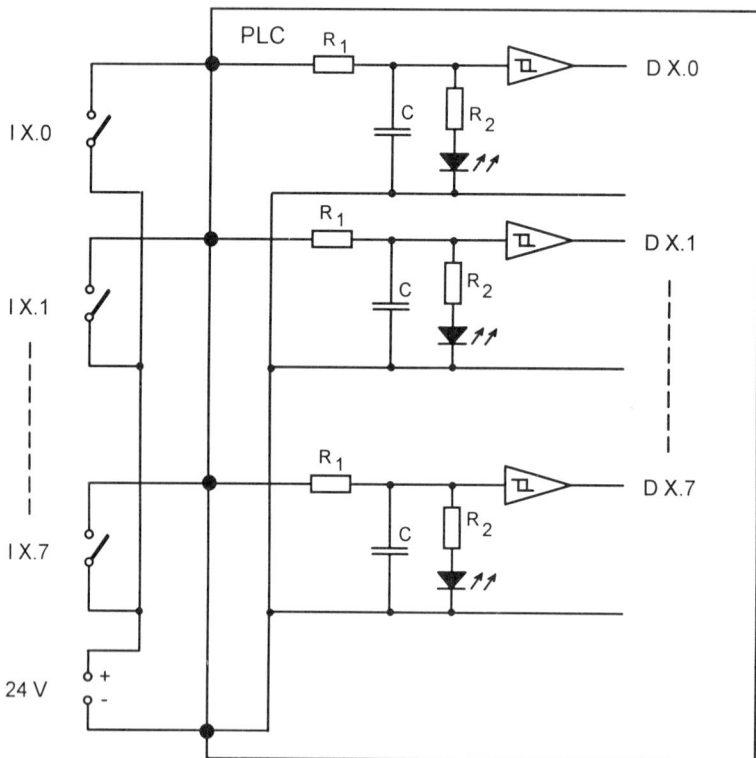

Figure 7.34 Digital input circuit.

As is frequently the case, the PLCs are organized in a network or at least connected to a programming station. As a rule, the PLC can execute a program and communicate over the serial channel simultaneously.

Typically, noise problems can be avoided if several precautions are taken against. First, shielding is used to block capacitive and inductive coupling. Furthermore, a correct grounding scheme should be organized. Finally, we admit that noise transients may send the program execution to a random address in the memory. In this situation, the system should recover as soon as possible. Practically, a watchdog timer is used to reset the PLC.

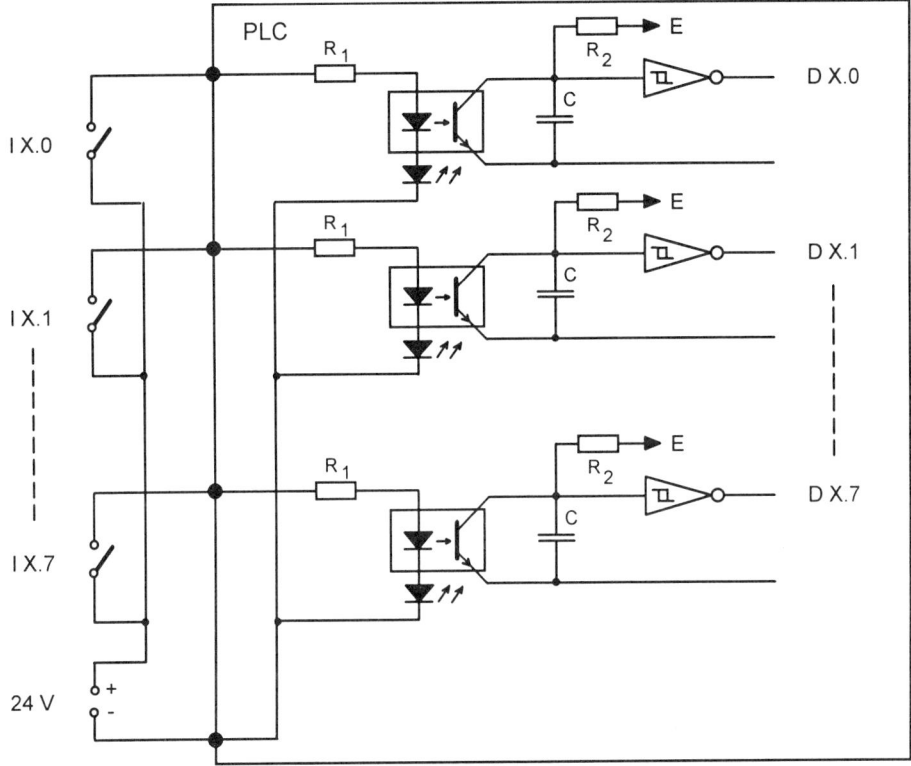

Figure 7.35 Digital input circuit with optocouplers.

In some applications, the type of the contacts indicated in the ladder diagram might be different from the actual implementation. What is certain is that the input values of the normally closed contacts in the diagram will be complemented before operation (as we did with the input I04 in a previous example). This case will be discussed in the following problem.

Problem 7.5

Use the ladder diagram language to specify a system which has two push-button inputs and one output. One of the inputs is a start input. The other is a stop input. Pressing the start button activates the output. Pressing the stop input disactivates the output. Write the microcontroller's instructions, which will be generated by the ladder diagram compiler.

Solution 7.5

Following the specification, we suggest a PLC wiring which can be seen in Figure 7.36. For safety reasons, we allocate a normally closed STOP push-button. Usually, the hazards are associated with active outputs. For example, if the output controls a motor, it will be more important to stop the motor rather than to start it. A normally closed STOP button would deactivate the output LOAD in case of malfunctions, such as loss of supply or cable fault.

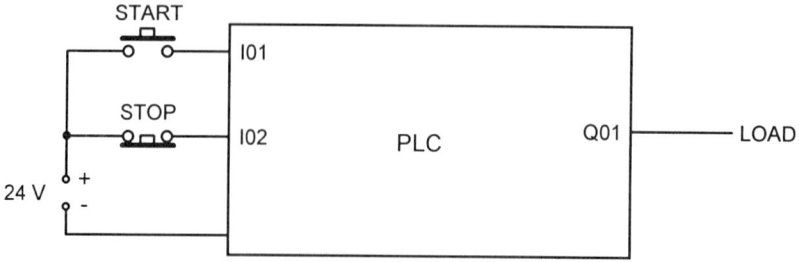

Figure 7.36 The PLC wiring.

The following logic equation is consistent with both the specification and the wiring:

$$LOAD = STOP(START \vee LOAD)$$

The equivalent equation for the PLC is

$$Q01 = I02(I01 \vee Q01)$$

As you can see, a bit from the PIQ, which is devoted to the output, keeps the output active when the START button is released.

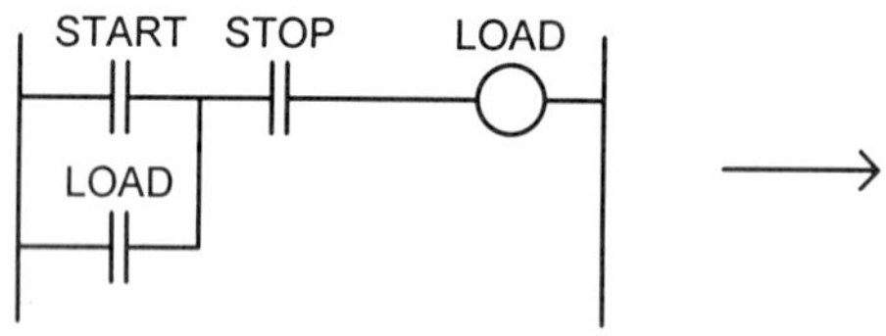

START STOP   LOAD
├─┤├──────┤├────────( )──────────┐
                                 │   MOV   C,START
   LOAD   │ ──→   ORL   C,LOAD
├─┤├─   │   ANL   C,STOP
                                 │   MOV   LOAD,C

Figure 7.37 The ladder diagram presentation and the corresponding instructions.

Figure 7.37 shows the ladder diagram presentation and the corresponding assembly language instructions.

7.12 Supplementary problems

Problem S7.1
Rewrite the solution of Problem 3.2 for the 83C552 microcontroller's Timer 2. Use the pin P4.0/CMSR0 for the pulse output. Assume that the input P1.5/RT2 is available and can be used to clear Timer 2.

Problem S7.2
Rewrite the solution of Problem 7.1 to use the complete 10-bit result from the A-to-D conversion.

Problem S7.3
Redesign the solution of Problem 7.1 with the possibility of selecting the analog channel from the PC.

Problem S7.4
Design a digital voltmeter based on the 83C552 built-in ADC. The input range must be 0 to 5.12 V, quantum 5 mV. For a display use the LCD LTN211 (see Figure 7.16 and Figure 7.28).

Problem S7.5
Write an assembly language program for the 83C552 microcontroller which serves as a driver for the keyboard with interfacing shown in Figure 7.38 [Thei 1995]. The program decodes the key pressed by measurement of the voltage.

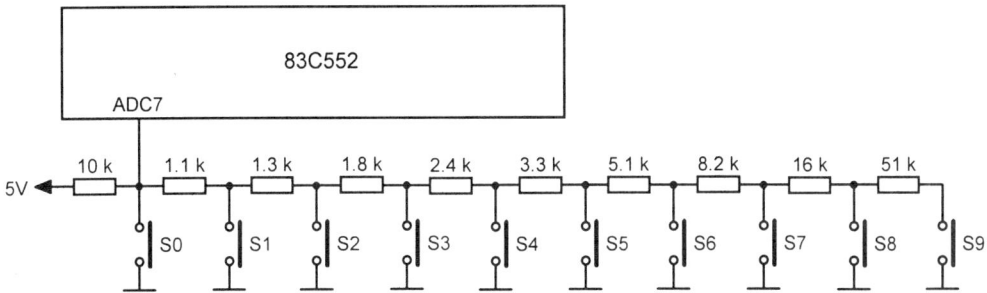

Figure 7.38 Interfacing a keyboard through an analog input.

Problem S7.6
Write an assembly language subroutine for the 8051 microcontroller, which takes as an entry condition the low nibble in the accumulator. The subroutine reloads the accumulator with the corresponding seven-segment pattern for LED displays. The codes between 0AH and 0FH are displayed as the letters A, b, c, d, E and F. The subroutine will be used for the interfacing approach illustrated in Figure 7.27.

Problem S7.7
Add an alarm feature to the digital clock design presented in Problem/Solution 7.4.

Problem S7.8
Write an assembly language program for the 8051 microcontroller which implements the ladder diagram example from Figure 7.39.

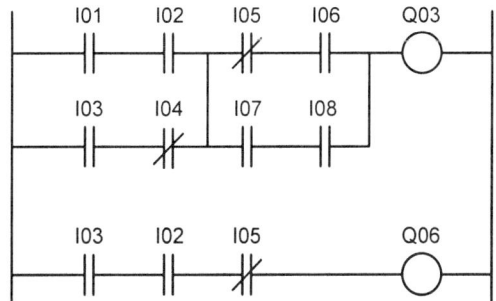

Figure 7.39 A two rung ladder diagram example.

7.13 References

Analog Devices, *Data Converter Reference Manual, Volume II*, 1992b.

Analog Devices, *Design - In Reference Manual*, 1994.

Hans Berger, *Programming of Control Systems in STEP 5, Volume 1 Basic Software*, Siemens Aktiengesellschaft, 1984.

Hewlett Packard, *Optoelectronics Designer's Catalog*, 1993.

Zdravko Karakehayov and Emil Saramov, *Applied Microcomputer Systems*, Technical University of Sofia, 1995a.

E. A. Parr, *Programmable Controllers*, Newnes, 1993.

Philips Components, *Liquid Crystal Displays and Driver ICs for LCD, Data Handbook*, 1991.

Philips Semiconductors, *80C51-Based 8-Bit Microcontrollers, Data Handbook IC20*, 1997a.

Philips Semiconductors, *Application Notes and Development Tools for 80C51 Microcontrollers, Data Handbook*, 1997b.

Philips Semiconductors, *I^2C Peripherals, Data Handbook IC12*, 1996.

Texas Instruments, *The TTL Data Book*, Volume 1, 1989.

Frank Theiner, Carsten Trapp and Bernd vom Berg, ""Geschlossene" Mikrocontroller - zuganglich gemacht", *Elektronik*, 22/1995, pp. 94-100.

I. G. Warnock, *Programmable Controllers - operation and application*, Prentice Hall, 1988.

Sencer Yeralan and Ashutosh Ahluwalia, *Programming and Interfacing the 8051 Microcontroller*, Addison-Wesley Publishing Company, 1995.

Information about 83C552 microcontrollers can be found at

    Philips Semiconductors   http://www.philips.com

Information about LCD manufacturers can be obtained from

                              http://www.eio.com/lcdmanuf.htm

Information about LED displays can be found at

    Hewlett-Packard Co.   http://www.hp.com

Chapter 8

SERIAL INTERFACES
FOR DISTRIBUTED EMBEDDED SYSTEMS

8.1 Introduction

In this chapter we discuss serial communications again. The shift from centralized to distributed control simplifies the wiring, introduces fault tolerance and streamlines the development process. After the brief introduction of the I^2C bus in section 7.1, the interface is now in focus. The I^2C bus is not yet another interface. It may give us cost-effective results which we would not otherwise be able to obtain, however, as with the other innovations in the computer industry, it requires wide acceptance from the market. Since the early 1980s, when the I^2C bus was introduced by Philips, more than 500 devices were developed and proved useful in applications, such as consumer electronics, telecommunications, industrial control and measurement equipment.

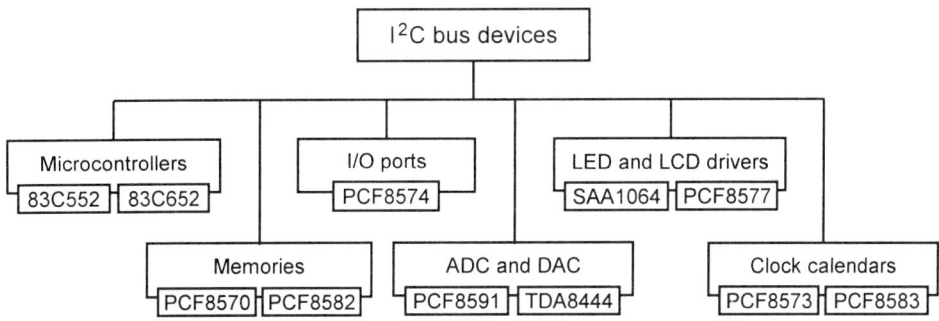

Figure 8.1 Survey of I^2C bus devices.

Figure 8.1 lays out a few representative devices produced by Philips [Phil 1996, Phil 1997a, Phil 1997b]. They possess I²C bus hardware. The 83C652 microcontroller is an 8051 pin compatible device with doubled memory. In addition, it has an I²C bus subsystem. The same applies, also, for the memories and peripherals included in Figure 8.1.

In this chapter we first discuss the basic concepts of the I²C interface. Next we explain the 83C552 I²C bus subsystem. Then, application examples and problems illustrate how I²C bus devices can be used. Finally, we introduce briefly the CAN bus, an alternative serial interface suitable for distributed embedded systems. The CAN bus was designed to improve the conventional electronic control devices in the motor vehicles. Also, section 8.6 introduces the 83C592 microcontroller, which is the mirror image of the 83C552 device for CAN applications.

8.2 I²C bus background

Serial communication is a common practice when a system interacts with the environment. The designers of the I²C bus looked at the microcomputer system itself as a set of components that could be connected by a serial link. Philips defined the I²C bus as a 2-wire bidirectional serial bus for communication between ICs located within a common chassis.

A general description of the I²C bus includes the following features:

• The I²C bus is a serial, 8-bit oriented bus.

• The bus is a multi-master bus with collision detection and arbitration.

• Each device connected to the bus has its own slave address.

• The data transfer rate can be up to 100 Kbps in the standard mode (S-mode), 400 Kbps in the fast mode (F-mode) or 3.4 Mbps in the high-speed mode (Hs-mode).

The I²C bus offers several advantages to the embedded systems designers. First of all, the system is completely software defined due to the fact that the bus interface is integrated on-chip. At a later phase, the system could be easily upgrade or enhanced by attaching new ICs.

The manufacturer benefits are also substantial. Naturally, the serial approach leads to ICs with fewer pins and smaller PCBs. Furthermore, there is no need for glue logic. Finally, the testing procedures are alleviated.

Figure 8.2 shows an example I²C bus system. The data line is called Serial DAta (SDA). The clock line is termed Serial CLock (SCL).

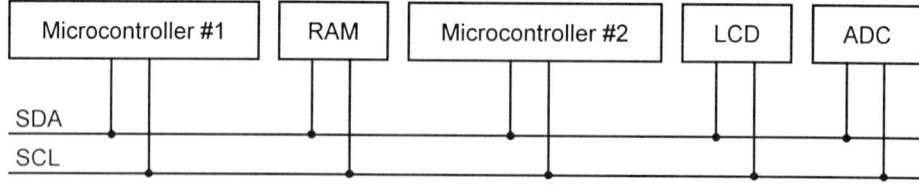

Figure 8.2 An example I²C bus system.

Figure 8.3 details the I/O buffers of the devices connected to the bus. Obviously, if one transmitter pulls an output low, the corresponding line will be brought low, regardless of the other outputs. A transmitter is a device which sends data to the bus. Likewise, a receiver is a device that receives data from the bus. A master is a device which initiates a transfer, provides the clock signal and terminates the transfer. The device addressed by the master is called slave.

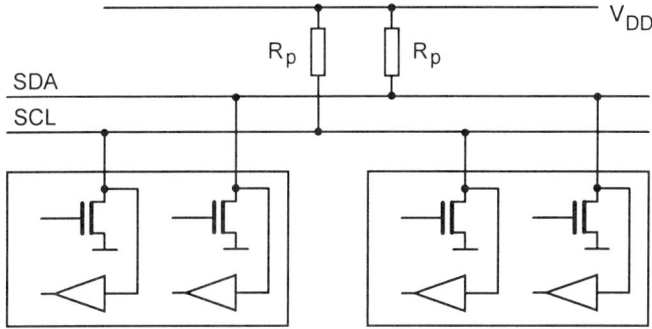

Figure 8.3 Two I²C bus devices and their I/O buffers.

As you might imagine, the data line signal must be stable when the clock line level is high. This can be seen in Figure 8.4. There are, however, two important exceptions. They are illustrated in Figure 8.5.

• First, we should be able to define a Start (S) condition. It is done by a high to low transition on the SDA line, when the SCL line is high.

• Second, a stoP (P) condition is established on the bus, when the SDA level is changed to high and the SCL line is hold high. As we mentioned earlier, the START and STOP conditions are generated by the master.

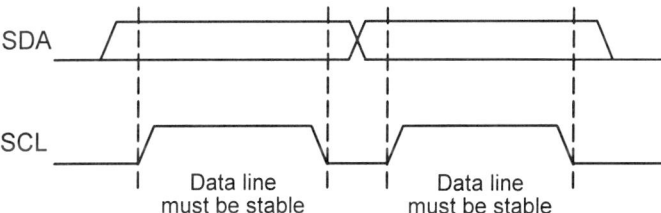

Figure 8.4 Bit transfer waveforms.

Now we move on to the byte format. One byte is the quantum used when data is transmitted over the bus. The number of bytes per transfer is not restricted. Naturally, the feedback between the transmitter and receiver makes the communication reliable. In this case, each byte is followed by an acknowledge slot. Figure 8.6 (on p.263) shows an example, when a master sends two bytes to a slave. A key point in understanding the operation of the bus is the difference which may occur between the output signals and the real logic levels on the bus. The first four waveforms attempt to show how the output buffers are driven. The last two waveforms present what actually can be seen on the I²C bus.

Figure 8.6 reveals certain key points:

• The most significant bit (MSB) is transferred first.

• When the master generates the acknowledge clock pulse (#9), it releases the SDA line and the slave pulls the SDA line down to confirm that a byte has been received.

Less obviously, if the receiver wants to slow down the data transfer, it could be done immediately after the acknowledge slot by holding the SCL line low. The transmitter will sense the SCL line and wait. Furthermore, it will be consistent with the I²C bus convention, if the rate is declined on the bit level. This is done by extending the low clock time.

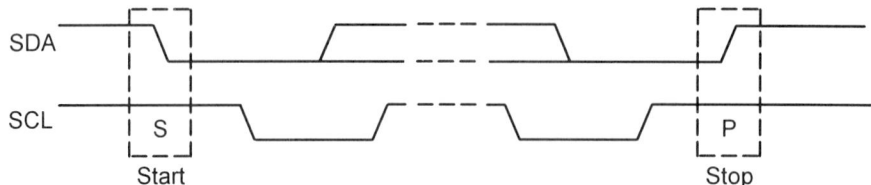

Figure 8.5 START and STOP situations.

The first byte which follows the START condition contains a 7-bit slave address. The eighth bit from the same byte defines the data flow direction. Slave address 0000 000 plus a low bit $R / \overline{W}$ is reserved for addressing every device connected to the I²C bus (general call). Figure 8.7 shows an example when a master transmits two bytes to a slave. The eighth bit of the address byte is zero to signal transmission.

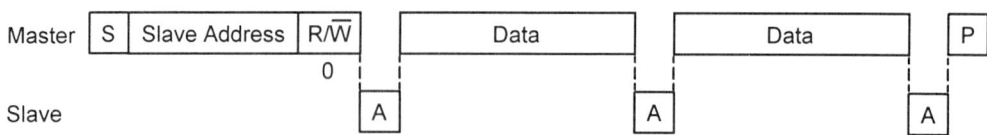

Figure 8.7 A master transmits two bytes to a slave, selected by a 7-bit address

Likewise, Figure 8.8 demonstrates a receive procedure on the bus. In this example, a master selects a slave and demands a read operation by a set bit $R / \overline{W}$. The slave sends back an acknowledgement bit (low) and the first byte. The master confirms the reception of the first byte. Again, the slaves transmits a byte and the master sends an acknowledgement bit. Finally, the master generates a STOP condition.

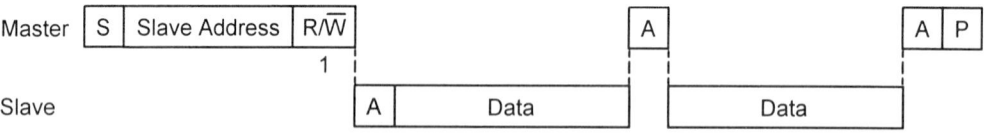

Figure 8.8 A master receives two bytes from a slave.

As yet another possibility, Figure 8.9 lays out the bus pattern when a master transmits one byte to a certain slave, generates a repeated START condition (S_R) and receives one byte from a slave. Inevitably, two or more masters could try to take over the bus simultaneously. The arbitration is done on the SDA line and the master, which transmits a high level while the SDA line is pulled down by another master, must switch off its output buffers and wait until the bus is released. The arbitration will continue in the data bytes if the masters are addressing a single slave and the desired direction of data flow is the same.

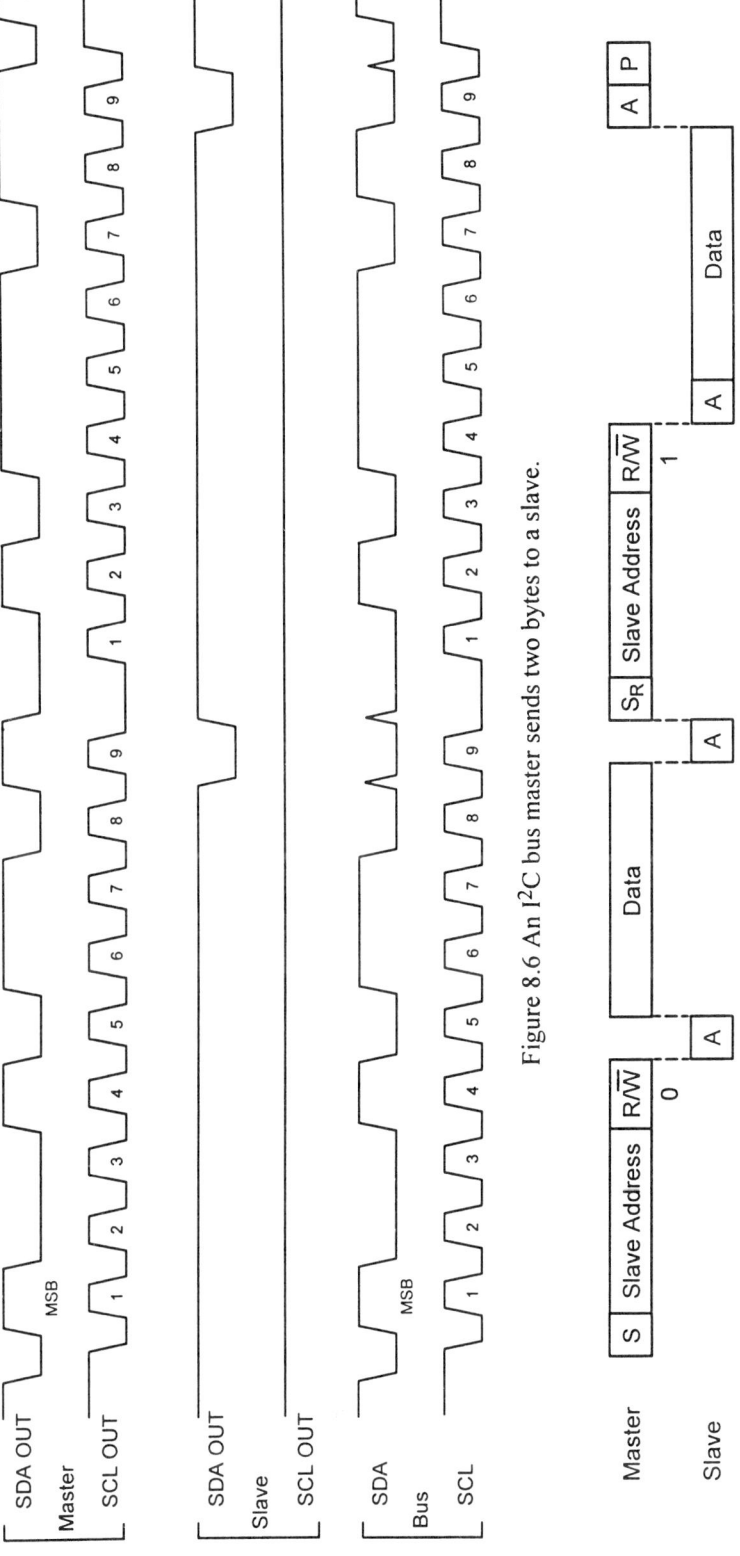

Figure 8.6 An I^{2}C bus master sends two bytes to a slave.

Figure 8.9 A master transmits one byte to a slave, generates a repeated START condition, and receives one byte.

8.3 The 83C552 microcontroller I²C bus subsystem

Now we are back to the 83C552 microcontrollers and our aim is to describe the embedded I²C bus hardware. Figure 8.10 lays out the involved registers. All register names start with S1 which means serial I/O #1. First of all, the microcontroller possesses a data register called S1DAT. The register contains a byte to be sent over the bus or a byte which has been received. As illustrated in Figure 8.10, the register S1DAT is used for both data and addresses.

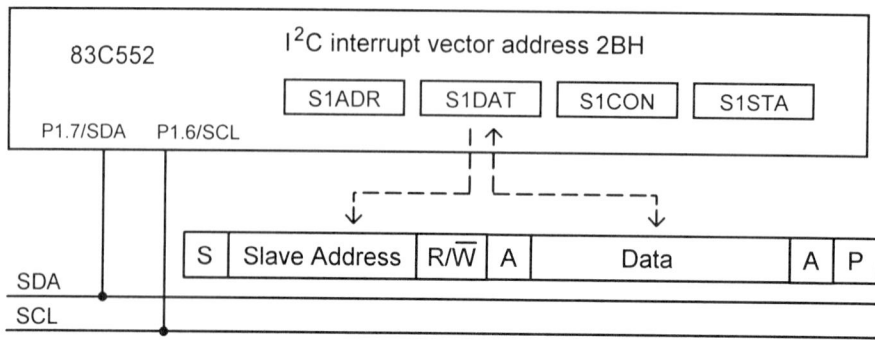

Figure 8.10 The 83C552 I²C bus related registers.

Furthermore, a register called S1ADR is used to hold the microcontroller 7-bit slave address. The address occupies the seven most significant bits. The least significant bit of the register S1ADR is a flag labelled General Call (GC). If the flag GC is set, the microcontroller will respond to the general call address (00H). In addition, the I²C bus on-chip logic includes a control register S1CON and a status register S1STA. And lastly, Figure 8.10 indicates that the SDA line is placed on P1.7 and the SCL pin is P1.6.

In a attempt to produce a clear picture of the interaction between the microcontroller's CPU, the interrupt and I²C bus subsystems, we propose Figure 8.11. The illustration will help us to draw the borderline between what is done automatically by hardware and what must be organized by software. The picture contains a piece of software, the I²C bus registers and the bus pattern. In this example the microcontroller transmits one byte to a certain slave.

The program starts with an initialization part (**INIT**). The first instruction loads the address register S1ADR with the microcontroller's own slave address. Although, this instruction is not a necessity for our example, it will be used in other programs when the microcontroller is addressed as a slave. The initialization goes on with programming the lines P1.6 and P1.7 as inputs. This is logical. The microcontroller must check if the bus is available before generating the START condition.

The next step is to define the place in the memory where the I²C bus interrupt subroutines are stored. We actually employ a temporary location called **HADD** to specify the high-order address byte **PAG1**. Consequently, the microcontroller will select a certain interrupt subroutine by means of the low-order address byte which comes from the status register S1STA. Moreover, the interrupt system is programmed to allow interrupts from the I²C bus. The initialization continues with the I²C bus clock rate. There are three bits in the control register called clock rate bits CR0, CR1 and CR2. The clock rate bits define the ratio between the oscillator frequency and the I²C bus clock rate. In this example, the oscillator frequency is divided by 120. As a result, the transfer rate is 100 Kbps.

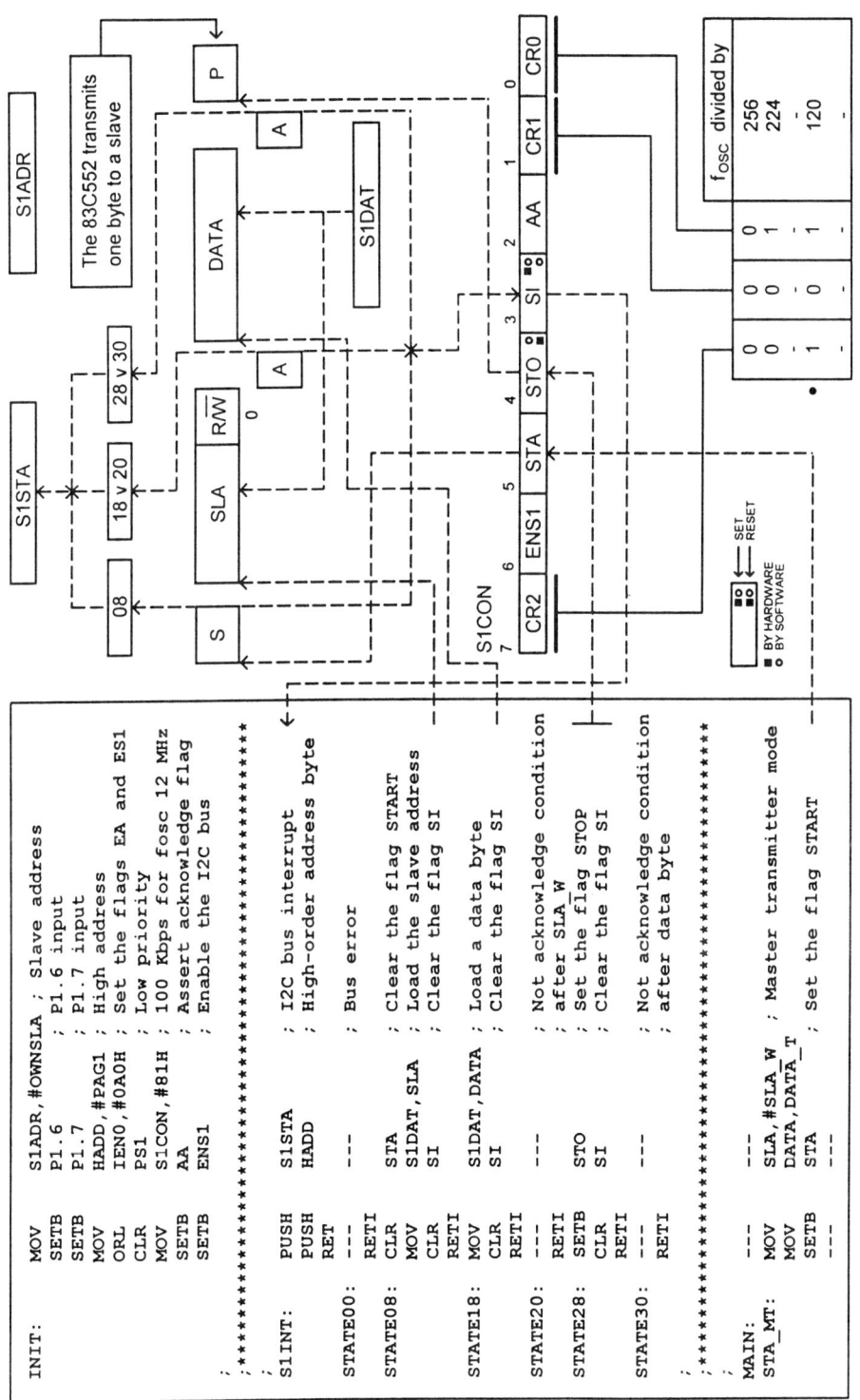

Figure 8.11 The interaction between the 83C552 hardware, software, and the I²C interface.

Furthermore, we set the bit Assert Acknowledge (AA) from the register S1CON in order to follow the normal procedure - an acknowledgement bit is returned during the corresponding clock pulse. And finally, the enable bit ENS1 is set to attach the I^{2}C bus hardware to the SDA and SCL lines. When the bit ENS1 is reset, the outputs P1.6 and P1.7 are in a high impedance state and the input signals are ignored.

Once the initialization instructions have been executed, the I^{2}C bus is ready for use. For example, at a certain point in the main program (**MAIN**) the microcontroller starts to transmit to a certain slave (**STA_MT**). The first instruction moves the slave address combined with a write bit (**SLA_W**) to a temporary location **SLA**. Likewise, the byte to be transmitted is moved to a location called **DATA**. Finally, the program must generate a START condition by setting the bit STA from the register S1CON. As soon as the bus is free, the I^{2}C unit will send the START condition over the bus. Then, the interrupt flag SI will be set and code 08H moved to the status register S1STA.

As can be seen in Figure 8.11, the interrupt subroutines share a common sequence of instructions labelled **S1INT**. The idea is to reach a certain interrupt subroutine by executing the instruction **RET**. First, we push the status code into the stack. Second, the high-order address byte **HADD** is also pushed to complete the target address. The START condition returns status code 08H. The corresponding subroutine is called **STATE08**. This subroutine clears the bit STA from the register S1CON, moves the slave address to the data register and clears the interrupt flag SI. Once the flag has been reset, the I^{2}C bus hardware is activated again and the slave address is transmitted. This phase can end up in two different ways. If an acknowledgement bit has been received, the status code will be 18H. A not acknowledge condition will be indicated by status code 20H. The reaction of the program when the slave does not return an acknowledgement bit will vary from application to application. One possibility is to set the flag STA and reset the interrupt flag SI. This will result in a repeated START condition over the bus. We recommend that the reader study the situations listed in the microcontroller data sheet, as space does not permit a full discussion [Phil 1997a].

When the slave address has been transmitted and an acknowledgement bit received, code 18H is moved to the register S1STA. The subroutine called **STATE18** writes a data byte to the register S1DAT and makes the I^{2}C interface active by clearing the interrupt flag SI. Naturally, acknowledge (status code 28H) and not acknowledge (status code 30H) are possible again.

With the assumption that the transfer has been successful, a STOP condition is generated by setting the bit STO. The stop flag will be reset automatically by hardware.

The status code 00H indicates bus error caused by a START or STOP condition occurred at an illegal position.

Thus, to summarize, the interrupt flag SI reactivates the I^{2}C bus logic for the next step, when it is reset.

8.4 I^{2}C bus application examples and problems

Interfacing I/O expanders

The examples and problems with I/O expanders can be viewed as an effort to systematically cover all the possibilities for I/O expansion while designing embedded systems. In the beginning of section 4.3 we mentioned that serial-to-parallel and parallel-to-serial conversion could be considered as well. At that time, however, we had in mind the 8051 microcontroller and did not discuss the details. Indeed, sacrificing the serial port for I/O expansion would fit a limited number of applications. Now, in light of the I^{2}C bus concept, the expansion by conversion is justified completely. The only parameter which is at stake is the speed, but this can be evaluated accordingly to the requirements.

The Philips engineers created an I^{2}C bus to parallel port expander which proved to be useful for I/O interfacing. Figure 8.12 shows the expander PCF8574 [Phil 1996]. The device possesses input and output register buffers. Furthermore, the illustration reveals the structure of the I/O expander slave address. The address can be broken down into three fields. The code of the first field (A6 A5 A4 A3) is fixed to 0100. In other words, all ICs have this pattern in their high-order address nibble. The second field (A2 A1 A0) is programmable. In this example, we have pulled down the inputs for the bits A2, A1 and A0. It is certainly implied that up to eight devices of this type can be used in a system. Finally, we distinguish between read and write operations by means of the LSB R / $\overline{\text{W}}$. As a result, the PCF8574 occupies the addresses 40H and 41H in the I^{2}C bus slave address memory map. Note that each I^{2}C bus device capable of receiving and transmitting occupies two addresses from the I^{2}C bus space.

The I/O expander is a slave device and is unable to initiate a transfer. Therefore, it has to be read for incoming data regularly. Alternatively, an interrupt output ($\overline{\text{INT}}$) can be connected to an interrupt input of the microcontroller. The open drain output $\overline{\text{INT}}$ is asserted by any rising or falling edge of the expander's lines programmed as inputs. The embedded systems designers might follow this opportunity in an attempt to avoid bus congestion.

Each line of the PCF8574 port can be individually used as an input or output. Similar to the microcontroller general purpose ports, if a certain port bit is used as an input, the corresponding output latch must be set.

Figure 8.13 shows the I^{2}C bus pattern, when the microcontroller writes to the I/O expander. The outputs are updated not later than 4 µs after the low-to-high transition of the SCL acknowledge clock pulse. Furthermore, Figure 8.14 explains the read operation. The required hold time for input data is 4 µs with regard to the low-to-high transition of the SCL acknowledge clock pulse. No acknowledgement from the microcontroller is an indication, that the read operation is over. The I/O expander leaves the line SDA high and the 83C552 generates a STOP condition.

As before with the LCD display, we are looking for a simple testing procedure. Again, we are going to apply single-stepping. In this case, there is no need to test the interrupt flag SI. Also, we assume that every byte sent is followed by an acknowledgement bit. Consequently, the program is simplified to the sequence of instructions below:

```
;*********************************************************
;        TEST_I2C.ASM                                    *
;        These instructions if single-step executed,     *
;        move a byte to the I/O expander                 *
;*********************************************************
SLA     EQU    05H     ; Write the I/O expander SLA_W (40H)
                       ; in location 05H
DATA    EQU    06H     ; Write a byte to be sent in location 06H
;
        ORG    2000H
INIT:   SETB   P1.6    ; P1.6 input
        SETB   P1.7    ; P1.7 input
        MOV    S1CON,#81H  ; 100 Kbps for 12 MHz crystal
        SETB   AA      ; Assert acknowledge
        SETB   ENS1    ; Enable I2C bus hardware
;
WRITE:  SETB   STA     ; Set the start flag
```

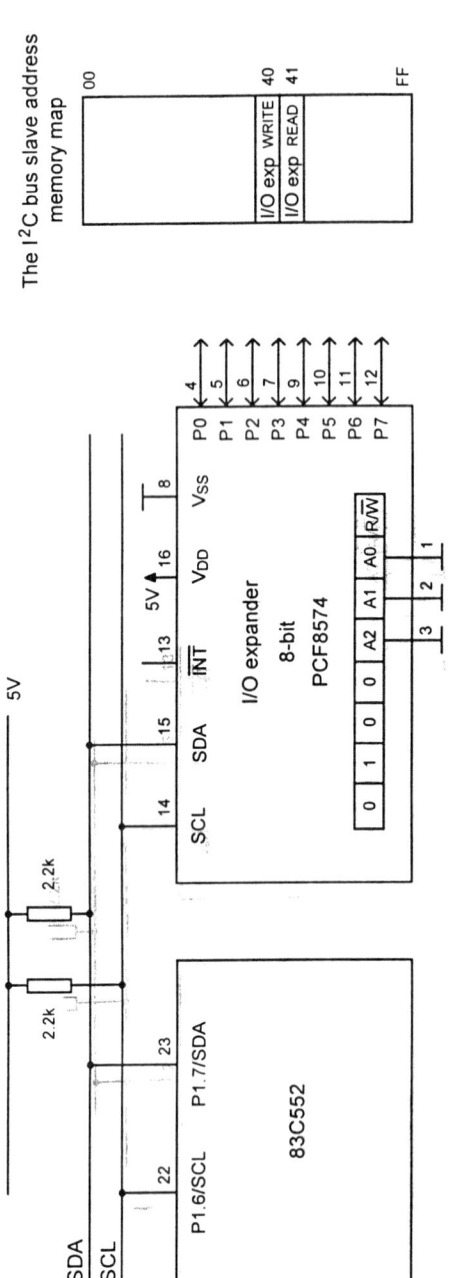

Figure 8.12 Interfacing an 8-bit I/O expander.

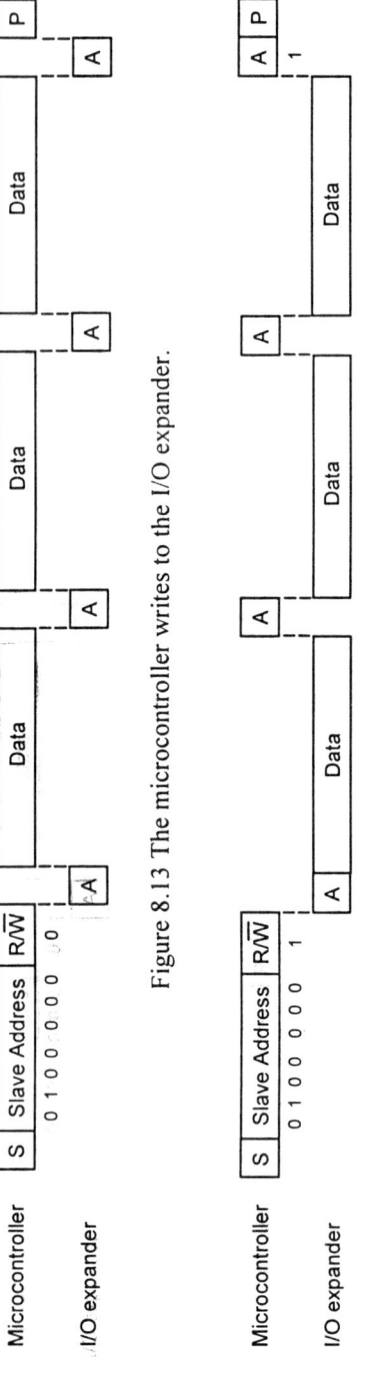

Figure 8.13 The microcontroller writes to the I/O expander.

Figure 8.14 The microcontroller reads from the I/O expander.

```
        CLR    STA          ; Clear the start flag
        MOV    S1DAT,SLA    ; Load the slave address
                           ; for write operations
        CLR    SI           ; Activate the I2C bus
;
        MOV    S1DAT,DATA   ; Load a data byte
        CLR    SI           ; Activate the I2C bus
                           ; update the outputs
        SETB   STO          ; Set the stop flag
        CLR    SI           ; Activate the I2C bus
;
        INC    DATA         ; Increment the data byte
        SJMP   WRITE        ; Loop
```

We store 40H in the internal Data Memory, address 05H, and single-step the code above. Each iteration in the loop moves the data from direct address 06H to the I/O expander. If you have more expanders in the system, you could modify the address in location 05H. Practically, it would be sufficient to indicate the state of the expander's output P0 (pin #4) by a LED lamp. Each iteration of the loop will alter the level at pin P0.

Once we have made sure that the hardware is functioning in single-step mode of execution, we can harness the I^2C bus by a perfect program.

Problem 8.1

Write an assembly language program for the 83C552 microcontroller which is connected to an I/O expander PCF8574 via I^2C bus. The program must turn the expander into a ring counter (rotating zero). The rotating rate is 1 bps.

Solution 8.1

Ring counters, alternatively labelled shift register counters, generate bit patterns that shift through all counter stages. The functionality of a ring counter implemented by a microcontroller can be broken down into three subtasks as shown in Figure 8.15. The subtasks T2 and T3 will be organized as subroutines. The program will execute an endless loop, which includes both subroutines **ROTAT** and **DEL_1S**. We assume that either the 83C552 microcontroller is the only master attached to the I^2C bus or if there are other masters, they will not take over the bus when the 83C552 is about to transmit.

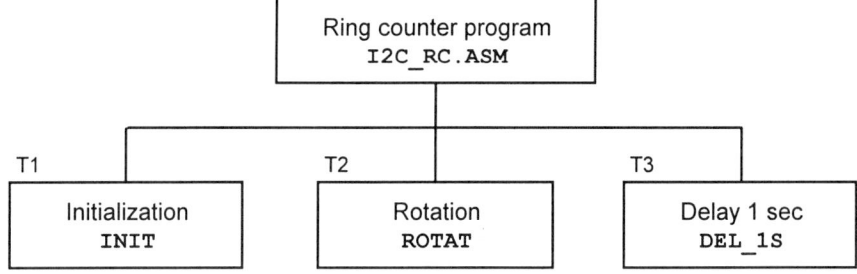

Figure 8.15 The subtasks in the ring counter program.

```
;******************************************************************
;*    I2C_RC.ASM                                                 *
;*    This program turns an I/O expander PCF8574                 *
;*       into a ring counter                                     *
;*    The rotating zero permits direct control of LED lamps      *
;******************************************************************
HADD       EQU    05H     ; A location for the high-order address
                          ; byte of the status sections in
                          ; the I2C interrupt subroutine S1INT
SLA        EQU    06H     ; A location for the I/O expander
                          ; slave address
DATA       EQU    07H     ; A location for data
IO_ADD_W   EQU    40H     ; The I/O expander slave address for
                          ; write operations
           ORG    0
           LJMP   INIT
;
           ORG    002BH
           LJMP   S1INT   ; The I2C bus interrupt subroutine
;
           ORG    0100H
INIT:      MOV    A,#7FH          ; Set-up the ring counter
           MOV    SP,#2FH         ; Initialize SP
;
           SETB   P1.6            ; P1.6 input
           SETB   P1.7            ; P1.7 input
           MOV    HADD,#02H       ; High-order address byte
           ORL    IEN0,#0A0H      ; Set the flags EA and ES1
           MOV    S1CON,#81H      ; 100 Kbps for 12 MHz crystal
           MOV    SLA,#IO_ADD_W   ; Select the I/O expander
           SETB   ENS1            ; Enable the I2C bus
;
START:     ACALL  ROTATE
           ACALL  DEL_1S
           SJMP   START
;
;****************************************************************
;    ROTATE : This subroutine rotates the accumulator         *
;    and moves the code to the I/O expander                   *
;****************************************************************
ROTATE:    RR     A       ; Rotate accumulator right
           MOV    DATA,A  ; Move the data byte
           SETB   STA     ; Start master transmitter mode
           RET
;
;***********************************************************
;    DEL_1S : This subroutine provides a delay        *
;    of 1 sec for fosc 12 MHz                         *
;***********************************************************
DEL_1S:    MOV    R1,#10
DEL1:      MOV    R0,#200
DEL2:      MOV    R2,#250
```

```
            DJNZ   R2,$
            DJNZ   R0,DEL2
            DJNZ   R1,DEL1
            RET
;
;***    The I2C bus interrupt subroutine   ***
S1INT:      PUSH   S1STA
            PUSH   HADD
            RET
;
            ORG    0208H
STATE08:    CLR    STA          ; Clear the flag START
            MOV    S1DAT,SLA    ; Load the slave address
            CLR    SI           ; Clear the flag SI
            RETI
;
            ORG    0218H
STATE18:    MOV    S1DAT,DATA   ; Load a data byte
            CLR    SI           ; Clear the flag SI
            RETI
;
            ORG    0228H
STATE28:    SETB   STO          ; Set the flag STOP
            CLR    SI           ; Clear the flag SI
            RETI
```

For simplicity, we do not take into account erroneous situations on the I^2C bus. Ring counters can be used in display devices where lights ripple on and off in a cyclical pattern.

Interfacing memory

When we investigated the memory design earlier in section 4.2, we discussed the possibility for nonvolatile memory solutions based on battery-backed CMOS RAM. Now, in an attempt to bring down the price and the RAM standby current, we replace the 8K x 8 RAM with a smaller I^2C RAM. Practically, there is no need to save 8K bytes. Usually, the vital information is a smaller amount. Therefore, the design has potential to decline the standby current and increase the battery life. In this situation, a low-voltage RAM with I^2C bus interface would be a good choice.

Figure 8.16 shows an example system. The RAM chip PCF8570 is organized as 256 bytes [Phil 1996]. The I^2C bus approach makes possible the IC to be produced in an 8-pin package. Furthermore, the maximum standby current is 15 µA.

The RAM contains an address counter which is incremented automatically after each write or read operation. As can be seen in Figure 8.16, the RAM address consists of a fixed part 1010 (A6 A5 A4 A3) and a programmable part (A2 A1 A0). In this example, the RAM occupies the address A0H for write operations and A1H for read operations. The operation type is defined with regard to the master involved, in our case the microcontroller.

Moreover, if we want to use two RAMs, placed on consecutive locations in the I^2C bus memory map, the input A0 (pin #1) of the second RAM must be pulled high. We can employ up to 8 ICs of this type in a system.

The input TEST, when brought high, resets the I^2C bus logic and the power saving mode is under way. Consequently, the supply current is reduced to a typical value of 50 nA.

Figure 8.17 lays out the I^2C bus pattern when the microcontroller writes to the RAM. After the RAM slave address, the 83C552 sends the word address to set up the internal address counter. The address counter will be incremented after each byte written to the RAM.

While in write operations a word address always follows the slave address; in a read operation two versions are possible. First, Figure 8.18 shows the bus pattern when the microcontroller reads from the current RAM address. The RAM transmits the first data byte immediately after the slave address. Second, Figure 8.19 illustrates a read operation from random address. Interestingly, it starts as a write operation ($R/\overline{W}$ is low). However, the microcontroller generates a repeated start condition after the word address. Then, the operation proceeds as a read one.

Finally, the microcontroller forces the RAM to terminate the transfer by generating no acknowledgement bit (high). The RAM leaves the line SDA high and the microcontroller yields a STOP condition.

Problem 8.2

Calculate the time the microcontroller will need to move 256 bytes to a RAM chip PCF8570 in the event of impending power loss. The I^2C bus frequency is 100 KHz.

Solution 8.2

If we neglect the time for START and STOP conditions, the desired result will be

$$T = (1 + 1 + 256) \, \text{bytes} \times 9 \, \text{bits} \times 10 = 23220 \, \mu s$$

Time duration in the area of 20 ms will fit to the inertia of the common power supply devices and the process of saving the data will be completed successfully.

As discussed earlier in section 4.2, EEPROMs were developed for upgrade of software or for configuration data. Serial EEPROMs and especially I^2C bus EEPROMs are also available. Figure 8.20 shows an EEPROM interfacing example. The National Semiconductor NM24C09 is an 1K x 8 EEPROM. The internal organization is four pages by 256 bytes.

Again, we can distinguish between fixed and programmable address fields. As shown in Figure 8.20 the high order nibble of the address is fixed to 1010. The next bit is compared with the value of a chip enable input. Therefore, we can have up to two devices of this type in a system. The page organization manifests itself in the next two bits (A0 and A1). They are termed page select bits. The pins #1 and #2 are unused, however the manufacturer requires both to be tied to ground. As you might expect, the last bit of the slave address determines the operation, read or write.

When the microcontroller writes to the EEPROM, it is similar to the RAM write operation. However, there is one difference. While the eight bit internal RAM address is incremented automatically, this is valid only for the four low order address bits of the EEPROM. This feature imposes a limitation of 16 consecutive data bytes. The previously written data will be overwritten if the number of bytes exceeds 16.

As far as the read operations are concerned, we distinguish between current address read and random read. The current address is the address of the last word accessed, incremented by one. However, if we want to work with any memory address different from the current one, we must use the procedure shown in Figure 8.19 for the I^2C bus RAM.

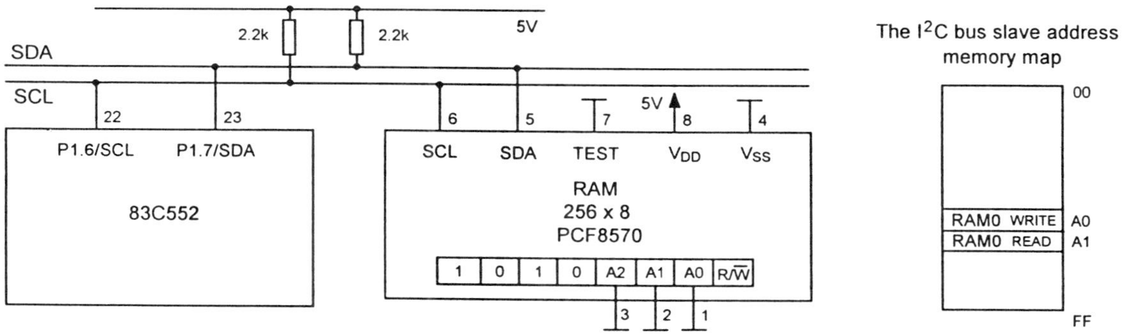

The I²C bus slave address memory map

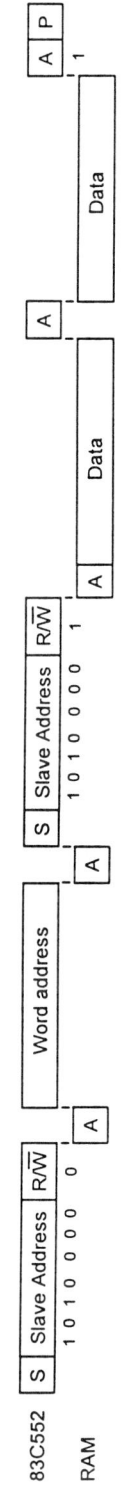

Figure 8.16 Interfacing a RAM.

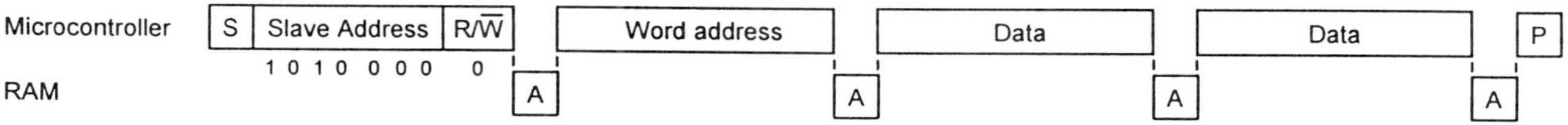

Figure 8.17 The microcontroller writes to the RAM.

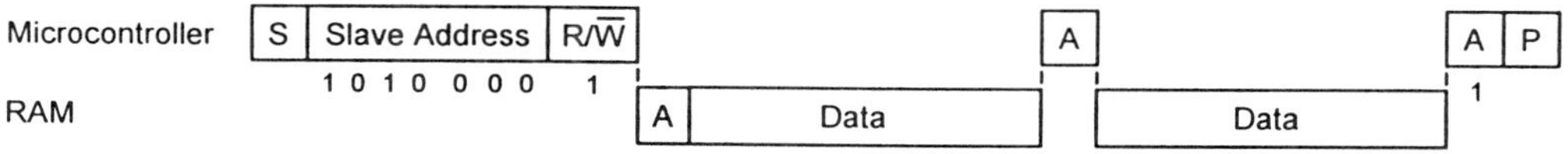

Figure 8.18 The microcontroller reads from the current RAM address.

Figure 8.19 The microcontroller reads from a random RAM address.

If the input Write Protection (WP) is pulled up, the write operations are blocked for the upper half of the memory (address 1FFH through 3FFH).

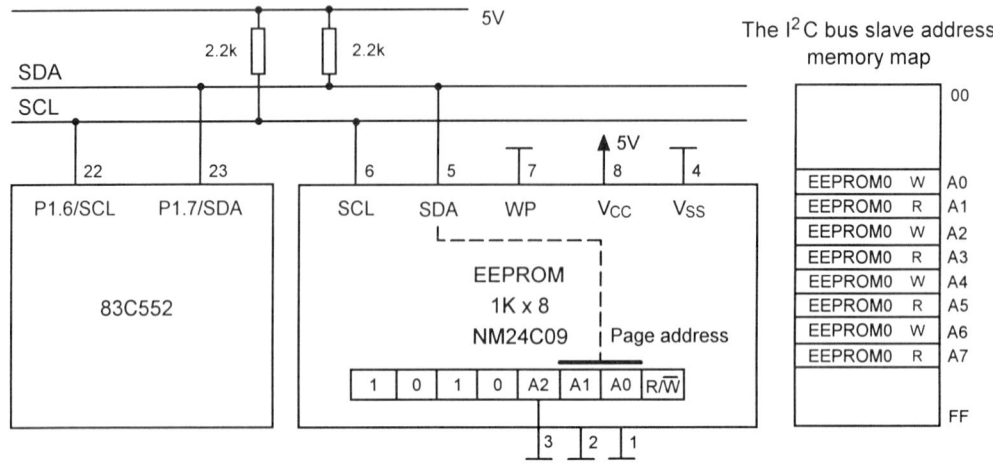

Figure 8.20 Interfacing an EEPROM.

So far, we have encountered two specifics of the I^2C bus memories:

• The word address is one byte in size and can be used for 256 locations. In case of bigger memories, the designers work around this limitation by paging.

• If we want to select a random address for read operations, we must include the word address immediately after the slave address. From the other hand, the memory will respond to the read operation as soon as the slave address is sent. One way to overcome the problem is to force the memory to wait for data rather than to transmit data. This is achieved by a "dummy" write operation.

8.5 Distributed embedded systems based on I^2C bus

When the performance of a microcontroller is not sufficient to cover the requirements, we can either allocate a more powerful chip or employ two or more devices of the same type. However, if the sensors and actuators are physically spread out in remote areas, a system with distributed intelligence would be the most natural solution. It is not necessarily all devices have computational facilities, however they must be able to exchange information over a common bus. The price of the system dictates the type of the interface and in most cases the designers use serial links between the nodes. Logically, the I^2C bus is a good candidate for a small scale distributed system.

The I^2C bus has been designed for a maximum capacitance of 400 pF when operated at a baud rate of 100 KHz. Inevitably, long cables result in bigger capacitive loads. We can mitigate the effect of the capacitive load by smaller pull-up resistors connected to the lines SDA and SCL. The limiting factor for the pull-ups is the maximum sink current permitted for I^2C bus devices, which is considered to be 3 mA.

The Philips engineers improved the situation by releasing a buffer for distributed I^2C bus applications. The IC named P82B715 is capable of driving a 4 nF capacitive load at 100 KHz [Phil 1997b]. Figure 8.21 shows two microcontrollers which communicate over a buffered cable. Other I^{2}C bus devices may reside either in the buffered or in unbuffered segments. The example in Figure 8.21 indicates that among the functions performed by microcontrollers are transfers between the ports P4 and P5. While the local I^{2}C buses are still characterized by sink capability of 3 mA, the buffered lines are able to sink 30 mA. Since the pull-ups are attached to the buffered lines, they are calculated on the base of 30 mA sink current. If several buffers P82B715 are used, pull-up resistors are spread over them. The maximum distance which can be covered vary with the type of the cable, but lengths in the area of one hundred meters should be reachable.

Also, there are two relatively new features, which improved the I^{2}C bus specification. First, a high-speed mode increases the rate to 3.4 Mbps. Second, a 10-bit addressing scheme allows 1024 addresses to be added to the basic functionality.

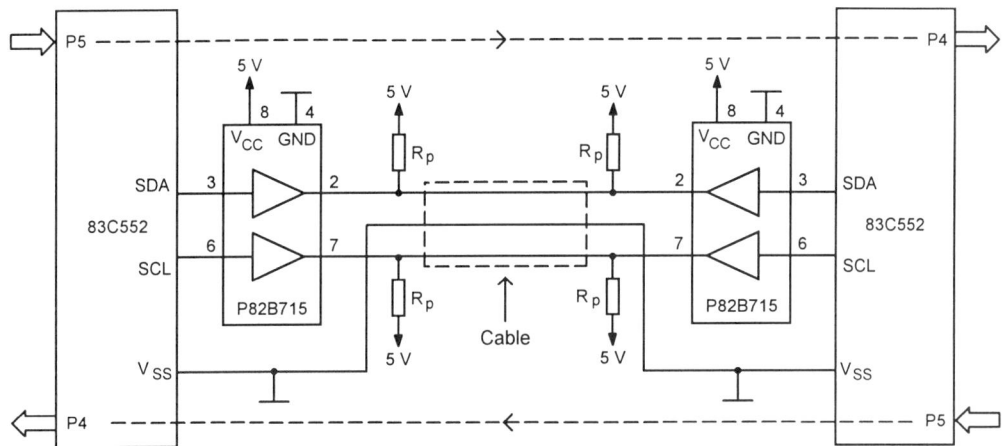

Figure 8.21 A distributed embedded system based on I^{2}C bus.

8.6 CAN bus basic concepts

There are numerous applications, such as automotive and industrial systems, where the designers must strike the balance between several conflicting requirements. First, fault tolerance in a noisy environment is a must for the distributed embedded systems in motor vehicles. Equally challenging is the demand for low cost and simple maintenance. Finally, the availability of standard solutions for a large spectrum of applications is very desirable.

A protocol that meets the requirements above is the Controller Area Network (CAN) [Phil 1994b]. The CAN bus was developed by Bosch for the field of motor vehicle systems. Currently, there is a massive penetration of the CAN bus in the area of industrial fieldbus systems. As is frequently the case, applications such as textile machines, smart elevators, medical equipment, agricultural and marine systems are based on the CAN protocol.

Any node of the CAN network can start a transmission at any time if it senses the bus is free. When one node transmits all other participants receive. A hallmark of the CAN convention is the lack of node addresses. Consequently, no addresses are associated with the CAN messages.

Figure 8.22 shows a classification of the CAN messages (packets). Data messages are used to broadcast up to eight data bytes. Remote Transmission Request (RTR) packets are generated by nodes which need information from other devices. Error messages are generated by receivers when they discover that the information in process of reception is erroneous. The error message forces the active node to retransmit the packet. Finally, overload messages delay the bus activity for a short period of time.

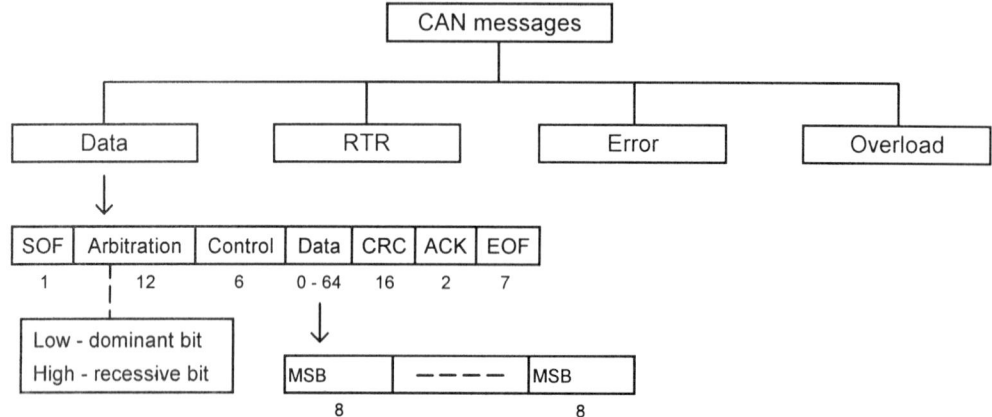

Figure 8.22 The CAN bus messages.

Furthermore, Figure 8.22 lays out the data message fine grain.

• The beginning of a frame is signalled by a Start Of Frame (SOF) bit. The start bit is a dominant one (low).

• The arbitration field includes an 11-bit identifier and a Remote Transmission Request (RTR) bit. The identifier characterizes the type of data transmitted. For example, temperature, pressure or revolutions could stand behind codes in this field. At the same time, the identifier dictates the message priority. The priority scheme is identical to the I^2C bus arbitration. A lower numbered identifier generates a dominant bit earlier over the bus and overrides the other transmitters. Since the identifier contains 11 bits, you might expect that 2048 different communication objects will be available. However, the identifiers 0 through 15 are not permitted and the actual number of unique messages is decreased to 2032. Newer versions of the protocol employ 29 bit identifiers.

The RTR bit is always dominant in data frames. In contrast, the RTR messages have recessive RTR bits. In other words, an RTR message is a frame very similar to the data frame, but possessing a recessive RTR bit. Theoretically, a node can demand information by a RTR message simultaneously with a transmitter which broadcasts the respective data. This case is easily solved when both nodes approach the RTR bit slot. The requester will recognize lost arbitration and will switch to reception.

• The control field includes six bits. The first two bits are reserved for future development. The reserved bits are sent dominant. The next four bits from the control field are termed Data Length Code (DLC). The allowed combinations are 0 through 8. Thus, the DLC specifies how many data bytes are encapsulated in the message.

• The core of the data frame is the data field. It contains a number of bytes which can vary between 0 and 8. The MSB of each byte is transferred first.

• The next 16 bits are devoted to the Cyclic Redundancy Check (CRC). The CRC sequence occupies 15 bits. The last bit in this field is named CRC delimiter. The CRC delimiter is a recessive bit.

• As indicated in Figure 8.22, the acknowledge field is two bits long. The transmitter sends both bits as recessive. The receiver reacts with a sent dominant bit in the first slot (ACK slot). The second acknowledgement bit is an ACK delimiter, which is a recessive bit.

• Finally, the End Of Frame (EOF) field contains seven recessive bits.

The CAN bit rate could be as high as 1 Mbps. The protocol does not specify the transmission medium. It is certainly implied that actual implementations must support the concept of dominant and recessive bits. In principle, the CAN systems can reach a bus length of 40 m while exchanging information at 1 Mbps.

There are five error checking schemes embedded in the CAN protocol. We show them in Figure 8.23. The error checking schemes can be classified in two hierarchy levels. At the message level we distinguish between CRC, frame check and ACK errors. The CRC scheme is based on the comparison between the check bits included in the message from the transmitter and the check bits recomputed by the receiver. If there is no match, an error is indicated. The frame check procedure verifies the structure of the frame over the bus. If a dominant bit is recognized instead of a recessive one or vice versa, an error is signalled. When a transmitter does not monitor a dominant bit during the ACK slot, it prompts an acknowledgement error. There are three possibilities in this case. First, the receivers detected an error in the message. Second, the acknowledge field has been corrupted. Finally, there might have been no receivers at all.

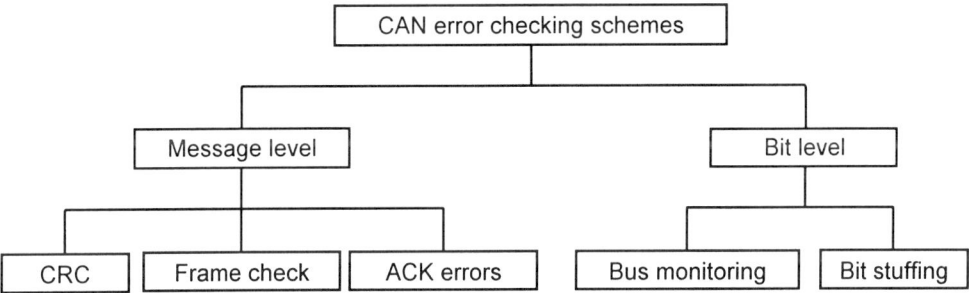

Figure 8.23 CAN error checking schemes.

Furthermore, the CAN convention includes two schemes effective at the bit level. The bus monitoring and the comparison between the bit sent and the bit received allows the transmitter to detect an error immediately when it occurs. The bit stuffing scheme is related to both code efficiency and reliability. Since the bit representation method is Non-Return-to-Zero (NRZ), it is unpredictable how many bits in a row with the same value will be transmitted. On the other hand, when there are no transitions the synchronization can be an issue. In seeking to balance conflicting requirements, the CAN designers adopted the bit stuffing method. As a result, after five consecutive equal bits the transmitter inserts a stuff bit with the complementary value. It is certainly implied that the stuff bit is removed by the receiver. However, the bit stuffing rule is a good opportunity to check the incoming stream. If six consecutive bits have equal values, the CAN nodes will recognize the error.

Figure 8.24 illustrates three different strategies which can underlie the CAN nodes design. When the main motivation is hardware and software reuse, a stand-alone CAN controller could be a natural addition to the microcontroller core (CAN node #1). New designs will probably bank on microcontrollers with a built-in CAN controller (CAN node #2). Finally, CAN compatible, Serial Linked I/O devices (SLIO) could be cost effective nodes in the network.

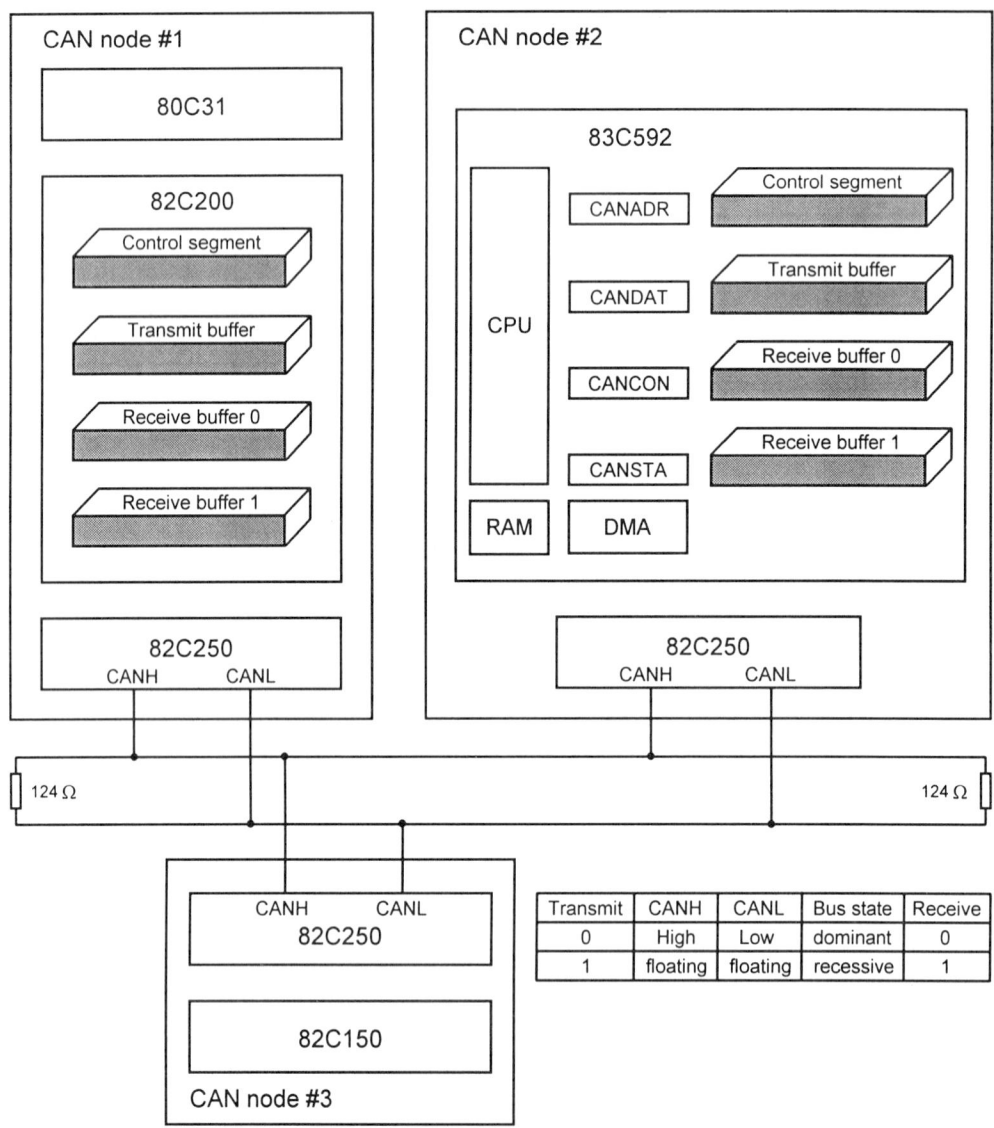

Figure 8.24 Designing CAN nodes - three views.

As shown in Figure 8.24, the stand-alone CAN controller 82C200 possesses a Transmit buffer and two Receive buffers [Phil 1997b]. The programming of the 82C200 and the control/status interaction with the microcontroller is organized through a set of registers named

Control segment. Each of the buffers and the control segment are 10 bytes long. From the microcontroller's point of view, the CAN controller appears as a block of 32 locations which belong to the external Data Memory address space. Both receive buffers share common addresses.

The CAN node #2 is based on an 83C592 microcontroller [Phil 1997a]. The device is a successor to the 83C552 microcontroller, however there are two differences. First, the internal program and data memory are doubled. Second, the I^2C bus interface is replaced by a CAN controller. In fact, the 256 bytes which are added to the internal data memory are part of the external Data Memory address space. Again, the built-in CAN controller includes a control segment, a transmit and two receive buffers organized identically to the 82C200's blocks. In addition, a Direct Memory Access (DMA) unit speeds-up the exchange of data between the CAN controller and the 83C552's style internal Data RAM (Main RAM). The interplay between the CPU and the on-chip CAN controller is implemented via four SFRs. A register named CANADR serves as an address register of the 32 locations (two of them unused) in the CAN controller. Respectively, a register called CANDAT is used as a data register for the interaction CPU - CAN controller. Furthermore, there are two extra SFRs (CANCON and CANSTA) which allow access to registers (locations) from the CAN control segment.

The CAN node #3 exhibits a piece of hardware which lacks the capability to execute a program. What the 82C150 can provide is 16 configurable I/O pins [Phil 1997b]. Some of the I/O lines can be either digital or analog. The Philips engineers built in a 6-input, 10-bit ADC and two quasi-analog outputs with 10-bit accuracy. For this cost-effective component the bit rate can vary between 20 Kbps and 125 Kbps.

All nodes in the example CAN system are equipped with an interface IC especially designed to link the controller and the physical bus. The 82C250 brings the noise immunity up by providing differential transmit/receive capability [Phil 1997b]. Inspection in the transceiver truth table shows that both outputs CANH and CANL are floating when a recessive bit is transmitted. In this way, what is gained is consistency between the differential approach and the concept of dominant and recessive bits.

8.7 Supplementary problems

Problem S8.1
An embedded system based on I^2C bus includes an 83C552 microcontroller and two I/O expanders PCF8574. Write a subroutine which reads one byte from I/O expander #1, increments the byte and move it to the I/O expander #2.

Problem S8.2
Write an assembly language subroutine which tests the RAM PCF8570 by a write-read-compare procedure for all addresses. Each location is checked once with code equal to its own address.

Problem S8.3
The distributed embedded system shown in Figure 8.21 exchanges data between the nodes. The byte which is read through Port 5 of each microcontroller is output by means of Port 4 of the other microcontroller. The transmissions are triggered when the input (Port 5) is changed. Assign slave addresses to both microcontrollers. Write the I^2C bus subroutine for the microcontroller with a lower slave address (higher priority). Figure 8.25 shows the possible status codes for this microcontroller while running in the slave receiver mode. You may wish to refer back to the status codes for the master transmitter mode illustrated in Figure 8.11.

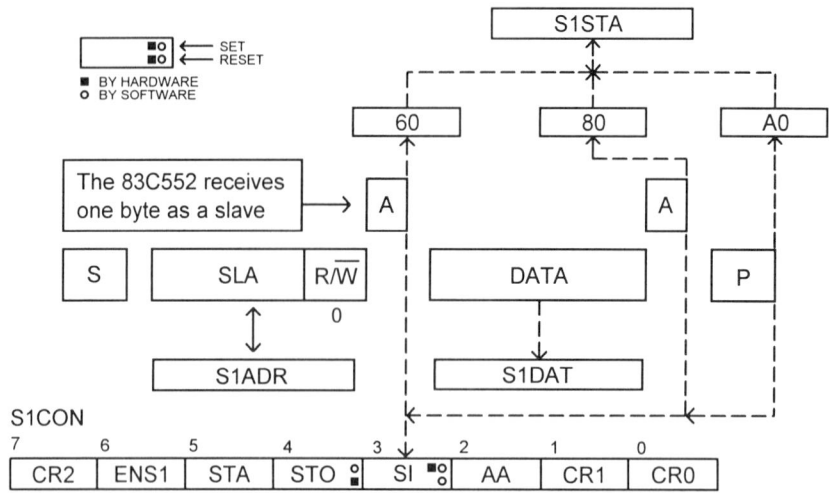

Figure 8.25 The 83C552 microcontroller receives one byte as a slave.

Problem S8.4

A node in a distributed embedded system based on CAN bus broadcasts 20 data bytes. Calculate how long the bus will be occupied for this task if the bit rate is 1 Mbps. Assume that the interframe space is three bit periods long.

8.8 References

Tim Brooksbank, "New development tools for I^{2}C bus", *Electronic Engineering*, February, 1998, p. 29.

Intel, *Embedded Microcontrollers*, 1996.

Philips Semiconductors, *CAN Bus Specification 2.0*, 1994b.

Philips Semiconductors, *The I^{2}C bus and how to use it*, 1995.

Philips Semiconductors, *I^{2}C Peripherals, Data Handbook IC12*, 1996.

Philips Semiconductors, *80C51-Based 8-Bit Microcontrollers, Data Handbook IC20*, 1997a.

Philips Semiconductors, *Application Notes and Development Tools for 80C51 Microcontrollers, Data Handbook*, 1997b.

Gerald Schickhuber and Oliver McCarthy, "Distributed fieldbus and control network systems", *Computing & Control Engineering Journal*, February, 1997, pp. 21-32.

SGS-Thomson Microelectronics, *Memory Products*, 1994.

Kenneth Terry, *Software Driver Routines for the Motorola MC68HC05 CAN Module*, Motorola Semiconductor, Application Note AN464, 1993.

Holger Zeltwanger, "An inside look at the fundamentals of CAN", *Control Engineering*, January, 1995, pp. 51-56.

Information about I^{2}C bus can be found at

    Philips Semiconductors   http://www-us2.semiconductors.philips.com/i2c

Information about CAN bus is available at

    Philips Semiconductors   http://www-us2.semiconductors.philips.com/can

Chapter 9

HIGH LEVEL LANGUAGES
FOR MICROCONTROLLERS

9.1 Why high level languages?

In this chapter we will discuss programming languages for embedded systems. Since we focus on small-scale, control-dominated embedded systems, we are interested in high level languages which are supported by compilers for microcontrollers.

While in areas such as database systems and scientific applications, the question is which high level language to be employed, in the field of microcontrollers both assembly language and high level languages may serve our needs. High level languages are characterized by the following advantages:

• They are much closer to the normal languages, and therefore probably easier to get a bug-free to a running program.

• In most cases, it is beneficial to work on higher level of abstraction and not use time and efforts to deal with hardware specifics. This not only frees you from an unnecessary burden, but also gives you programs with no bugs.

• Also, software written in a high level language is more portable. There is a learning curve associated with each instruction set and having to master the details of a new language represents too much time lost. There might be some alternations in a high level language due to different platforms, but not as significant.

On the other hand, the assembly language has more immediate appeal for people who pursue the design of fast and limited size programs. Furthermore, an assembly language will give you a better understanding what is going on, when you study a new microcontroller.

In this chapter, after the outline of the major points, we use an actual C language, SYS51C, for a sequence of examples. A timer example demonstrates the capability of this language to work with interrupt handlers. A keyboard scanner written in C presents a solution for a typical embedded system's entry. The testing procedure for this example employs the terminal emulator programs created in section 6.4. Next, we deal with programming of the serial port. Once we have obtained high level language support at both ends of the communication link, we can efficiently shift from exchanging characters to exchanging strings. Finally, we discuss a speech machine example which as a time critical application issues a challenge to the high level language approach.

9.2 When to use high level languages?

Most of the high level languages are originally written for purposes completely different from embedded applications. It might therefore seem impractical or even impossible to use such languages with small embedded controllers. In practice, only a few additions or modifications could be sufficient for generating very useful code in most cases.

Of course, you should not write your programs in a high level language unless you have access to a compiler for the actual language and processor. But even with that access, there might be situations where to use assembler instead of, or as a supplement to using high level language. If your problem is very time critical, your compiler may be unable to make code efficient enough or close enough to some special hardware.

Normally you will have access to memory, interrupts and ports directly in your high level language, so this will not be a reason to include assembler code. Indeed, we experienced that in section 6.4, when designed interrupt-driven communication programs for the PC. In addition, we illustrated direct access to the I/O address space of the PC's microprocessor.

Also, in case of heavy computational tasks, a high level language will help reduce the time to market.

9.3 Which language to choose?

Whatever language you choose, the selected compiler should be able to access whatever is necessary in writing your embedded application. For instance, you should be able to write interrupt routines and port access directly in the high level language. It is also essential that the generated code is ROM-able. Usually, it is not expected that an operating system presents. However, in some cases that could be convenient or even necessary.

For use together with the 8051 microcontroller, at least three potential possibilities exist. They all fulfil the above requirements.

• Traditional and oldest is the Intel PLM51 language, which is a subset of the language PLI, changed a little to meet the requirements of the 8051. A compiler for that language is supported by Intel itself. It generates very efficient code, but has limited facilities: only bits, bytes and 16-bit words are supported together with arrays and structures.

• Another language which has been successfully used for embedded controllers is the language Pascal. The syntax is slightly modified to meet the requirements of the microcontrollers. Consequently, it is possible to include interrupts directly in the language. Moreover, the language allows port access and access to individual bits in the ports. Also, ROM-able code and code not depending on a specific operating system can be generated.

• At the moment, the high level language C is most widely used for embedded systems. Being originally designed as a high level, portable, machine close general assembler language, it is well suited for the purpose. However, there are some drawbacks and pitfalls in using C for the 8051 microcontrollers. For example, the widespread and integrated use of pointers may lead to unreliable programs. Furthermore, the way the libraries are declared and work makes the process of global optimization difficult. Finally, the 8051 memory architecture with several address spaces requires the pointers to be bound to a chosen address space or generalized by an additional byte. However, three byte long pointers could make the code slow and impractical.

Here is an example of a small C program:

```
/*********************************************/
/*        compl_p1.c                      */
/*        This program complements Port 1   */
/*********************************************/
#pragma code=0x2000   /* Starting address of the code */
main()
{
   do
     {
     if (P3.4)
        P1=0xFF;
     else
        P1=0x00;
     }
   while (1);
}
```

The first C example is more or less self-explanatory. We assume that the pin P3.4 of the 8051 microcontroller is driven by a pushbutton connected to ground. When the pushbutton is not pressed, the input P3.4 is high (there is an internal pull-up) and code FFH is moved to Port 1. In the case of a pressed pushbutton, the input P3.4 is low and code 00H is moved to Port 1. Along with the simple action which is very easy to observe, we deliberately specified the code to start from address 2000H. If you try out this example by a platform, such as, a single board computer and the address 2000H is not covered by RAM, you could simply modify the address in the compiler directive. In this way, you will be able to download the code generated from the compiler to the RAM and run the program from there.

9.4 The SYS51C language

As our aim is to program the 8051 software compatible microcontrollers in high level language C, we must choose a compiler designed for this purpose. The compiler translates programs written in C into machine language which is the only language a microcontroller can understand. Also, the compiler could generate an assembly language file. Inevitably, the implementation of a particular compiler will result in a C language different from ANSI C.

In this section we outline the language SYS51C which is a C language especially designed for the 8051 compatible microcontrollers [Chri 1995]. As code efficiency is a major goal in this implementation, the following differences from ANSI C should be pointed out:

• The functions in SYS51C are not reentrant.

• In SYS51C the arithmetic operations on variables type **char** are 8 bit by default. However, the compiler directive

```
#pragma 8 bit calc=16
```

may be used to force conversion to 16 bit before **char** calculations, as specified in ANSI C.

• The variable type **char** is processed by the SYS51C compiler as unsigned. The type **signed char** does not exist.

• The next feature affects the pointers. There are three memory models embedded in the SYS51C compiler. In model small (default), pointers point to internal Data Memory if not otherwise specified. Next, in model medium the pointers are related to the external Data Memory. Finally, the model large is associated with three bytes long pointers which are capable of pointing either to the internal Data Memory, external Data Memory or the Program Memory.

 • Since the 8051 microcontroller possesses a non-von Neumann memory architecture, different memory types should be defined, which is not normally necessary in C. This approach will be illustrated through examples.

9.5 Application examples and problems

Timer program

In the C example `compl_pl.c` we assigned values to Port 1. This feature, direct access, applies to all SFRs and flags. For example, if you want Timer/Counter 0 to operate as a timer in mode 2, you could include the following statement in the C program:

```
TMOD=2    /* Load the register TMOD with 2 */
```

Likewise, the statement

```
EA=1    /* Set the flag EA */
```

will enable all interrupts.

Furthermore, if you need an interrupt handler named `inc_pl`, which is activated by the Timer/Counter 0 interrupts, you should insert in the program

```
void inc_pl() interrupt TIMER0
{
    P1++;     /* This interrupt handler increments Port 1 */
              /* when Timer/Counter 0 overflows */
}
```

Recall that we used an interrupt handler in the C communication programs in section 6.4. Even though in principle it is the same, the actual descriptions differ due to the specifics of the compilers and the target hardware. Now, we are ready to approach the following problem.

Problem 9.1

Write a C program to blink the output P1.0 of the 8051 microcontroller at intervals of 1 second. Use interrupts from Timer 0.

Solution 9.1

We assume that the oscillator frequency is 12 MHz. As a result, the Timer 0 input frequency is 1 MHz. The required frequency for the output P1.0 is 1 Hz. The auto-reload mode can provide a maximum ratio of 256. Obviously, this is not sufficient and we are forced to add a software counter. A convenient balance between both counters would be a 250 µs delay generated by Timer 0 and a software counter with module 4000.

Here is the program:

```
/****************************************************************/
/*      blink_pl.c                                             */
/*      This program blinks P1.0 at intervals of 1 second      */
/****************************************************************/
```

```
unsigned int sec,frac;

/***    Interrupt handler for Timer 0     *******************/
void increment() interrupt TIMER0
{
   if (frac++ == 4000)
      {
      sec++;
      frac=0;
      }
}

/***********************************************************/
main()
{
sec=0;
frac=0;
TMOD=2;     /* Timer/Counter 0 as a timer in mode 2 */
TH0=6;      /* Auto-reload code for interrupts every 250 mcs */
ET0=1;      /* Enable interrupts from Timer 0 */
EA=1;       /* Enable all interrupts */
TR0=1;      /* Start Timer 0 */
while (1)
   P1.0=sec & 1;    /* Blink P1.0 every second */
}
```

As you can see, using a C program to obtain a frequency of 1 Hz is much simpler compared with the assembly language version in Problem/Solution 4.4 and 7.4.

Keyboard scanner

Keyboards are popular peripheral devices for all types of computers. While the PC's keyboard includes approximately one hundred keys, a big number of embedded computers could be interfaced by a keyboard which has just 16 buttons. In both cases a microcontroller scans the keyboard and communicates with the other units. We will concentrate on a small matrix keyboard. For better understanding, we will rewrite Problem/Solution 3.1 first. Here is the alternative program written in C:

```
/*****************************************/
/*    inc_p1.c                        */
/*    This program increments P1      */
/*    when the input P3.5 is toggled  */
/*****************************************/
#pragma code=0x2000  /* We download the code in RAM */
                     /* and execute from address 2000H */
/***********************************************************/
main()
{
do
   {
   while (P3.5);       /* Wait for P3.5 low */
```

```
    while (!P3.5);       /* Wait for P3.5 high */
    P1++;
    }
while (1);
}
```

Both `while` loops consist only of test expressions. They are terminated when the input P3.5 is changed. Again, we assume that the pushbutton has been debounced by hardware.

A more useful data entry for embedded systems can be organized by a 16 buttons keyboard as discussed in the following problem.

Problem 9.2

Write a C program for the 8051 microcontroller which scans a keyboard as shown in Figure 9.1. The program decodes the pressed key and sends out its ASCII code through the serial port.

Solution 9.2

Figure 9.2 shows the functionality of the keyboard scanner decomposed into three tasks: "Detect", "Wait" and "Decode". The task T1 detects if a key is being pressed. The task T2 must overcome the problem of bouncing contacts, which is typical for mechanical switches. In this solution, we wait the bouncing to settle down. Finally, the program should decode the pressed key and generate its ASCII code. The code is transmitted through the serial port.

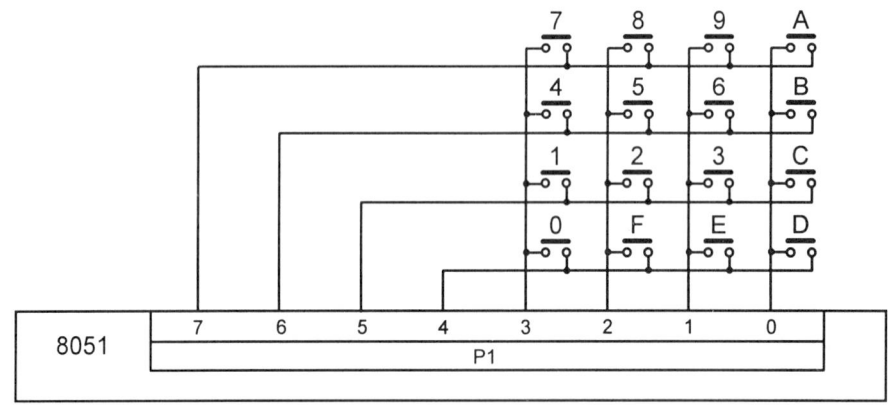

Figure 9.1 Interfacing a keyboard.

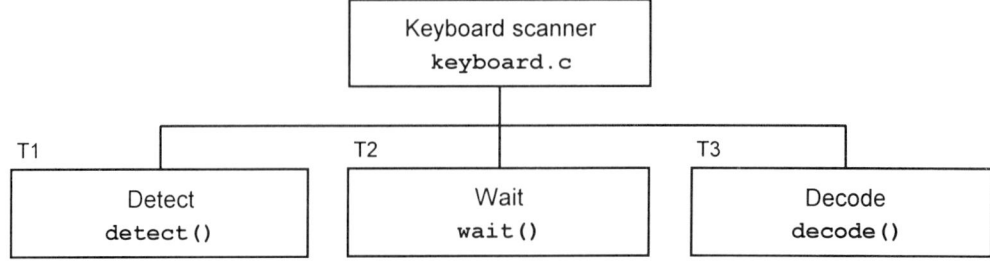

Figure 9.2 The keyboard scanner program decomposed into three tasks.

The C program appears here:

```
/***********************************************************/
/*    keyboard.c                                        */
/*    This program scans a cross-bar switch keyboard    */
/*    Debounces the key pressed                         */
/*    Translates the code to ASCII and sends it out     */
/***********************************************************/
#pragma code=0x2000

/***   Detect if a key is pressed    ******************/
detect()
{
   char i;
   P1=0x0F;                /* Pull all rows down */
   while (P1 != 0x0F);   /* Wait here if the previous key */
                          /* is still pressed */
   while (P1 ==  0x0F); /* Wait until a key is pressed again */
}

/***   Wait for debounce time    ***********************/
wait()
{
   char i;
   for ( i=0 ; i<=100 ; i++ );
}

/***   Decode the key which is pressed   ***************/
char decode()
{
   char output,input,i;
   output=0xEF;
   P1=output;
   for ( i=1 ; i=4 ; i++ )
     {
     input=P1;
     if (input != output)
        {
        switch (input)
          {
          case 0xE7:    /* P1.4 low */
          return '0';
          break;
          case 0xEB:
          return 'F';
          break;
          case 0xED:
          return 'E';
```

```
            break;
            case 0xEE:
            return 'D';
            break;
            case 0xD7:      /* P1.5 low */
            return '1';
            break;
            case 0xDB:
            return '2';
            break;
            case 0xDD:
            return '3';
            break;
            case 0xDE:
            return 'C';
            break;
            case 0xB7:      /* P1.6 low */
            return '4';
            break;
            case 0xBB:
            return '5';
            break;
            case 0xBD:
            return '6';
            break;
            case 0xBE:
            return 'B';
            break;
            case 0x77:      /* P1.7 low */
            return '7';
            break;
            case 0x7B:
            return '8';
            break;
            case 0x7D:
            return '9';
            break;
            case 0x7E:
            return 'A';
            break;
            }
        }
    output = output << 1 | 1;
    P1 = output;
    }
    return '?'; /* Error, two or more keys have been pressed */
}
```

```
/*************************************************************/
main()
{
do
   {
   do
      {
      detect();
      wait();
      }
   while ( P1 == 0x0F );
   SBUF=decode();
   }
while (1);
}
```

In contrast to the previous example (inc_p1.c), it is not sufficient to follow a transition for this program. The pressed key must be decoded and this can only be done when the key is still pressed. Consequently, the program is designed to leave the function detect() when a key is pressed again.

If two or more keys have been pressed, the program transmits over the serial port the character question mark to indicate an error. Even though we failed in our attempt to produce a question mark, it is theoretically possible.

A logical way to try out the program is to connect the microcontroller to a PC. A basic test can be accomplished by the program DEBUG. As we demonstrated in section 6.3, we can configure a PC's communication port by the DOS command MODE. For example,

```
MODE COM2:9600,N,8,1↵
```

Then, we are ready to use the function Input (I) from DEBUG. We start DEBUG, press a key on the keyboard which is scanned by the microcontroller and type on the PC's keyboard

```
-I 02F8↵
```

The PC displays on the following line the last character received and we can compare.

A more reliable test can be organized if you run on the PC the terminal emulator program term_1.c (see section 6.4). The program term_1.c displays on the screen all characters arrived from the selected communication port. Again, we are in position to compare the key pressed with the character received. This test is more appropriate since we can trap errors, such as two or more characters generated from a single keystroke.

Serial port programming

Using a high level language, we can work efficiently with data structures, such as files and strings. In section 11.2, the EPROM programmer case study, we design a function which downloads Intel HEX files in the RAM. While Intel HEX files are generated from compilers and linkers, strings would be likely sent by communication programs which interact with the embedded system.

The following problem/solution demonstrates how the serial communications could be organized by strings.

Problem 9.3

Write a C program for the 8051 microcontroller which receives a string through the serial port, converts the string and sends it out. The conversion rule requires a space character to be inserted after every two characters in the received string. Organize an endless loop to test the program.

Solution 9.3

The specification prompts that we could break down this example into four tasks. Figure 9.3 shows a natural decomposition. The design is based on the polling method.

In the beginning of the program, we define the size of the string buffers. Also, we declare the pointers `*ptr_rec` and `*ptr_send` which are related to the strings.

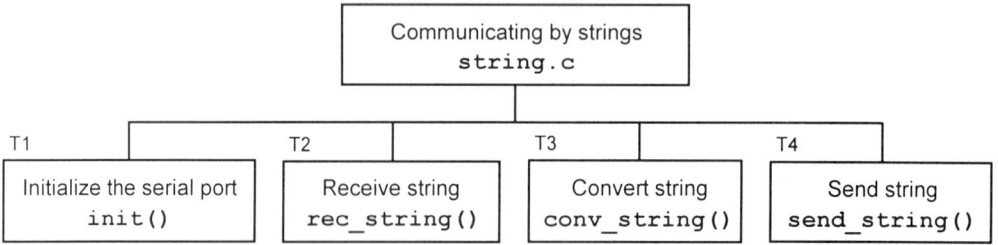

Figure 9.3 The string example consists of four tasks

```
/*******************************************************************/
/*      string.c                                                  */
/*      This program executes a loop which                        */
/*         includes the following actions:                        */
/*      Receive a string through the serial port                  */
/*      Convert the received string into a new one                */
/*      Send out the new string through the serial port           */
/*      The conversion rule requires a space character            */
/*         to be inserted after every two characters              */
/*         in the received string                                 */
/*******************************************************************/
#pragma code=0x2000      /* Download and execute the program */
                         /* from RAM, address 2000H */
#include <stdio.h>       /* This header file includes */
                         /* the function fopen(stdout) */
#define REC_BUFFER_SIZE 20
#define SEND_BUFFER_SIZE 20
char rec_buffer[REC_BUFFER_SIZE], *ptr_rec ;
char send_buffer[SEND_BUFFER_SIZE], *ptr_send ;
char ch;

/***    Initialize the serial port ***********************/
init()
{
fopen(stdout) ;        /* This function initializes the serial */
```

```
                        /* port in mode 1, rate 9600 bps for */
                        /* an 11.0592 MHz crystal */
}
/***    Receive string    *********************************/
rec_string()
{
ptr_rec = rec_buffer ;
do
   {
   while (!RI);     /* Wait for a set receive interrupt flag */
   RI = 0 ;
   if ((SBUF == '\0') || (SBUF == 0x1A))
/* We expect the incoming string to be terminated by code 0 */
/* We are also prepared for Ctrl Z */
      *ptr_rec = '\0' ;
   else
      *ptr_rec = SBUF ;
   ptr_rec++ ;
   }
while (!(SBUF == '\0') && !(SBUF == 0x1A));
}
/***    Send string    *********************************/
send_string()
{
ptr_send = send_buffer ;
do
   {
   while (!TI);     /* Wait for a set transmit interrupt flag */
   TI = 0;
   SBUF = *ptr_send ;
   ch = *ptr_send;  /* Remember that SBUF is a common name */
                    /* for two different registers */
   ptr_send++ ;
   }
while (ch != '\0');
}
/***    Convert string    *********************************/
conv_string()
{
ptr_rec = rec_buffer ;          /* Receive buffer, first address */
ptr_send = send_buffer ;        /* Send buffer, first address */
while (1)
   {
   *ptr_send = *ptr_rec ;  /* Move one character */
   if ( *ptr_rec == '\0')  /* NULL character, the send string */
      return ;             /* is ready for transmission */
   else
      {
      ptr_rec++ ;
      ptr_send++ ;
      *ptr_send = *ptr_rec ;  /* Move one more character */
      if ( *ptr_rec == '\0')  /* NULL character, the send */
```

```
      return ;                  /* string is ready */
                                /* for transmission */
   else
      {
      ptr_send++ ;
      *ptr_send = ' ' ;        /* Insert the space character */
      ptr_send++ ;
      ptr_rec++ ;
      }
   }
}
}
/**************************************************************/
main()
{
init() ;
while (1)
   {
   rec_string() ;
   conv_string() ;
   send_string() ;
   }
}
```

Again, we can test the program either by the terminal emulator program **term_1.c** or by the improved version **term_2.c** (see section 6.4). Both programs must be activated by a command line parameter 96 (9600 bps). In this way, we achieve consistence with the function **fopen(stdout)**.

Speech machine

Let's turn now to an application which can be viewed as a DAS. The sensor is a microphone and the actuator is a speaker. The microphone converts the sound waves into electrical signals. The 83C552's built-in ADC provides the digital equivalent of a continuous signal sampled at a certain rate. The digital results are stored in the embedded system memory and can be involved in calculations. In straightforward tasks, the recorded data simply stays in the memory until a playback is under way. We will start our discussion with the analog output subsystem.

Problem 9.4

Write a C program for the 83C552 microcontroller which can harness the hardware shown in Figure 9.4 to the task of generating sound. The sound starts when the input P3.4 is pulsed down. The speaker is activated for five seconds. The sound frequency should be in the area of 1 KHz for a 12 MHz crystal.

Solution 9.4

A few comments about the analog output subsystem are in order. The subsystem consists of a DAC, output signal conditioner and a very widespread actuator, a speaker. The DAC AD557 is an 8-bit device produced by Analog Devices [Anal 1994]. The AD557 possesses a precision voltage reference and an output amplifier. The output range is 0 to 2.56 V, quantum

10 mV. The typical output settling time is 0.8 μs. In addition, the DAC has an input register buffer. In the schematic diagram shown in Figure 9.4, both control inputs $\overline{CS}$ and $\overline{CE}$ are pulled down to make the register buffer transparent.

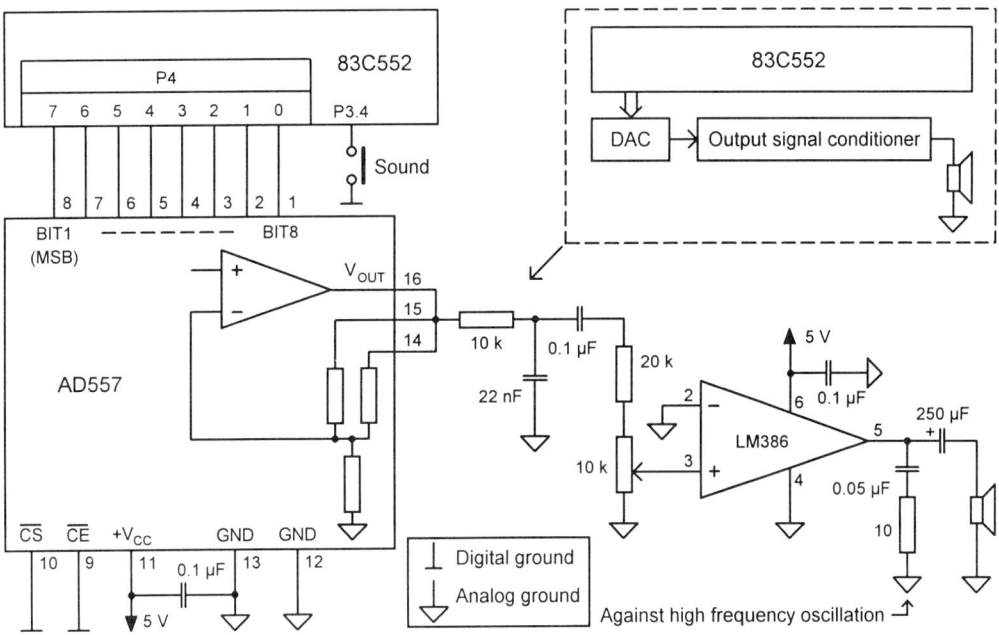

Figure 9.4 An analog output for generating sound.

The output signal conditioner consists of a low-pass filter and an amplifier. The amplifier, LM386, is a low voltage, audio power amplifier [Nati 1995].

The program **sound.c** is based on two functions. In the function **main()** the system is initialized. Thereafter, the program remains in a self loop. The second function is an interrupt handler for Timer 0. Each interrupt signals that a value should be sent to the DAC through Port 4. An array named **wave[]** contains 32 characters which represent the sine wave points. Therefore, the interrupts should occur 32 times per cycle or every 32 μs.

The program appears here:

```
/********************************************************************/
/*      sound.c                                                     */
/*      This program for the 83C552 microcontroller                 */
/*      generates sound when the input P3.4 is pulsed down          */
/*      The sound duration is 5 seconds                             */
/*      The desired frequency is 1 KHz for a 12 MHz crystal         */
/********************************************************************/
#pragma code=0x2000       /* We need this offset to download */
                          /* the code in RAM from address 2000H */
                          /* Then, the RAM is switched to */
                          /* address 0000H and the 83C552 is */
                          /* restarted */
                  /* If you will move the code to an EPROM from */
```

```
                    /* address 0000H, discard the directive pragma */
#define CYCLE_NUMBER 5000 /* 5 seconds x 1000 cycles = 5000 */
char point = 0 ;
int cycle = 1 ;
const char wave[ ] = { 127, 152, 176, 199, 218, 234, 244, 253,
255, 253, 244, 234, 218, 199, 176, 152, 127, 103, 79, 56, 37,
21, 9, 2, 0, 2, 9, 21, 37, 56, 79, 103 } ;
/*** Interrupt handler for Timer 0  ********************/
void clock() interrupt TIMER0
{
P4 = wave[point] ;     /* Output a point from the sine wave */
point++ ;
if ( point == 32)
   {
   point = 0 ;
   cycle++ ;
   if ( cycle == CYCLE_NUMBER )
      {
      ET0 = 0 ;          /* Disable interrupts from Timer 0 */
      cycle = 1 ;
      }
   }
}
/*************************************************************/
main()
{
   TMOD = 2 ;           /* Timer/Counter 0 as a timer in mode 2 */
   TH0 = 224 ;          /* 1 msec/32 points = 32 mcsec */
                        /* 256 - 32 = 224 */
   EA = 1 ;             /* Enable all interrupts */
   TR0 = 1 ;            /* Start Timer 0 */
   do
      {
      while ( P3.4 ) ;   /* Press and realise the */
                         /* push-button to go on */
      while ( !P3.4 ) ;
      ET0 = 1 ;          /* Enable interrupts from Timer 0 */
      }
   while (1) ;
}
```

Closing the last parenthesis in the program above you might think that success would be undeniable. Indeed, the system generates sound when the input P3.4 is pulsed down. However, we realized that the duration of the sound is longer than five seconds. Not surprisingly, people lose the feeling for execution time when programming in a high level language. An important detail is the execution time of all instructions in the interrupt handler in that it imposes limitations on the sound frequency. In particular, it should be less than 32 μs. Otherwise, the interrupt rate is declined and the curve is stretched. The resulting frequency is lower than expected. While the translation to machine instructions may vary from compiler to compiler, the minimal execution times for a certain oscillator frequency will be revealed by interrupt handlers written in assembly language.

Once we have made sure the output channel of the system is operational, we can focus on the input subsystem.

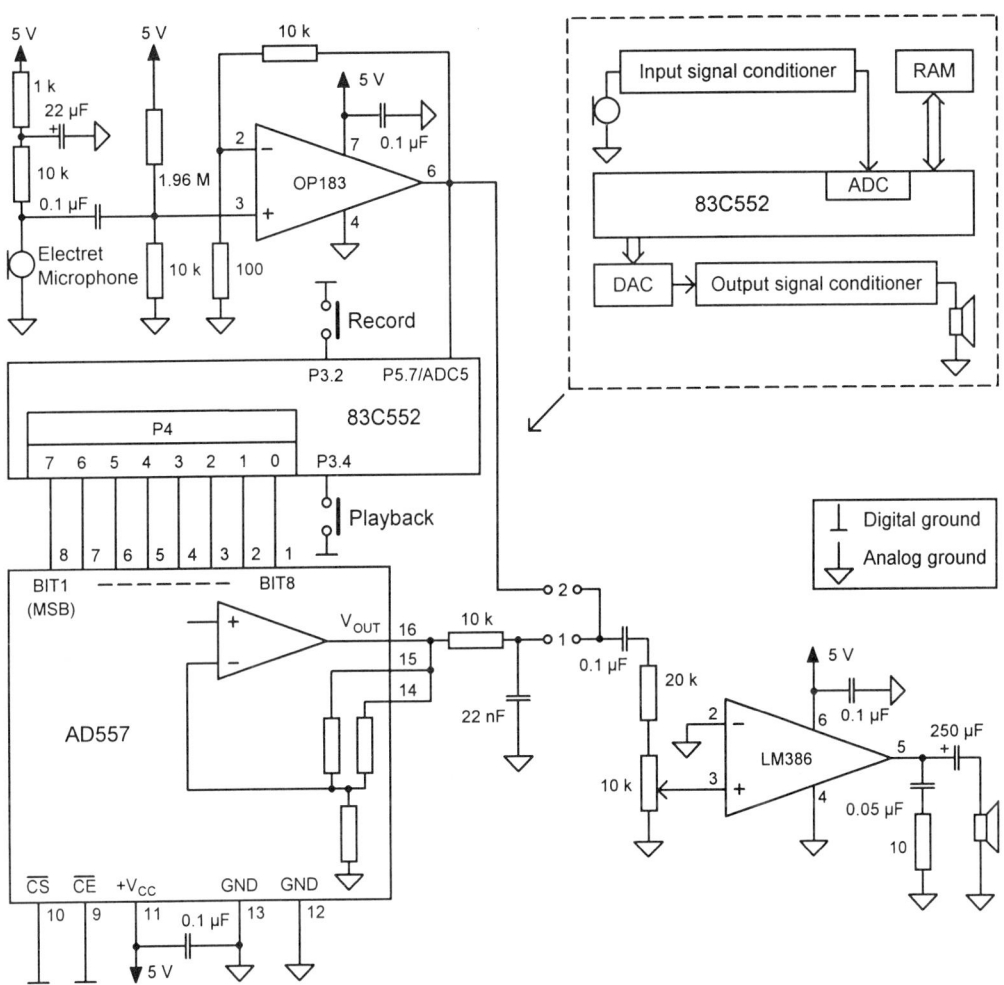

Figure 9.5 The speech machine.

Problem 9.5

Write a C program for the 83C552 single chip microcomputer which controls the simple speech machine shown in Figures 9.5. The program:
- Implements a speech recording for four seconds when the input P3.2 is pulsed down
- Plays back the recorded speech when the input P3.4 is pulsed down.

Write the program for a sampling rate of 8 KHz.

Solution 9.5

The main goal we pursue with this example is to demonstrate a C program written for a time-critical application, the performance of the analog subsystem is not essential. The input signal conditioner is simplified to a noninverting amplifier. The amplifier has a gain of 100. The amplifier's input circuit has been designed with the assumption that the amplitude of the microphone's output could be up to 25 mV. So, when we do not speak into the microphone, the amplifier's output level is 2.5 V, just in the middle of the ADC input range. The device OP183 is a low noise amplifier produced by Analog Devices [Anal 1994]. While the jumper #1 is normally wired, it is also possible to bypass the microcontroller and try out the analog subsystem connecting jumper #2.

Again, we use interrupts from Timer 0 to clock the speech recording and play back. The SYS51C image of the external Data Memory is an array named xmem[]. Furthermore, the language allows specification of the starting address which in our example is 8000H.

The timing constrains in this application are not so overwhelming as in Problem 9.4. The backbone of the interrupt handler is an if-else decision-making structure. The most critical part is the body of the if statement. What tips the balance is the ADC conversion time. This extra burden equals 25 average microcontroller's instructions. Even though the ADC runs in parallel with the CPU, we can not benefit from that.

A program with the following layout meets the requirements:

```
/********************************************************/
/*      rec_play.c                                    */
/*      This program for the 83C552 microcontroller   */
/*      Records speech for four seconds if the input P3.2  */
/*         is pulsed down                             */
/*      Plays back the recorded speech if the input P3.4   */
/*         is pulsed down                             */
/*      The sampling rate is 8 KHz                      */
/********************************************************/
#pragma code=0x2000      /* We need this offset to download */
                         /* the code in RAM from address 2000H */
                         /* Then, the RAM is switched to */
                         /* address 0000H and the 83C552 is */
                         /* restarted */
                  /* If you will move the code to an EPROM from */
                  /* address 0000H, discard the directive pragma */
#define SAMPLE_NUMBER 32768    /* The number of samples */
                              /* for 4 seconds */
bit record, playback ; /* Flags to indicate the current task */
xdata char xmem[] at 0x8000 ;    /* External Data Memory */
                              /* from address 8000H */
unsigned sample = 0 ;

/***   Interrupt handler for Timer 0   ********************/
void clock() interrupt TIMER0    /* The execution time */
          /* of this handler should be less than 122 mcs */
{
  if ( record )                      /* Record */
    {
    ADCON = ADCON | 0x0D ;           /* Start ADC, select ADC5 */
    while ( ! ( ADCON & 0x10 ) ) ;
    /* Wait for ADC to complete the conversion */
    /* The conversion time is 50 mcs for a 12 MHz crystal */
```

```
      /* The register ADCON is not a bit-addressable register */
      ADCON = ADCON & 0xEF ;          /* Reset the flag ADCI */
      xmem[sample] = ADCH ;           /* 8-bit conversion */
      sample++ ;                      /* Increment the address */
                                      /* in the external RAM */

      if ( sample == SAMPLE_NUMBER )
        {
        ET0 = 0 ;        /* Disable the interrupts from Timer 0 */
        sample = 0 ;               /* Initialize the system */
        record = 0 ;
        }
      }
   else                             /* Play back */
      {
      P4 = xmem[sample] ;           /* Output a sample */
      sample++ ;                    /* Increment the address */
                                    /* in the external RAM */

      if ( sample == SAMPLE_NUMBER )
        {
        ET0 = 0 ;                   /* Disable the interrupts */
                                    /* from Timer 0 */
        sample = 0 ;               /* Initialize the system */
        playback = 0 ;
        }
      }
   }

/****************************************************************/
main()
{
TMOD = 2 ;            /* Timer/Counter 0 as a timer in mode 2 */
TH0 = 134 ;          /* 1sec/8K = 1/8192 = 122 mcsec */
                     /* TH0 = 256 - 122 = 134 */
EA = 1 ;             /* Enable all interrupts */
TR0 = 1 ;            /* Start Timer 0 */
while ( P3.2 ) ;     /* Wait the input P3.2 to be pulsed down */
while ( !P3.2 ) ;
record = 1 ;         /* Record speech */
ET0 = 1 ;            /* Enable interrupts from Timer 0 */
while ( record ) ;   /* Wait the record task to complete */
do
   {
   if ( !P3.2 )
      {
      while ( !P3.2 ) ;
      record = 1 ;            /* Record speech */
      ET0 = 1 ;              /* Enable interrupts from Timer 0 */
      while ( record ) ;    /* Wait the record task to complete */
      }
   if ( !P3.4 )
      {
      while ( !P3.4 ) ;
```

```
        playback = 1 ;          /* Play back */
        ET0 = 1 ;               /* Enable interrupts from Timer 0 */
        while ( playback ) ;    /* Wait the play back task */
                                /* to complete */
        }
    } while (1) ;
}
```

While the first recording is obligatory, from then on the program enters a self loop in which the sequence of recordings and playbacks is selected from the user.

Along with the speech machine real-time requirements, equally challenging is the needed memory capacity. As the microcontroller is capable of addressing up to 64K external Data Memory locations, we have to work around this obstacle when longer speech intervals are demanded.

9.6 Supplementary problems

Problem S9.1
Write a C program for the 8051 microcontroller which increments Port 1 once per second.

Problem S9.2
Write a C program for the 8051 microcontroller which increments or decrements Port 1 every second. Initially, Port 1 is incremented. When the input P3.5 is pulsed low, Port 1 starts to count down. When the input P3.5 is pulsed down again, Port 1 begins to count up. Use interrupts from Timer 0.

Problem S9.3
Write a C program for the 83C552 microcontroller which scans a four key keyboard that might be used for a TV set control. The program decodes the pressed key and takes an appropriate action as specified in Figure 9.6.

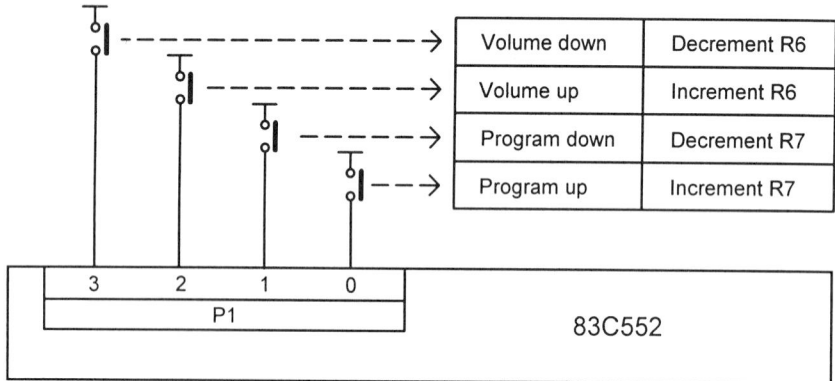

Figure 9.6 Interfacing a four key input which might be used for a TV set control.

Problem S9.4
Rewrite the solution of Problem 5.5 in high level language SYS51C. Calculate the temperature values rather than referencing a table.

Problem S9.5
Write a C program for the 8051 microcontroller which performs the following actions:
- Receives a string from the serial port
- Converts the received string into a new one
- Sends out the new string through the serial port.

The conversion rule requires a dash character to be inserted after every character in the received string. In both directions the data travels as ASCII strings.

Problem S9.6
Rewrite the program for the microcontroller in Problem 7.1 using the high level language SYS51C.

Problem S9.7
Design a 96K bytes external Data Memory subsystem for an embedded computer based on the 83C552 microcontroller. Rewrite the solution of Problem 9.5 for this system and for recording/playback intervals of 10 seconds.

Problem S9.8
A differential pressure flow meter is based on an 83C552 microcontroller. Figure 9.7 shows a block diagram of the flow meter. The measurement procedure use the fact that the pressure difference on both sides of the orifice plate is proportional to the flow rate [Parr 1993, Tomp 1988]. The flow

$$F = B\sqrt{P_D}$$

where P_D is the differential pressure and B is a constant. Use the language SYS51C to code a part of the program that converts the input voltage U_{IN} and calculates the flow F.

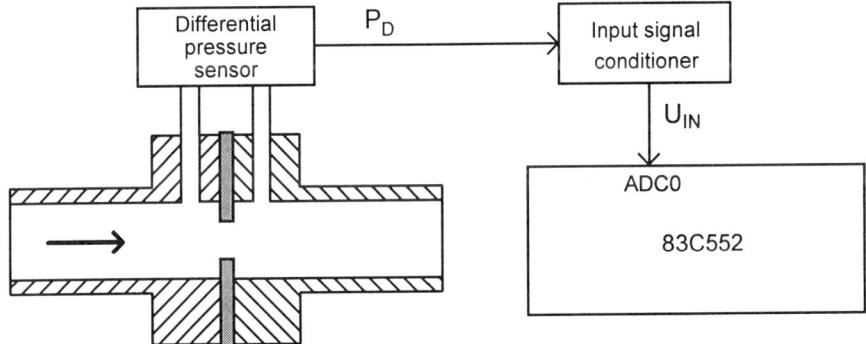

Figure 9.7 Differential pressure flow meter

9.7 References

Analog Devices, *Design-in Reference Manual*, 1994.

Atmel, *Microcontroller Data Book*, 1997.

Knud Christensen, "A Pascal-like portable, interactive development system for small microcontrollers", *Microprocessing and Microprogramming*. The Euromicro Journal, December, 1986.

Knud Smed Christensen, *User Manual for The System51*, KSC Software Systems, 1995.

Anders Gezelius, "A simulator simplifies design verification of control software", *EDN*, September, 1989a, pp.131-140.

Anders Gezelius, "Simulation helps you debug I/O software for embedded systems", *EDN*, October, 1989b, pp.161-170.

ISD, *Data Book, Voice Record and Playback ICs*, 1996.

Zdravko Karakehayov, Knud Smed Christensen and Ole Winther, *Embedded Systems*, Technical University of Denmark, Department of Applied Electronics, 1995.

Peter Marwedel and Gert Goossens, *Code Generation for Embedded Processors*, Kluwer, 1995.

P. E. Melsen and B. T. Madsen, *Microcontrollere*, Teknisk Forlag, 1992.

E. A. Parr, *Programmable Controllers*, Newnes, 1993.

National Semiconductor, *National Application Specific Analog Products, Databook*, 1995.

Philips Semiconductors, *80C51-Based 8-Bit Microcontrollers, Data Handbook IC20*, 1997a.

Philips Semiconductors, *Application Notes and Development Tools for 80C51 Microcontrollers, Data Handbook*, 1997b.

Thomas W. Schultz, *C and the 8051 Programming for Multitasking*, PTR Prentice Hall, 1993.

Thomas W. Schultz, *C and the 8051 Hardware, Modular Programming and Multitasking*, Volume I, Prentice Hall PTR, 1998.

Robert W. Sebesta, *Concepts of Programming Languages*, Addison-Wesley, 1996.

Kenneth L. Short, *Embedded Microprocessor Systems Design*, Prentice-Hall, 1998.

James W. Stewart, *The 8051 Microcontroller, Hardware, Software and Interfacing*, Prentice Hall, 1993.

Willis J. Tompkins and John G. Webster, *Interfacing sensors to the IBM PC*, Prentice Hall, 1988.

Information about C compilers can be found at

2500AD Software	http://www.2500ad.com
Archimedes Software	http://www.archimedesinc.com
Avocet Systems	http://www.avocetsystems.com
Franklin Software	http://www.fsinc.com
Hitex	http://www.hitex.com
IAR Systems	http://www.iar.com
Intel	http://www.intel.com
Keil Software	http://www.keil.com
KSC Software Systems	http://www.ksc-softsys.com/c.htm
Nohau	http://www.nohau.com

A limited demo version of a C compiler can be downloaded from

KSC Software Systems	http://www.ksc-softsys.com/c.htm

Chapter 10

EMBEDDED SYSTEMS DESIGN

10.1 Introduction

There are two important trade-offs which underlie the design of embedded systems. First, the designers have to strike the balance between the target system parameters and the design process in terms of resources. Unfortunately, our efforts to optimize the design object, the embedded system, result in longer and more expensive development cycles. The second trade-off comes with the borderline which the design process puts between the embedded system's hardware and software. The hardware-software trade-off has a significant impact on the system performance. Consequently, it is vitally important how the design methods will tackle the balance between hardware and software. The traditional design methodology splits the development into two separate flows: hardware design and software design. Thus, the hardware-software trade-off is predefined on the basis of previous experience. It is changed only if the designers are forced to approach a new iteration due to unsatisfactory results. This method could be sufficient when applied to the design of simple embedded systems. Also, the designers needs are confined to widespread conventional tools. However, when complex embedded systems are designed under tight deadlines, the most probable results will be overspecified solutions. This in turn brings the product's price up and puts the market penetration at stake.

In seeking to improve the situation, researchers are actively attempting to develop methods for concurrent design of hardware and software (hardware-software co-design). The hardware-software co-design will give us optimized designs in terms of performance, economics and development time, which we would not otherwise be able to obtain. A distinctive feature of the co-design methodology is the dynamic balance between hardware and software throughout the design process.

In this chapter we first present a summary of the hardware design. Next, we discuss ideas beneficial for the software design. Then, we explain debugging tools. Section 10.5 attempts to outline the hardware-software co-design approach. Later, we explain a method for I/O interface co-design of distributed embedded systems. Finally, section 10.7 is an overview of hardware-software co-design systems.

10.2 Hardware design

Reduced to its fundamental principles, the hardware design process is an interconnection of existing blocks. Less frequently, new blocks are designed and combined with existing units. The term block does not necessarily mean IC. The designers may connect gates and flip-flops within programmable VLSI devices. Furthermore, a single chip microcomputer can be designed by linking a certain CPU core with peripheral devices available in the libraries or designed especially for the project.

The benefits of a single chip system are not only weight and size. In addition, the price will be lower in high volumes. Furthermore, the single chip embedded system is basically more reliable and demands less power. Ultimately, the single chip approach will help protect the design. The internal Program Memory lock system discussed in section 2.11 is an example which justifies this view.

Alternatively, a board-level integration could be the only feasible option. As is frequently the case, a board level system is cheaper due to the fact that standard components are cost effective. Moreover, we can expect shorter manufacturing time and the system will be easily adjustable to changing conditions. If we want to compare the reliability of a single chip system versus a board-level integration, we will need many technological details. However an attempt to design a highly safe system will probably select a physically distributed approach [Erns 1995].

An interesting idea is to allocate an area on the microcontroller chip which could be programmed by the user [Wolf 1994b]. In this way, additional peripherals can be added according to the specific needs.

Regardless of the approach taken, single chip system or board-level integration, the hardware design process is more efficient when begun at an abstraction level higher than gates and flip-flops. The embedded system hardware is decomposed into subsystems. Inherently, some of the subsystems are built in the microcontroller. The interaction between them is predefined. As far as the interplay with the other subsystems is concerned, appropriate interfaces must be organized. Examples of memory design, parallel and serial interfaces are discussed in Chapter 4. In addition, serial interfaces are in focus in Chapter 6 and 8. Timing analysis, such as the one performed in section 4.2, also, is an integral part of the hardware design.

When we are considering component allocation we can use the basic taxonomy shown in Figure 10.1. Standard logic with different parameters has been widely used for many years. However, the design of embedded systems must be consistent with a set of requirements such as reaction time, power consumption and physical space. The above demands, when combined with economic justification, lead to Application Specific Integrated Circuit (ASIC) implementations [Pell 1991, Herb 1996]. As the term implies an ASIC is an IC, which is designed or programmed for a specific application.

The ASICs could be broken down into three major groups. The first option is custom logic. The user supplies the chip maker with specifications and the IC is designed and fabricated to meet the requirements. Not surprisingly, the design time and the price will be acceptable for few projects.

A step forward in the development of ASICs is the involvement of the user in the design process. The manufacturer provides specialized CAD tools and as soon as the design is completed, launches the IC into production phase.

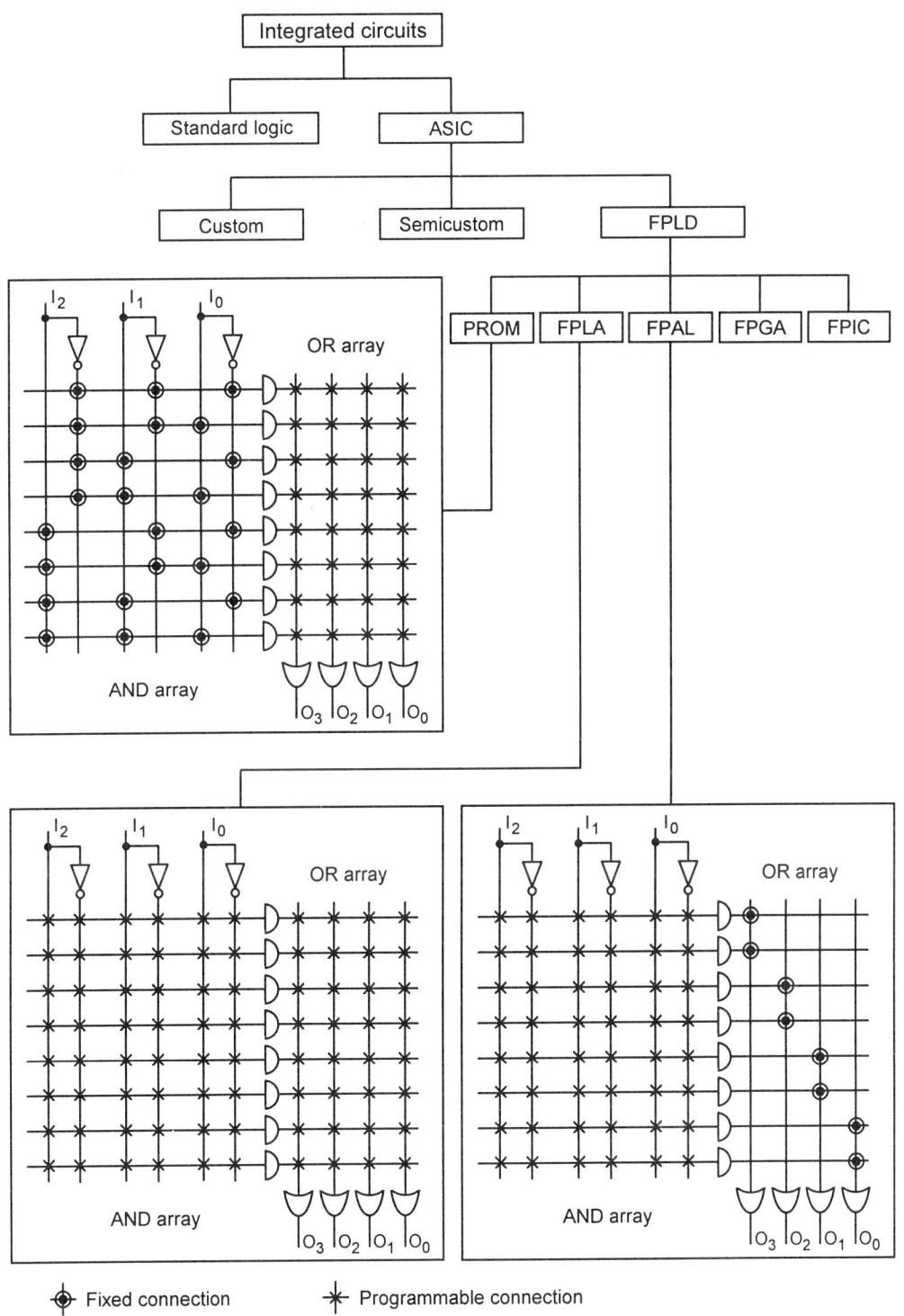

Figure 10.1 Basic hardware implementations.

The semicustom logic might be a more beneficial implementation in terms of design time and cost, but still remains unrealistic for small and medium scale projects.

What really tipped the balance in favor of ASICs was the prospect of the devices being programmed and reprogrammed by the end-user (in the field). The ICs called Field Programmable Logic Devices (FPLD) possess this feature and manifest a good trade-off between performance and price. The first three subgroups - PROM, Field Programmable Logic Array (FPLA), and Field Programmable Array Logic (FPAL), are based on an AND array and an OR array which are programmed to implement a specific function. As can be seen in Figure 10.1, we distinguish between fixed connections and programmable connections. The balance between both types of connections defines the degree of programmability.

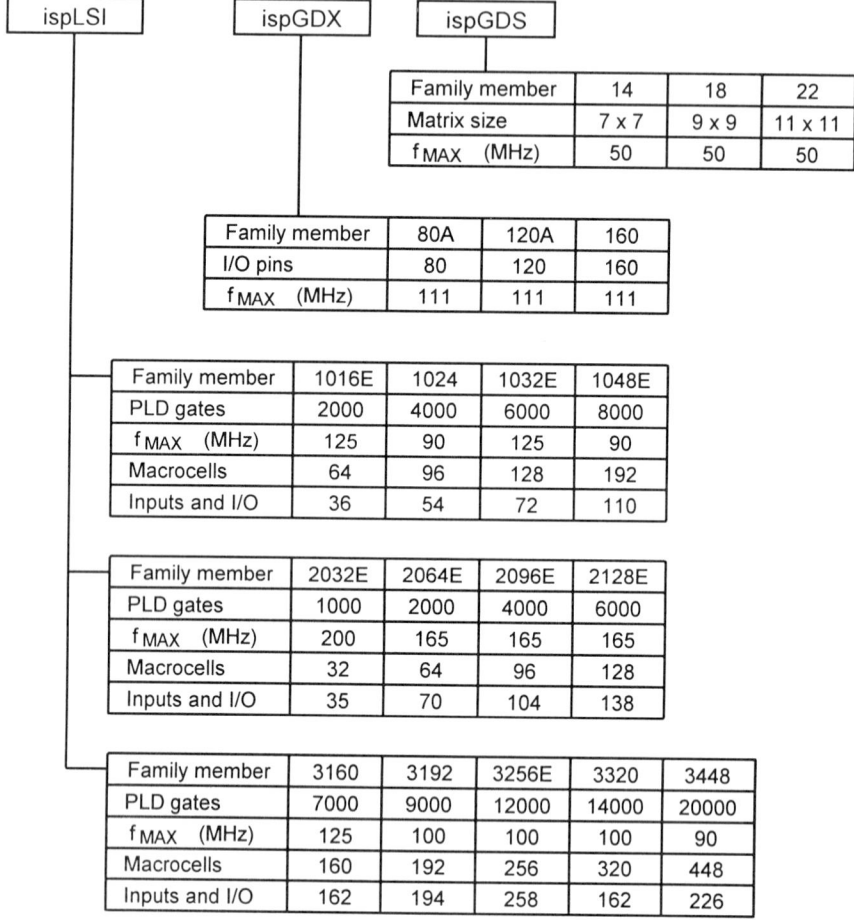

Figure 10.2 Parameters of FPLDs produced by Lattice.

The Field Programmable Gate Array (FPGA) logic is currently on the crest of the wave. In principle, the FPGA consists of configurable cells which are programmed and interconnected. Field Programmable Interconnect Circuits (FPIC) replace solid-state and mechanical switches and jumpers. Figure 10.2 summarizes the parameters of some FPGAs produced by Lattice. The table includes ispLSI families 1000, 2000 and 3000. The abbreviations stand for in-system programmable Large Scale Integration (ispLSI). The in-system programming is an impressive feature which allows modifications of the design without removing the IC from the board. In this way, we can implement multifunctional hardware blocks which are reconfigured by software. Moreover, Figure 10.2 shows the ispGDX (Generic Digital Crosspoint) and ispGDS (Generic Digital Switch) FPIC devices, which are especially good for system-level signal routing and prototyping.

The programmable devices design entry can utilize either text or graphics. The abstraction level may vary from high level specification languages to schematics. The flow goes on with validation followed by the actual implementation.

The computers that use FPGAs to provide better performance are termed netlist computers [Mang 1997]. The name stems from the fact that the gates and flip-flops in the FPGAs are connected to form the desired configuration.

In an attempt to utilize better the silicon, new architectures are based on more complex functional units such as ALUs and multipliers. They may be referred to as chunky function unit architectures. Along with the advantage of using less IC area, the chunky function unit architectures shift the design to a higher level of abstraction.

You could use PROM, FPLA or FPAL to organize decoders, register buffers, counters and other circuits that surround the microcontroller. The FPGA logic you can employ to design and implement your own single chip microcomputer.

An imperative goal for many applications is to minimize the power consumption. The designers could substantially improve the situation by calculating the optimal oscillator frequency. Two strategies make sense in that regard. First, the microcontroller saves power by running at a lower speed. Second, the system is running in fast outbursts, staying mostly asleep. Inherently, the dynamic logic imposes a limitation on the minimal oscillator frequency. In contrast, static devices, such as the microcontroller 80CL31 can reduce the oscillator frequency to zero [Phil 1997a].

Problem 10.1

Calculate the most beneficial low power design oscillator frequencies for the 8051 microcontrollers with minimal oscillator frequencies 0, 0.5 MHz, 1.2 MHz or 3.5 MHz if the serial port requirements are baud rates 1200, 2400, 4800, 9600 or 19200 bps. Assume that the serial port mode 1 or 3 and the Timer 1 mode 2 will be used. Define also the auto-reload codes for the register TH1.

Solution 10.1

In contrast to Problem 6.1, now our aim is the lowest possible oscillator frequency which meets the serial port requirements. Another difference is that we use one baud rate from the spectrum 1200 through 19200 bps, rather than organizing programmable baud rates.

If the serial port mode 1 or 3 and the Timer 1 mode 2 are employed, the correspondence baud rate to oscillator frequency is given in the following equation:

$$f_{OSC} = 12 (256 - TH1)(32 - 16(SMOD)) BAUD RATE$$

Observe that we can achieve minimal oscillator frequencies if the auto-reload code TH1 = 255 and the bit SMOD is set. Therefore, auto-reload code 255 will be our starting point. However, if the resulting oscillator frequency is lower than the minimum, indicated by the manufacturer, we must decrease the auto-reload value.

Figure 10.3 presents a set of results for the oscillator frequency and the corresponding auto-reload code (after the slash). For example, if our target is baud rate 1200 bps and the microcontroller's minimal oscillator frequency is zero, we can run it at 0.2304 MHz and the register TH1 must be initialized to 255. Similarly, if the same baud rate must be achieved by a microcontroller with oscillator frequency floor 0.5 MHz, the lowest possible oscillator frequency will be 0.6912 MHz combined with code 253 in register TH1.

f_{OSC} (MHz)/TH1		Baud rate (bps)				
		1200	2400	4800	9600	19200
$f_{OSC\ MIN}$ (MHz)	0	0.2304 / 255	0.4608 / 255	0.9216 / 255	1.8432 / 255	3.6864 / 255
	0.5	0.6912 / 253	0.9216 / 254	0.9216 / 255	1.8432 / 255	3.6864 / 255
	1.2	1.3824 / 250	1.3824 / 253	1.8432 / 254	1.8432 / 255	3.6864 / 255
	3.5	3.6864 / 240	3.6864 / 248	3.6864 / 252	3.6864 / 254	3.6864 / 255

Figure 10.3 The minimum oscillator frequencies and the corresponding auto-reload codes.

In addition, Figure 10.4 graphically represents the results and also indicates by links the possibilities for programmable baud rates. For instance, if we select oscillator frequency 1.8432 MHz, we will be able to choose between baud 4800 and 9600 by software.

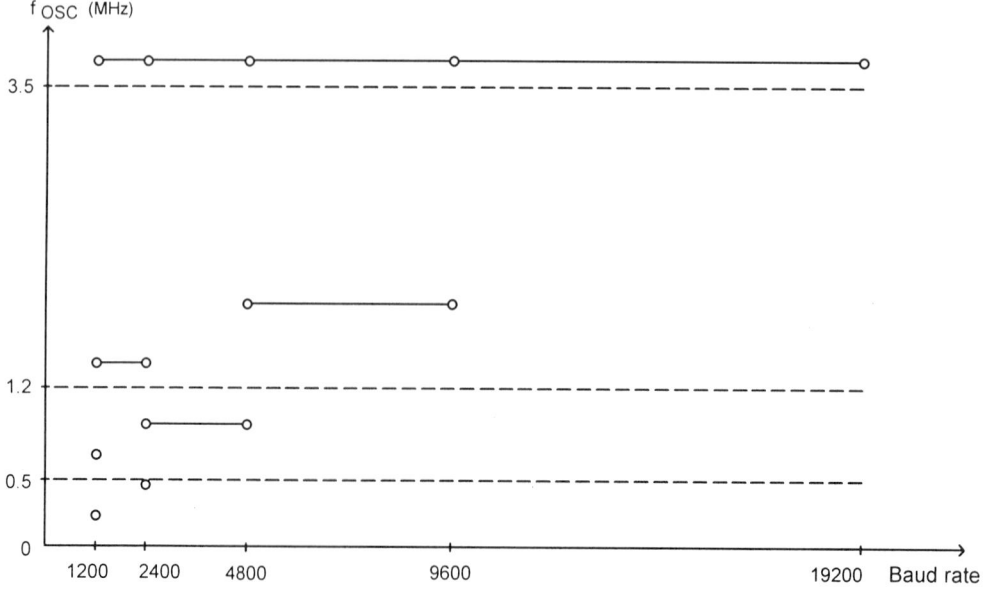

Figure 10.4 The minimum oscillator frequencies.

Finally, the low power design of the serial interface certainly implies the use of sophisticated interface buffers, capable of reducing significantly the power consumption when the transmitter is off.

Since embedded systems are frequently linked to analog peripherals, analog interfacing is an essential part of the hardware design. Examples of utilizing the built-in analog subsystems and external components are discussed in Chapter 5, sections 7.7 and 7.8. As the price of the embedded computer is vitally important, overspecified analog subsystems could decline the market penetration. A subtle effect of the information exchange between the digital and analog domains is that the memory capacity requirements are increased (see Problem/Solution 9.5).

Once the embedded system hardware has been implemented, we can start real test procedures. Preferably, the subsystems are examined separately first, as we frequently suggested throughout the text. While no one will dispute that tests are indispensable, the intensity of the tests is a controversial issue. In some cases, the manufacturer could afford to minimize the tests and offer a free replacement. It is certainly implied that Anti-lock Braking Systems (ABS), which come with most cars, can not be a subject for replacement policy.

The efficiency of the tests is improved if the designers introduce testability at the development phase. A classical approach is to insert test points. The signals can be measured at the test points and the errors located. By means of sophisticated test equipment each signal of a PCB can be sampled and the efforts shifted to development of test programs.

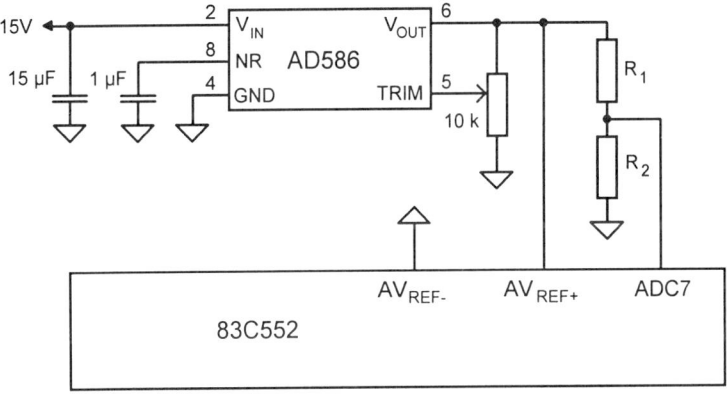

Figure 10.5 An example of a self-test design.

Moreover, the embedded systems can be programmed to perform self-tests in the field. A typical example of a self-test design is a system with ADC and a dedicated analog input which measure the reference voltage. Figure 10.5 illustrates this approach.

10.3 Software design

Typically, the embedded system software is of modest size, however the code must be highly optimized. In the majority of cases, the programmers succeed in making their programs work. Even then, the following questions are relevant [Jone 1988]:
• Are the execution times consistent with the timing requirements?
• Are the programs easy to understand?

• Are the programs easy to alter?
• Are the programs protected against inconsistent input data and hardware faults?

During the design of embedded systems, errors are inevitable. By decomposing a complex task into a set of subtasks we minimize the risk of design errors. In the field of software design this approach is known as modular programming [Jack 1975]. Apparently, the basic advantages of working with modules are:

• The modules can be separately compiled and tested.
• By processing small and simple modules, we can more efficiently use the development tools and available software.
• The modular approach allows better software reuse.

Over the years, it was proved that this method does not automatically produce good results. What is imperative in the software design process is how to split the program into smaller subtasks. A substantial improvement came with a method termed structured programming. The idea of designing a hierarchical structure has been very popular for twenty years. The intended functionality is decomposed into a set of subtasks. The next step is to break down the subtasks into smaller components and the depth of this process will be defined by the criteria for manageable subtasks. Once we have a clear structure of the program we can start coding. This concept is also known as stepwise-refinement. Figure 10.6 illustrates the structured programming method. The task T is broken down into subtasks T1, T2 and T3. The subtask T1 is divided further into subtasks T11 and T12.

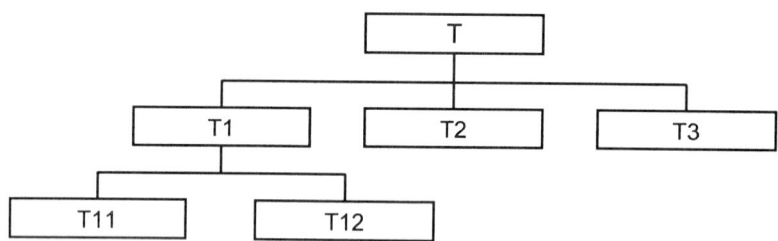

Figure 10.6 The task T is decomposed into smaller subtasks.

We usually seek to find a convenient correspondence between the software structure and the language building blocks. For example, Figure 10.7 shows the way we handled the subtasks in Problem/Solution 9.5. The functionality of the speech machine is decomposed into four subtasks. The subtasks are mapped onto the set of building blocks the language SYS51C offers. While the function `main()` and the other user defined functions are pure software, the interrupt handlers are software only on the surface. They are activated by events in the hardware subsystems of the microcontroller. In all cases, an interrupt flag is set by hardware. It is also possible for two flags to be set in one subsystem and the interrupt handler must check the flags in order to produce an appropriate reaction. For instance, an interrupt from the serial port is generated when either the receive buffer is full or the transmit buffer is empty. On the other hand, two or more subtasks can be directed to one interrupt handler, as we did in the speech machine example. Figure 10.7 shows the decomposition and the mapping for this case. User defined flags are used in the interrupt handler to select and execute one subtask.

Finally, the graphical symbols introduced in Figure 1.20 will help us to handle the special function flags properly.

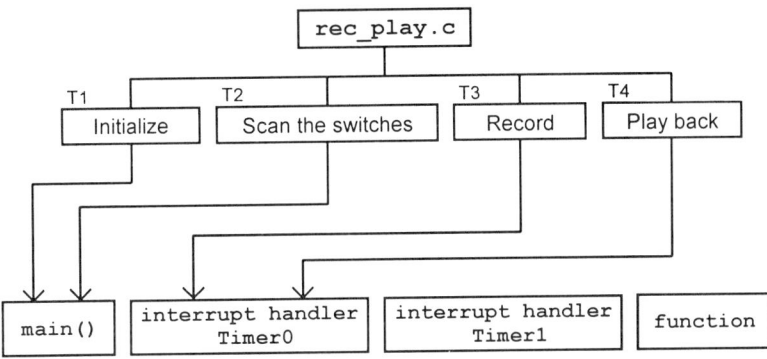

Figure 10.7 Mapping the software structure onto the language building blocks.

While discussing the hardware design earlier in this chapter, we mentioned that it is important to introduce testability and recoverability in the design object. Similarly, testability and recoverability must be considered at the phase of software design. An example of recoverability is the completion of the I^{2}C bus subroutines for the cases, when the arbitration is lost or no acknowledgement has been received. We also aim to achieve a certain degree of recoverability, when the 8051 serial port is used in modes 2 and 3 and the additional bit D8 is employed for parity check.

Moreover, some precautions must be taken against exceptional situations. For instance, noise, inconsistent input data, unstable power supply voltage and some other problems can be overcome if the system has been designed for recoverability. There are many circumstances in which design for recoverability will prove useful. Think, for example, of an embedded system based on the 8051 microcontroller which is interrupted by an external event and the input is programmed for low level mode. Imagine that the interrupt source output is stuck-at-zero. Consequently, the system will hang-up. A test of the interrupt input at the end of the interrupt subroutine would solve this problem.

The supervisory circuit discussed in section 4.2 and the watchdog timer described in section 7.6 are typical examples for introducing recoverability in the system. Exception handling is usually done by both hardware and software and it is also an argument that the idea to split the design into hardware design and software design at early stage might not be beneficial.

Furthermore, designers of embedded systems look for hardware components, especially analog ones, which can be replaced by software. For example, when analog signals are sampled at regular intervals, the correspondence between the sampling rate and the highest-frequency component of the signal is given by the Nyquist sampling theorem. The theorem states that the analog information can be recovered if the sampling rate is at least twice as high as the frequency of the highest-frequency component of the analog signal. It is a standard practice to decrease the spectral content of the sampled analog signal to make it consistent with the selected sampling rate. Low-pass filters are employed to eliminate the high frequency components. Alternatively, the microcontroller can be harnessed to do the filtering. If the microcontroller processes the samples in the memory and moves to the DAC a weighted sum of previous measured values, the result will be similar to passing the signal through a filter. This approach is known as digital filtering. In essence, using software instead of hardware is a method to achieve cost-effective solutions. It should be pointed out, however, that some time-critical applications will demand a diametrically opposite approach of using hardware instead of software.

10.4 Debugging tools

In the introduction of this chapter we mentioned that the first trade-off that the designers should consider involves the development process. The verification is an essential part of the design process and it can be organized only by appropriate debugging tools. Different tools can significantly differ in capabilities and prices. The market offers a large spectrum of tools released by microcontroller's manufacturers and independent companies. The application of the debugging tools is not confined to detecting errors. A big share of the tools sold are for educational purposes. Of course, a convenient user interface is crucial in this field.

Figure 10.8 illustrates a debugging tools taxonomy. At a high level we distinguish between tools for simulation/emulation and prototype related debugging tools. While the debugging tools from the first group can be used at an early stage of verification, the tools from the second group require a prototype. It is certainly implied that we can harness two different tools for the task of debugging. For example, an EPROM emulator could be connected to an evaluation board which will serve as an approximation of the prototype.

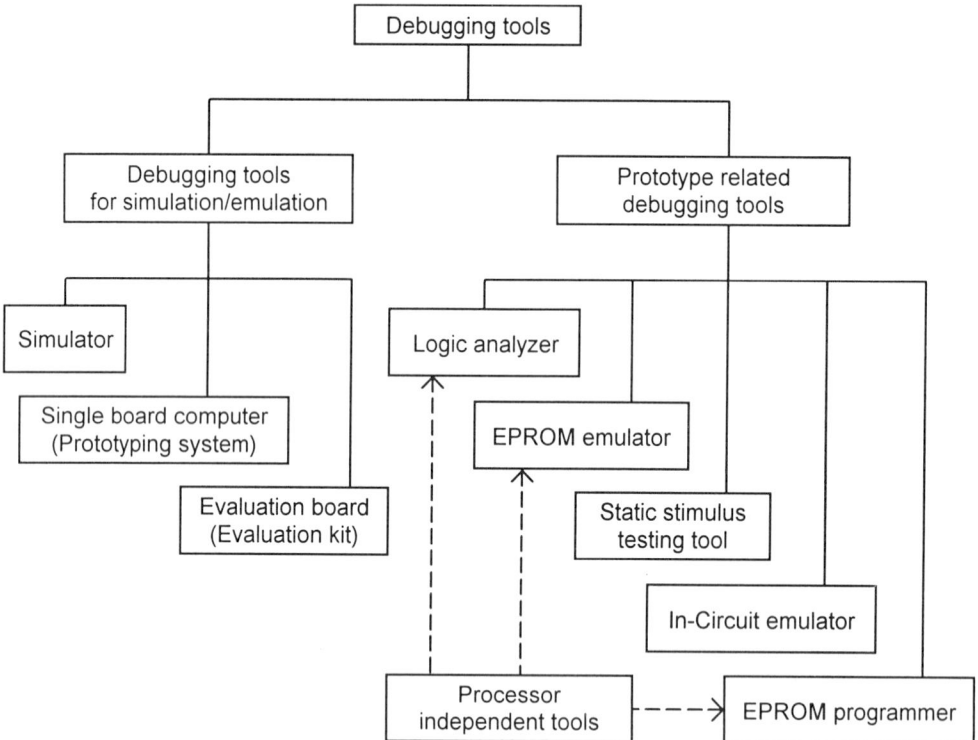

Figure 10.8 A debugging tools taxonomy.

A pure software tool that can be applied to debugging and evaluation is termed a simulator. As a rule, simulators run on PCs. The PC employs its own processor instruction set to simulate the instruction set of the microcontroller. As with many other debugging tools, the simulator has two basic modes: command mode and execution mode. In the command mode you can load the object code of your program, set breakpoints and start execution. Alternatively, you could execute a single instruction at a time. In both cases the simulator will be back in the command mode when the execution stops. The registers and memory are examined and modified. Usually, the simulators display the execution time and you can feel whether the timing constrains are met.

The cheapest piece of equipment is a single board computer, which is also referenced as a prototyping system. If we focus on 8051 microcontrollers, each prototyping system will resemble more or less the one shown in Figure 4.26. As is frequently the case, the systems have some free area to accommodate additional memory or peripheral components.

Although it is hard to define the borderline between the numerous systems available on the market, we could term the next, more sophisticated tool as an evaluation board (evaluation kit). The evaluation boards can be either stand-alone or connected to a personal computer. In the second case, a serial link is established between the kit and the PC. The programs are downloaded and then debugged by single stepping and breakpoints. The microcontroller internal registers and memory are examined and modified. The evaluation boards are useful for fast prototyping and education.

It is normal practice, both prototyping systems and evaluation boards possess a program called Monitor. The Monitor controls the communication with the PC and takes care of the debugging functions. The communication at the PC's end is handled by a terminal emulator program, such as the versions we designed in Chapter 6, or by more sophisticated programs capable of exchanging files. Since the Monitor requires a certain amount of EPROM and RAM in the evaluation board, we often try to move as many functions from the resident Monitor (on the evaluation board) to the remote debugger program (on the PC) [Shor 1998].

The second group of tools, classified as prototype related, can be used for design verification when a prototype is available. We can further subdivide them into processor independent tools and tools which are related to a specific machine. A tool which is particularly valuable when we emphasise the digital interface and its history, is called a logic analyzer [Clem 1987]. The basic principle behind the logic analyzer is recording the signals over a certain time interval and then displaying them on the screen. Hence, the logic analyzer has two modes: an acquisition mode and a display mode. The samples are stored in the memory as digital signals, regardless of the fact that the real voltages could be somewhere in the middle. The collected data can be displayed either as a table or by waveforms. In addition, if the target address and data buses are sampled, a mnemonic presentation would be possible. In this case, the analyzer must possess a module for the target microcontroller. This will allow the analyzer to disassemble the code.

The next debugging tool is termed EPROM emulator. It is based on a promising idea, that the target EPROM is replaced by RAM. Of course, the RAM will function as an EPROM from the embedded system's point of view. The EPROM emulator works as follows. An EPROM from the target system is removed and the emulator cable plug is inserted in the socket. Now, it is much more convenient to modify the code and as a result, the errors in the target will be easily detected and eliminated. Usually, the EPROM emulator is connected to a PC via the Centronics port or a RS-232 communication port. The EPROM emulators debugging technique follows the classical approach in that breakpoints are inserted and the vital information is examined and modified.

The biggest advantage of the EPROM emulators (the memory emulators), is that the development tool and the target are linked via a well documented and stable interface - the EPROM socket. Unfortunately, the internal memory implementations are a case when the memory emulation does not work. Section 11.3 deals with an example EPROM emulator design.

Naturally, the next step is to emulate the most significant IC in the target - the processor (microcontroller). Two methods have taken root in this field. The first approach is called Static Stimulus Testing (SST) [Coff 1983]. The hallmark of the SST is that the target system hardware can be tried out independently from software. Thus, the method is still applicable in cases of total breakdown or for prototypes, which have never worked. The equipment is termed the SST switch panel due the fact it contains mainly switches and LED lamps. The switches are used to define the logical levels of the address, data and control signals. Similar to EPROM emulation, the target microcontroller is removed and the SST cable plug is inserted in the socket. Then, a certain pattern is arranged manually in order to watch the LEDs, which close the feedback indicating the target status. The SST tool is a good trade-off between price and performance.

The second debugging technique is more aggressive. The target microcontroller is replaced by a microcontroller residing in the development system. This tool is known as In-Circuit Emulator (ICE) [Taba 1995, Shor 1998]. Normally, the ICE is based on a PC. The debugging process starts with most of the hardware allocated in the ICE. If the target is off and running, its hardware is mapped to the address space step by step. Unfortunately, the ICE is an expensive development tool and their purchase might not be justified as an investment for small projects.

A big step forward in the design of microcontrollers is the implementation of some debugging and run-control blocks on the chip. An approach termed single-wire Background Debug Mode (BDM) uses a dedicated pin to exchange information while the target embedded system is running.

Ultimately, the development process requires an EPROM programmer. The tool is produced as a stand alone unit or as a part of a more complex system. An example of an EPROM programmer design is discussed in section 11.2.

10.5 Hardware-software co-design

The hardware-software co-design process goes through several phases and each one of them is a subject of its own. Figure 10.9 outlines the design flow. The design process starts with specifying the desired functionality. Usually, a native language, such as English or Chinese, could be used for the initial specification. For example, a market research team informs the development engineers that customers need a certain product. Since this type of specification is at conceptual level, it is not bound to any physical implementation. It is also possible the first attempt for system specification to include more details. In this case, a Hardware Description Language (HDL) could be a natural choice. Describing a target system at a lower hierarchy level avoids ambiguity, but might impose frames for the following design steps. Once we have captured the embedded system functionality by an executable language, we can apply synthesis tools and simulators. This in turn will help us to verify the correctness and reduce the design time. Specification languages such as VHDL, Verilog, Hardware C, CSP, Statecharts and SDL are widespread and suitable for a large spectrum of applications [Gajs 1994, Berg 1997].

At this phase, the system specification is usually combined with a particular model. We distinguish between state-oriented, activity-oriented, structure-oriented and data-oriented models [Gajs 1994]. Examples of state-oriented models are the Mealy and Moor FSMs discussed in section 1.4. If the specification of an embedded system is based on a FSM model, the functionality is introduced frequently by a state transition table or a state transition graph. An example of a state transition table is used in the counter design in Figure 1.11. A state transition graph can be seen in the EPROM emulator design in section 11.3. The state-oriented models have always been candidates for specification of control dominated systems where the real-time parameters are essential. Extended FSM models for co-design form the basis of the POLIS system [Bala 1997].

The activity-oriented models are a good starting point when computational processes dominate the functionality. An activity-oriented model, a data flow graph, illustrates the partitioning phase in Figure 10.9. Another popular activity-oriented model is termed Control-Flow Graph (CFG). This model, alternatively labelled flowchart, was used to specify the clock's functionality in Figures 4.36 and 7.31.

There are cases when we find it easiest to think about the target system as a structure of interconnected modules. Structure-oriented models will help us to approach the design in this style. Finally, data-oriented models emphasize the data attributes and are more appropriate for the specification of a database system.

The next phase is to find an architecture which is the most suitable for the specific application. In the field of hardware-software co-design, three basic architectural decisions dominate the research.

The first target architecture can be specified as a single processor architecture. The single processor system could consist of a main processor and Application-Specific Integrated Circuits (ASIC) [Erns 1993, Bala 1997]. The majority of research teams use commercial FPGAs to implement the ASIC.

The second candidate architecture is the multiprocessor or parallel architecture. The taxonomy is completed by distributed architectures. The distributed systems are parallel systems which are physically spread out in the control object. Consequently, communication is organized, mostly based on serial links.

After the system specification, the design process goes on with partitioning, mapping and scheduling [Wolf 1994b, Gajs 1994]. Cases of partitioning were discussed in many example designs. Now, the functionality is decomposed into smaller tasks with the assumption that both hardware and software solutions are possible. The partitioning example included in Figure 10.9 displays the embedded system function broken down into six tasks. The data flow graph also indicates that Task 3 must wait until Task 1 and Task 2 are competed. Likewise, Task 4 can start when the results from Tasks 3, 5 and 6 are available. Partitioning the function to be implemented into smaller pieces can be done at different levels of granularity [Erns 1993, Gajs 1994]. Fine-grain partitioning may give us results which we would not otherwise be able to obtain.

In the field of embedded control, the system specifications are frequently given as individual algorithms for each output, rather than byte- or word-oriented procedures used in classical data processing. Consequently, it is probably natural to implement fine-grain partitioning where the quantum is a single output function. Apparently, this approach could lead to unrealistic search space for larger systems. Less obviously, it would be more difficult to bind some software modules to one hardware module in an attempt to improve the time properties or other system parameters.

In the phase of allocation (hardware partitioning), the designers select basic system components. The type and number of components will depend on which parameters of the embedded system dominate the design. As far as parallel and distributed systems are concerned it would be natural to use hardware modules. For example, Figure 10.9 shows three hardware modules. The first module is based on an 80C31 microcontroller. The second module employs an 80C552 single chip microcomputer. Lastly, an ASIC will be designed to improve the system performance.

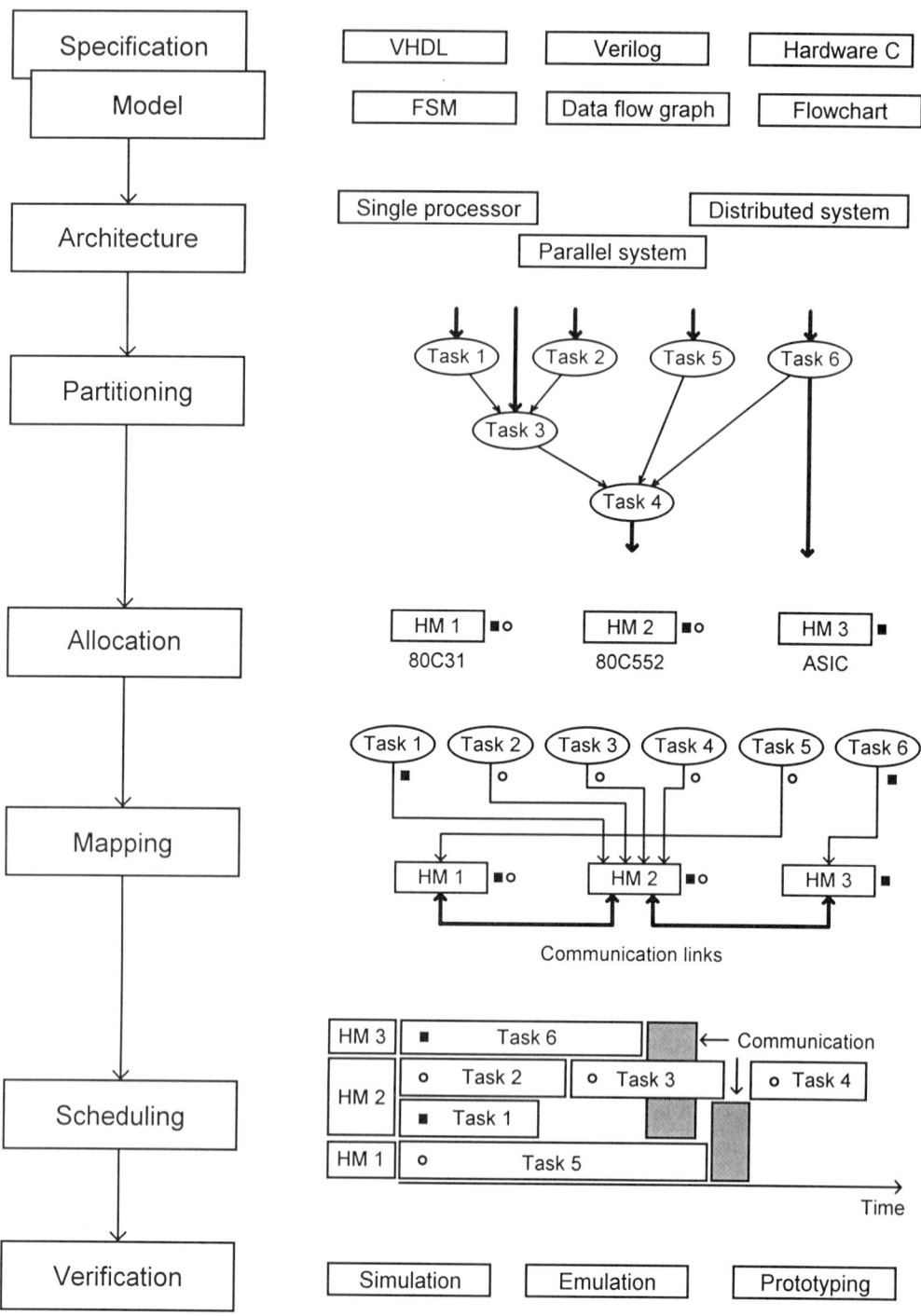

Figure 10.9 Hardware-software co-design of embedded systems with microcontrollers.

Hardware modules which contain CPUs are marked by **O**. Pure hardware modules, such as, ASICs are indicated by **■**. Microcontrollers have a CPU and parallel running hardware blocks (counters, timers, serial ports and so on) and are marked by **■O**. Typically, a hardware module based on a processor may possess a certain amount of memory and I/Os. Embedded systems built from identical hardware modules are termed homogeneous systems. The designers may also consider a heterogeneous system which is composed from different modules.

Once we have the system functionality partitioned into tasks (processes) and the allocation completed, we can map the tasks onto the set of hardware modules [Wolf 1994b]. In fact, the mapping is organized in two steps. First, for each task either a software (symbol **O**) or a hardware (symbol **■**) implementation is chosen. Second, each task is bound to a certain hardware module. When the mapping is completed, we use the data flow graph again to identify the required communication links. For this particular example, the hardware module HM2 must receive results from subtasks #5 and #6 which have been mapped to other modules. Consequently, communication between the module HM2 and the other two modules is organized.

At the phase of scheduling, the designers concentrate on the order in which the tasks can be processed [El-R 1994, Ferr 1994, Bala 1997]. The goal of scheduling is to find an arrangement which optimizes the system timing parameters. As you might expect, it is not always an easy job. If a task includes conditional branches and loops, the execution time might be unknown until the task is off and running. Consequently, the approaches applied for deterministic scheduling, when the information about the tasks is completely known in advance, will differ from the methods used for nondeterministic scheduling. Figure 10.9 reflects the scheduling scheme chosen in our example. The CPU embedded in Hardware Module 2 (HM2) executes Task 2 first. Task 4 is scheduled last.

Naturally, microcontrollers introduce parallelism in the frame of one hardware module. Thus, it is possible for Task 1 to be executed concurrently with Task 2. Moreover, some single chip microcomputers possess not only parallel running hardware blocks, but also more than one CPUs [Moto 1994]. The scheduling fragment of Figure 10.9 shows that the communication between hardware modules brings the total execution time up. Unfortunately, in some circumstances the communication penalty takes a big share of the total time. In the end of the scheduling phase the designers might be able to feel if new iterations of the previous three phases should be approached. The degree of confidence will depend of how sophisticated the scheduling tools are.

The verification is the last phase of the design flow. A classical way to verify the design is to implement one or more systems and to carry out tests. This approach, which is termed prototyping, makes it possible for the system to be evaluated in a real environment. However, constructing a system with all the technological operations stands in contrast with the requirement for a short development cycle. Also, it is hard to predict how many systems should be produced and test to eliminate all errors.

We can save time and money using emulation [Rose 1997]. Practically, we can test a large amount of the embedded system functionality by a board with the target microcontroller and FPGAs. The FPGAs are linked by Field Programmable Interconnect Circuits (FPICs), such as the ispGDX and ispGDS, to form the system hardware. Since all components of the emulation board are programmable, the designers can easily manipulate gates and flip-flops to obtain the functional equivalent of the target system with timing parameters in the close vicinity of the final implementation. In many cases, simulation may precede the emulation tests. For example, gate-level simulators have been used for many years.

10.6 I/O interface co-design for distributed systems

The data flow graph in the partitioning segment of Figure 10.9 shows the computational process of an embedded system. When the I/O interface is designed, we must take into account the connections to the environment. As shown in Figure 10.10, the typical peripheral devices are sensors and actuators.

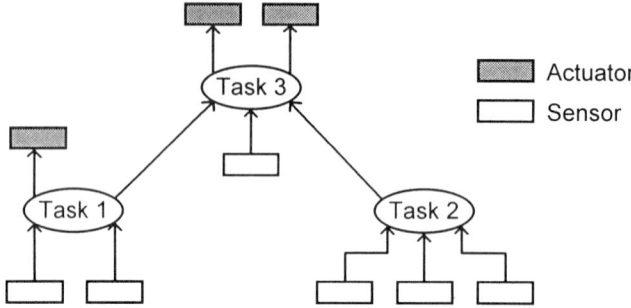

Figure 10.10 The embedded system data flow graph and I/O interface.

We assume that the sensors and actuators are physically spread out in different areas of the control object and a distributed architecture is the only efficient approach. The embedded system I/O design becomes of primary importance. The I/O capacity requirements can be met by enhanced hardware modules [Chou 1992]. An alternative solution provides I/O expansion by overlapping the interface and the computation [Kara 1995b].

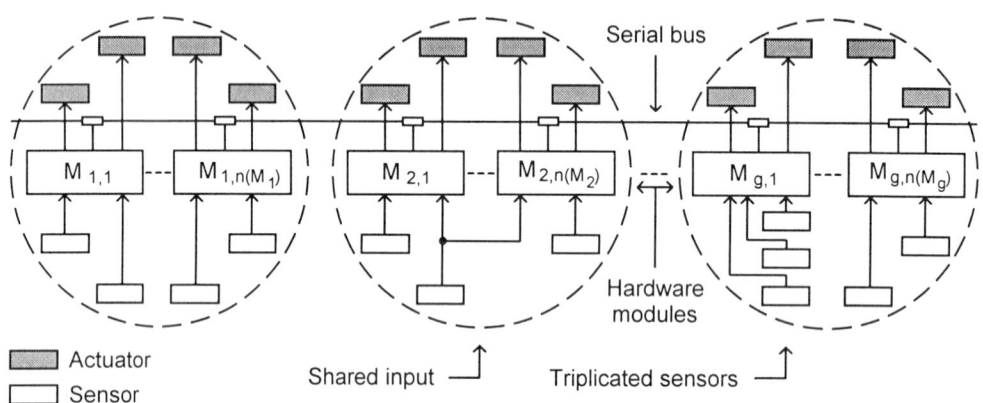

Figure 10.11 The embedded system architecture.

Figure 10.11 shows the target architecture. The sensors and actuators form g groups which are distributed among the control object. The embedded system has a set of hardware modules M_i in each group $(1 \le i \le g)$. Distributed architectures require communication. A serial bus seems to be the most convenient way for communication between the groups. In an attempt to design a homogeneous and flexible embedded system, all hardware modules are attached to the network. Due to the distribution of the work load among the hardware modules some input variables become shared (see Figure 10.11). We assume that every sensor can reach every hardware module in the frame of the corresponding group. In most cases shared analog input is not allowed and we exclude it as a possibility. The overlap could be used for diagnostic purposes as well.

It is also desirable to provide diagnostics to the field device level where the majority of all failures typically occur. One reason for input expansion could be the use of triplicated sensors and the suggested approach is consistent with this tendency.

Let the following sets be a description of the embedded system interface.

$$I = \bigcup_{1 \leq i \leq g} I_i = \bigcup_{1 \leq i \leq g} \left(\bigcup_{1 \leq j \leq n(I_i)} I_{i,j} \right)$$	(10-1)	Digital inputs
$$A = \bigcup_{1 \leq i \leq g} A_i = \bigcup_{1 \leq i \leq g} \left(\bigcup_{1 \leq j \leq n(A_i)} A_{i,j} \right)$$	(10-2)	Analog inputs
$$P = \bigcup_{1 \leq i \leq g} P_i = \bigcup_{1 \leq i \leq g} \left(\bigcup_{1 \leq j \leq n(P_i)} P_{i,j} \right)$$	(10-3)	Parallel running hardware inputs
$$Q = \bigcup_{1 \leq i \leq g} Q_i = \bigcup_{1 \leq i \leq g} \left(\bigcup_{1 \leq j \leq n(Q_i)} Q_{i,j} \right)$$	(10-4)	Digital outputs
$$C = \bigcup_{1 \leq i \leq g} C_i = \bigcup_{1 \leq i \leq g} \left(\bigcup_{1 \leq j \leq n(C_i)} C_{i,j} \right)$$	(10-5)	Analog outputs
$$F = \bigcup_{1 \leq i \leq g} F_i = \bigcup_{1 \leq i \leq g} \left(\bigcup_{1 \leq j \leq n(F_i)} F_{i,j} \right)$$	(10-6)	Flags

The interface is viewed at two levels. First, we investigate how the I/Os are organized in groups (we vary with i). Second, we look at the individual I/Os in each group (we point by j). The number of digital inputs in group i is $n(I_i)$.

Parallel running hardware inputs are used as a separate set to introduce some specific, interface oriented blocks in the hardware modules. These blocks run in parallel with the processor core. In this way, the system architecture has more potential to optimize the design process, taking into account the need of counters, timers, interrupt inputs and so on. Flags are considered to be internal variables.

Once we have the system architecture and the concept about the hardware modules accepted, we can split the partitioning into software partitioning and hardware partitioning. The real-time requirements should be met by free running hardware units. We assume that the possibility to process software modules on different nodes concurrently leads to reduced execution time.

Software partitioning

Software partitioning is organized in two steps. First, all output functions are attached to a certain group. The goal is reduction of swapping. The data closeness criterion is a maximum number of input variables in the group. In other words, we want to compute an output function in the group where the majority of its input variables are sensed. Since we distinguish between different types of inputs, we must decide how to count. We check for each output function if it has inputs of the P type. If there are any, we determine which group holds the majority of them and attach the function there. If the function lacks parallel running inputs (type P), we check for analog inputs. Also, if the type P inputs are equally distributed in two or more groups we go on with the analog inputs. Finally, if an output function lacks type P and A inputs or if they are balanced over the groups, we take a decision on the base of the digital inputs (type I). In essence, when an output function is attached to a group, the set P has the highest priority and the set I the lowest.

Second, the output functions are clustered in every group. At this point, our design approach has to offer sufficient potential to form the embedded system as a reactive system. In order to improve the timing parameters the criteria for clustering become dependent on common input variables from the subsets parallel running hardware P_i inputs and analog A_i inputs. Analog input variables can not be employed as shared input and they are involved in the clustering criteria in order to decline the communication penalty.

Exploiting the fact that each output function $v \in Q \cup C \cup F$ defines corresponding subsets $I(v) \subset I$, $A(v) \subset A$, $P(v) \subset P$ and $F(v) \subset F$, we recursively merge the output functions until we finally obtain partitioning of all functions into $n(O_i)$ clusters $O_{i,j}$ in each group i.

$$Q \cup C \cup F = \bigcup_{1 \le i \le g} (Q_i \cup C_i \cup F_i) = \bigcup_{1 \le i \le g} \left(\bigcup_{1 \le j \le n(O_i)} O_{i,j} \right) \tag{10-7}$$

If in the group i, we move from one cluster to another $(1 \le j \le n(O_i))$, we can see the subsets of parallel running hardware inputs and analog inputs and no two of them share a common input P or A.

$$\forall 1 \le i \le g, \quad P_{O_{i,j}} = \bigcup_{v \in O_{i,j}} P(v), \quad P_{O_{i,j}} \cap P_{O_{i,q}} = \varnothing \text{ if } j \ne q, 1 \le j,q \le n(O_i) \tag{10-8}$$

Similarly,

$$\forall 1 \le i \le g, \ A_{O_{i,j}} = \left(\bigcup_{v \in O_{i,j}} A(v) \right) \cap A_i, \quad A_{O_{i,j}} \cap A_{O_{i,q}} = \varnothing \ \text{ if } \ j \ne q, 1 \le j, q \le n(O_i)$$

$$(10\text{-}9)$$

The result of the software partitioning is that all output functions are distributed over the groups for calculation and clustered.

Hardware partitioning

The hardware partitioning can be viewed as an allocation phase (see section 10.5). We assume that the embedded system is homogeneous as all hardware modules are identical. The minimal input/output capacity of one hardware module is defined by the five most demanding clusters.

$$IM = \max_{1 \le i \le g} \left(\max_{1 \le j \le n(O_i)} n\left(I_{O_{i,j}} \cap I_i \right) \right) \tag{10-10}$$

$$AM = \max_{1 \le i \le g} \left(\max_{1 \le j \le n(O_i)} n\left(A_{O_{i,j}} \cap A_i \right) \right) \tag{10-11}$$

$$PM = \max_{1 \le i \le g} \left(\max_{1 \le j \le n(O_i)} n\left(P_{O_{i,j}} \right) \right) \tag{10-12}$$

$$QM = \max_{1 \le i \le g} \left(\max_{1 \le j \le n(O_i)} n\left(Q_{O_{i,j}} \cap Q_i \right) \right) \tag{10-13}$$

$$CM = \max_{1 \le i \le g} \left(\max_{1 \le j \le n(O_i)} n\left(C_{O_{i,j}} \cap C_i \right) \right) \tag{10-14}$$

Note that we take into account the intersection between all inputs/outputs of a certain type associated to a cluster and all inputs/outputs of this type which belong to the current group (physically connected to the group). In this way, we eliminate input/output values which travel between the groups. They are exchanged through the serial communication subsystems of the microcontrollers, such as I^2C bus or CAN bus, and alter the burden of the parallel interface.

Denote by I_M, A_M, P_M, Q_M and C_M the sets of digital inputs, analog inputs, parallel running hardware inputs, digital outputs and analog outputs as attributes of a certain hardware module. Denote also by

$$IN = n(I_M) \quad AN = n(A_M) \quad PN = n(P_M) \quad QN = n(Q_M) \quad CN = n(C_M) \quad (10\text{-}15)$$

the number of elements in the corresponding set. Then, the minimal hardware module frame which is consistent with the clustering approach is given in inequality (10-16).

$$IN \geq IM, \quad AN \geq AM, \quad PN \geq PM, \quad QN \geq QM, \quad CN \geq CM \qquad (10\text{-}16)$$

Now we move on to the lower bound of the number of hardware modules. Suppose the type of hardware module has been chosen according to (10-16) and our aim is a solution based on a minimal number of hardware modules. A lower bound of this number, for a certain group, is given in (10-17),

$$n\left(M_i^{MIN}\right) = \max\left(\left\lceil \frac{n(I_i)}{IN} \right\rceil, \left\lceil \frac{n(A_i)}{AN} \right\rceil, \left\lceil \frac{n(P_i)}{PN} \right\rceil, \left\lceil \frac{n(Q_i)}{QN} \right\rceil, \left\lceil \frac{n(C_i)}{CN} \right\rceil\right) \qquad (10\text{-}17)$$

where $\lceil \alpha \rceil$ denotes the ceiling function, i.e., the smallest integer $\geq \alpha$. However, the microcontroller's port lines can be used as either input or output. Therefore, a more precise estimation for the minimal number we derive from (10-18).

$$n\left(M_i^{MIN}\right) = \max\left(\left\lceil \frac{n(I_i) + n(Q_i)}{IN + QN} \right\rceil, \left\lceil \frac{n(A_i)}{AN} \right\rceil, \left\lceil \frac{n(P_i)}{PN} \right\rceil, \left\lceil \frac{n(C_i)}{CN} \right\rceil\right) \qquad (10\text{-}18)$$

Problem 10.2

Calculate the required minimal number of hardware modules based on the 83C592 microcontroller for a group where the demand for I/O interface is as follows:

Digital inputs plus digital outputs $n(I_i) + n(Q_i) = 40$

Analog inputs $n(A_i) = 8$

Interrupt inputs $n(P_i) = 2$

Analog outputs $n(C_i) = 2$

Solution 10.2

The hardware module can be designed in two different styles:
- The microcontroller employs Port 0 and Port 2 to access the external memory.
- The microcontroller does not interact with external memory and two more ports are available for I/O interface.

For the first case,

$$IN + QN = 24, \quad AN = 8, \quad PN = 4, \quad CN = 2 \qquad (10\text{-}19)$$

and we get for the minimal number of modules

$$n\left(M_i^{MIN}\right) = \max\left(\left\lceil \frac{40}{24} \right\rceil, \left\lceil \frac{8}{8} \right\rceil, \left\lceil \frac{2}{2} \right\rceil, \left\lceil \frac{2}{2} \right\rceil\right) = \max(2,1,1,1) = 2 \qquad (10\text{-}20)$$

For the second case (without external memory),

$$IN + QN = 40, \quad AN = 8, \quad PN = 4, \quad CN = 2 \tag{10-21}$$

and we obtain

$$n\left(M_i^{MIN}\right) = max\left(\left\lceil \frac{40}{40} \right\rceil, \left\lceil \frac{8}{8} \right\rceil, \left\lceil \frac{2}{2} \right\rceil, \left\lceil \frac{2}{2} \right\rceil\right) = max(1,1,1,1) = 1 \tag{10-22}$$

Mapping

When we distribute the workload over the hardware modules we could use our software modules represented by the clustered output functions. However, it might be wise to transform the clusters into sets called reorganized clusters. For instance, the cluster $O_{i,j}$ generates the corresponding reorganized cluster $R_{i,j}$.

$$R_{i,j} = \left(I_{O_{i,j}} \cap I_i\right) \cup \left(A_{O_{i,j}} \cap A_i\right) \cup \left(P_{O_{i,j}}\right) \cup \left(Q_{O_{i,j}} \cap Q_i\right) \cup \left(C_{O_{i,j}} \cap C_i\right) \tag{10-23}$$

The important thing is that the reorganized cluster $R_{i,j}$ includes all outputs of the cluster $O_{i,j}$ and the associated inputs except the inputs and outputs that are exchanged over the network. The type P inputs are not allowed to travel between the groups.

The distribution can be done by evaluation of all possible mappings:

$$\varphi_{i,m} \qquad : \quad R_{i,j} \quad \rightarrow \quad \{i,1 \quad i,2 \quad i,3 \quad ... \quad i,n(M_i)\} \tag{10-24}$$
$$\scriptstyle 1 \le i \le g, \ 1 \le m \le \left(n\left(M_i^{MIN}\right)\right)^{n(O_i)} \qquad \scriptstyle 1 \le j \le n(O_i)$$

starting with

$$n(M_i) = n\left(M_i^{MIN}\right) \tag{10-25}$$

and completing the procedure as soon as an acceptable mapping is found. Each row of the table in Figure 10.12 reflects one mapping. The decision of whether to accept or reject a certain row in Figure 10.12 is made on the basis of two tests.

First, each input and output must find its place in the hardware modules. For example, the reorganized cluster $R_{i,j}$ is directed to hardware module $M(i,2)$ and occupies a certain number of type P inputs. As can be seen in Figure 10.12, there is an overlap of type P inputs between reorganized clusters $R_{i,2}$ and $R_{i,j}$. As is frequently the case, the set P_M is a subset of the set I_M. This fact motivates us to start the procedure with type P inputs. We define the same priority for the type A inputs and type C outputs. Furthermore, the reorganized cluster $R_{i,j}$ has several digital inputs which use up a share of the set I_M.

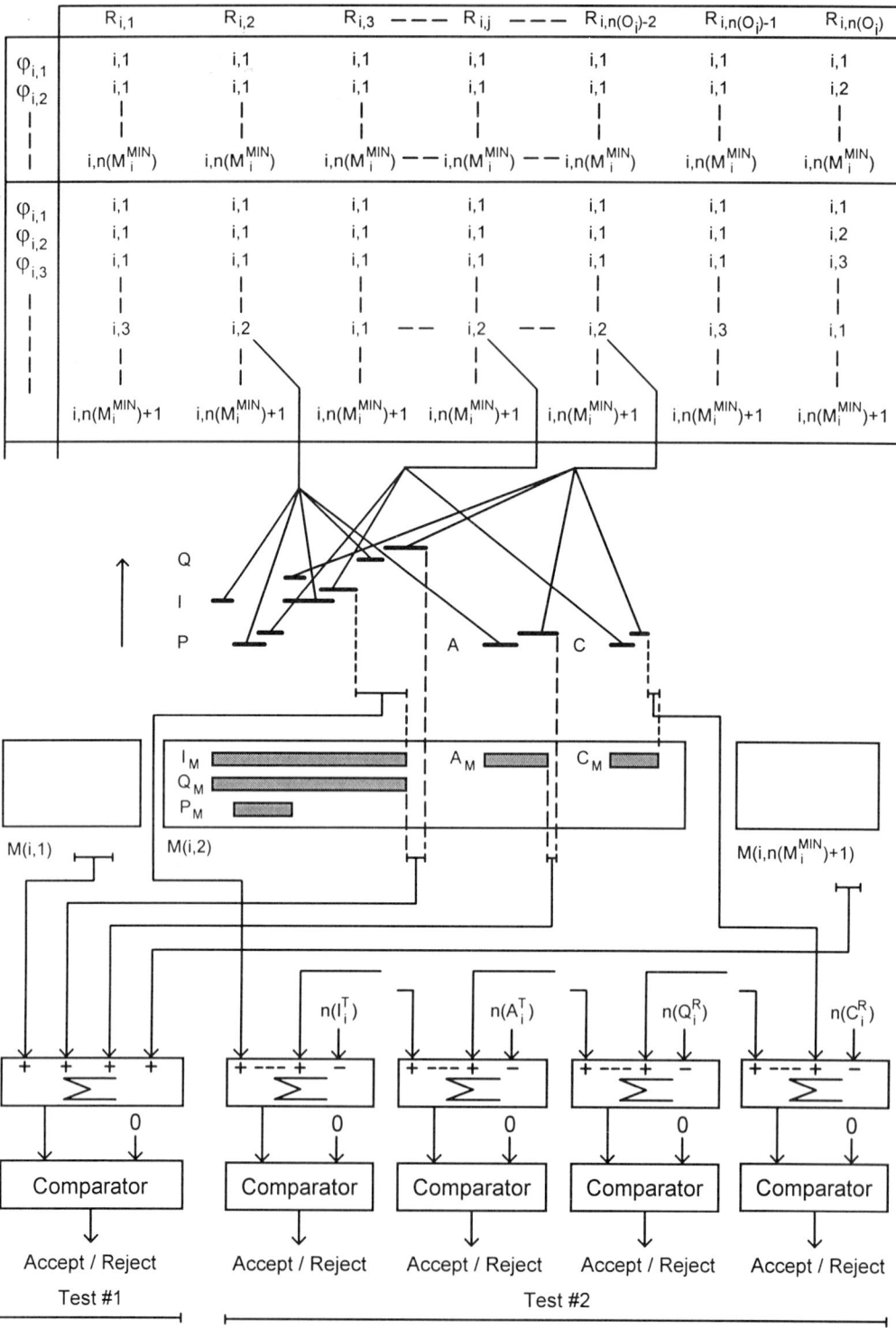

Figure 10.12 The mapping function evaluation.

The outputs of the reorganized cluster $R_{i,j}$ (the outputs are type C in this example) consume a subset from the module M(i,2), set C_M. If all inputs and outputs are successfully placed in the hardware modules we mark "Accept" and test #1 is over.

Second, the inputs I_i^T and A_i^T which are not involved in calculations in the current group, however, must be transmitted over the network, are compared with unused inputs from the corresponding type. Similarly, the outputs Q_i^R and C_i^R, which are received from the network and emitted in the current group, are compared with unused outputs. If test #2 is also successful, the row is accepted and the mapping evaluation procedure is terminated.

It is certainly implied that the rows containing less hardware modules than the corresponding number for the current section of the table are skipped.

A straightforward implementation of the algorithm will result in up to $M_i^{n(O_i)}$ row tests per table section. In fact, the number of the relevant mappings in the frame of one table section is given by the Stirling numbers of the second kind [Bala 1995].

10.7 Hardware-software co-design systems

A variety of methodologies for hardware-software co-design of embedded systems are capturing the attention of the academic community and industry. Several hardware-software co-design systems are targeted on control dominated applications.

POLIS is a hardware-software co-design environment developed in the University of California, Berkeley, Magneti Marelli, Torino Italy and the Politecnico di Torino, Italy [Chio 1996, Bala 1997, Hsie 1997]. The specification languages used are Esterel, graphical CFSM and subsets of Verilog and VHDL. The model which underlies the system is Co-design Finite State Machines (CFSMs). The CFSMs are capable of implementing arithmetic, relational and logical operations on a set of integer values. The communication between the CFSMs is asynchronous and based on events. The target architecture is composed of microcontrollers and ASICs. In addition, a library of microcontroller's peripherals can be used for performance evaluation, co-simulation and verification [Hsie 1997]. Initially, the embedded system is prototyped by FPGA. As soon as the ASIC of the microcontroller's CPU plus the built-in peripherals is available, the co-synthesis system is switched from FPGA to peripheral subsystems.

Cosyma is a co-synthesis system for embedded control applications developed in The Technical University of Braunschweig, Germany [Erns 1993, Henk 1994, Stau 1997]. The specification language, C^x, is an extension of C which allows communication between parallel processes. In principle, the target architecture is based on a RISC processor and an automatically generated application specific coprocessor. Both processors communicate by means of shared memory. A recent version of Cosyma can handle target architectures which include heterogeneous processors, such as SPARC and 8051 [Stau 1997]. The Cosyma system starts with all basic blocks directed to software. The basic blocks are obtained by partitioning process. The granularity is a compromize between statement level and process level. The system extracts hardware components by simulated annealing until the timing parameters are met. In essence, the Cosyma is a platform which can be used for a variety of applications from digital signal processing to embedded control.

LYCOS is a co-synthesis system created in the Department of Information Technology, The Technical University of Denmark, Lyngby [Knud 1996, Stau 1997]. The input specification can be written in VHDL. The model used is Control/Data Flow Graph (CDFG). The target architecture includes a processor and an ASIC. A software library contains technology files for different processors. Furthermore, the ASIC is designed as data path - controller architecture. The trade-off between the control and data path area is evaluated. The LYCOS hardware-software co-design system has been used for complex image processing applications.

COSMOS is a hardware-software co-design environment developed in the TIMA Laboratory, National Polytechnical Institute of Grenoble, France [Isma 1995, Stau 1997]. The design process starts with specification in SDL (Specification and Description Language). The basic model is an extended FSM capable of representing hierarchy and parallelism. The target architecture is a general concept of hardware, software and communication modules. The hardware modules are represented in VHDL. The software modules are represented in C. The COSMOS system has proved useful for the design of embedded applications such as robot controllers [Stau 1997].

ZEBRA is a system for automatic hardware-software co-design which is under development in the Computer Systems Department of The Technical University of Sofia, Bulgaria [Kara 1995b, Kara 1996]. The system is based on the methodology described in section 10.6.

10.8 References

James R. Armstrong and F.Gail Gray, *Structured Logic Design with VHDL*, PTR Prentice Hall, 1993.

V. K. Balakrishnan, *Theory and Problems of Combinatorics*, McGraw-Hill, 1995.

Felice Balarin, Massimiliano Chiodo, Paolo Giusto, Harry Hsieh, Attila Jurecska, Luciano Lavagno, Claudio Passerone, Alberto Sangiovanni-Vincentelli, Ellen Sentovich, Kei Suzuki and Bassam Tabbara, *Hardware-Software Co-Design of Embedded Systems, The POLIS Approach*, Kluwer, 1997.

A.Berger, "Following simple rules lets embedded systems work with μP emulators", *EDN*, April 13, 1989.

Jean-Michel Berge, Oz Levia and Jacques Rouillard, *Hardware/Software Co-Design and Co-Verification*, Kluwer, 1997.

Jean Paul Calvez, "A CoDesign Case Study with the MCSE Methodology", *Design Automation for Embedded Systems*, vol 1, No. 3, July 1996.

Massimiliano Chiodo, Daniel Engels, Paolo Giusto, Harry Hsieh, Attila Jurecska, Luciano Lavagno, Kei Suzuki and Alberto Sangiovanni-Vincentelli, "A case study in Computer-Aided Co-design of embedded controllers", *Design Automation for Embedded Systems*, vol. 1, 1996, pp. 51-67.

P.Chou, R.Ortega and G.Borriello, "Synthesis of the Hardware/Software interface in microcontroller-based systems", *Proc. IEEE/ACM Int. Conference on Computer-Aided Design*, Santa Clara, 1992, pp. 488-495.

Alan Clements, *Microprocessor Systems Design*, PWS Publishers, 1987.

James W. Coffron, *Z80 Applications*, SYBEX, 1983.

Data I/O Corporation, *SYNARIO, ABEL-HDL Reference*, 1996.

Hesham El-Rewini, Theodore G. Lewis and Hesham H. Ali, *Task Scheduling in Parallel and Distributed Systems*, PTR Prentice Hall, 1994.

Rolf Ernst, Jorg Henkel and Thomas Benner, "Hardware-software cosynthesis for microcontrollers", *Design & Test of Computers*, December 1993, pp. 64-75.

Rolf Ernst, "Target architectures", EUROCHIP Course on Hardware/Software Codesign, DTU, Lyngby, Denmark, 1995.

Alberto Daniel Ferrari, "Real-time scheduling algorithms", *Dr. Dobb's Journal*, December 1994, pp. 60-65.

Daniel D. Gajski, Frank Vahid, Sanjiv Narayan and Jie Gong, *Specification and Design of Embedded Systems*, PTR Prentice Hall, 1994.

Jorg Henkel, Rolf Ernst, Ullrich Holtmann and Thomas Benner, "Adaptation of partitioning and high-level synthesis in hardware/software co-synthesis", *Proc. IEEE/ACM Int. Conference on Computer-Aided Design*, San Jose, California, 1994, pp. 96-100.

L. J. Herbst, *Integrated Circuit Engineering*, Oxford University Press, 1996.

Harry Hsieh, Luciano Lavagno, Claudio Passerone, Claudio Sansoe, Alberto Sangiovanni-Vincentelli, "Modeling micro-controller peripherals for high-level co-simulation and synthesis", *Proc. 5th International Workshop on Hardware/Software Co-Design, Codes/CASHE'97*, March 1997, pp. 127-130.

Tarek Ben Ismail and Ahmed Amine Jerraya, "Synthesis steps and design models for codesign", *Computer*, February, 1995, pp. 44-52.

M. A. Jackson, *Principles of Program Design*, Academic Press Inc., 1975.

Gwyn Jones and Mike Headon, *Structured program design with Pascal*, Paradigm, 1988.

Zdravko Karakehayov and Tzvetan Ostromsky, "Distributed control using shared input", *Proc. IASTED Int. Conference on APPLIED INFORMATICS*, Annecy, France, 1994, pp. 276-278.

Zdravko Karakehayov, "A fine-grain approach to distributed embedded systems design", *Proc. IASTED/ISMM Int. Conference on Parallel and Distributed Computing and Systems*, pp. 376-380, Washington, D.C., 1995b.

Zdravko Karakehayov and Emil Saramov, "A fuzzy geography approach to hardware-software co-design of distributed embedded systems", *IEEE International Workshop on Embedded Fault-Tolerant Systems*, Dallas, USA, 1996.

Peter Voigt Knudsen and Jan Madsen, "PACE: A dynamic programming algorithm for hardware/software partitioning", *Proc. 4th International Workshop on Hardware/Software Co-Design, Codes/CASHE'96*, March 1996, pp. 85-92.

Sanjaya Kumar, James H. Aylor, Barry W. Johnson and Wm. A. Wulf, *The Codesign of Embedded Systems : A Unified Hardware/Software Representation*, Kluwer, 1996.

Ganesh Lakshminarayana and Niraj K. Jha, "High-level synthesis of power-optimized and area-optimized circuits from hierarchical data-flow intensive behaviors", *IEEE Trans. Computer-Aided Design of Integrated Circuits and Systems*, vol. 18, No. 3, March 1999, pp. 265-281.

Lattice Semiconductor, *Data Book*, 1994.

William H. Mangione-Smith, Brad Hutchings, David Andrews, Andre DeHon, Carl Ebeling, Reiner Hartenstein, Oskar Mencer, John Morris, Krishna Palem, Viktor K. Prasanna and Henk A.E. Spaanenburg, "Seeking solutions in configurable computing", *Computer*, December, 1997, pp. 38-43.

Motorola Semiconductor, *Neuron® Chip Distributed Communications and Control Processors*, 1994.

Zainalabedin Navabi, *VHDL: Analysis and Modeling of Digital Systems*, McGraw-Hill, 1993.

John Novellino, "Emulators race to keep up with new devices", *Electronic Design*, pp. 151-164, November 20, 1995.

John B. Peatman, *Design with Microcontrollers*, McGraw-Hill, 1988.

David Pellerin and Michael Holley, *Practical Design Using Programmable Logic*, Prentice Hall, 1991.

Philips Semiconductors, *CAN Bus Specification 2.0*, 1994b.

Philips Semiconductors, *High-speed CMOS Logic family, Data Handbook IC06*, 1994a.

Philips Semiconductors, *80C51-Based 8-Bit Microcontrollers, Data Handbook IC20*, 1997a.

Wolfgang Rosenstiel, "Prototyping and emulation", Jorgen Staunstrup and Wayne Wolf (eds.), *Hardware/Software Co-Design: Principles and Practice*, pp. 75-112, Kluwer, 1997.

Michael A. Schuette and John R. Barr, "Embedded systems design for low-energy consumption", *Proceedings, ICCAD-94*, 1994.

Sol M. Shatz, *Development of Distributed Software: Concepts and Tools*, Macmillan Publishing Company, 1993.

Kenneth L. Short, *Embedded Microprocessor Systems Design*, Prentice-Hall, 1998.

S. Srinivasan and N.K. Jha, "Hardware-software co-synthesis of fault-tolerant real-time distributed embedded systems", *Proc. European Design Automation Conference*, pp. 334-339, Brighton, U.K., 1995.

Jorgen Staunstrup and Wayne Wolf, *Hardware/Software Co-Design: Principles and Practice*, Kluwer, 1997.

Daniel Tabak, *Advanced Microprocessors*, McGraw-Hill, 1995.

Malcolm Wallace, *Functional Programming and Embedded Systems*, Ph.D. Thesis, University of York, January 1995.

Wayne Wolf, *Modern VLSI Design*, Prentice Hall, 1994a.

Wayne Wolf, "Hardware-software co-design of embedded systems", *Proc. IEEE*, vol. 82, No. 7, July 1994b, pp. 967-989.

Andrew Wolfe, "A case study in low-power system design", *Proceedings, ICCD-95*, IEEE Computer Society, 1995, pp. 332-338.

S. Yajnik, S. Srinivasan and N.K. Jha, "TBFT: A task-based fault tolerance scheme for distributed systems", *Proc. ISCA Int. Conference Parallel and Distributed Computing Systems*, Las Vegas, USA, 1994.

Information about FPGAs can be found at
Lattice	http://www.latticesemi.com
Xilinx	http://www.xilinx.com
QuickLogic	http://www.quicklogic.com

Information about development tools for 8051 microcontrollers can be found at
Ceibo	http://www.ceibo.com
Intel	http://www.intel.com
Keil Software	http://www.keil.com
KSC Software Systems	http://www.ksc-softsys.com/c.htm
Nohau	http://www.nohau.com

Information about 8051 soft cores written in Verilog and VHDL can be obtained from
| Mentor Graphics | http://www.mentorg.com/inventra |

Chapter 11

DESIGN EXAMPLES

11.1 Introduction

In this chapter we work through two design problems: an EPROM programmer and an EPROM emulator. Both examples are typical embedded systems based on microcontrollers. While the EPROM programmer project can be split into hardware design and software design, a successful design of EPROM emulator requires a co-design approach. Designing the emulator architecture we first investigate the application range in case of software control. The result suggests that the architecture must be reinforced by an ASIC. Using FPGAs to implement the ASIC makes the design affordable and provides the capacity to adapt the hardware to different target processors.

11.2 EPROM programmer

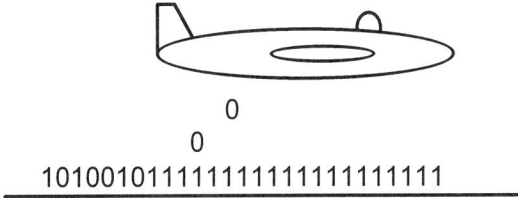

In this case study, the embedded system under consideration is an EPROM programmer. A new EPROM has high in all bits. The programming process resets bits to achieve the desired pattern. The only way to change a bit from low to high is to expose the device to ultraviolet light. However, the exposition affects all bits. The EPROMs are produced in two different packages. First, packages with a transparent lid (UV EPROM) allow the user to erase the content by ultraviolet light. Second, packages without a transparent lid (OTP ROM) are produced for applications where the pattern is programmed only one time. In both cases, we need a device termed EPROM programmer to implement the programming procedure.

The programmer has the following specification:

• The device is a part of a microprocessor development system based on a PC and a single board computer.

• The device is capable of burning components type 32K x 8 (27C256).

• The EPROM programmer possesses 32K byte on-board RAM.

• The programmer is linked to the PC via the RS-232 interface.

• The format of the file which is written to the EPROM is Intel HEX.

Figure 11.1 shows the operation modes of a typical 32K x 8 EPROM. The programming mode requires a few pulses to be applied to the input $\overline{CE}$. Note that the 8K x 8 EPROM, discussed in section 4.2, was programmed by pulsing the input $\overline{PGM}$. Keeping the 28 pins package for 32K x 8 EPROMs could be possible if, for example, pin #20 is used for both $\overline{CE}$ and programming pulse input. Normally, the programming pulse should be 100 µs long. The 8051 microcontroller could generate this pulse either by a subroutine or using a timer. Since no other timing constrains are involved, it would be justified to break down the design into hardware design and software design.

		Pins				
		$\overline{CE}$	$\overline{OE}$	V_{PP}	V_{CC}	D0 - D7
Mode	Program	⎍	1	12.75 ± 0.25 V	6.25 ± 0.25 V	D_{IN}
	Verify	1	0	12.75 ± 0.25 V	6.25 V	D_{OUT}
	Read	0	0	5 V	5 V	D_{OUT}
	Standby	1	x	5 V	5 V	OFF

Figure 11.1 EPROM modes of operation.

Hardware design

Figure 11.2 outlines the flow of data on the base of a memory map as discussed earlier in section 4.5. The shaded area in the RAM is the place where the file is downloaded. The correspondent shaded area in the EPROM address space is the actual segment that is programmed.

The pin's arrangement for 32K x 8 EPROMs can be seen in Figure 11.3. The single board computer designed in section 4.5 provides sufficient I/O interface to cover the EPROM address and data lines. Figure 11.3 shows only this part of the single board computer which supports the interface to the EPROM socket. Also, Port 1 has ample lines to meet the control inputs of the

EPROM. Figure 11.1 indicates that the pins V_{PP} and V_{CC} must be switched between different voltage levels. This can be successfully handled if Port 1 is reinforced by extra switching circuitry. Again, looking for a simple solution we select reed relays to switch the voltages. The relays are housed in a standard 14 pin IC package with built-in diode protection.

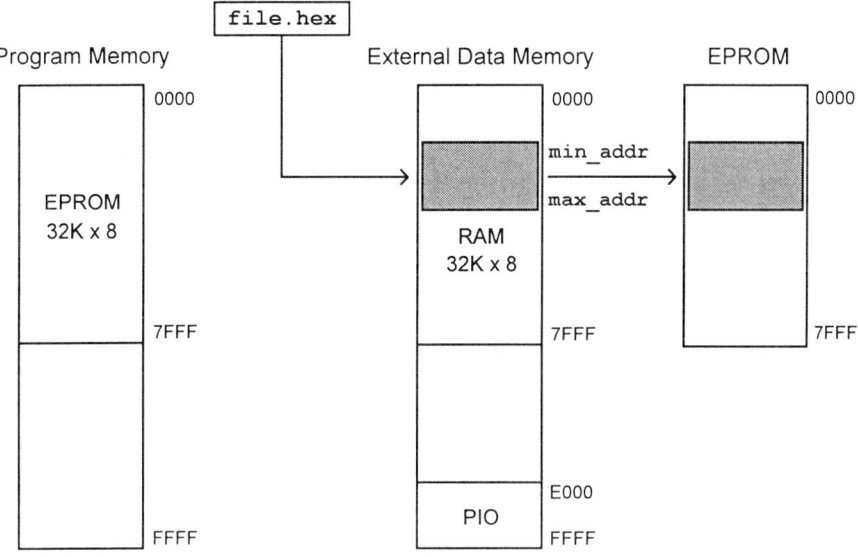

Figure 11.2 The EPROM programmer memory map.

The maximum values of the supply (pin V_{CC}) and programming (pin V_{PP}) currents are in the area of 50 mA. The contacts of the relays are closed when the driving open collector gate has a low output.

Software design

In the interest of simplicity, we decompose the EPROM functionality into two clusters of subtasks. Figure 11.4 shows the subtasks and presents the borderline between the PC and the single board computer in that regard. Thus, we have a clear picture of the interaction between the PC and the 8051-based embedded system. It is logical to map more tasks to the PC where the resources are practically unlimited. However, the drawback of a PC dominated control would be a complicated communication procedure. A straightforward implementation can be obtained by downloading the code to the single board computer's RAM and copying the contents of the RAM to the EPROM. This is the motivation for 32K bytes on-board RAM. Consequently, the major task of the PC is to download the file to the EPROM programmer's RAM. Once the code has been downloaded, the PC's role can be simplified to printing messages on the screen.

The 8051's operation covers two essential tasks: "Download" and "Program" the EPROM. The download is further subdivided into a few actions which are reflected in Figure 11.4 by the task receive a "Line" (see the Intel HEX file example in section 3.10). The task receive a line calls the task receive a "Byte". Finally, receive a "Nibble" is the task which operates at the lowest level.

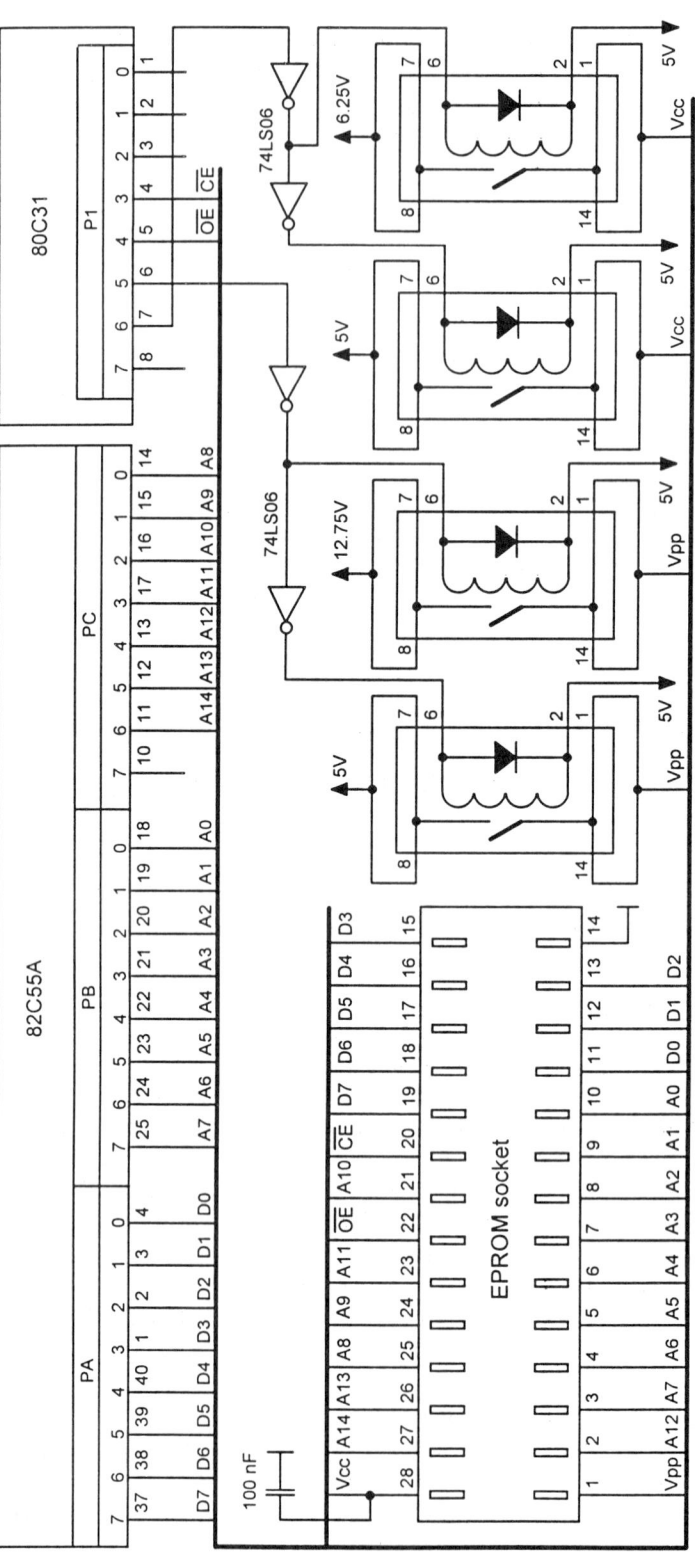

Figure 11.3 The EPROM programmer.

The task "Program" the EPROM, can be broken down into initialization, an essential core ("Program and verify"), and final check ("Check all bytes"). The program and verify operation can be further decomposed into two, byte grain subtasks. The only time critical subtask is delay 100 µs which defines the programming pulse duration. The subtask "Delay 10 ms" is used to debounce the relays and do not need a precise interval. Both delay subtasks are related to Timer 0. Since we have chosen the software to be written in C for both the PC and the embedded system, Figure 11.4 also reveals the names of the C functions which correspond to the accepted decomposition scheme.

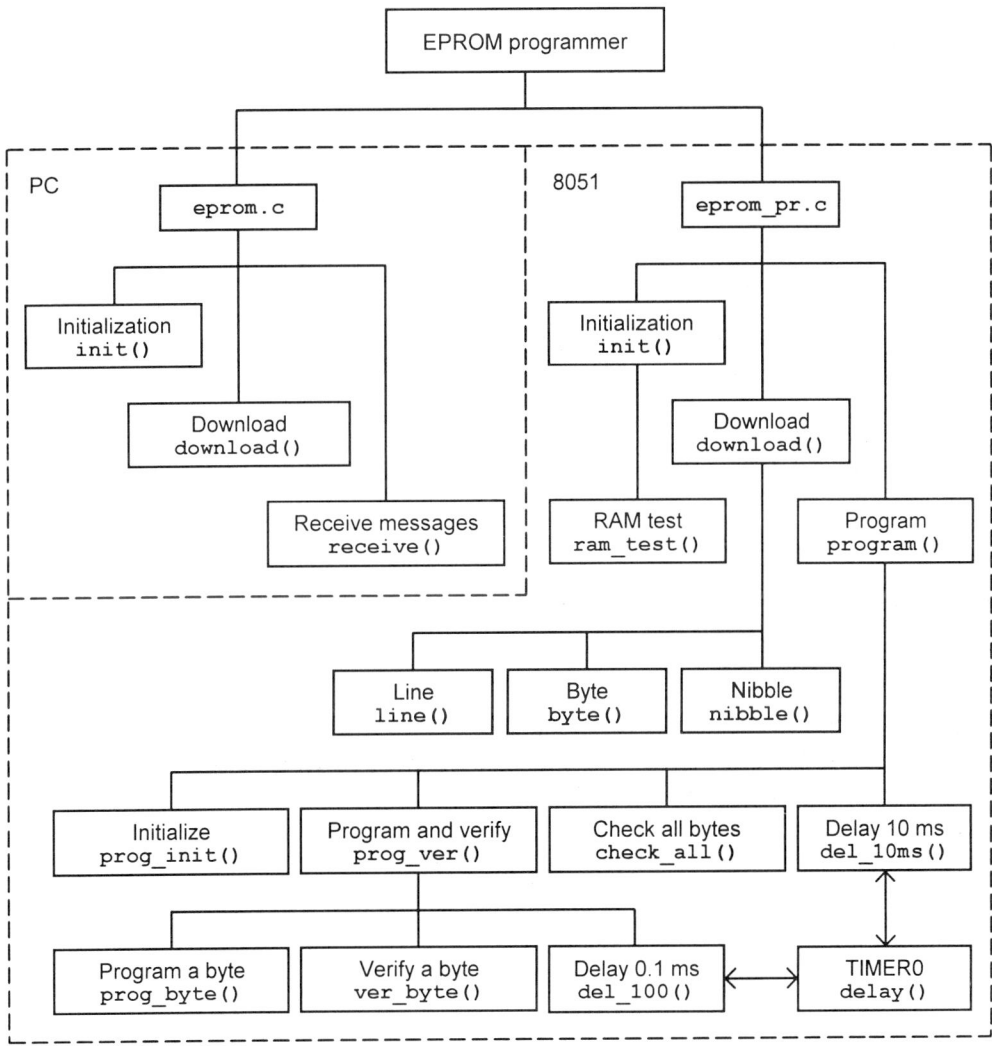

Figure 11.4 The EPROM programmer functionality decomposed into subtasks.

Here is the program which runs on the PC:

```
/****************************************************************/
/*      eprom.c                                               */
/*      This program for the PC                               */
/*      Downloads an Intel HEX file through a serial port     */
/*      Reads messages from the same port and prints them     */
/*         on the screen                                      */
/****************************************************************/
#include <stdio.h>
#include <conio.h>
#include <process.h>
#include <bios.h>

#define ESC                '\x1B'
#define COM1               0
#define COM2               1
#define COM_INIT           0
#define COM_SEND           1
#define COM_RECEIVE        2
#define COM_STAT           3
#define COM_DATA_READY     0x0100
#define COM_TxHOLD         0x2000

int com = COM2;
FILE *fptr;
char ch;
int stat;

/*** Initialize the communication port *********************/
init()
  {
  char data;
  data = (0x03 | 0x00 | 0x00 | 0xE0);
          /* 8 data bits, 1 stop bit, no parity, 9600 bps */
  bioscom(COM_INIT,data,COM2);
  }

/*** Download the specified file ***************************/
download()
  {
  puts("Downloading");
  ch = getc(fptr);
  while ( ch != EOF)
    {
    stat = bioscom(COM_STAT, 0, com);
    while ( !(stat & COM_TxHOLD ) )
      stat = bioscom(COM_STAT, 0, com);
    bioscom(COM_SEND, ch, com);          /* Send a character */
    if (stat & COM_DATA_READY)
```

```
        {
        ch = bioscom(COM_RECEIVE, 0, com);
        putch(ch);
        }
     ch = getc(fptr);
     }
  }

/*** Receive messages from the single board computer ********/
receive()
   {
   do
      {
      stat = bioscom(COM_STAT, 0, com);
      if (stat & COM_DATA_READY)
         {
         ch = bioscom(COM_RECEIVE, 0, com);
         putch(ch);
         }
      if ( kbhit() )
         ch=getch();
      } while ( ch != ESC);
   }

/*************************************************************/
main(int argc, char *argv[])
   {
   init();
   if (argc != 2)
      {
      puts("Format : eprom filename");
      exit(0);
      }
   if ( ( fptr = fopen(argv[1], "r") ) == NULL )
      {
      printf("Can't open file %s.", argv[1] );
      exit(0);
      }
   download();
   receive();
   }
```

The EPROM programming algorithm requires a sequence of 100 μs ± 5 μs pulses to be applied to the input $\overline{CE}$ until a correct verify occurs. If 25 attempts to program a byte are unsuccessful, the IC is considered failed.

The following is a C program for the EPROM programmer:

```c
/******************************************************/
/*      eprom_pr.c                                    */
/*      This program for the 8051 microcontroller     */
/*      implements an EPROM programmer                */
/******************************************************/
#include <stdio.h>

#define EPROM_R_CR    0x90       /* PIO CR value for read */
#define EPROM_W_CR    0x80       /* PIO CR value for write */
#define RAM_MIN       0x0000     /* Available RAM start address */
#define RAM_MAX       0x7FFF     /* Available RAM end address */

xdata char eprom_data at 0xE000; /* PIO at address 0xE000 */
xdata char eprom_a_l at 0xE001;
xdata char eprom_a_h at 0xE002;
xdata char pio_cr at 0xE003;

char ch, c_sum, number, n, frac=0;
unsigned int addr, min_addr, max_addr;
bit dl_error, dl_end, f_min_addr, f_100, f_10, ready;

/***    Interrupt handler for Timer 0    ********************/
delay() interrupt TIMER0
   {
   if ( f_100 )
      {
      TR0 = 0;        /* Stop Timer 0 */
      f_100 = 0;      /* Reset the flag */
      }
   else
      {
      if ( frac++ == 40 )
         {
         TR0 = 0;     /* Stop Timer 0 */
         f_10 = 0;    /* Reset the flag */
         frac = 0;
         }
      }
   }

/*** Test of the available RAM *****************************/
ram_test()
   {
   addr = RAM_MIN;
   do
      XMEM[addr++] = 0x00;         /* Reset all locations */
   while ( addr <=  RAM_MAX );
   addr = RAM_MIN;
```

```
   do
     if ( XMEM[addr++] != 0x00 )
        {
        puts("\n\r RAM error");
        while (1);
        }
   while ( addr <= RAM_MAX );
   addr = RAM_MIN;
   do
     XMEM[addr++] = 0xFF;          /* Store 0xFF in all locations */
   while ( addr <= RAM_MAX );
   addr = RAM_MIN;
   do
     if ( XMEM[addr++] != 0xFF )
        {
        puts("\n\r RAM error");
        while (1);
        }
   while ( addr <=  RAM_MAX );
}

/*** Initialization ***************************************/
init()
   {
   P1.5 = 0;      /* Vpp = 5 V */
   P1.6 = 0;      /* Vcc = 5 V */
   P1.4 = 1;      /* Set *OE */
   P1.3 = 1;      /* Set *CE */

   fopen(stdin); /* Initialize the serial port for 9600 bps */
                 /* Crystal 11.0592 MHz */
   f_min_addr = 1;
   ram_test();
   }

/*** Receive a character and convert it to a nibble *********/
char nibble()
   {
   if ( ( ch = getchar( ) ) > '9' )
     return ch - 55 ;
   else
     return ch - 48 ;
   }

/*** Form a byte from two nibbles ***************************/
char byte()
   {
   char value ;
```

```c
   value=(nibble()<<4)+nibble();
   c_sum+= value ;          /* Update the checksum */
   return value ;
   }

/*** Download a line ***************************************/
line()
   {
   char count;
   c_sum = 0 ;
   number = byte() ;       /* Number of code bytes */
   addr = byte() ;
   addr = ( addr << 8 ) + byte();
   if ( f_min_addr )
      {
      min_addr = addr;     /* Save the start address */
      f_min_addr = 0;
      }
   ch = byte() ;
   if ( ch == 0x01 )       /* Final line? */
      dl_end = 1 ;         /* Set the flag download end */
   if ( number > 0 )
      {
      for ( count = 1 ; count <= number ; count++ )
         XMEM[addr++] = byte();  /* Store the code in the RAM */
      max_addr = addr - 1;       /* Update the end address */
      }
   ch = byte() ;           /* Add the checksum byte */
   if ( c_sum != 0 )
      {
      dl_error = 1 ;       /* Set the flag download error */
      addr = addr - 1 ;
      for ( count = 1 ; count <= number ; count++ )
         XMEM[addr--] = 0xFF ;   /* Set the RAM locations for */
                                 /* the last line */
      }
   }

/*** Store the incoming Intel HEX file in the RAM ************/
download()
   {
   do
      {
      while ( ( ch = getchar( ) ) != ':' );
      line();
      if ( dl_error )
         puts("\n\r Download error");
      } while ( !dl_end );
```

```
   }

/*** Delay 10 ms **********************************************/
del_10ms()
   {
   TH0 = 6;
   TR0 = 1;          /* Start Timer 0 */
   f_10 = 1;
   while (f_10) ;    /* Wait for interrupt */
   }

/*** Delay 0.1 ms *********************************************/
del_100()
   {
   TH0 = 170;        /* Crystal 11.0592 MHz */
   TR0 = 1;          /* Start Timer 0 */
   f_100 = 1;
   while (f_100) ;   /* Wait for interrupt */
   }

/*** Initialize for programming ******************************/
prog_init()
   {
   puts("\n\r Programming");
   printf("\n\r min_addr = 0x%x     max_addr = 0x%x",min_addr,
max_addr);
   f_100=0;
   P1.6 = 1;   /* Vcc = 6.25 V */
   P1.5 = 1;   /* Vpp = 12.75 V */
   TMOD=TMOD | 0x02;     /* Timer 0, mode 2 */
   TMOD=TMOD & 0xF2;
   ET0 = 1;                   /* Enable interrupts from Timer 0 */
   EA = 1;                    /* Enable all interrupts */
   del_10ms();
   }

/*** Program a byte ******************************************/
prog_byte()
   {
   pio_cr = EPROM_W_CR;       /* PIO Port A output */
   eprom_a_l = (char)addr;    /* Take the low byte */
   eprom_a_h = (char)(addr >> 8);
   eprom_data = XMEM[addr];   /* Output a byte to the EPROM */
   P1.3 = 0;                  /* Pull down the input *CE */
   del_100();                 /* Wait 0.1 ms */
   P1.3 = 1;                  /* End of the pulse */
   }
```

```
/*** Verify a byte *****************************************/
ver_byte()
   {
   ready = 0;
   pio_cr = EPROM_R_CR;      /* PIO Port A input */
   eprom_a_l = (char)addr;  /* Take the low byte */
   eprom_a_h = (char)(addr >> 8);
   P1.4 = 0;                 /* Reset the input *OE */
   if ( eprom_data == XMEM[addr] )
      ready = 1;             /* The byte has been programmed OK */
   P1.4  = 1;                /* Set the input *OE */
   }

/*** Program and verify the specified area *****************/
prog_ver()
   {
   for ( addr=min_addr ; addr<=max_addr; addr++ )
      {
      n = 0;                    /* Reset the iteration counter */
      do
         {
         prog_byte();        /* Apply one programming pulse */
         ver_byte();         /* Verify */
         if ( ++n == 25 )
            {
            puts("\n\r Device failed");
            while (1);
            }
         } while ( !ready );
      }
   }

/*** Enter the read mode and verify all bytes ***************/
check_all()
   {
   P1.5 = 0;                          /* Vpp = 5 V */
   P1.6 = 0;                          /* Vcc = 5 V */
   del_10ms;
   P1.3 = 0;                          /* Reset the input *CE */
   P1.4 = 0;                          /* Reset the input *OE */
   for ( addr=min_addr ; addr<=max_addr ; addr++ )
      {
      eprom_a_l = (char)addr; /* Take the low byte */
      eprom_a_h = (char)(addr >> 8);
      if (eprom_data != XMEM[addr])
         {
         printf("\n\r Final check error at address 0x%x",addr);
         while (1) ;
```

```
        }
      }
    puts("\n\r EPROM programmed successfully");
    }

/*** Program the EPROM ************************************/
program()
    {
    prog_init();
    prog_ver();
    check_all();
    }

/***********************************************************/
main()
    {
    init();
    download();
    program();
    while (1);
    }
```

If you want to make experiments with the program **eprom_pr.c**, you might need to adapt the code to a specific hardware platform. Manipulating the RAM start address (**RAM_MIN**) and RAM end address (**RAM_MAX**) you could compromise between the Intel HEX file parameters and the available RAM. Likewise, you could move the PIO 82C55 to a more suitable place in the memory map.

As discussed earlier in section 4.3, the PIO's output registers are reset when the code in the control register is changed. Therefore, the program must rewrite the EPROM address to ports B and C each time the direction of Port A is altered.

Finally, do not be surprised if the EPROM socket is empty and the program is capable of passing successfully the subtask "Verify a byte". The Port A capacitive load could keep the voltage levels for a certain period of time if there is no IC to drive the lines.

The EPROM programmer design can be viewed as a starting point for the development of sophisticated systems, capable of programming a cluster of memory components.

11.3 EPROM emulator

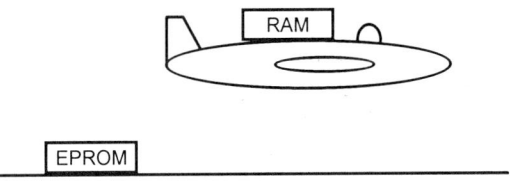

Now our design goal is an EPROM emulator. We are going to use the EPROM socket in an attempt to design a processor independent debugging tool. We intend to control the EPROM emulator by an 8051 microcontroller.

A microcontroller architecture

A good starting point can be Figure 11.5 which presents the overall system architecture viewed as a distributed system. The embedded system under consideration, the emulator, is a single processor system. So is the system that will be tested, the target system.

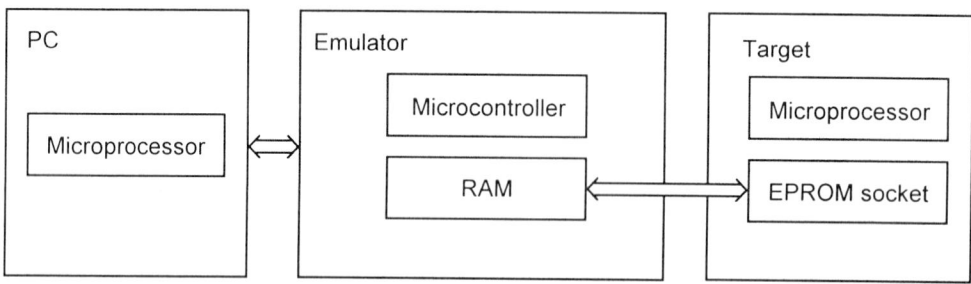

Figure 11.5 The EPROM emulator in conjunction with a PC and a target system.

The EPROM emulator can be used when the target system has one or more EPROMs. We remove an EPROM from the target and plug in the emulator cable. Thus, the emulator RAM substitutes the target EPROM. The personal computer downloads the code to the RAM. Surely no one would dispute that a program is modified much easier when it resides in a RAM instead of in an EPROM.

The emulator single chip microcomputer controls the debugging process according to the instructions from the PC.

Figure 11.6 shows the EPROM emulator functionality broken into six subtasks. The subtask "Set breakpoints" marks a certain number of addresses. When the target system reaches a breakpoint address, the execution of the user program is suspended and a program called Monitor is activated. The Monitor controls the communication between the PC and the target system. Consequently, the emulator emerges with two extra RAMs as shown in Figure 11.7. Essentially, RAMs are addressed in parallel. As a result, each byte in the user program RAM (USER RAM) has an attached location in the breakpoint (BP) RAM.

The size of the BP RAM cell could be just one bit - a breakpoint flag. However, it would be wise to use byte sized locations and implement, for example a trace buffer function which will require one start and one stop bit. Nevertheless, a debugging approach based on a high level language might require additional bits in the BP RAM.

The three-RAM architecture is motivated by the following considerations. The simplest manner to organize a breakpoint is to replace the pattern of a certain instruction with a jump to the Monitor (relevant to Figure 11.5). Inevitably, this solution imposes a limitation. The breakpoint instruction must be at least two bytes long. As far as the 8051 family is concerned, the method is not feasible for 44% of the instructions. We overcome the problem by introducing the breakpoint RAM (see Figure 11.7). In this case, we still need to cary out the jump to the Monitor. The second extra RAM accommodates the Monitor (RAM MON). Of course, we could cut down the emulator architecture to two RAMs and place the Monitor in the user RAM. However, this will limit the accessible size of the user program.

Apparently, the subtask "Stop the user program on a breakpoint" requires the fastest reaction which will be a challenge for the selected architecture.

The design flow goes on with the following assumption:

The target is based on an 8051 microcontroller.

When the target microcontroller runs a user program, the emulator checks the breakpoint bit BP from the BP RAM. As far as the bit BP is not set, the emulator moves the code from the USER RAM to the EPROM socket. When a set bit BP occurs, the emulator emits the three bytes of the instruction **LCALL** to the EPROM socket. The destination address of this instruction is an entry point in the Monitor. Thus, the execution of the user program is suspended and the target microcontroller runs the Monitor program. Beginning now, the emulator redirects the code from the MON RAM to the EPROM socket.

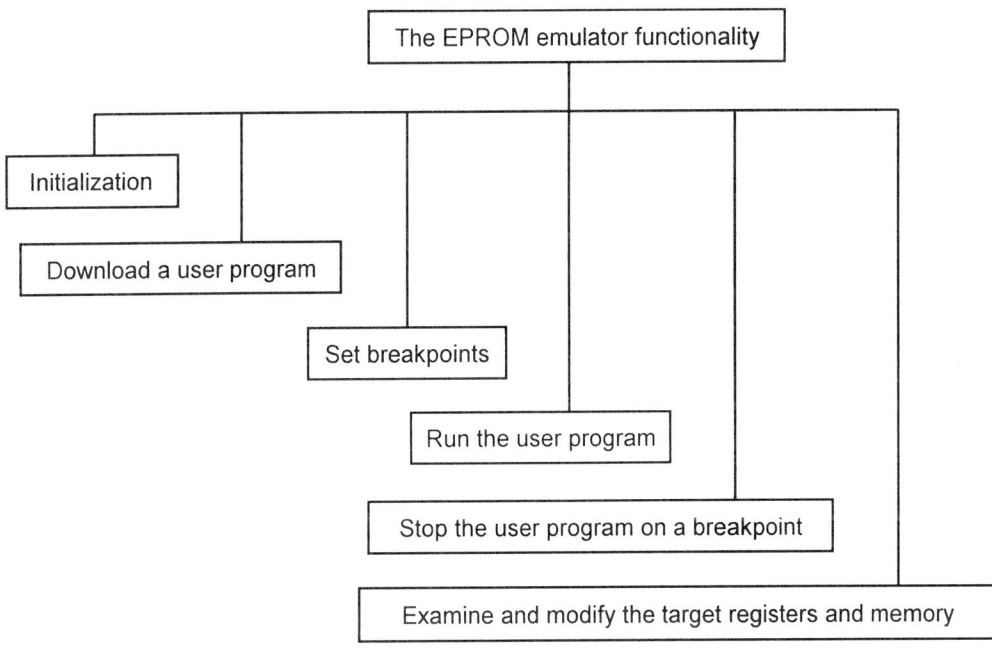

Figure 11.6 The EPROM emulator functionality shown in subtasks.

Naturally, the emulator will make a decision of whether to insert a breakpoint or to continue the execution of the user program within a certain period of time which we call reaction time (t_R). The reaction time determines a target oscillator frequency, which must not be exceeded.

The emulator microcontroller must test the signals $\overline{CE}, \overline{OE}$ and BP. Also, the change of the address line A0 is taken into account. The emulator must know if the signal A0 has just been altered. In this way, the design will be consistent with the 8051 feature to read bytes from the Program Memory ahead (see section 2.5).

For example, there is an instruction **INC A** in the program. We would like to insert a breakpoint immediately after the instruction. Therefore, the first byte of the following instruction (it is a strict rule) is marked by setting the correspondent bit BP in the BP RAM. The execution of the instruction **INC A** includes a fetch of two consecutive bytes. The first byte is the instruction opcode. The second byte is ignored and the Program Counter is not incremented.

However, the emulator microcontroller receives an active bit BP and substitutes the first byte of the instruction following **INC A** with the **LCALL** opcode. The same opcode (12H) must be maintained until the next read which is a real transfer. The number of redundant read actions will vary from instruction to instruction and we need a criterion of when to emit the second byte of the instruction **LCALL**. The solution is to follow the signal A0 and pass the second byte of **LCALL** when A0 is altered.

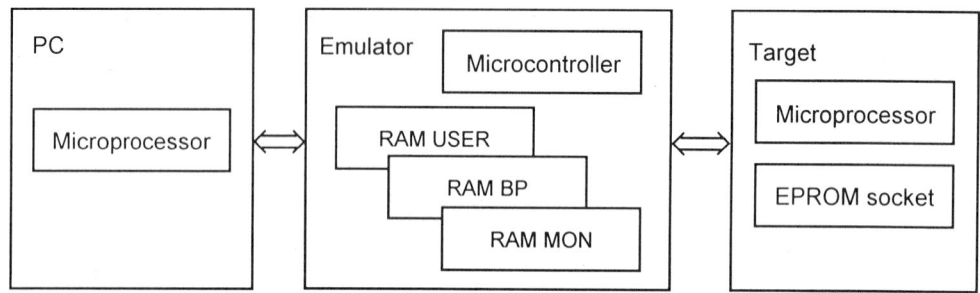

Figure 11.7 The EPROM emulator with two extra RAMs.

As you can see, we were forced to move away from the concept "processor independent interface". In particular, the 8051 manner to read bytes from the Program Memory ahead must be taken into account.

Unfortunately, there is another case when it seems impossible to find a solution at a reasonable price. Assume that we would like to insert a breakpoint immediately after a **JB** (Jump if bit) instruction. This instruction performs a redundant read from the first address of the following instruction. The code is ignored, but the breakpoint scheme is activated regardless of the result produced by the conditional jump. As a consequence for our emulator:

> The user is not allowed to insert breakpoints immediately after the program control instructions (**ACALL**, **LCALL**, **RET**, **RETI**, **AJMP**, **LJMP**, **SJMP**, **JMP**, **JZ**, **JNZ**, **JC**, **JNC**, **JB**, **JNB**, **JBC**, **CJNE** and **DJNZ**).

The user could work around this problem by adding **NOP** instructions where necessary.

We have reached the point when we must allocate a specific microcontroller for the emulator. As you might predict:

> The emulator microcontroller is selected to be a member of the 8051 family.

We assume that **JB** (**JNB**) instructions will be used to test the signals $\overline{CE}, \overline{OE}$ and BP. Thus, the reaction time becomes 8 cycles (96 oscillator periods).

When we put a few calculations into a table, the correspondence between the target and emulator clocks immediately become obvious (Figure 11.8). For instance, a target microcontroller running at 1 MHz will require an emulator microcontroller with oscillator frequency 20 MHz. The calculations are based on BP RAM access time 50 ns and a correction of 20 ns which covers the delay introduced from buffers, the cable and the decoding circuit in the target.

Target oscillator frequency f_{OSC} (MHz)	0.2	0.4	0.6	0.8	1.0	1.2
$T_{OSC} = 1/f_{OSC}$ (ns)	5000	2500	1666.7	1250	1000	833.3
The reaction time in case of the 80CL31 target microcontroller $t_R = 5T_{OSC} - 115 - t_{ACC}^{RAM\_BP} - 20$ (ns) $t_{ACC}^{RAM\_BP} = 50$ ns	24815	12315	8148	6065	4815	3981
The emulator microcontroller oscillator frequency $f_{OSC\_E} = 96/t_R$ (MHz)	3.869	7.796	11.783	15.829	19.938	24.115

Figure 11.8 The correspondence between the target and emulator microcontroller clocks.

Furthermore, Figure 11.9 illustrates the design process when the subtask "Stop the user program on a breakpoint" is moved from software to hardware. The transition point is oscillator frequency of 1 MHz. The hardware implementation of this time critical subtask would demand an ASIC.

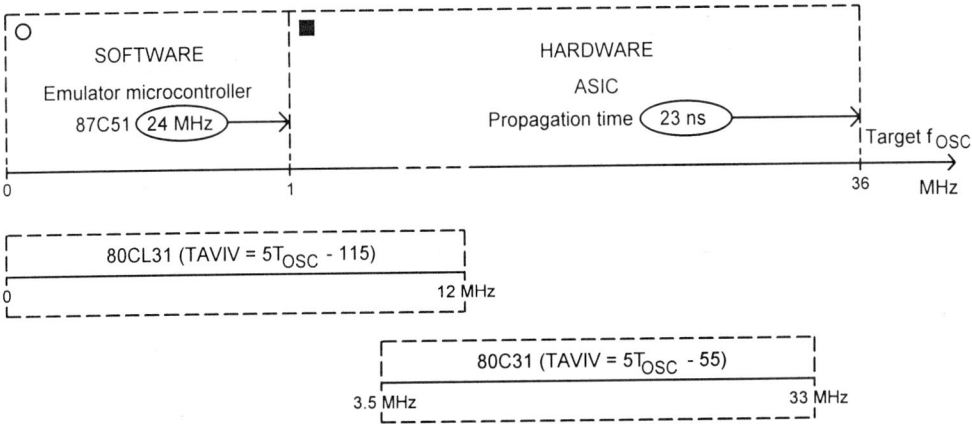

Figure 11.9 The implementation of the subtask "Stop the user program on a breakpoint".

Two different target microcontrollers are involved in this example - 80CL31 and 80C31. Figure 11.9 indicates the difference in the timing parameter TAVIV (Address to valid instruction in), which may vary from one device to another. The period of time TAVIV is used in the equation for the reaction time t_R (Figure 11.8).

While the emulator architecture based on an 87C51 microcontroller with a 24 MHz crystal is suitable for target systems running up to approximately 1 MHz, a microcontroller plus ASIC architecture expands the application range up to 36 MHz , if the ASIC propagation time is 23 ns and the BP RAM access time is 40 ns.

The overall conclusion is that the EPROM emulator design requires a hardware-software co-design approach and a microcontroller plus ASIC architecture is the only solution which provides sufficient application range.

A microcontroller plus ASIC architecture

The results illustrated in Figure 11.9 prove that a pure software implementation of the subtask "Stop the user program on a breakpoint" will cover a small segment of the frequency range even for slow microcontrollers.

Figure 11.10 outlines a microcontroller plus an ASIC architecture. The picture concentrates on the address and data paths. In order to distinguish between the emulator and the target signals we add a prefix "Target" or "T" to all EPROM I/Os. The target microcontroller reset input is named TRST as well.

As can be seen in Figure 11.10, an ASIC links the target with the emulator microcontroller and the RAMs. The ASIC gives us timing properties which we would not otherwise be able to obtain. The ASIC passes the code from the user RAM to the target and the user program is executed. In parallel, the breakpoint RAM emits the BP bit. When a set BP bit occurs, the ASIC generates the three bytes long instruction **LCALL**. From then on the ASIC conveys the code from the Monitor RAM to the target. The control inputs of the RAMs can not be seen in Figure 11.10. This is deliberate. We would like to find the right balance between hardware (ASIC) and software (microcontroller) at a later stage.

At this point, we have the knowledge to rewrite the emulator functionality previously given in Figure 11.6. The new specification, which is done for the sake of design clarity, appears in Figure 11.11. In other words, if the subtasks in Figure 11.6 specify what the emulator should do, the functionality presented in Figure 11.11 emphasizes the way it could be achieved. While migrating from the user oriented specification (Figure 11.6) to the one that better supports the design process (Figure 11.11), two points are essential.

First, the transition must not decline the functionality. On the contrary, we can expect some new features to be involved. For example, if we introduce testability or rectify omissions which become visible during the design process, the second version will be richer. For simplicity, we will not discuss the communication emulator microcontroller - PC.

Second, the introductory design phase may indicate that a certain number of subtasks must be bound into one subtask, helping the design to meet timing constrains or to solve other problems. In our particular case, we formed the subtask "Run a user program, switch to the Monitor if there is a breakpoint" on the basis of the subtasks "Run the user program" and "Stop the user program on a breakpoint". Also, the set in Figure 11.11 includes a new subtask which allows interrupt subroutines to be activated on the background of the Monitor program.

While Figure 11.11 shows the intended functionality broken into subtasks, it does not describe the temporal behavior of the emulator. Using a FSM model, you may recognize the basic emulator states from the decomposition scheme in Figure 11.11. Figure 11.12 shows a state transition graph for the EPROM emulator.

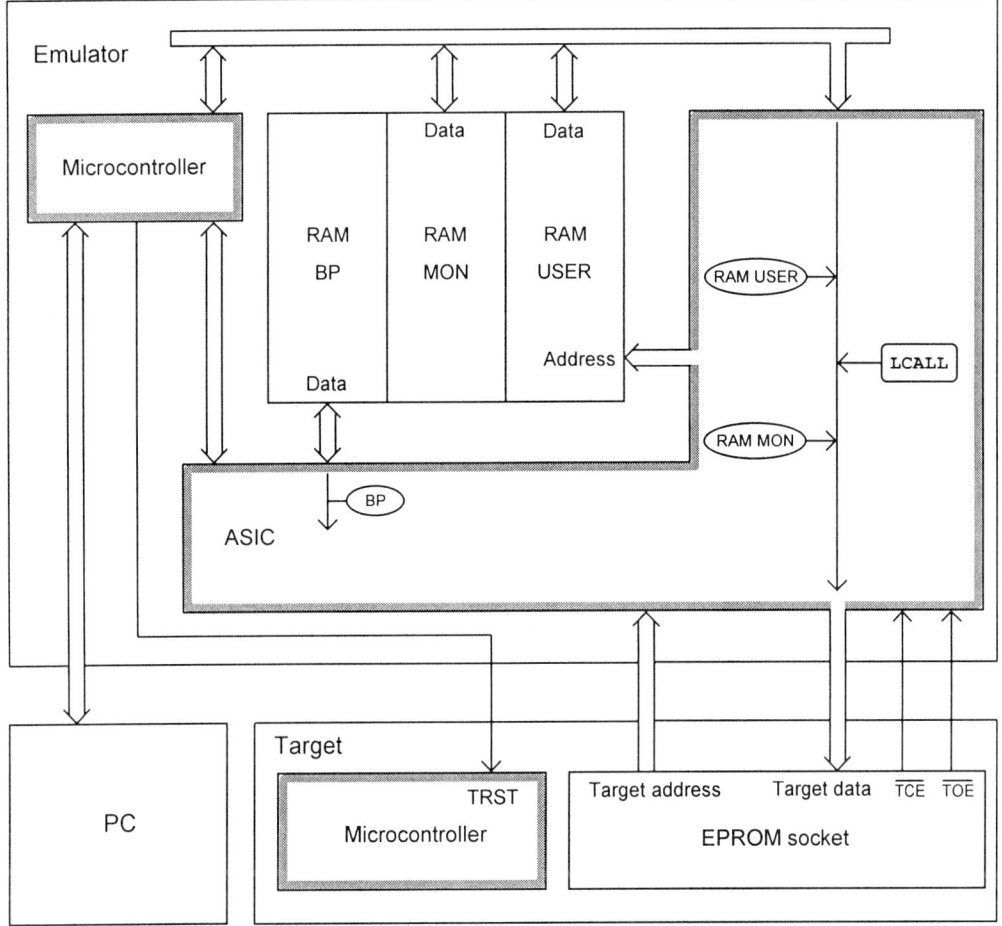

Figure 11.10 The EPROM emulator architecture in outline.

The FSM model is especially good for small-scale, control-dominated embedded systems. The emulator state transition graph from Figure 11.12 gives us a rough idea of how it works. After initialization the emulator unconditionally goes to the state "Run User". Depending on the breakpoint flag BP the emulator can either keep the state or switch to the state "Run Monitor". As soon as the condition User is asserted, the emulator will be back in the state "Run User". Indeed, we use the FSM model for the emulator with some leeway. It is important to start out simply and gradually detail the design. The exact values of the emulator outputs are replaced by more complex actions: "Initialization", "Run User" and "Run Monitor". Moreover, instead of real inputs which control the transition from one state to another, we use the internal variables BP and USER. You may think of the BP flag as a real input, whose influence on the emulator depends on the current address. Similarly, the USER flag is set when the PC instructs the emulator to switch back to the user program.

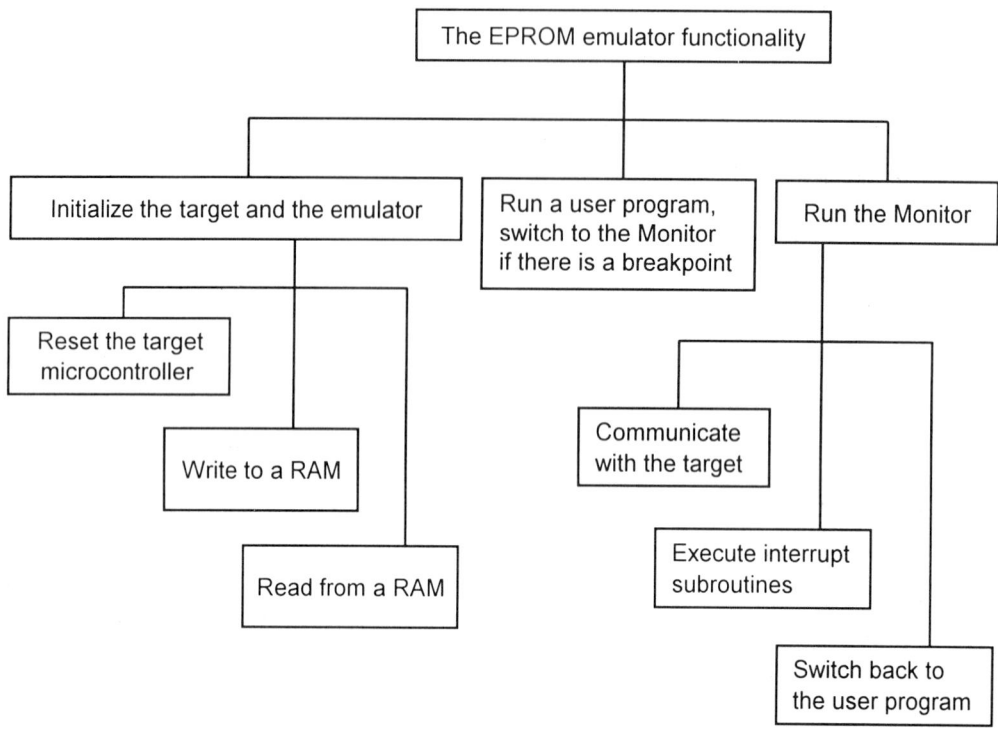

Figure 11.11 The EPROM emulator functionality which reflects the designer point of view.

Moving down in the emulator hierarchy, we present the ASIC state transition graph in Figure 11.13. The picture gives more details of the transition User mode - Monitor mode. Also, Figure 11.13 manifests what we have already stated, that the second byte of the instruction LCALL must be passed to the target when the address line TA0 is altered (A0_ALTER). We include a state "Wait" in the graph due to the redundant read in the end of the instruction LCALL.

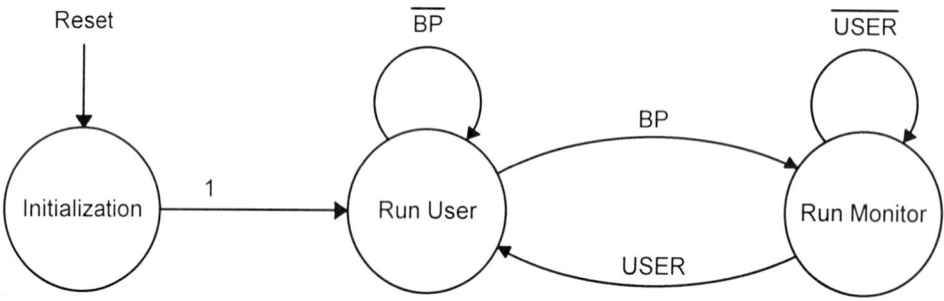

Figure 11.12 A state transition graph for the emulator.

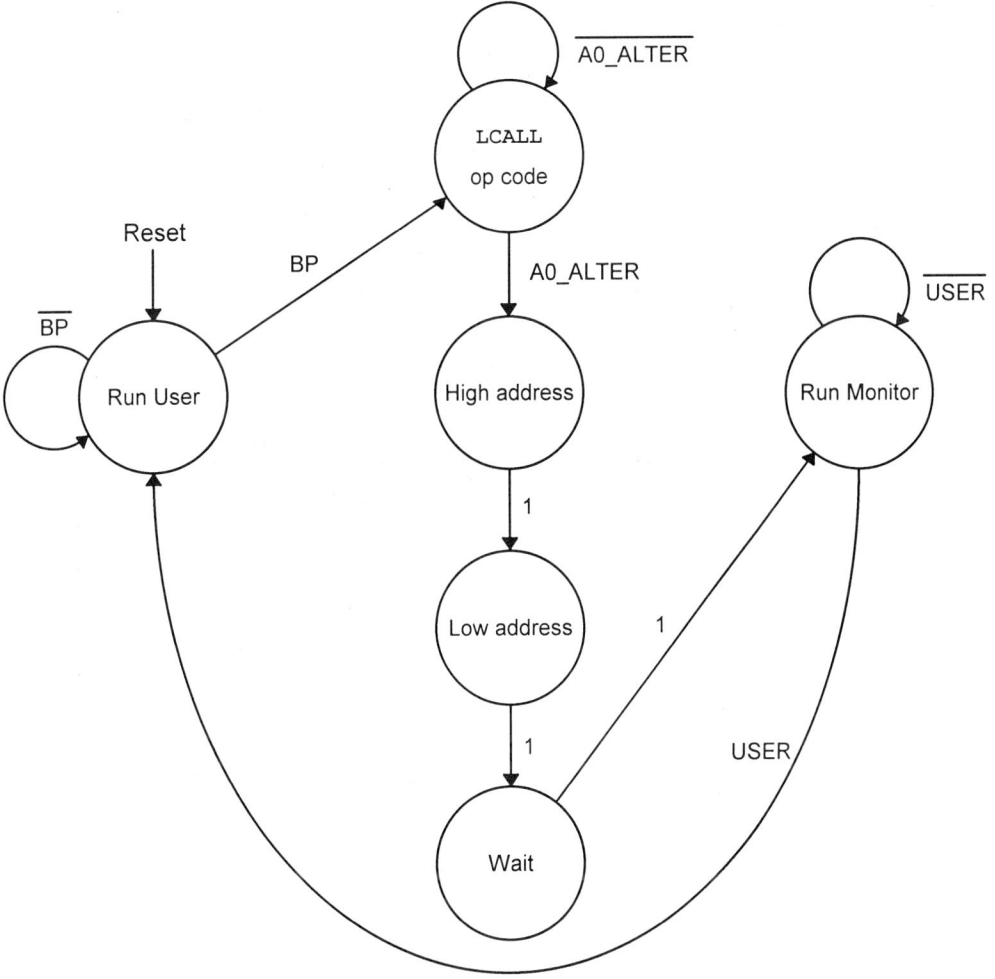

Figure 11.13 State transition graph for the ASIC.

A more detailed emulator architecture can be seen in Figure 11.14. The design refinement is determined by two chief factors:

First, the classical hardware-software trade-off driven mainly by timing requirements is influenced by the number of pin limitations. We must minimize the interconnections between the components. For example, the architecture in Figure 11.10 signals that we might run into problems when implementing an ASIC with too many pins. Only the address lines will occupy 32 I/O pins. The same problem affects the microcontroller interface as well.

In order to alleviate the situation we decided that the target address lines should bypass the ASIC when a program is executed. Also, we introduce three-state buffers to be able to cut the connection target - RAMs when the emulator microcontroller communicates with the RAMs. Assume that the ASIC possesses two 8-bit registers which output the two address bytes. This approach leads to reduction of the ASIC address devoted pins from 32 (Figure 11.10) to 16

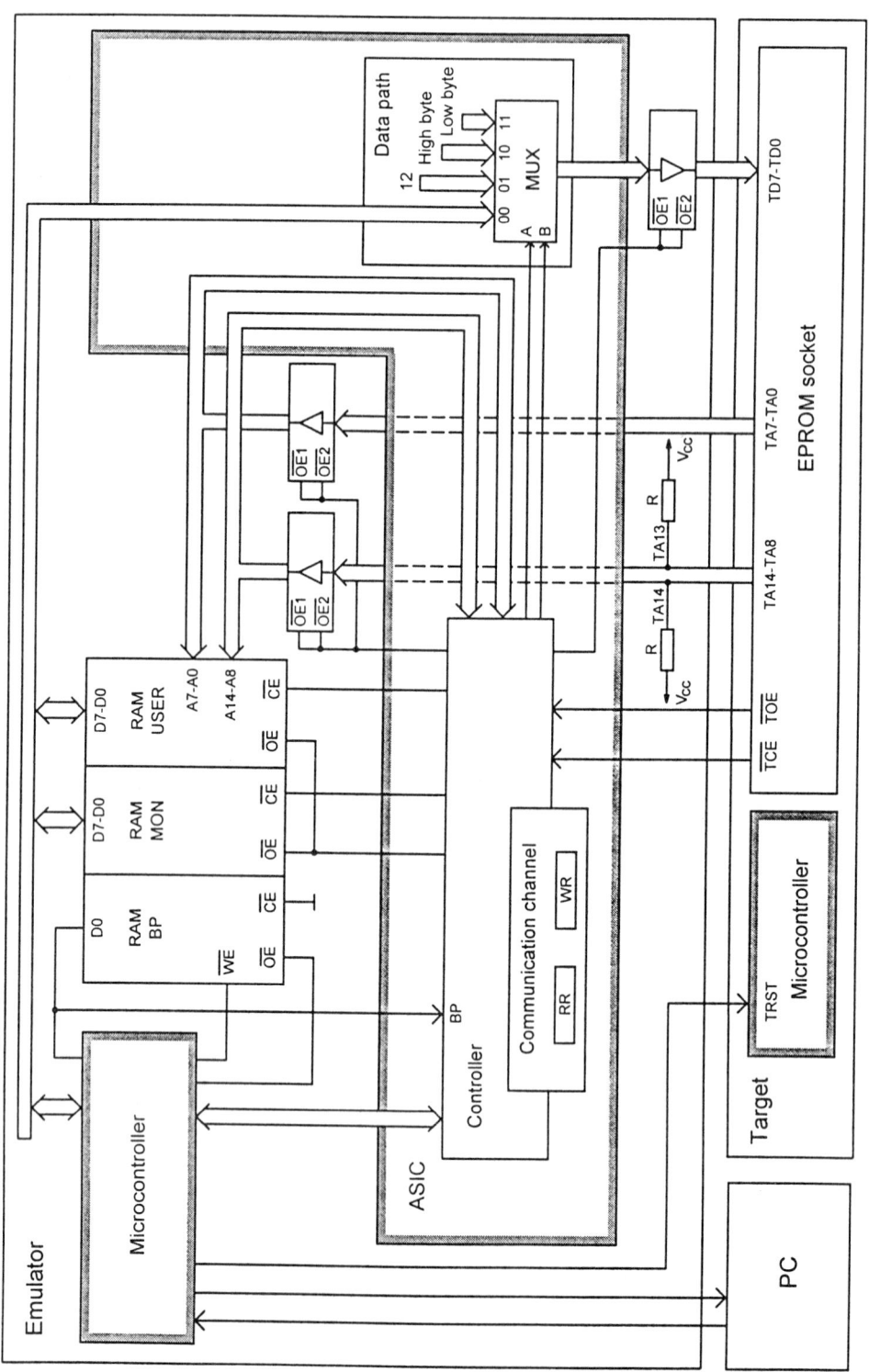

Figure 11.14 An emulator structure with a data path - controller architecture for the ASIC.

(Figure 11.14). Indeed, the 16 lines in the latest version are bi-directional, but it is a common practice and could be easily achieved. The emulator microcontroller takes care to update the two address registers in the ASIC.

Also, there is another reason to introduce buffers. The emulator, when attached to the target will influence it in terms of output current and load capacitance. Inevitably, the EPROM emulator and the target system are connected by a cable. The effect of this complex load to the target must be brought down as much as possible. Moreover, the shape of the signals, which the emulator receives, must be restored. The buffers (line drivers) perform this function.

Furthermore, the emulator outputs which go through the cable must be buffered as well. Therefore, we need isolation, driving capability and correct input values on both sides of the cable. It could be achieved by using line drivers.

Second, the ASIC architecture is also an important factor in the emulator design. We break it down into data and control. The idea behind the data path-controller architecture combines intuitive approach with formal design methods [Wolf 1994a]. First, we allocate and link standard functional blocks such as registers, adders and counters. Second, we design a FSM to control them. Thus, if the application is too complex for a direct FSM approach, we could intuitively design the data processing path and add a formally synthesized control unit.

In our particular case, the data path is a multiplexer. The multiplexer has four input groups of eight lines. According to the select inputs A and B (MSB), a certain group of inputs is switched to the outputs. Check if the multiplexer functionality is consistent with the state transition graph for the ASIC (Figure 11.13).

The byte 12H is the opcode of the instruction **LCALL**. As soon as the address bit TA0 is altered the controller sets the input B and resets the input A. The high order address byte (High byte) is sent to the target. Waiting for the next time when both $\overline{\text{TCE}}$ and $\overline{\text{TOE}}$ inputs are low, the ASIC generates the low order address byte (Low byte). The byte read in the last transition state, which is named "Wait", is ignored by the target microcontroller.

Along with the data path - controller hardware we introduce a communication channel. The communication channel is based on two mailbox registers: Read Register (RR) and Write Register (WR). The registers are named with regard to the emulator microcontroller. The emulator microcontroller writes to the WR and reads from the RR. The target microcontroller writes to the RR and reads from the WR.

A few remarks about the buffers (line drivers) in Figure 11.14 are in order. You might have experience with the SN74LS244 octal buffers and line drivers with three-state outputs [Texa 1989]. Recently, a new series of advanced bus interface ICs, based on combined bipolar and CMOS technology (BiCMOS) have shown excellent features. Figure 11.15 shows the BiCMOS replacement for the 74LS244 [Texa 1993]. Higher drive output capability and significantly reduced power dissipation are the hallmarks of the BiCMOS technology. Two control inputs are available (pins 1 and 19). The outputs are in the ON state when the control input of the corresponding group is low.

Figure 11.14 also indicates that our design deals with EPROMs not larger than 32K byte - the address line A15 has been discarded. Due to the fact that for smaller than 32K byte EPROMs either A14 or A14 and A13 will flow (there will be no source signal to drive them), we must set them high by resistors. Consequently, we reach the situation illustrated in Figure 11.16. The two pull-up resistors automatically align the User RAM in the emulator memory map to the end address of 7FFFH. Naturally, the resulting address will appear back through the cable on the EPROM socket. As you might remember from section 4.2, the address inputs A14 and A13 (for smaller EPROMs) will be either NC (no internal connection) or $\overline{PGM}$ (tied up in the target for read mode).

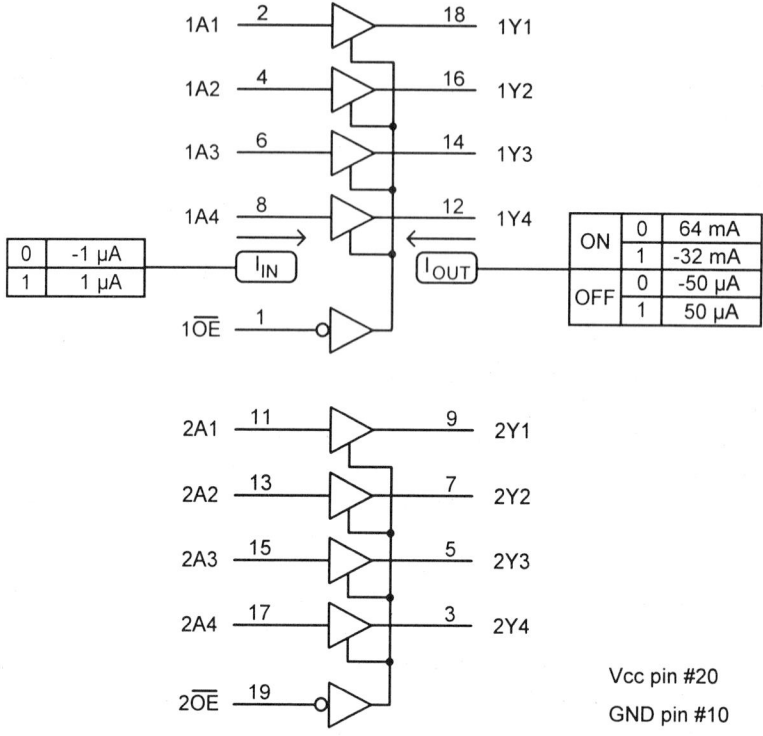

Figure 11.15 The octal buffer SN74ABT244 - pins and maximal input/output currents.

Furthermore, Figure 11.17 gives the correspondence target - Monitor RAM map. The shaded area is the Monitor program which occupies 1K byte. An essential feature of our design is that the unshaded area (7K, 15K or 31K in size) contains a big part of the user program. In other words, the user program is downloaded in the Monitor RAM as well. Only 1K byte is not available because it is substituted by the Monitor. As a result, the target is capable of executing interrupt subroutines on the background of the Monitor. Obviously, the following limitation is in order:

> The emulator is capable of executing interrupt subroutines while in the Monitor if they do not overlap the last 1K byte of the target EPROM.

Since the interrupt vectors are in the beginning of the Program Memory, another restriction sounds logical:

> The target EPROM must start from address 0000H.

The different EPROM options and the correspondence between the address pattern generated by the target and the actual address in the emulator have a significant impact on the design.

Essentially, the target memory map in Figure 11.17 proves that it is impossible to use a single address as an entry point for the Monitor (there is no overlap between the shaded areas). Thus, the address 1C00H will be the entry point for 8K byte EPROMs, the address 3C00H for 16K byte memory and finally, 7C00H will be relevant for the 32K byte option.

In addition, Figure 11.17 indicates a few specific addresses used by the Monitor for interaction with the target. We will discuss them a little later.

At this point, we are ready to approach the final emulator architecture shown in Figure 11.18. For simplicity, we impose the following specification:

> The emulator is capable of working with 8K and 32K byte EPROMs.

By eliminating the 16K byte EPROMs, we simplify the data path block in the ASIC and the corresponding control (the Monitor entry points are different).

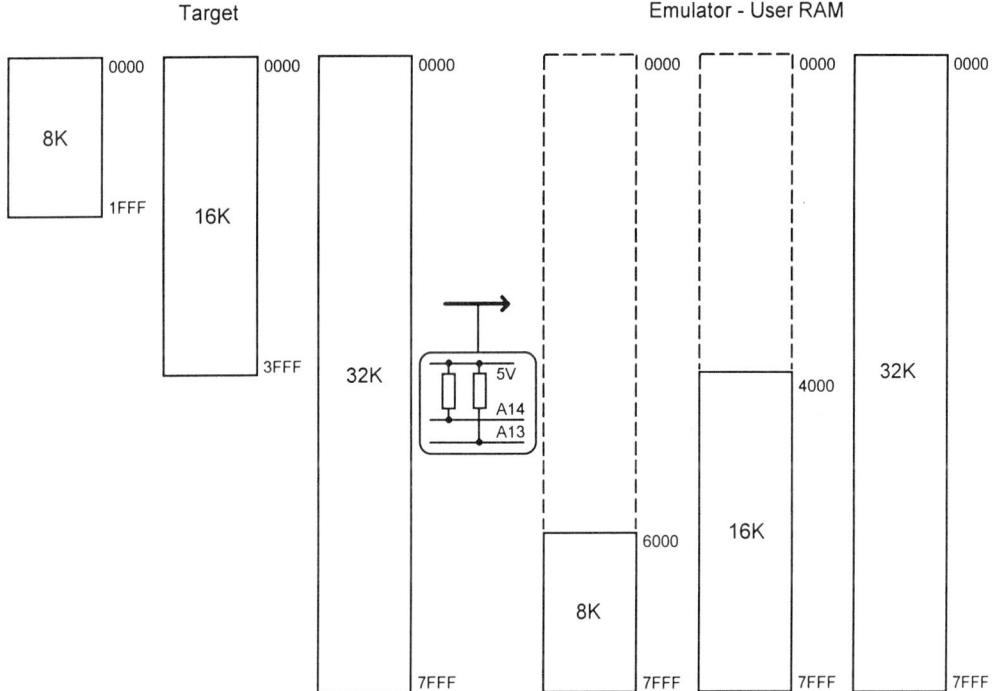

Figure 11.16 The target - user RAM address translation.

The ASIC architecture shown in Figure 11.18 includes a couple of registers, flip-flops, decoders (DEC), multiplexers (MUX), Combinational Logic Circuits (CLC) and a Sequential Logic Circuit (SLC). We have allocated flip-flops and registers which update their outputs with the rising edge of the clock input. The flip-flop's asynchronous inputs CL (Clear) are asserted when they are high.

A register called Mode Register (MR) adjusts the ASIC to a certain mode of operation. In fact, the emulator possesses two fundamental modes of operation:
- Access to the RAMs (either write or read)
- Run a program (either a user program or the Monitor).

The mode register value is set by the emulator microcontroller. Figure 11.19 explains how the four flags in the register are used. A set bit M32 adapts the emulator to work with 32K byte EPROMs. If 8K byte EPROMs are used, the flag M32 is reset.

When the flag RUN is set, the emulator is prepared to run a user program. If the RUN flag is reset, which means access to the RAMs, four combinations are relevant. We use a bit MON to select a RAM and a bit RD to define the direction.

The communication channel introduced in Figure 11.14 is based on an 8-bit bus which links the emulator microcontroller and the ASIC. All ASIC registers are attached to the bus. The communication approach is identical to the one discussed in Problem/Solution 4.3. In addition to the eight I/O lines there is one microcontroller output that drives the clock inputs of the flip-flops Q0 through Q3.

There is one register, Receive Register (RR), which is not related to the common synchronization circuit. The RR clock input is driven from the decoder DEC1.

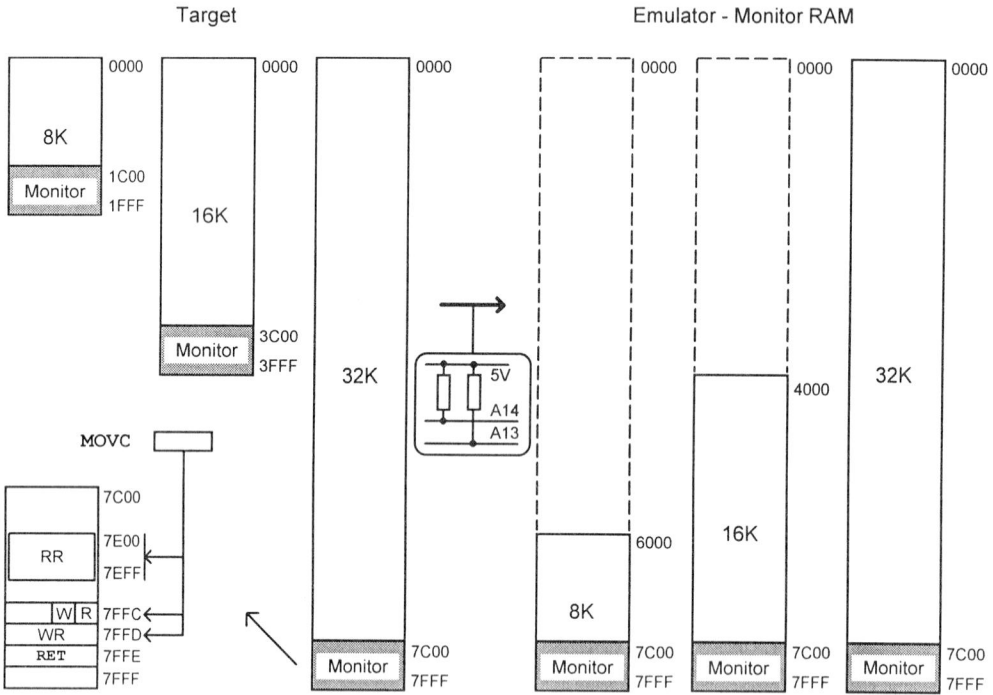

Figure 11.17 The target - Monitor RAM address translation.

We can look at the emulator architecture in Figure 11.18 from different angles. However, the most logical way is to make sure that this phase of the design is consistent with the tasks listed in Figure 11.11. Consequently, we will discuss the proposed emulator architecture in a functional manner task by task.

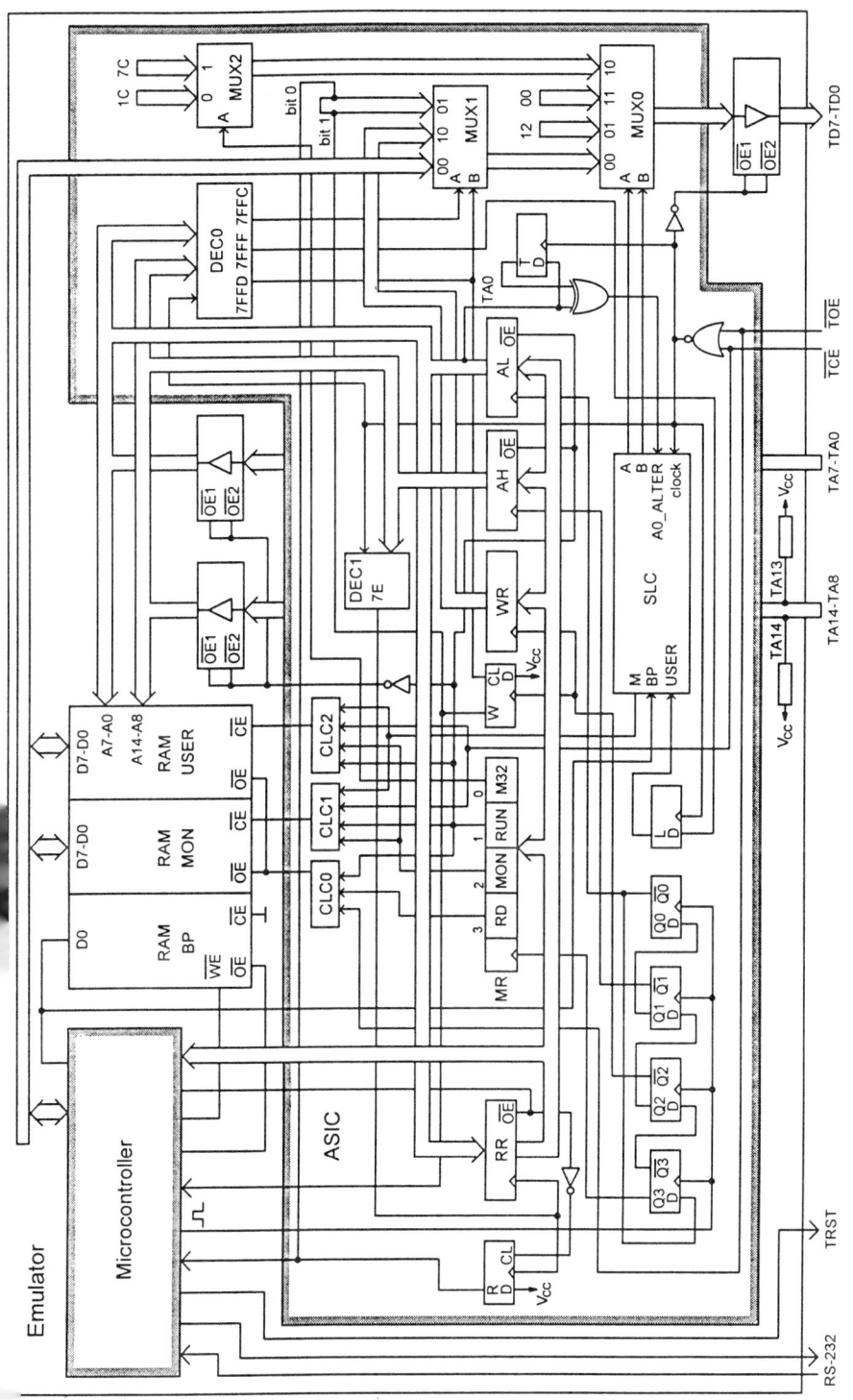

Figure 11.18 The final emulator architecture.

Reset the target microcontroller

This task is very simple. The emulator microcontroller sets a dedicated output and the targ processor is reset. While the target microcontroller is in this condition data exchange betwe the emulator microcontroller and the RAMs can be organized.

Write to the RAMs

After you have had a clear idea how to debug your program, you should download code the RAMs.

The emulator microcontroller resets the bits RUN and RD. You could follow the connecti between the flip-flop RUN and an inverter which controls the address buffers. The buffers a in the OFF state. Contrary to this, the outputs of the ASIC address registers AH and AL are the ON state (the output enable inputs $\overline{OE}$ are low). The microcontroller moves the address the registers AH and AL. Both registers provide the address for the RAMs.

Furthermore, the desired RAM is selected by the flag MON. The emulator microcontrol outputs the code for the RAMs and clears a dedicated output which controls the RAM's inp $\overline{WE}$.

At first glance it seems that the task "Write to the RAMs" must be performed three tim (the emulator includes three RAMs). It should be pointed out, however, that the $\overline{WE}$ signal common for all RAMs. Inevitably, two RAMs are involved in the write procedu simulteneously. As this solution is to some extent controversial, we could go a little further a write to the three RAMs without changing the address. For instance, we organize the followi actions:

• The microcontroller loads a certain address into the registers AH and AL.

• The user RAM is selected by the mode register MR, the emulator microcontroller outpu the byte to be writen and pulls down the inputs $\overline{WE}$.

• The Monitor RAM is selected, the emulator microcontroller emits the byte together w the BP bit (D0), and the inputs $\overline{WE}$ are asserted.

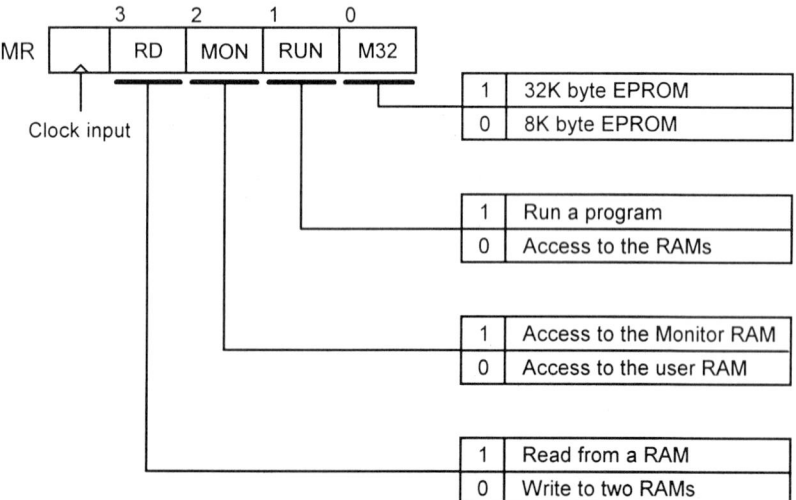

Figure 11.19 The emulator mode register MR.

Read from a RAM

Similarly, we must take care of the flags M32 and RUN. Now the flag RD is set. Naturally, the emulator microcontroller port, which is linked to the RAMs, must be programmed as an input port.

Run a user program, switch to the Monitor if there is a breakpoint

Once we have reset the target microcontroller and downloaded code to the RAMs, we are ready to start a program. The RUN flag from the register MR has been set. The address buffers are in the ON state. As soon as the emulator microcontroller pulls the signal TRST down, the user program will be off and running.

We should make a check for the path user RAM - emulator data buffer (outputs TD7-TD0). If you compare the ASIC architecture in Figure 11.14 and the one in Figure 11.18, you will see that we have two extra multiplexers in Figure 11.18. They are 8-bit word multiplexers again. We need the multiplexer MUX2 to insert different high order address bytes when the emulator is switched to the Monitor. As you might predict, the multiplexer MUX2 is controlled by the flag M32. Likewise, we need the multiplexer MUX1 to read the register WR and the flags R and W.

There is no doubt that the current task requires the MUX1 input word indicated by 00 to be passed to the MUX0 when the emulator runs the user program. Is this done? Not quite. If the address is 7FFCH the multiplexer MUX1 selects the input word marked as 01. In addition, the address 7FFDH through decoder DEC0 will pull up input B of the MUX1 and link the register WR with the multiplexer MUX0.

On the other hand, we need the multiplexer MUX1 and the corresponding control scheme to establish the connection between the communication channel blocks (register WR, flag W and flag R) and the target. By means of the following limitation we can restore the correctness of the design:

> When a user program is executed it must not pass the addresses 1FFCH and 1FFDH for 8K byte EPROMs. The corresponding addresses for 32K byte EPROMs are 7FFCH and 7FFDH.

Practically, the limitation has no impact on the debugging facilities of the emulator.

		$\overline{\text{TCE}}$	$\overline{\text{TOE}}$	$\overline{\text{CE}}$ User RAM	$\overline{\text{CE}}$ Mon RAM	$\overline{\text{OE}}$
Emulator memories	User RAM	0	0	0	1	0
	Mon RAM	0	0	1	0	0
Target memories	Other Program Memory	1	0	1	1	1
	External Data Memory	1, 0	1	1	1	1

Figure 11.20 The RUN mode relationships.

When we design the SLC unit, we will take care of the appropriate outputs A and B. Furthermore, we would like to state another feature of the SLC. Initially, when a user program is in progress, the SLC output M is low. If a set bit BP (breakpoint) occurs, it will result in high level for the output M. The output M dictates which RAM is to be the active one. If the output M is high the current program memory for the target is the Monitor RAM.

Before we show all the details of the combinational circuits CLC0, CLC1 and CLC2, we should clear up the correspondence between the signals $\overline{TCE}$ and $\overline{TOE}$ from one side and the RAMs control inputs from the other side (in the RUN mode). Figure 11.20 will help us to distinguish between the emulator RAMs and the other target memories.

When the target runs a user program it will start from the user RAM (the reset vector is there). However, at a certain moment the target might fetch instructions from another EPROM, different from the one we substituted by a RAM. If that is the case, the emulator input $\overline{TCE}$ will be high. As you can see in the same line of the table, the circuits CLC0, CLC1 and CLC2 are under obligation to pull their outputs high.

We can use either truth tables or logic expressions to specify functions. Experience shows that constructing a truth table is the prudent course of action. It will help us to avoid errors. The next step is simplification. We are interested to find logically equivalent expressions which lead to the simplest implementation. We intend to employ an FPGA for the ASIC. Normally, the manufacturers provide CAD tools capable of minimizing logic expressions and implementing technology dependent optimization. In this situation, it will be not only sufficient but advisable to construct truth tables and go on with CAD tools.

The truth tables for the functions CLC0, CLC1 and CLC2 are shown in Figure 11.21. The flag RUN from the mode register MR influences all three functions. In the first half of the tables (RUN is low) the emulator is in the memory access mode and the variables RD and MON are important.

In the second half of the tables (RUN is high) the emulator is in the RUN mode. The inputs $\overline{TOE}$ and $\overline{TCE}$ together with the internal variable M define the output value.

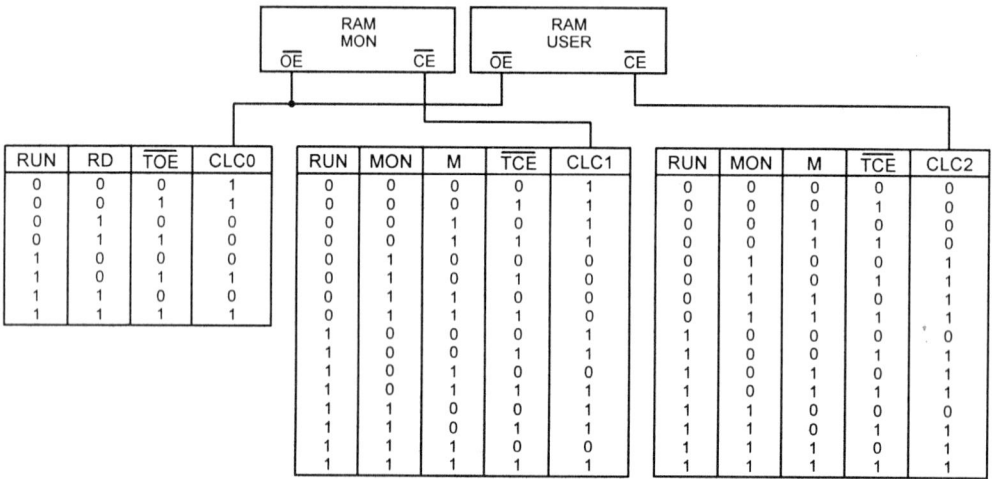

Figure 11.21 Truth tables for the functions CLC0, CLC1 and CLC2.

Now we move on to the design of the sequential circuit SLC. We have already presented the first cut of the ASIC state transition graph in Figure 11.13. As the role of the SLC is to control the data path unit, the ASIC state transition graph is the specification to be followed. The design of the SLC can be done either by hand or by a CAD tool. What is more important is to make sure that the foundation of the design, the state transition graph, is correct.

In our particular case, it turned out that the state transition graph must be modified as shown in Figure 11.22. The design validation would not be successful if we did not use timing diagrams. Figure 11.23 shows an example of a timing diagram for the transition user program - Monitor program. Furthermore, Figure 11.24 presents a timing diagram for the other key transition: Monitor - user program. Both timing diagrams are used in the refinement of the state transition graph.

We are going to discuss what is different in the final state transition graph (Figure 11.22) and why the change has been made.

As far as the current task is concerned, we discarded the state "Wait" from Figure 11.13. We knew that it could be dropped, however we did not know that it would be a more reliable solution. The timing diagram in Figure 11.23 displays a set bit BP under the $\overline{\text{TOE}}$ pulse #2. The SLC is moved from the state S0 to the state S1. The SLC output M is set. Consequently, the Monitor RAM becomes the active one. The transition is marked by an ellipse. The SLC output A is set and the multiplexer MUX0 generates the code 12H. In this example of a timing diagram we assume that the byte at #2 is ignored and a real fetch takes place at #3. In the end of the state S1 the SLC input A0_ALTER is low. As a result, the sequential machine keeps the state S1 and maintains the input word 01 of the multiplexer MUX0 as a selected one. The code 12H is available for the target microcontroller. At #4, the input A0_ALTER indicates a change in the address bit TA0 and the SLC goes to the state S2. After the rising edge of the clock signal at #5, the SLC moves to the state S3 unconditionally. There is a redundant read (xx) at #6. Logically, we use the transition at #6 to prepare the multiplexer MUX0 for the RUN mode by clearing inputs A and B. If we make calculations for the RAM's timing requirements in both cases (Figure 11.13 and Figure 11.22), they would be practically the same, but as a principle the final state transition graph is a more natural implementation.

Switch back to the user program

The new states S5, S6 and S7 in the state transition graph are related to this task. Again, we combine a timing diagram (Figure 11.24) with the FSM model. The idea is to switch back to the user program by executing an instruction **RET** from address 7FFEH (see also Figure 11.17). This instruction fits for the purpose in two ways. It allows the jump to the user program to be organized by the stack. In addition, the **RET** instruction has three redundant read actions (#3, 4 and 5) which give sufficient time for the emulator to switch to the user RAM.

As you can see in the timing diagram (Figure 11.24), the flag BP is set again when the emulator is back on the breakpoint address (#6). Following the state transition graph in Figure 11.13, the emulator would jump to the Monitor and the loop would be endless. In view of this problem, we introduce states called "Return 0", "Return 1" and "Return 2". In these states, the emulator starts executing the user program, but ignores the input BP. As can be seen in Figure 11.18, the SLC input USER is set when the address 7FFFH is accessed.

A few remarks about the read #7 are in order. The question is if the emulator could react properly to a set flag BP at #7. If the flag BP is really set, we can conclude that the instruction at #6 (the first after the Monitor), is a one byte instruction. Hence, the byte read at #7 is discarded, the address is not incremented and the flag BP appears again at #8 when the SLC is already sensitive to the BP flag. Thus, the emulator will not miss a breakpoint at #7.

Incidentally, the implementation of the SLC will require three flip-flops in both cases (Figure 11.13 and Figure 11.22).

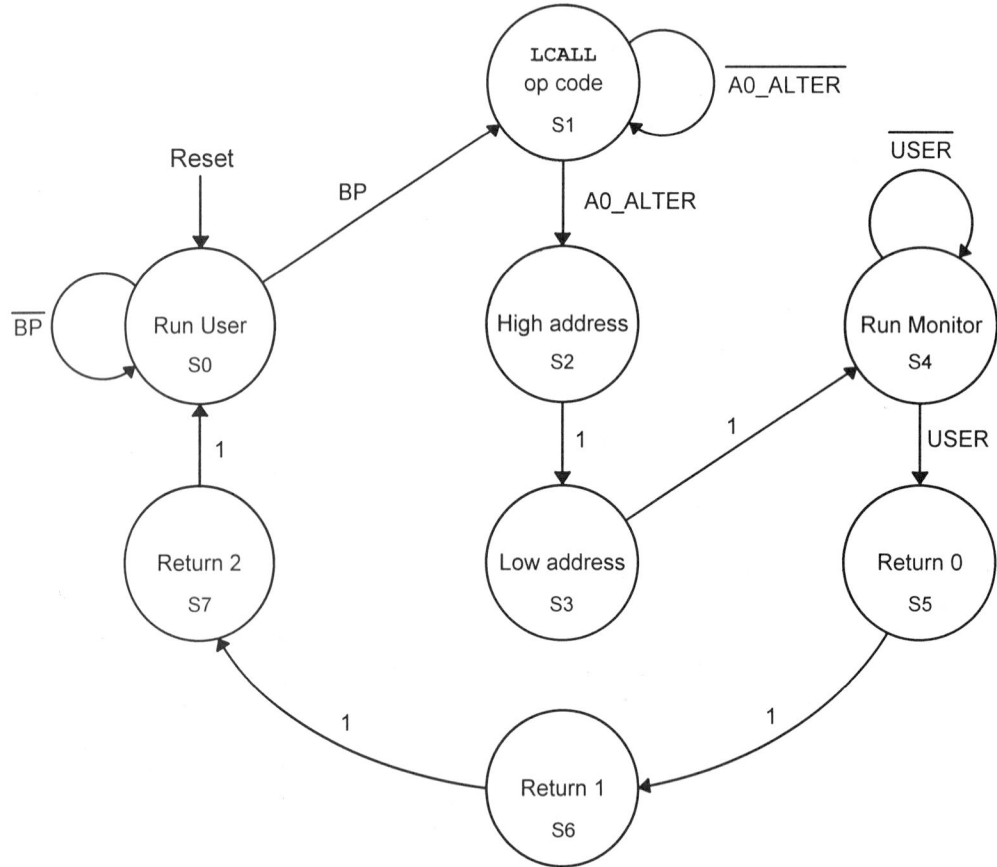

Figure 11.22 The final state transition graph for the ASIC in RUN mode.

Execute interrupt subroutines

Figure 11.17 shows the code in the Monitor RAM. As far as the interrupt subroutines are outside the last 1K byte area of the user program they can be executed. The corresponding state in Figure 11.22 is S4. The flag BP is not checked. Therefore, if you insert breakpoints in an interrupt subroutine they will be ignored. It would be possible for the emulator microcontroller to disable the interrupts in the target system, if it is beneficial to the debugging process.

Communicate with the target

In the beginning of this example project we described the overall system as a distributed one. The system includes three nodes: the personal computer, the emulator and the target machine.

The current task focuses on the communication between the emulator microcontroller and the target. Just to stop on a breakpoint is not sufficient. We have to inspect certain registers and memory locations in the target. Furthermore, we might need to modify data in the target. Finally, we should be able to go on with the user program. All these actions require communication.

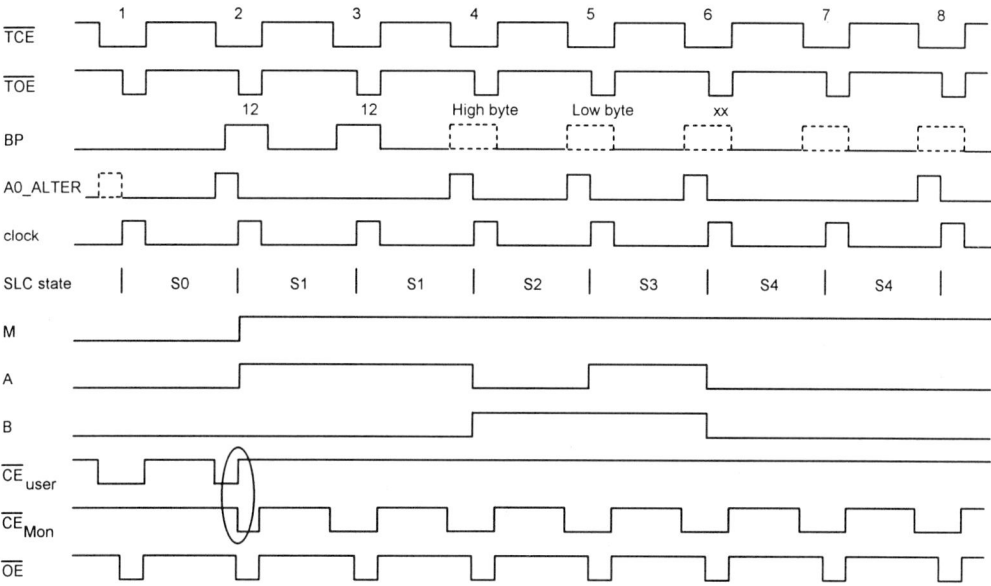

Figure 11.23 Timing diagram for the transition user program - Monitor.

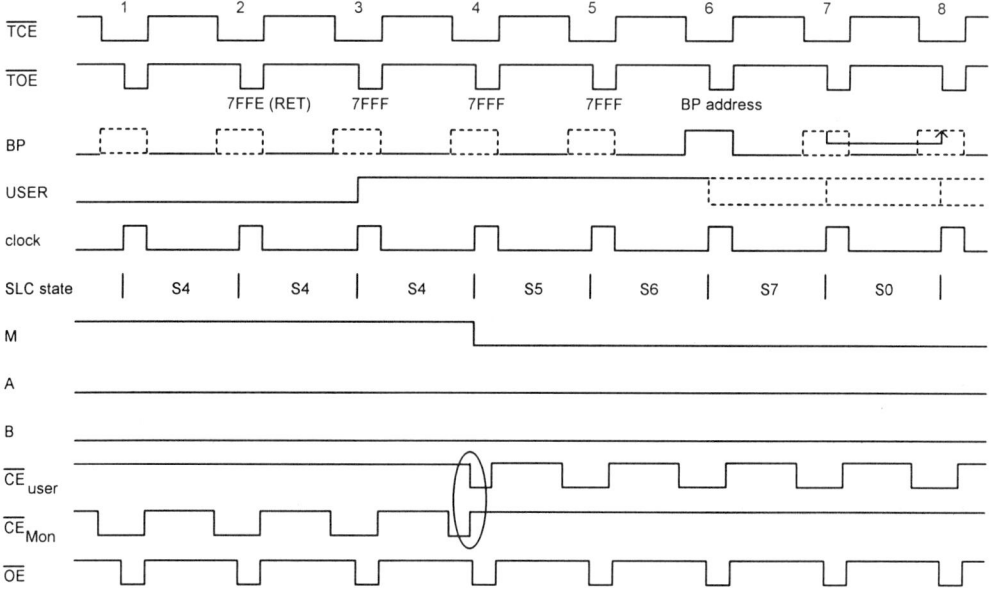

Figure 11.24 Timing diagram for the transition Monitor - user program.

The communication channel was introduced in Figure 11.14. The core of this block includes two mail-box registers: the write register WR and the read register RR. An asynchronous communication style fits best to our specific application. We organize handshaking by two flags - R and W (Figure 11.18). When the emulator microcontroller throws a byte to the write register WR, the flag W is set as well. The flag W can be read from both microcontrollers. It signals to the target microcontroller that there is a byte waiting to be read.

In order to check the flag, the target performs the following instructions:

```
CLR     A
MOV     DPTR,#1FFCH     ; 8K byte EPROM
MOV     DPTR,#7FFCH     ; 32K byte EPROM
MOVC    A,@A+DPTR       ; ACC, bit1 contains the flag W
```

Regardless of the base address loaded in the register DPTR (1FFCH or 7FFCH), the MOVC instruction will generate address 7FFCH in the emulator (see also Figure 11.17). However, we need the right address in the register DPTR to guarantee that the target hardware will assert the signal TCE. The ASIC internal decoder DEC0 pulls up the input A of the multiplexer MUX1. The MUX1 input word 01 (in fact, only two bits are important) is switched to the outputs. After execution of the instructions above, the accumulator's bit 1 contains the flag W value. Upon a set flag W, the target microcontroller can read the register WR. The approach is analogous to the one we have just used for the flag. The only difference is in the address of the MOVC instruction which now is 1FFDH/7FFDH. The input word 10 of the multiplexer MUX1 becomes the selected one and the code from the register WR passes through MUX1, MUX0 and the output buffer. In parallel, the flag W is cleared by the input CL. A cleared flag W is an indication for the emulator microcontroller that the register WR could be updated.

The data stream from the target to the emulator microcontroller seems more difficult to organize. The problem stems from the EPROM interface where only the address lines are outputs. Hence, we are forced to convert addresses into data. We must sacrifice a certain number of addresses from the Monitor quota and use them for this purpose. Practically, we can choose between the following options:
- Convert 16 address combinations into 4 data bits (decode a nibble)
- Convert 256 address combinations into 8 data bits (decode a byte).

We have decided to allocate 256 addresses from the Monitor in order to read a complete byte from the target. This approach is faster and more convenient.

As you might expect, the MOVC instruction will help us again. There is no harm in reading from the memory. It is certainly implied that the byte moved to accumulator will be discarded. However, the low byte address pattern will be captured in the read register RR. For example, if the byte to be sent is 00H, the MOVC instruction is combined with address 1E00H/7E00H. Likewise, if the target microcontroller sends FFH, the address will be 1EFFH/7EFFH. The decoder DEC1 not only clocks the register RR, but also sets the flag R. The emulator microcontroller scans the flag R and when the flag is set reads the byte from the register RR. It is done by pulling down the output enable input OE of the register. In parallel, the flag R is reset.

Exchanging bytes is the lowest level in the communication hierarchy. We can group one or more bytes and form commands. In our two node subsystem the emulator microcontroller is a master and the target is a slave. Consequently, the emulator microcontroller generates the commands and the target interprets them. Figure 11.25 shows an example set of communication commands.

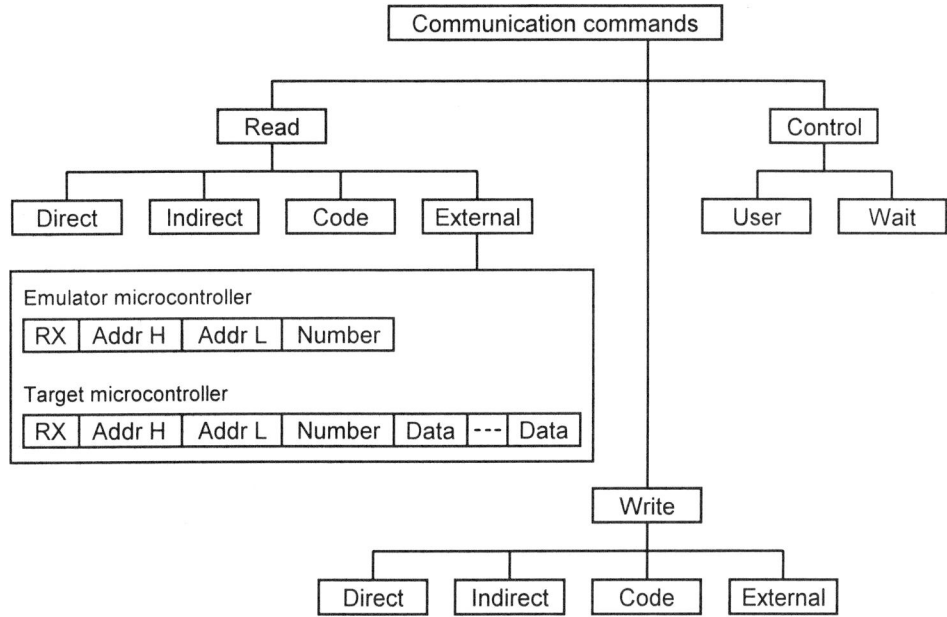

Figure 11.25 A set of essential communication commands.

We distinguish between read, write and control commands. We must be able to access locations in all memory address spaces. For example, the "Indirect" read/write communication command covers the internal Data Memory in the range 80H through FFH (see Figure 7.4). Using microcontrollers such as the 83C552, we need this option.

The target microcontroller echoes the command received from the emulator microcontroller and appends data as the case requires. Figure 11.25 details a read from the external Data Memory (External). You can see the strings generated from both microcontrollers. The target microcontroller repeats the first four bytes and adds the requested number of bytes. The command "User" tells the target to resume the user program. The command "Wait" is self-explanatory.

At this point, the list of tasks is over and we are ready to approach the final schematic diagram. We must select a specific microcontroller and make a decision about the ASIC.

As far as the microcontroller is concerned, we need to estimate if a member of the previously discussed families (the 8051 and the 83C552) is good enough for this application. In fact, we must count the wires connecting the microcontroller and the other components. Moreover, we can choose between internal or external program memory. The advantage of using external program memory is that we could employ an available EPROM emulator to debug the system. This might remind you of "the chicken or the egg dilemma". Finally, we have allocated an 80C552 microcontroller, as shown in Figure 11.26.

The personal element can never be ignored and on the basis of our experience with the Latice FPGAs, we decided to try them for the ASIC. As you might remember, Figure 10.2 contains a few typical parameters of these programmable devices. Two features are crucial - the number of I/Os and functional capacity (the type and number of macrocells). You may find some other FPGAs which are also suitable for this project.

It turned out that the ASIC can be implemented by two FPGAs type ispLSI 1016. Of course, we could use a larger FPGA, however the ispLSI 1016 are very popular and fit better to our example design. In addition, we would like to take this opportunity to demonstrate partitioning of the architecture into smaller units.

The emulator processor core consists of an 80C552 microcontroller, a 32K byte EPROM and a register buffer. The limited space forced us to skip the pin numbers, however, they can be obtained from Figure 4.26, 7.8 and 7.9. The pulse width modulated outputs $\overline{PWM0}$ and $\overline{PWM1}$ are used as general purpose outputs.

An interface buffer (MAX232A) connects the microcontroller and the PC (see again Figure 4.26).

Regarding the ASIC design we start from the final architecture (Figure 11.18). If we count the pins we will have 51 I/Os. A reset input will increase the number to 52. In any case, we need a reset condition for the ASIC when we restart a user program.

Once we have selected the ispLSI 1016 circuits for the ASIC implementation, we can test different partitioning schemes by trying to meet the number of pins demand. Inevitably, when architecture is decomposed into blocks the communication between the blocks will influence the final result in terms of delay and pin numbers. A classical approach used to optimize the timing parameters is replication. Some blocks can be replicated so that they reside in more than one partition. Consequently, the communication is minimized. Unfortunately, some inputs might also be replicated and the number of pins would put at stake the practical implementation. For simplicity, Figure 11.26 does not show the internal connections in both FPGAs. They can be seen in Figure 11.18.

The ASIC design goes on with mapping the partitioned architecture onto the FPGA's hardware. It is a vital test for the FPGA's functional capacity and interconnection capability.

We used a HDL (Hardware Description Language) called ABEL (Advanced Boolean Expression Language) as a design entry language for both FPGA circuits [Pell 1991, Data 1996]. The Synario CAD tool was employed for logic optimization, mapping and routing. The tool produced JEDEC files for both FPGAs. The files contain the device programming data. The format of the files is standardized by the Joint Electron Device Engineering Council (JEDEC). Consequently, we can have compatible design tools and ASIC programmers. As is frequently the case, the programmers unite a PC, a software package and a piece of hardware connected to the parallel port. An important feature of the JEDEC files is that they also possess test information (test vectors). The test code specifies the correspondence inputs - outputs for a certain number of input combinations. Thus, the FPGA programmers become capable of verifying the actual functionality of the programmed devices.

A part of the design entry, related to FPGA0, is given below. Using ABEL-HDL we first specify the names and attributes of external signals.

```
Declarations
"Inputs
BP,USER pin; "SLC
D7..D0,R,MUX1_A,MUX1_B pin; "MUX1, set operator (..)
TCE_N,TOE_N pin; "Logic
TA0 pin; "Flip-flop T
SR_CLK pin; "Shift register
IO7..IO0 pin; "Registers WR and MR

"Outputs
M pin istype 'com'; "SLC, combinational output
```

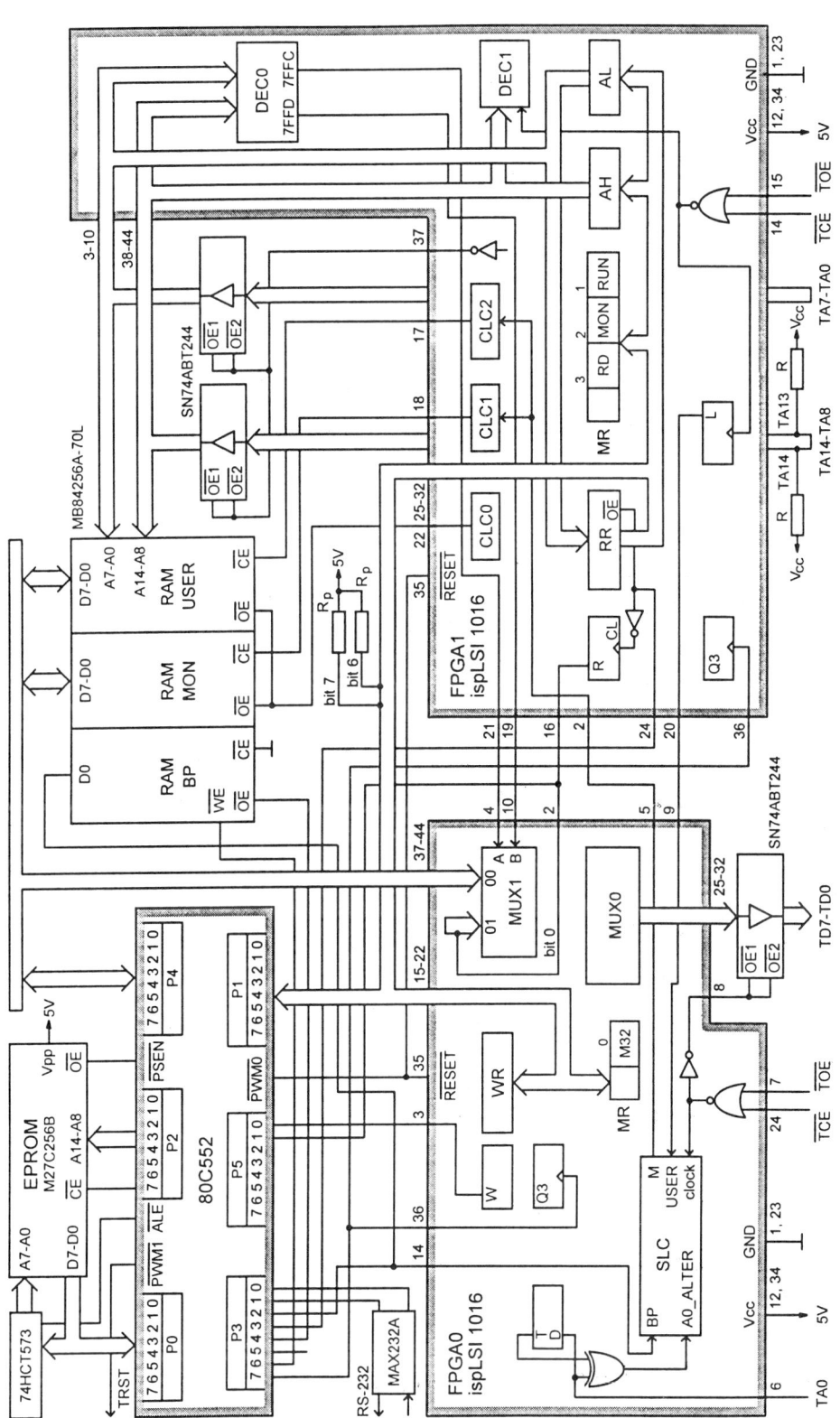

Figure 11.26 The EPROM emulator schematic diagram.

```
TD7..TD0 pin istype 'com'; "MUX0, combinational outputs
OED_N pin istype 'com'; "Logic, combinational output
W pin istype 'reg_d,buffer'; "Flip-flop W, flip-flop type D
```

In ABEL-HDL comments begin with a double question mark ("). Next we apply node statements to indicate internal signals. Use both Figure 11.18 and Figure 11.26.

```
"Nodes
A,B,A0_ALTER node istype 'com'; "SLC
P2,P1,P0 node istype 'reg_d,buffer'; "SLC, flip-flops, 8 states
clock node istype 'com'; "Logic
T node istype 'reg_d,buffer'; "Flip-flop T
Q3,Q2,Q1,Q0 node istype 'reg_d,buffer'; "Shift register
WR7..WR0 node istype 'reg_d,buffer'; "Register WR
M32 node istype 'reg_d,buffer'; "Flip-flop M32
```

In the end of the current specification phase we define names.

```
"Names
DATA_IN=[D7..D0];
DATA_OUT=[TD7..TD0];
WR=[WR7..WR0];
F_W_R=[0,0,0,0,0,0,W,R];
LCALL=[0,0,0,1,0,0,1,0]; "^H12 is the LCALL op code
AHB=[0,0,0,0,0,0,0,0]; "Address high-byte
ALB_1C=[0,0,0,1,1,1,0,0]; "Address low-byte ^H1C
ALB_7C=[0,1,1,1,1,1,0,0]; "Address low-byte ^H7C
IO=[IO7..IO0];
SLC=[P2,P1,P0];
S0=0; S1=1; S2=2; S3=3; S4=4; S5=5; S6=6; S7=7; "State values
```

The second phase of the logic design is termed equations. In particular, the FPGA0 intended hardware can be described as follows:

```
Equations
DATA_OUT=(((!B & !A & !MUX1_B & !MUX1_A) & DATA_IN) #
((!B & !A & MUX1_B & !MUX1_A) & WR) #
((!B & !A & !MUX1_B & MUX1_A) & F_W_R) #
((!B & A) & LCALL) # ((B & A) & AHB) #
((B & !A & !M32) & ALB_1C) #
((B & !A & M32) & ALB_7C); "NOT operator (!), AND operator (&)

OED_N=TCE_N # TOE_N; " Data buffers enable, OR operator (#)

Q3:=!Q0.FB; " Shift register
Q3.CLK=SR_CLK;

Q2:=!Q3.FB; " Shift register
Q2.CLK=SR_CLK;

Q1:=Q2.FB; " Shift register
Q1.CLK=SR_CLK;

Q0:=Q1.FB; " Shift register
```

```
Q0.CLK=SR_CLK;

M32:=IO0; " Flip-flop M32
M32.CLK=Q3;

W:=1; " Flip-flop W
W.CLK=!Q2;
W.AR=MUX1_B; " Asynchronous reset (CL)

WR:=IO; " Register WR
WR.CLK=!Q2;

T:=TA0; " Flip-flop T
T.CLK=clock;

clock=!(TCE_N # TOE_N); " SLC
[P2,P1,P0].CLK=clock;
A0_ALTER=T $ TA0; "XOR operator ($)

State_diagram SLC
State S0: M=0; " Run User
          A=0;
          B=0;
          if (BP) then S1 else S0;
State S1: M=1; " LCALL op code
          A=1;
          B=0;
          if (A0_ALTER) then S2 else S1;
State S2: M=1; " Address high-byte
          A=0;
          B=1;
          goto S3;
State S3: M=1; " Address low-byte
          A=1;
          B=1;
          goto S4;
State S4: M=1; " Run Monitor
          A=0;
          B=0;
          if (USER) then S5 else S4;
State S5: M=0; " Return 0
          A=0;
          B=0;
          goto S6;
State S6: M=0; " Return 1
          A=0;
          B=0;
          goto S7;
State S7: M=0; " Return 2
          A=0;
          B=0;
          goto S0;
```

The states of the SLC are described by the values of the corresponding outputs. The output signals are maintained as far as the SLC keeps the state. In addition, the description contains the relationship of the present state and inputs - next state (Figure 11.22).

The operation of the sequential circuit SLC begins in state S0 and remains there as long as the input BP is low. The SLC enters this state when the microcontroller's output $\overline{PWM0}$ is low and the input $\overline{RESET}$ is asserted. Both FPGAs must be reset simultaneously as shown in Figure 11.26. The sequential machine has three outputs (A, B and M) and all of them are low in state S0. When the input BP is asserted, the SLC advances to state S1. In this state both outputs A and M are pulled high. The transition from one state to another is synchronized by the low-to-high change of the input clock.

The Synario tool reported successful design process for both FPGAs. The ispLSI 1016 possesses 16 basic units of logic called Generic Logic Block (GLB). Each GLB contains four flip-flops plus an associated logic array. Figure 11.27 shows the utilization of the FPGAs calculated by the CAD tool. The net utilization indicates the level of interconnectivity between the internal elements achieved for this project.

	GLB utilization	I/O utilization	Net utilization
FPGA0	62%	100%	56%
FPGA1	75%	100%	67%

Figure 11.27 The FPGA0 and FPGA1 utilization.

A specific feature of the synthesis system is that the pins may change position from run to run. At the same time, the description allows signals to be locked to pins which makes the results predictable.

The EPROM emulator software resides in both the emulator microcontroller program memory and the Monitor RAM. The ASIC includes four registers which can be loaded from the emulator microcontroller: MR, WR, AH and AL. We organize an output buffer for them in the emulator microcontroller internal Data Memory. When we want to update a register, we move the new value to the buffer and copy the buffer to the ASIC registers. The following is a subroutine for this operation:

```
;*********************************************************************
;    E_WRITE.ASM                                                    *
; This subroutine copies the register buffer from the               *
; emulator microcontroller internal Data Memory to the              *
; ASIC registers                                                    *
; The labels used for direct addresses are MR, WR, AH and AL *
;*********************************************************************
E_WRITE:  CLR     P3.7
          MOV     P1,MR   ; Output the value for register MR
          SETB    P3.7    ; Copy the value for register MR
          CLR     P3.7
          MOV     P1,WR   ; Output the value for register WR
```

```
        SETB    P3.7    ; Copy the value for register WR
        CLR     P3.7
        MOV     P1,AH   ; Output the value for register AH
        SETB    P3.7    ; Copy the value for register AH
        CLR     P3.7
        MOV     P1,AL   ; Output the value for register AL
        SETB    P3.7    ; Copy the value for register AL
        RET
```

When a byte is moved to the ASIC write register WR, the Monitor program will transfer it to the target microcontroller by means of the subroutine below:

```
;*****************************************************************
;       T_READ.ASM                                              *
;       This subroutine moves a byte from the ASIC WR register  *
;       to the accumulator of the target microcontroller        *
;       Set the flag MEM32 for 32K byte EPROMs and test         *
;       the flag W beforehand                                   *
;*****************************************************************
T_READ: CLR     A
        MOV     DPTR,#7FFDH     ; Assume 32K byte EPROM
        JB      MEM32,T_READ1
        MOV     DPTR,#1FFDH     ; Correct for 8K byte EPROM
T_READ1: MOVC   A,@A+DPTR       ; Read the register WR and
        RET                     ; clear the flag W
```

The subroutine **T_READ.ASM** sets the decoder DEC0's output 7FFDH, which in turn adjusts the data path to emit the code in register WR. In parallel, the subroutine clears the flag W. A prototype of the EPROM emulator is shown in Figure 11.28.

Figure 11.28 The EPROM emulator.

Finally, if we summarize what has been done in this case study, we could mark the following features:

- The conceptualization used in the beginning of the design process gradually matured in rigorous specification. The reason for this, though it may be obvious, is that we wanted to unroll alternatives and directions for the reader to follow if the EPROM emulator functionality is modified and different components or CAD tools are used.

- We moved a task from software to hardware and from then on continued the project by concurrent design of the hardware and software parts.

- The ASIC implementation based on in-circuit programmable FPGAs demonstrates an efficient design of a small-scale embedded system. The FPGA not only provides the required timing parameters, but also allows the hardware to be reconfigured for different target processors.

The EPROM emulator example is readily transferable to other applications when microcontrollers and FPGAs are combined to improve performance.

11.4 References

Data I/O Corporation, *SYNARIO, ABEL-HDL Reference*, 1996.

Fujitsu, *Static RAM Products, Data Book*, 1991.

Lattice Semiconductor, *Data Book*, 1994.

David Pellerin and Michael Holley, *Practical Design Using Programmable Logic*, Prentice Hall, 1991.

Philips Semiconductors, *High-speed CMOS Logic family, Data Handbook IC06*, 1994a.

Philips Semiconductors, *80C51-Based 8-Bit Microcontrollers, Data Handbook IC20*, 1997a.

SGS-Thomson Microelectronics, *Memory Products*, 1994.

Texas Instruments, *The TTL Data Book*, Volume 1, 1989.

Texas Instruments, *Advanced BiCMOS ABT Bus Interface Logic, Data Book*, 1993.

Wayne Wolf, *Modern VLSI Design*, Prentice Hall, 1994a.

Wayne Wolf, "Hardware-software co-design of embedded systems", *Proc. IEEE*, vol. 82, No. 7, July 1994b, pp. 967-989.

The following URLs can be used for additional information:

Data I/O Corporation	http://www.dataio.com
Lattice	http://www.latticesemi.com
Philips Semiconductors	http://www.philips.com
SGS-Thomson	http://www.st.com
Texas Instruments	http://www.ti.com

Appendix A

SURVEY OF MICROCONTROLLERS

Microcontroller: MC68HC11A8   Vendor: Motorola

http://www.mot.com

- 8-bit
- ROM 8K bytes
- RAM 256 bytes
- EEPROM 256 bytes
- Enhanced 16-bit timer system
- 8-bit pulse accumulator circuit
- Serial communications interface
- Serial peripheral interface
- Eight channel, 8-bit ADC
- Watchdog system
- Pins 48, DIP

Reference: Motorola, *MC68HC11A8, HCMOS Single - Chip Microcontroller*, 1991.

Microcontroller: COP8640C   Vendor: National Semiconductor

http://www.national.com

- 8-bit
- ROM 2K bytes
- RAM 64 bytes
- EEPROM 64 bytes
- I/O pins 24
- Execution time per instruction 1 microsecond
- Fully static design

- Supply voltage 4.5 to 6.0 V
- MICROWIRE PLUS^{TM} serial I/O
- 16-bit timer/counter with auto reload and capture register
- Pins 28

Reference: National Semiconductor, *COP8^{TM} Microcontroller Databook*, 1994.

Microcontroller: PIC16C73   Vendor: Microchip Technology

http://www.microchip.com

- 8-bit, high-performance RISC-like CPU
- EPROM 4K bytes (OTP)
- RAM 192 bytes
- 14-bit wide instruction set
- Programmable I/O lines 22
- Three timer/counters
- Two capture/compare/PWM modules
- Two serial ports
- I^{2}C bus
- Five channel, 8-bit ADC, conversion time 1.6 µs
- Watchdog timer
- Power-down mode
- Clock speed DC - 20 MHz
- Supply voltage 3.0 - 6.0V
- Pins 28

Reference: Microchip, *PIC16/17 Microcontroller Data Book*, 1995.

Microcontroller: AT89C52   Vendor: Atmel

http://www.atmel.com

- 8-bit, compatible with the 8051
- In-system reprogrammable flash memory 8K bytes, endurance: 1000 write/erase cycles
- Three-level Program Memory lock
- RAM 256 bytes
- Programmable I/O lines 32
- Three 16-bit timer/counters
- Eight interrupt sources
- Programmable serial channel
- Low power Idle and Power Down modes
- Fully static operation: 0 Hz to 24 MHz

Reference: Atmel, *Microcontroller Data book*, 1997.

Microcontroller: AT90S8515   Vendor: Atmel

                                              http://www.atmel.com

- 8-bit, based on AVR® enhanced RISC architecture
- 120 powerful instructions - most single clock cycle execution
- In-system reprogrammable downloadble flash memory 8K bytes
- EEPROM 512 bytes
- RAM 512 bytes
- Programmable I/O lines 32
- Programmable serial UART
- SPI (Serial Peripheral Interface)
- Supply voltage 2.7 to 6.0 V
- Fully static operation: 0 Hz to 20 MHz
- Instruction cycle time: 50 ns @ 20 MHz
- Dual PWM
- Programmable Watchdog timer
- On-chip analog comparator

Reference: Atmel, *8-Bit RISC Microcontrollers, Data book*, 1997.

Microcontroller: SAB80C515   Vendor: Siemens

                                              http://www.siemens.com

- 8-bit
- ROM 8K bytes
- RAM 256 bytes
- Binary-code compatible with the 8051
- Three 16-bit timer/counters
- 16-bit reload, compare, capture capability
- ADC, 8 multiplexed analog inputs, programmable reference voltages
- 16-bit watchdog timer
- 12 interrupt sources , 4 priority levels
- Pins 68

Reference: Siemens, *Microcomputer Components, SAB 80515 / SAB 80C515, 8-Bit Single-Chip Microcontroller Family,* 1992.

Microcontroller: DS80C320   Vendor: Dallas Semiconductor

                                              http://www.dalsemi.com

- 8-bit, 80C32 compatible
- 4 clocks per bus cycle
- Runs DC to 33 MHz clock rates
- Three 16-bit timer/counters
- Dual data pointers
- Programmable Watchdog timer
- Two full-duplex serial ports

- Precision power-fail reset
- Early warning power-fail interrupt
- 13 total interrupt sources with 6 external
- Variable length MOVX to access fast/slow peripherals

Reference: Dallas Semiconductor, *High-Speed Micro User's Guide*, 1994.

Microcontroller: 80C151SA   Vendor: Intel

http://www.intel.com

- 8-bit
- ROM/OTPROM 8K bytes
- RAM 256 bytes
- Software compatible with the 8051 and the performance is improved three to six times
- 16-bit instruction fetch
- A static design
- Three 16-bit timer/counters
- Program Counter Array (PCA)
- Watchdog timer
- Programmable I/O lines 32
- Serial port with framing error detection and automatic address recognition
- Pins 40/44

Reference: Intel, *8XC151SA/SB High-Performance CHMOS Microcontroller*, 1996.

Microcontroller: 83C251SB   Vendor: Intel

http://www.intel.com

- 8-bit, 16-bit and 32-bit arithmetic and logic instructions
- ROM 16K bytes
- RAM 1K bytes
- Binary-code compatible with the 8051
- 16-bit internal code fetch
- Oscillator frequency 0 - 16 MHz
- Three flexible 16-bit timer/counters
- Program Counter Array (PCA)
- Watchdog timer
- Programmable I/O lines 32
- Serial port with framing error detection and automatic address recognition
- Pins 44

Reference: Intel, *Embedded Microcontrollers*, 1996.

Microcontroller: 83930AE   Vendor: Intel
                                  http://www.intel.com

- Register-based MCS® 251 architecture
- Complete Universal Serial Bus Specification 1.0 compatibility
- Serial bus Interface Engine (SIE)
- Four transmit FIFOs
- Four receive FIFOs
- ROM 16K bytes
- RAM 1K bytes
- Four I/O ports
- Program Counter Array (PCA)
- Hardware Watchdog timer

Reference: http://www.intel.com/design/usb/datashts/

Microcontroller: MC68HC16Z2   Vendor: Motorola
                                   http://www.mot.com

- 16-bit
- ROM 8K bytes, 16-bit wide array
- RAM 2K bytes
- Program memory address space 1M bytes
- Data memory address space 1M bytes
- Eight channels, 8/10-bit ADC
- Queued serial module
- General - purpose timer
- Fully static operation
- Two pulse width modulation outputs

Reference: Motorola, *MC68HC16Z2, 16-Bit Modular Microcontroller*, 1992.

Microcontroller: P51XAG23JB   Vendor: Philips
                                    http://www.philips.com

- 16-bit, a member of the Philips Extended Architecture (XA) family
- ROM 16K bytes
- RAM 512 bytes
- Source code compatible with the 8051
- Address range 20-bit, 1M byte each program and data space
- Oscillator frequency 0 - 25 MHz
- Eight 16-bit CPU registers each capable of performing all arithmetic and logic operations as well as acting as memory pointers
- Three counter/timers with enhanced features
- Watchdog timer
- Four 8-bit I/O ports with 4 programmable output configurations
- Two enhanced UARTs

- Supply voltage 2.7V to 5.5V
- Pins 44

Reference: Philips, *16-bit 80C51XA Microcontrollers (eXtended Architecture), Data Handbook IC25*, 1996.

Microcontroller: 80386EXTB   Vendor: Intel

http://www.intel.com

- 32-bit internal architecture, 16-bits wide external data bus
- 26-bit address bus, 64M bytes of memory address space and 64K bytes of I/O address space
- Integrated peripherals
    - PC - compatible peripherals
        - Interrupt control unit
        - Timer/counter unit
        - Asynchronous serial I/O
        - Direct memory access controller
    - Embedded application-specific peripherals
        - System management mode architectural extension to the Intel386 CPU
        - Clock and power management unit
        - Synchronous serial I/O unit
        - Chip-select unit
        - Refresh control unit
        - Parallel I/O ports
        - Watchdog timer unit
        - JTAG test - logic unit
- On-chip debugging support including breakpoint registers
- Supply voltage 2.7V to 3.6V
- Operating frequency 25 MHz at 3.0V to 3.6V
- Two package types - 132 or 144 pins

Reference: Intel, *Intel386^{TM}EX Embedded Microprocessor User's Manual*, 1996.

Appendix B

THE 8051 MICROCONTROLLER
SPECIAL FUNCTION REGISTERS

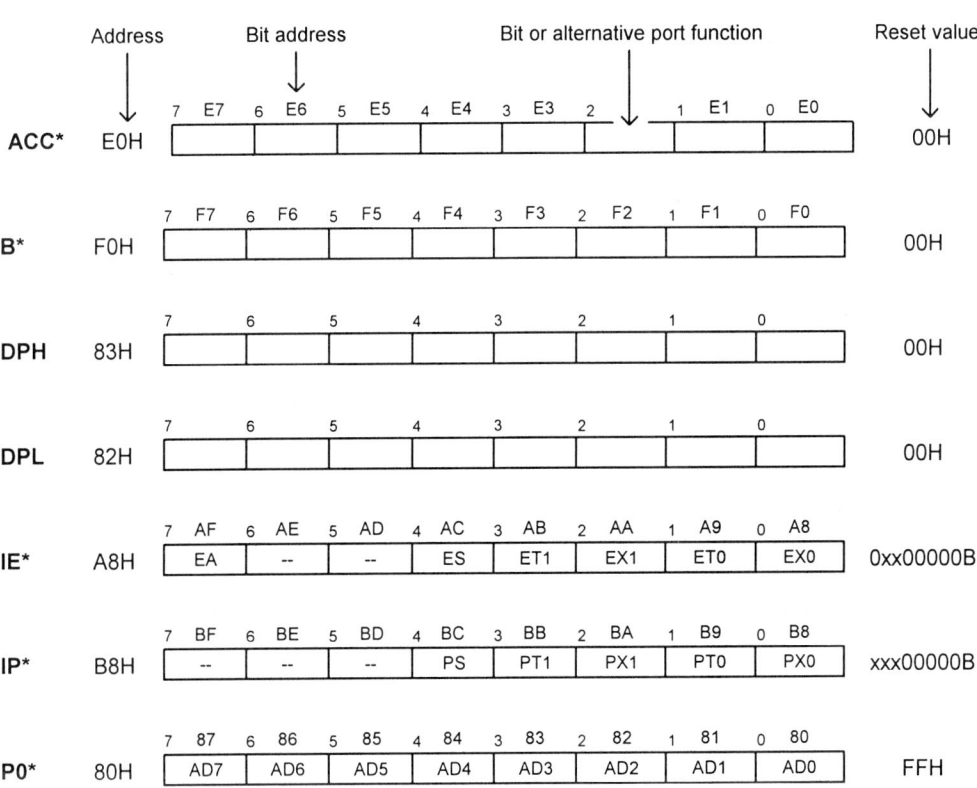

P1* 90H

7 97	6 96	5 95	4 94	3 93	2 92	1 91	0 90

FFH

P2* A0H

7 A7	6 A6	5 A5	4 A4	3 A3	2 A2	1 A1	0 A0
A15	A14	A13	A12	A11	A10	A9	A8

FFH

P3* B0H

7 B7	6 B6	5 B5	4 B4	3 B3	2 B2	1 B1	0 B0
$\overline{RD}$	$\overline{WR}$	T1	T0	$\overline{INT1}$	$\overline{INT0}$	TXD	RXD

FFH

PCON 87H

7	6	5	4	3	2	1	0
SMOD	--	--	--	GF1	GF0	PD	IDL

0xxx0000B

PSW* D0H

7 D7	6 D6	5 D5	4 D4	3 D3	2 D2	1 D1	0 D0
CY	AC	F0	RS1	RS0	OV	--	P

00H

SBUF 99H

7	6	5	4	3	2	1	0

xxxxxxxB

SCON* 98H

7 9F	6 9E	5 9D	4 9C	3 9B	2 9A	1 99	0 98
SM0	SM1	SM2	REN	TB8	RB8	TI	RI

00H

SP 81H

7	6	5	4	3	2	1	0

07H

TCON* 88H

7 8F	6 8E	5 8D	4 8C	3 8B	2 8A	1 89	0 88
TF1	TR1	TF0	TR0	IE1	IT1	IE0	IT0

00H

TH0 8CH

7	6	5	4	3	2	1	0

00H

TH1 8DH

7	6	5	4	3	2	1	0

00H

TL0 8AH

7	6	5	4	3	2	1	0

00H

TL1 8BH

7	6	5	4	3	2	1	0

00H

TMOD 89H

7	6	5	4	3	2	1	0
GATE	$C/\overline{T}$	M1	M0	GATE	$C/\overline{T}$	M1	M0

00H

Appendix C

THE 83C552 MICROCONTROLLER SPECIAL FUNCTION REGISTERS

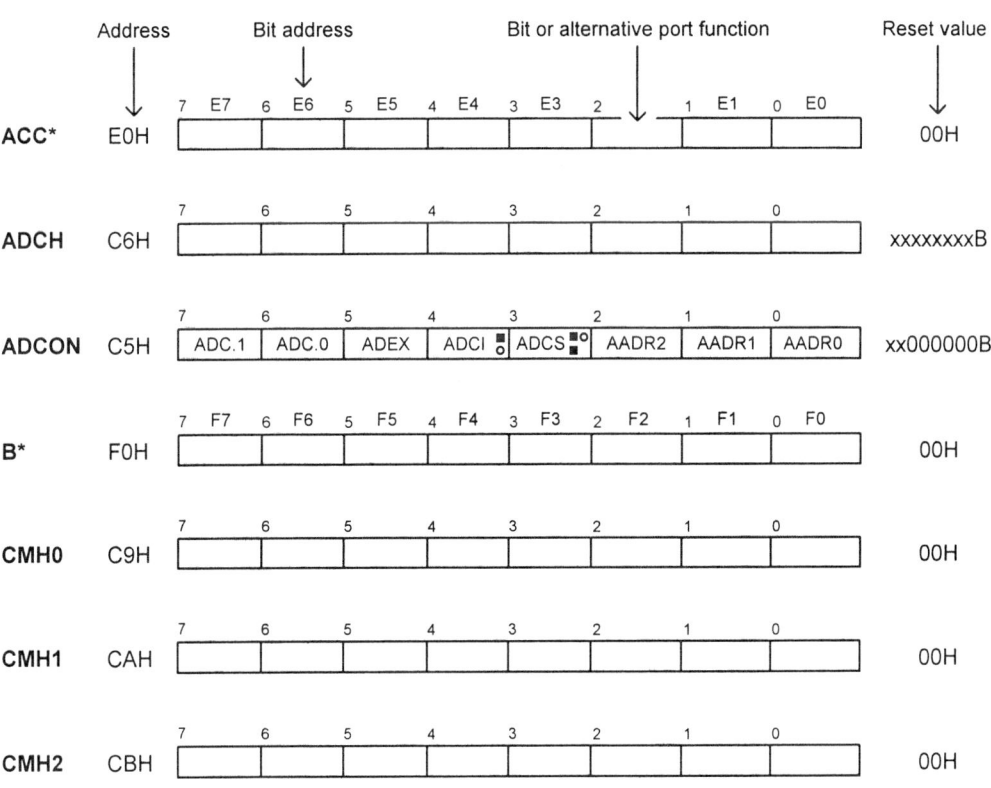

Register	Addr	7	6	5	4	3	2	1	0	Reset
CML0	A9H									00H
CML1	AAH									00H
CML2	ABH									00H
CTCON	EBH	CTN3	CTP3	CTN2	CTP2	CTN1	CTP1	CTN0	CTP0	00H
CTH0	CCH									xxxxxxxxB
CTH1	CDH									xxxxxxxxB
CTH2	CEH									xxxxxxxxB
CTH3	CFH									xxxxxxxxB
CTL0	ACH									xxxxxxxxB
CTL1	ADH									xxxxxxxxB
CTL2	AEH									xxxxxxxxB
CTL3	AFH									xxxxxxxxB
DPH	83H									00H
DPL	82H									00H

IEN0* A8H

7 AF	6 AE	5 AD	4 AC	3 AB	2 AA	1 A9	0 A8	
EA	EAD	ES1	ES0	ET1	EX1	ET0	EX0	00H

IEN1* E8H

7 EF	6 EE	5 ED	4 EC	3 EB	2 EA	1 E9	0 E8	
ET2	ECM2	ECM1	ECM0	ECT3	ECT2	ECT1	ECT0	00H

IP0* B8H

7 BF	6 BE	5 BD	4 BC	3 BB	2 BA	1 B9	0 B8	
--	PAD	PS1	PS0	PT1	PX1	PT0	PX0	x0000000B

Register	Addr	7	6	5	4	3	2	1	0	Reset
IP1*	F8H	FF PT2	FE PCM2	FD PCM1	FC PCM0	FB PCT3	FA PCT2	F9 PCT1	F8 PCT0	00H
P0*	80H	87 AD7	86 AD6	85 AD5	84 AD4	83 AD3	82 AD2	81 AD1	80 AD0	FFH
P1*	90H	97 SDA	96 SCL	95 RT2	94 T2	93 CT3I	92 CT2I	91 CT1I	90 CT0I	FFH
P2*	A0H	A7 A15	A6 A14	A5 A13	A4 A12	A3 A11	A2 A10	A1 A9	A0 A8	FFH
P3*	B0H	B7 $\overline{RD}$	B6 $\overline{WR}$	B5 T1	B4 T0	B3 $\overline{INT1}$	B2 $\overline{INT0}$	B1 TXD	B0 RXD	FFH
P4*	C0H	C7 CMT1	C6 CMT0	C5 CMSR5	C4 CMSR4	C3 CMSR3	C2 CMSR2	C1 CMSR1	C0 CMSR0	FFH
P5	C4H	ADC7	ADC6	ADC5	ADC4	ADC3	ADC2	ADC1	ADC0	xxxxxxxxB
PCON	87H	SMOD	--	--	WLE	GF1	GF0	PD	IDL	0xx00000B
PSW*	D0H	D7 CY	D6 AC	D5 F0	D4 RS1	D3 RS0	D2 OV	D1 F1	D0 P	00H
PWM0	FCH									00H
PWM1	FDH									00H
PWMP	FEH									00H
RTE	EFH	TP47	TP46	RP45	RP44	RP43	RP42	RP41	RP40	00H
S0BUF	99H									xxxxxxxxB
S0CON*	98H	9F SM0	9E SM1	9D SM2	9C REN	9B TB8	9A RB8	99 TI	98 RI	00H
S1ADR	DBH			SLAVE ADDRESS					GC	00H

		7 DF	6 DE	5 DD	4 DC	3 DB	2 DA	1 D9	0 D8	
S1CON*	D8H	CR2	ENS1	STA	STO	SI	AA	CR1	CR0	00H

		7	6	5	4	3	2	1	0	
S1DAT	DAH									00H

		7	6	5	4	3	2	1	0	
S1STA	D9H	SC4	SC3	SC2	SC1	SC0	0	0	0	F8H

		7	6	5	4	3	2	1	0	
SP	81H									07H

		7	6	5	4	3	2	1	0	
STE	EEH	TG47	TG46	SP45	SP44	SP43	SP42	SP41	SP40	C0H

		7	6	5	4	3	2	1	0	
T3	FFH									00H

		7 8F	6 8E	5 8D	4 8C	3 8B	2 8A	1 89	0 88	
TCON*	88H	TF1	TR1	TF0	TR0	IE1	IT1	IE0	IT0	00H

		7	6	5	4	3	2	1	0	
TH0	8CH									00H

		7	6	5	4	3	2	1	0	
TH1	8DH									00H

		7	6	5	4	3	2	1	0	
TL0	8AH									00H

		7	6	5	4	3	2	1	0	
TL1	8BH									00H

		7	6	5	4	3	2	1	0	
TM2CON	EAH	T2IS1	T2IS0	T2ER	T2BO	T2P1	T2P0	T2MS1	T2MS0	00H

		7 CF	6 CE	5 CD	4 CC	3 CB	2 CA	1 C9	0 C8	
TM2IR*	C8H	T2OV	CMI2	CMI1	CMI0	CTI3	CTI2	CTI1	CTI0	00H

		7	6	5	4	3	2	1	0	
TMH2	EDH									00H

		7	6	5	4	3	2	1	0	
TML2	ECH									00H

		7	6	5	4	3	2	1	0	
TMOD	89H	GATE	$C/\overline{T}$	M1	M0	GATE	$C/\overline{T}$	M1	M0	00H

Appendix D

THE 8051 AND 83C552 MICROCONTROLLERS INSTRUCTION SET

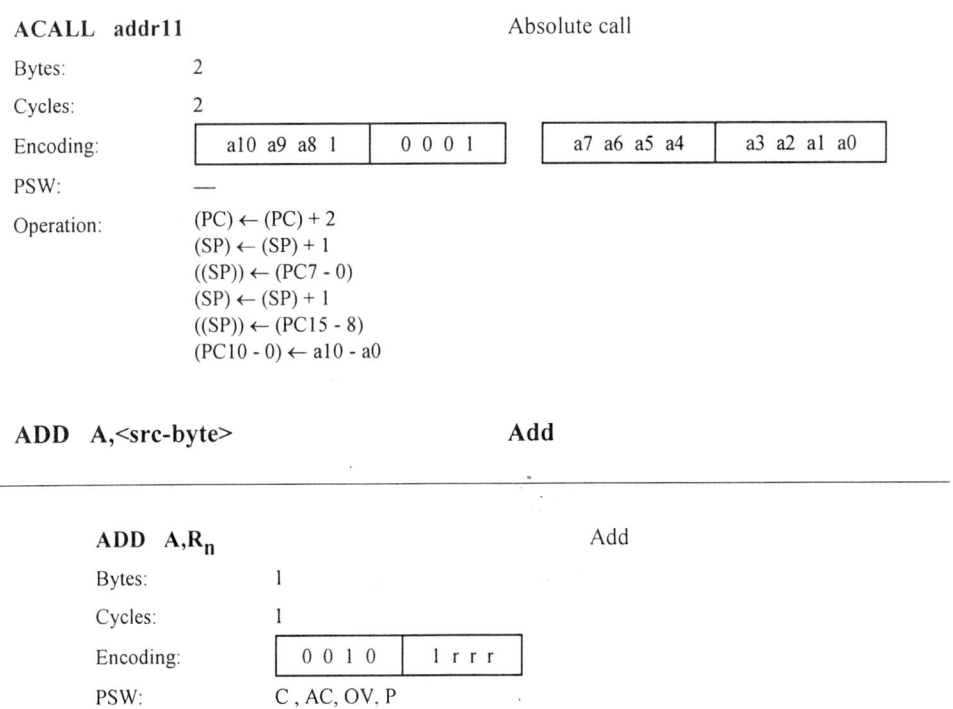

ACALL addr11   Absolute call

Bytes:	2
Cycles:	2

Encoding:

a10 a9 a8 1	0 0 0 1		a7 a6 a5 a4	a3 a2 a1 a0

PSW: —

Operation:
(PC) ← (PC) + 2
(SP) ← (SP) + 1
((SP)) ← (PC7 - 0)
(SP) ← (SP) + 1
((SP)) ← (PC15 - 8)
(PC10 - 0) ← a10 - a0

ADD   A,<src-byte>   **Add**

ADD   A,R$_n$ Add

Bytes:	1
Cycles:	1

Encoding:

0 0 1 0	1 r r r

PSW: C , AC, OV, P

Operation: (A) ← (A) + (R$_n$)

381

ADD   A,dir   Add

Bytes: 2

Cycles: 1

Encoding: | 0 0 1 0 | 0 1 0 1 | | dir |

PSW: C, AC, OV, P

Operation: (A) ← (A) + (dir)

ADD   A,@R$_i$ Add

Bytes: 1

Cycles: 1

Encoding: | 0 0 1 0 | 0 1 1 i |

PSW: C, AC, OV, P

Operation: (A) ← (A) + ((R$_i$))

ADD   A,#data   Add

Bytes: 2

Cycles: 1

Encoding: | 0 0 1 0 | 0 1 0 0 | | #data |

PSW: C, AC, OV, P

Operation: (A) ← (A) + #data

ADDC   A,<src-byte>   **Add with carry**

ADDC   A,R$_n$ Add with carry

Bytes: 1

Cycles: 1

Encoding: | 0 0 1 1 | 1 r r r |

PSW: C, AC, OV, P

Operation: (A) ← (A) + (C) + (R$_n$)

ADDC   A,dir   Add with carry

Bytes: 2

Cycles: 1

Encoding: | 0 0 1 1 | 0 1 0 1 | | dir |

PSW: C, AC, OV, P

Operation: (A) ← (A) + (C) + (dir)

ADDC A,@R$_i$ Add with carry

Bytes: 1

Cycles: 1

Encoding: | 0 0 1 1 | 0 1 1 i |

PSW: C, AC, OV, P

Operation: $(A) \leftarrow (A) + (C) + ((R_i))$

ADDC A,#data   Add with carry

Bytes: 2

Cycles: 1

Encoding: | 0 0 1 1 | 0 1 0 0 | | #data |

PSW: C, AC, OV, P

Operation: $(A) \leftarrow (A) + (C) + \#data$

AJMP addr11   Absolute jump

Bytes: 2

Cycles: 2

Encoding: | a10 a9 a8 0 | 0 0 0 1 | | a7 a6 a5 a4 | a3 a2 a1 a0 |

PSW: —

Operation: $(PC) \leftarrow (PC) + 2$
              $(PC10 - 0) \leftarrow a10 - a0$

ANL <dest-byte> , <src-byte>   **AND logical**

ANL A,R$_n$ AND logical

Bytes: 1

Cycles: 1

Encoding: | 0 1 0 1 | 1 r r r |

PSW: P

Operation: $(A) \leftarrow (A) \wedge (R_n)$

ANL A,dir   AND logical

Bytes: 2

Cycles: 1

Encoding: | 0 1 0 1 | 0 1 0 1 | | dir |

PSW: P

Operation: $(A) \leftarrow (A) \wedge (dir)$

ANL A,@R$_i$ AND logical

Bytes: 1

Cycles: 1

Encoding: | 0 1 0 1 | 0 1 1 i |

PSW: P

Operation: $(A) \leftarrow (A) \wedge ((R_i))$

ANL A,#data   AND logical

Bytes: 2

Cycles: 1

Encoding: | 0 1 0 1 | 0 1 0 0 | | #data |

PSW: P

Operation: $(A) \leftarrow (A) \wedge$ #data

ANL dir,A   AND logical

Bytes: 2

Cycles: 1

Encoding: | 0 1 0 1 | 0 0 1 0 | | dir |

PSW: —

Operation: $(dir) \leftarrow (dir) \wedge A$

ANL dir,#data   AND logical

Bytes: 3

Cycles: 2

Encoding: | 0 1 0 1 | 0 0 1 1 | | dir | | #data |

PSW: —

Operation: $(dir) \leftarrow (dir) \wedge$ #data

ANL C,<src-bit>   AND logical for bit variables

ANL C,bit   AND logical for bit variables

Bytes: 2

Cycles: 2

Encoding: | 1 0 0 0 | 0 0 1 0 | | bit |

PSW: C

Operation: $(C) \leftarrow (C) \wedge$ bit

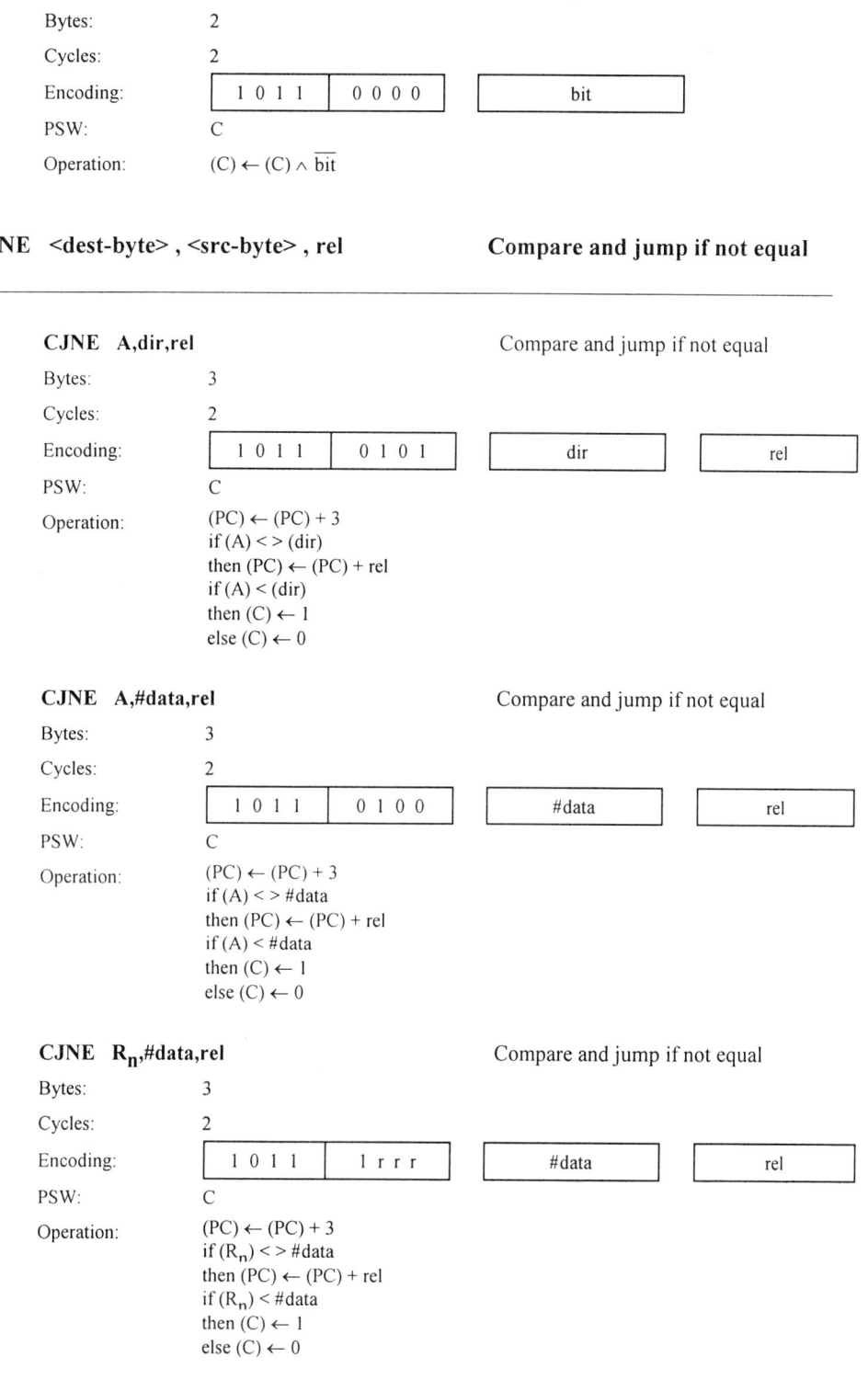

ANL C,/bit   AND logical for bit variables

Bytes: 2

Cycles: 2

Encoding: | 1 0 1 1 | 0 0 0 0 | | bit |

PSW: C

Operation: $(C) \leftarrow (C) \wedge \overline{bit}$

CJNE <dest-byte> , <src-byte> , rel   **Compare and jump if not equal**

CJNE A,dir,rel   Compare and jump if not equal

Bytes: 3

Cycles: 2

Encoding: | 1 0 1 1 | 0 1 0 1 | | dir | | rel |

PSW: C

Operation: $(PC) \leftarrow (PC) + 3$
if $(A) < > (dir)$
then $(PC) \leftarrow (PC) + rel$
if $(A) < (dir)$
then $(C) \leftarrow 1$
else $(C) \leftarrow 0$

CJNE A,#data,rel   Compare and jump if not equal

Bytes: 3

Cycles: 2

Encoding: | 1 0 1 1 | 0 1 0 0 | | #data | | rel |

PSW: C

Operation: $(PC) \leftarrow (PC) + 3$
if $(A) < > \#data$
then $(PC) \leftarrow (PC) + rel$
if $(A) < \#data$
then $(C) \leftarrow 1$
else $(C) \leftarrow 0$

CJNE R_n,#data,rel   Compare and jump if not equal

Bytes: 3

Cycles: 2

Encoding: | 1 0 1 1 | 1 r r r | | #data | | rel |

PSW: C

Operation: $(PC) \leftarrow (PC) + 3$
if $(R_n) < > \#data$
then $(PC) \leftarrow (PC) + rel$
if $(R_n) < \#data$
then $(C) \leftarrow 1$
else $(C) \leftarrow 0$

CJNE   @R$_i$,#data,rel   Compare and jump if not equal

Bytes: 3

Cycles: 2

Encoding: | 1 0 1 1 | 0 1 1 i | | #data | | rel |

PSW: C

Operation: $(PC) \leftarrow (PC) + 3$
if $((R_i)) < > $#data
then $(PC) \leftarrow (PC) + $rel
if $((R_i)) < $#data
then $(C) \leftarrow 1$
else $(C) \leftarrow 0$

CLR   A   Clear accumulator

Bytes: 1

Cycles: 1

Encoding: | 1 1 1 0 | 0 1 0 0 |

PSW: P

Operation: $(A) \leftarrow 0$

CLR   bit   **Clear bit**

CLR   C   Clear carry

Bytes: 1

Cycles: 1

Encoding: | 1 1 0 0 | 0 0 1 1 |

PSW: C

Operation: $(C) \leftarrow 0$

CLR   bit   Clear bit

Bytes: 2

Cycles: 1

Encoding: | 1 1 0 0 | 0 0 1 0 | | bit |

PSW: P (if bit in accumulator)

Operation: $(bit) \leftarrow 0$

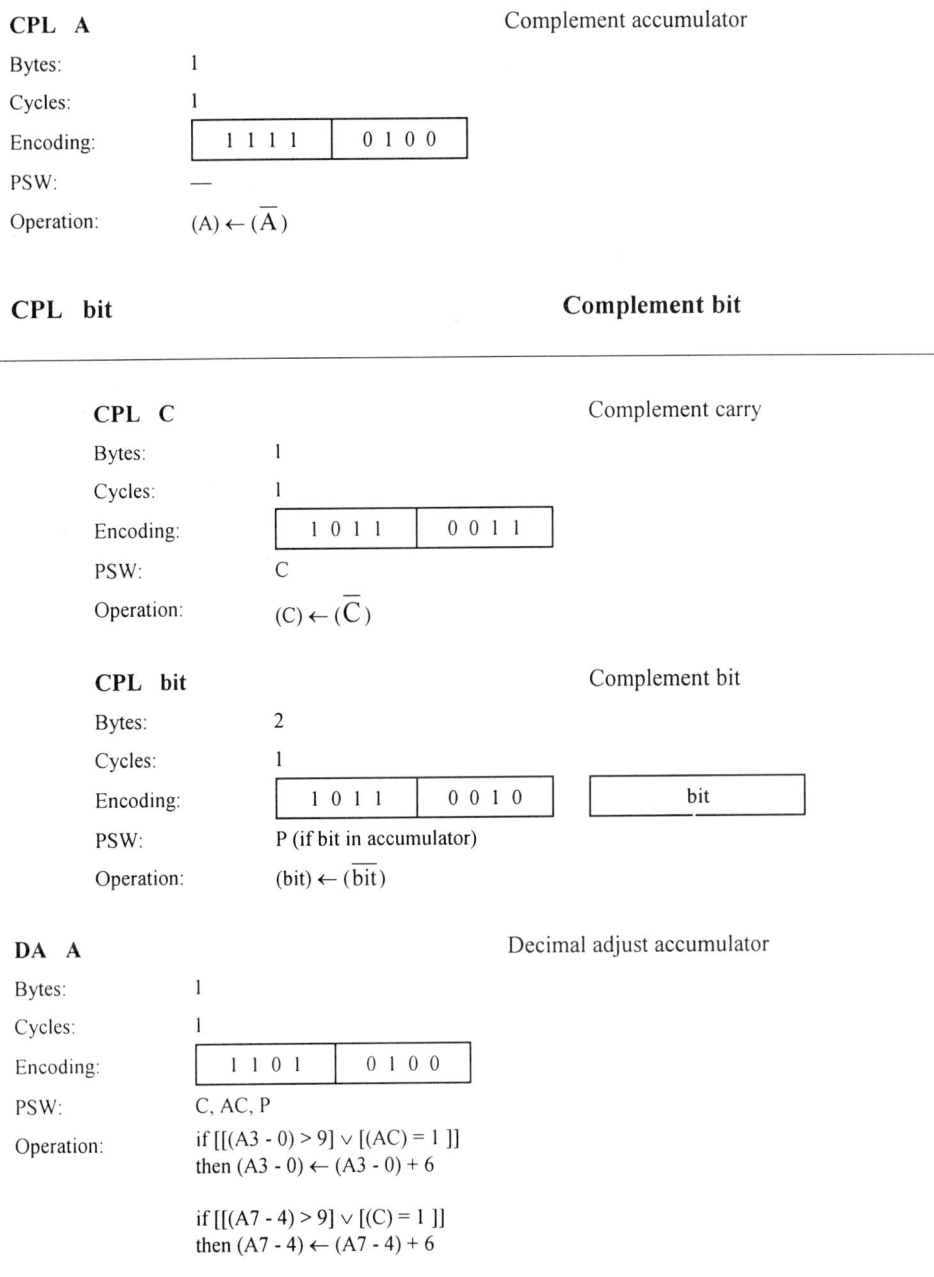

CPL A   Complement accumulator

Bytes:	1
Cycles:	1
Encoding:	1 1 1 1 0 1 0 0
PSW:	—
Operation:	$(A) \leftarrow (\overline{A})$

CPL bit   **Complement bit**

CPL C   Complement carry

Bytes:	1
Cycles:	1
Encoding:	1 0 1 1 0 0 1 1
PSW:	C
Operation:	$(C) \leftarrow (\overline{C})$

CPL bit   Complement bit

Bytes:	2
Cycles:	1
Encoding:	1 0 1 1 0 0 1 0 bit
PSW:	P (if bit in accumulator)
Operation:	$(bit) \leftarrow (\overline{bit})$

DA A   Decimal adjust accumulator

Bytes:	1
Cycles:	1
Encoding:	1 1 0 1 0 1 0 0
PSW:	C, AC, P
Operation:	if $[[(A3 - 0) > 9] \vee [(AC) = 1]]$
	then $(A3 - 0) \leftarrow (A3 - 0) + 6$
	if $[[(A7 - 4) > 9] \vee [(C) = 1]]$
	then $(A7 - 4) \leftarrow (A7 - 4) + 6$

DEC byte   **Decrement**

DEC A

		Decrement
Bytes: 1

Cycles: 1

Encoding: | 0 0 0 1 | 0 1 0 0 |

PSW: P

Operation: $(A) \leftarrow (A) - 1$

DEC R_n Decrement

Bytes: 1

Cycles: 1

Encoding: | 0 0 0 1 | 1 r r r |

PSW: —

Operation: $(R_n) \leftarrow (R_n) - 1$

DEC dir   Decrement

Bytes: 2

Cycles: 1

Encoding: | 0 0 0 1 | 0 1 0 1 | | dir |

PSW: P (if dir = E0H)

Operation: $(dir) \leftarrow (dir) - 1$

DEC @R_i Decrement

Bytes: 1

Cycles: 1

Encoding: | 0 0 0 1 | 0 1 1 i |

PSW: P (if (R_i) = E0H)

Operation: $((R_i)) \leftarrow ((R_i)) - 1$

DIV AB   Divide

Bytes: 1

Cycles: 4

Encoding: | 1 0 0 0 | 0 1 0 0 |

PSW: C, OV, P

Operation: (A)15 - 8

                                    $\leftarrow$ (A)/(B)
                  (B)7 - 0 remainder

DJNZ <byte>,<rel>   **Decrement and jump if not zero**

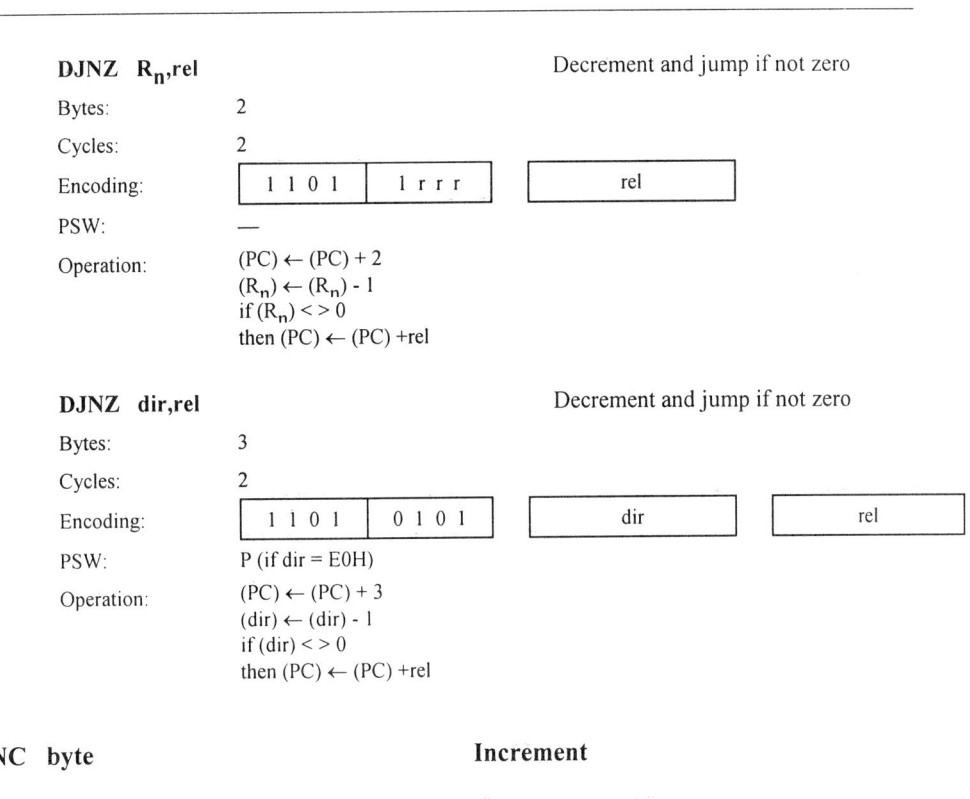

DJNZ R_n,rel   Decrement and jump if not zero

Bytes: 2

Cycles: 2

Encoding: | 1 1 0 1 | 1 r r r | | rel |

PSW: —

Operation: $(PC) \leftarrow (PC) + 2$
$(R_n) \leftarrow (R_n) - 1$
if $(R_n) <> 0$
then $(PC) \leftarrow (PC) + rel$

DJNZ dir,rel   Decrement and jump if not zero

Bytes: 3

Cycles: 2

Encoding: | 1 1 0 1 | 0 1 0 1 | | dir | | rel |

PSW: P (if dir = E0H)

Operation: $(PC) \leftarrow (PC) + 3$
$(dir) \leftarrow (dir) - 1$
if $(dir) <> 0$
then $(PC) \leftarrow (PC) + rel$

INC byte   **Increment**

INC A   Increment

Bytes: 1

Cycles: 1

Encoding: | 0 0 0 0 | 0 1 0 0 |

PSW: P

Operation: $(A) \leftarrow (A) + 1$

INC R_n Increment

Bytes: 1

Cycles: 1

Encoding: | 0 0 0 0 | 1 r r r |

PSW: —

Operation: $(R_n) \leftarrow (R_n) + 1$

INC dir   Increment

Bytes: 2

Cycles: 1

Encoding: | 0 0 0 0 | 0 1 0 1 | | dir |

PSW: P (if dir = E0H)

Operation: (dir) ← (dir) + 1

INC @R$_i$ Increment

Bytes: 1

Cycles: 1

Encoding: | 0 0 0 0 | 0 1 1 i |

PSW: P (if (R$_i$) = E0H)

Operation: ((R$_i$)) ← ((R$_i$)) + 1

INC DPTR   Increment

Bytes: 1

Cycles: 2

Encoding: | 1 0 1 0 | 0 0 1 1 |

PSW: —

Operation: (DPTR) ← (DPTR) + 1

JB bit,rel   Jump if bit

Bytes: 3

Cycles: 2

Encoding: | 0 0 1 0 | 0 0 0 0 | | bit | | rel |

PSW: —

Operation: (PC) ← (PC) + 3
                    if (bit) = 1
                    then (PC) ← (PC) + rel

JBC bit,rel   Jump if bit and clear

Bytes: 3

Cycles: 2

Encoding: | 0 0 0 1 | 0 0 0 0 | | bit | | rel |

PSW: —

Operation: (PC) ← (PC) + 3
                    if (bit) = 1
                    then   (bit) ← 0
                           (PC) ← (PC) + rel

JC rel   Jump if carry

Bytes: 2

Cycles: 2

Encoding: | 0 1 0 0 | 0 0 0 0 | | rel |

PSW: —

Operation: (PC) ← (PC) + 2
                if (C) = 1
                then (PC) ← (PC) + rel

JMP @A+DPTR   Jump indirect

Bytes: 1

Cycles: 2

Encoding: | 0 1 1 1 | 0 0 1 1 |

PSW: —

Operation: (PC) ← (A) + (DPTR)

JNB bit,rel   Jump if no bit

Bytes: 3

Cycles: 2

Encoding: | 0 0 1 1 | 0 0 0 0 | | bit | | rel |

PSW: —

Operation: (PC) ← (PC) + 3
                if (bit) = 0
                then (PC) ← (PC) + rel

JNC rel   Jump if no carry

Bytes: 2

Cycles: 2

Encoding: | 0 1 0 1 | 0 0 0 0 | | rel |

PSW: —

Operation: (PC) ← (PC) + 2
                if (C) = 0
                then (PC) ← (PC) + rel

JNZ rel   Jump if accumulator is not zero

Bytes: 2

Cycles: 2

Encoding: | 0 1 1 1 | 0 0 0 0 | | rel |

PSW: —

Operation: (PC) ← (PC) + 2
                if (A) < > 0
                then (PC) ← (PC) + rel

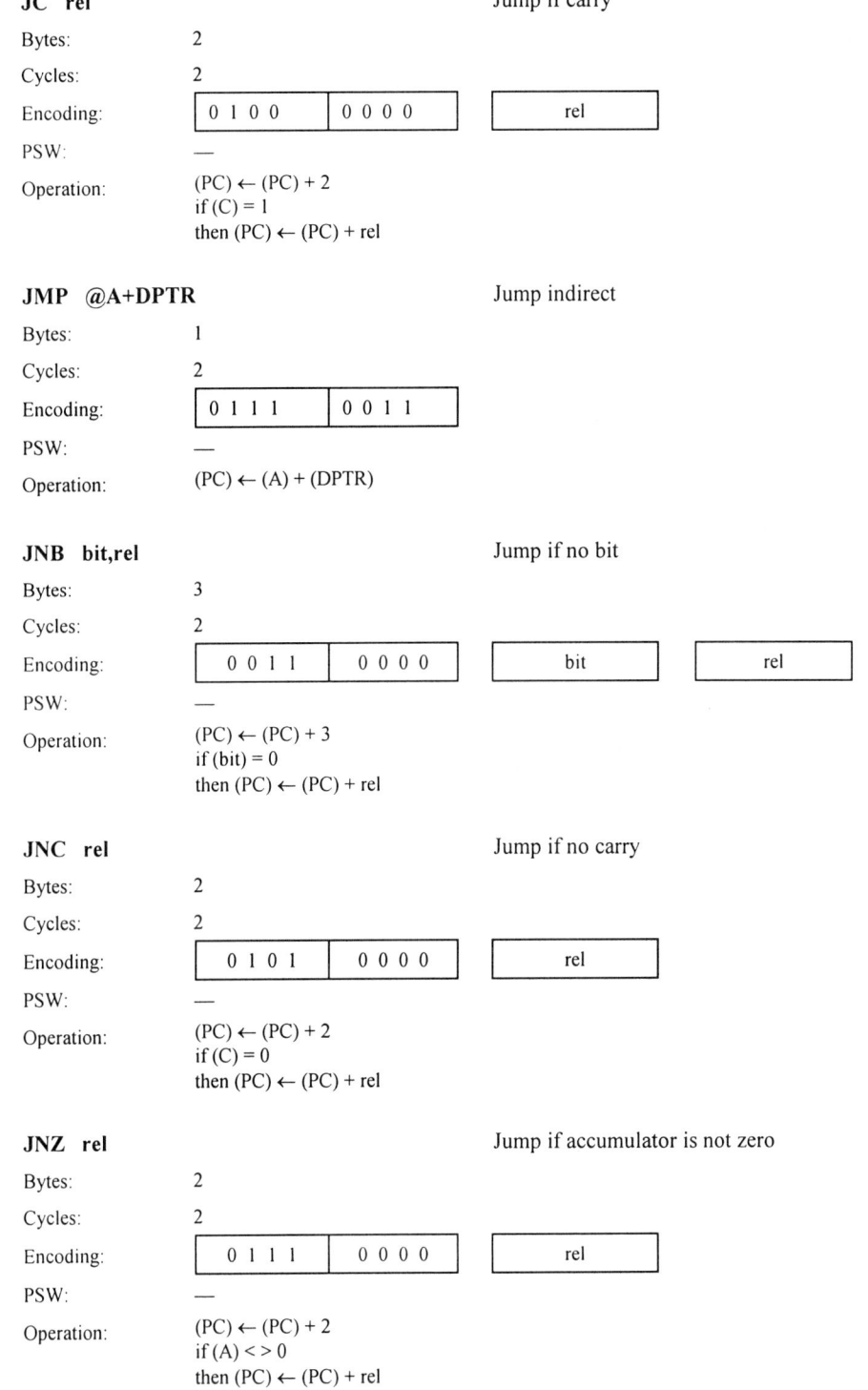

JZ rel   Jump if accumulator is zero

Bytes: 2

Cycles: 2

Encoding: | 0 1 1 0 | 0 0 0 0 | | rel |

PSW: —

Operation: (PC) ← (PC) + 2
              if (A) = 0
              then (PC) ← (PC) + rel

LCALL addr16   Long call

Bytes: 3

Cycles: 2

Encoding: | 0 0 0 1 | 0 0 1 0 | | addr15 - addr8 | | addr7 - addr0 |

PSW: —

Operation: (PC) ← (PC) + 3
              (SP) ← (SP) + 1
              ((SP)) ← (PC7 - 0)
              (SP) ← (SP) + 1
              ((SP)) ← (PC15 - 8)
              (PC) ← addr15 - 0

LJMP addr16   Long jump

Bytes: 3

Cycles: 2

Encoding: | 0 0 0 0 | 0 0 1 0 | | addr15 - addr8 | | addr7 - addr0 |

PSW: —

Operation: (PC) ← addr15 - 0

MOV   <dest-byte> , <src-byte>   **Move**

MOV A,R$_n$ Move

Bytes: 1

Cycles: 1

Encoding: | 1 1 1 0 | 1 r r r |

PSW: P

Operation: (A) ← (R$_n$)

MOV A,dir   Move

Bytes: 2

Cycles: 1

Encoding: | 1 1 1 0 | 0 1 0 1 | | dir |

PSW: P

Operation: $(A) \leftarrow (dir)$

MOV A,@R$_i$ Move

Bytes: 1

Cycles: 1

Encoding: | 1 1 1 0 | 0 1 1 i |

PSW: P

Operation: $(A) \leftarrow ((R_i))$

MOV A,#data   Move

Bytes: 2

Cycles: 1

Encoding: | 0 1 1 1 | 0 1 0 0 | | #data |

PSW: P

Operation: $(A) \leftarrow \#data$

MOV R$_n$,A   Move

Bytes: 1

Cycles: 1

Encoding: | 1 1 1 1 | 1 r r r |

PSW: —

Operation: $(R_n) \leftarrow (A)$

MOV R$_n$,dir   Move

Bytes: 2

Cycles: 2

Encoding: | 1 0 1 0 | 1 r r r | | dir |

PSW: —

Operation: $(R_n) \leftarrow (dir)$

MOV R$_n$,#data   Move

Bytes: 2

Cycles: 1

Encoding:

| 0 1 1 1 | 1 r r r |

| #data |

PSW: —

Operation: (R$_n$) ← #data

MOV dir,A   Move

Bytes: 2

Cycles: 1

Encoding:

| 1 1 1 1 | 0 1 0 1 |

| dir |

PSW: —

Operation: (dir) ← (A)

MOV dir,R$_n$ Move

Bytes: 2

Cycles: 2

Encoding:

| 1 0 0 0 | 1 r r r |

| dir |

PSW: —

Operation: (dir) ← (R$_n$)

MOV dir,dir   Move

Bytes: 3

Cycles: 2

Encoding:

| 1 0 0 0 | 0 1 0 1 |

| dir (src) |

| dir (dest) |

PSW: —

Operation: (dir) ← (dir)

MOV dir,@R$_i$ Move

Bytes: 2

Cycles: 2

Encoding:

| 1 0 0 0 | 0 1 1 i |

| dir |

PSW: —

Operation: (dir) ← ((R$_i$))

MOV   dir,#data   Move

Bytes: 3

Cycles: 2

Encoding: | 0 1 1 1 | 0 1 0 1 | | dir | | #data |

PSW: —

Operation: (dir) ← #data

MOV   @R$_i$,A   Move

Bytes: 1

Cycles: 1

Encoding: | 1 1 1 1 | 0 1 1 i |

PSW: —

Operation: ((R$_i$)) ← (A)

MOV   @R$_i$,dir   Move

Bytes: 2

Cycles: 2

Encoding: | 1 0 1 0 | 0 1 1 i | | dir |

PSW: —

Operation: ((R$_i$)) ← (dir)

MOV   @R$_i$,#data   Move

Bytes: 2

Cycles: 1

Encoding: | 0 1 1 1 | 0 1 1 i | | #data |

PSW: —

Operation: ((R$_i$)) ← #data

MOV   <dest-bit> , <src-bit>   **Move bit data**

MOV   C,bit   Move bit data

Bytes: 2

Cycles: 1

Encoding: | 1 0 1 0 | 0 0 1 0 | | bit |

PSW: C

Operation: (C) ← (bit)

MOV   bit,C   Move bit data

Bytes: 2

Cycles: 2

Encoding: | 1 0 0 1 | 0 0 1 0 | | bit |

PSW: —

Operation: (bit) ← (C)

MOV   DPTR,#data16   Move

Bytes: 3

Cycles: 2

Encoding: | 1 0 0 1 | 0 0 0 0 | | #data15 - 8 | | #data7 - 0 |

PSW: —

Operation: DPH ← #data15 - 8
              DPL ← #data7 - 0

MOVC   A,@A + <base-reg>   **Move code**

MOVC   A,@A + DPTR   Move code

Bytes: 1

Cycles: 2

Encoding: | 1 0 0 1 | 0 0 1 1 |

PSW: P

Operation: (A) ← ((A) + (DPTR))

MOVC   A,@A + PC   Move code

Bytes: 1

Cycles: 2

Encoding: | 1 0 0 0 | 0 0 1 1 |

PSW: P

Operation: (PC) ← (PC) + 1
              (A) ← ((A) + (PC))

MOVX <dest-byte> , <src-byte>   **Move external**

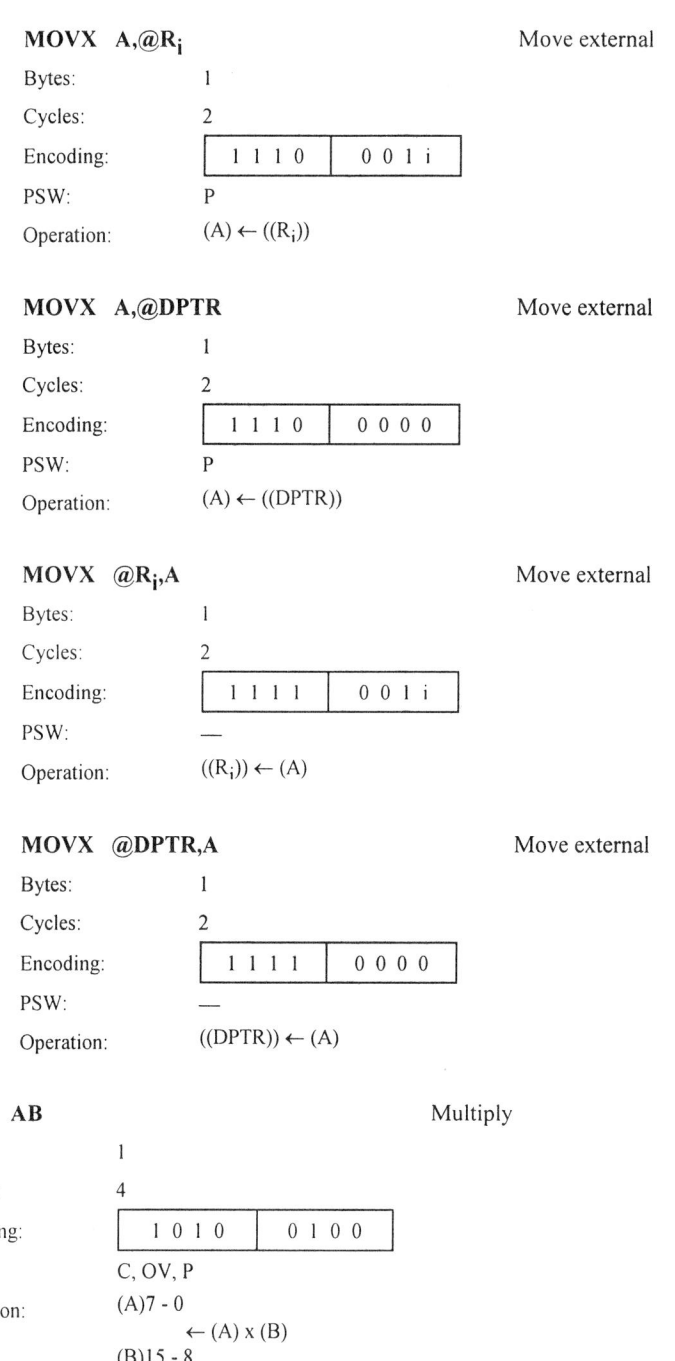

MOVX A,@R$_i$ Move external

Bytes: 1

Cycles: 2

Encoding: | 1 1 1 0 | 0 0 1 i |

PSW: P

Operation: $(A) \leftarrow ((R_i))$

MOVX A,@DPTR   Move external

Bytes: 1

Cycles: 2

Encoding: | 1 1 1 0 | 0 0 0 0 |

PSW: P

Operation: $(A) \leftarrow ((DPTR))$

MOVX @R$_i$,A   Move external

Bytes: 1

Cycles: 2

Encoding: | 1 1 1 1 | 0 0 1 i |

PSW: —

Operation: $((R_i)) \leftarrow (A)$

MOVX @DPTR,A   Move external

Bytes: 1

Cycles: 2

Encoding: | 1 1 1 1 | 0 0 0 0 |

PSW: —

Operation: $((DPTR)) \leftarrow (A)$

MUL AB   Multiply

Bytes: 1

Cycles: 4

Encoding: | 1 0 1 0 | 0 1 0 0 |

PSW: C, OV, P

Operation: (A)7 - 0
                        $\leftarrow (A) \times (B)$
              (B)15 - 8

NOP   No operation

Bytes: 1
Cycles: 1
Encoding: | 0 0 0 0 | 0 0 0 0 |
PSW: —
Operation: (PC) ← (PC) + 1

ORL   \<dest-byte\> , \<src-byte\>   **OR logical**

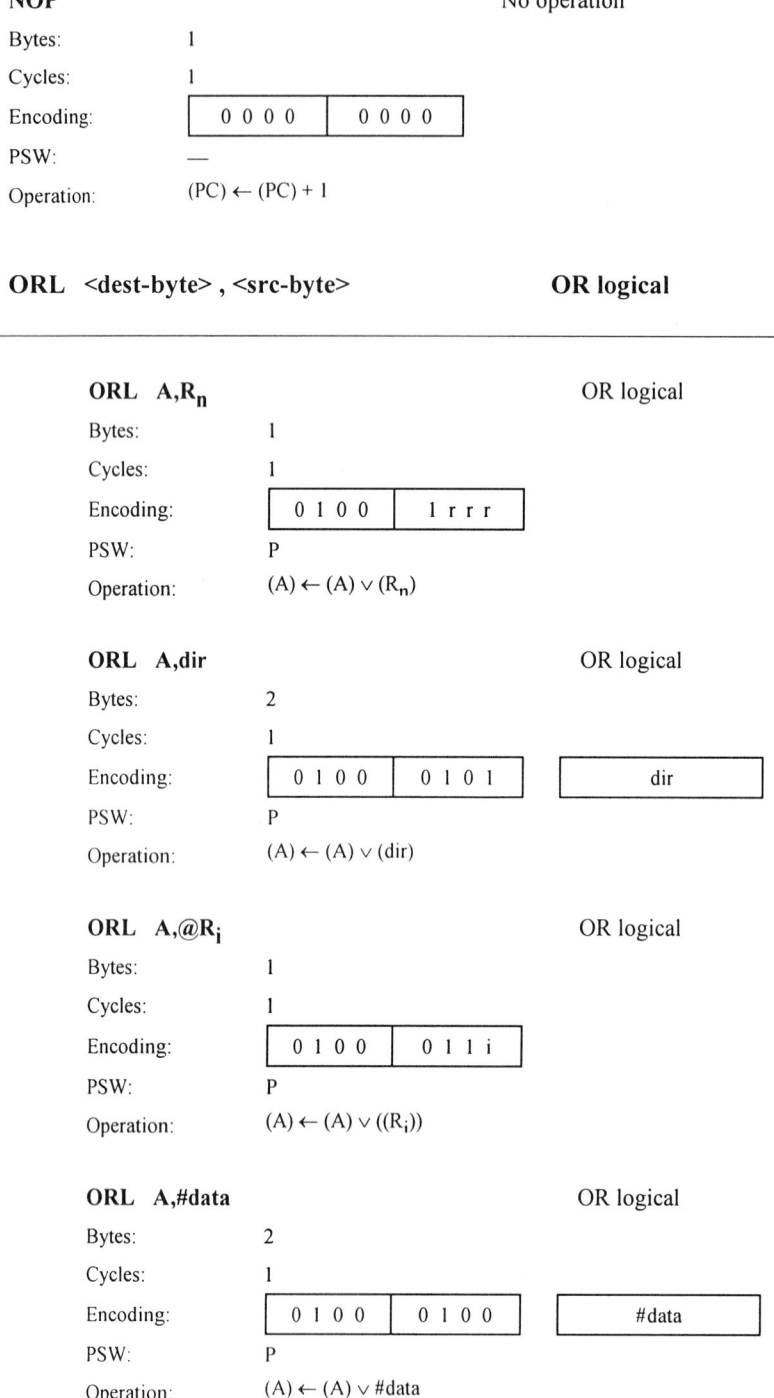

ORL   A,R$_n$ OR logical

Bytes: 1
Cycles: 1
Encoding: | 0 1 0 0 | 1 r r r |
PSW: P
Operation: (A) ← (A) ∨ (R$_n$)

ORL   A,dir   OR logical

Bytes: 2
Cycles: 1
Encoding: | 0 1 0 0 | 0 1 0 1 | | dir |
PSW: P
Operation: (A) ← (A) ∨ (dir)

ORL   A,@R$_i$ OR logical

Bytes: 1
Cycles: 1
Encoding: | 0 1 0 0 | 0 1 1 i |
PSW: P
Operation: (A) ← (A) ∨ ((R$_i$))

ORL   A,#data   OR logical

Bytes: 2
Cycles: 1
Encoding: | 0 1 0 0 | 0 1 0 0 | | #data |
PSW: P
Operation: (A) ← (A) ∨ #data

ORL dir,A   OR logical

Bytes: 2

Cycles: 1

Encoding: | 0 1 0 0 | 0 0 1 0 | | dir |

PSW: —

Operation: $(dir) \leftarrow (dir) \vee A$

ORL dir,#data   OR logical

Bytes: 3

Cycles: 2

Encoding: | 0 1 0 0 | 0 0 1 1 | | dir | | #data |

PSW: —

Operation: $(dir) \leftarrow (dir) \vee \#data$

ORL C, <src-bit>   OR logical for bit variables

ORL C,bit   OR logical for bit variables

Bytes: 2

Cycles: 2

Encoding: | 0 1 1 1 | 0 0 1 0 | | bit |

PSW: C

Operation: $(C) \leftarrow (C) \vee (bit)$

ORL C,/bit   OR logical for bit variables

Bytes: 2

Cycles: 2

Encoding: | 1 0 1 0 | 0 0 0 0 | | bit |

PSW: C

Operation: $(C) \leftarrow (C) \vee (\overline{bit})$

POP dir   Pop from stack

Bytes: 2

Cycles: 2

Encoding: | 1 1 0 1 | 0 0 0 0 | | dir |

PSW: P (if dir = E0H)

Operation: $(dir) \leftarrow ((SP))$
                $(SP) \leftarrow (SP) - 1$

PUSH dir   Push onto stack

Bytes: 2

Cycles: 2

Encoding: | 1 1 0 0 | 0 0 0 0 | | dir |

PSW: —

Operation:
$(SP) \leftarrow (SP) + 1$
$((SP)) \leftarrow (dir)$

RET   Return from subroutine

Bytes: 1

Cycles: 2

Encoding: | 0 0 1 0 | 0 0 1 0 |

PSW: —

Operation: $(PC15 - 8) \leftarrow ((SP))$
$(SP) \leftarrow (SP) - 1$
$(PC7 - 0) \leftarrow ((SP))$
$(SP) \leftarrow (SP) - 1$

RETI   Return from interrupt

Bytes: 1

Cycles: 2

Encoding: | 0 0 1 1 | 0 0 1 0 |

PSW: —

Operation: $(PC15 - 8) \leftarrow ((SP))$
$(SP) \leftarrow (SP) - 1$
$(PC7 - 0) \leftarrow ((SP))$
$(SP) \leftarrow (SP) - 1$

RL A   Rotate left

Bytes: 1

Cycles: 1

Encoding: | 0 0 1 0 | 0 0 1 1 |

PSW: —

Operation: $(An+1) \leftarrow (An)$ $n = 0 - 6$
$(A0) \leftarrow (A7)$

RLC A   Rotate left through carry flag

Bytes: 1

Cycles: 1

Encoding: | 0 0 1 1 | 0 0 1 1 |

PSW: C, P

Operation: $(An+1) \leftarrow (An)$ $n = 0 - 6$
$(A0) \leftarrow (C)$
$(C) \leftarrow (A7)$

RR A   Rotate right

Bytes: 1

Cycles: 1

Encoding: | 0 0 0 0 | 0 0 1 1 |

PSW: —

Operation: (An) ← (An + 1) n = 0 - 6
                    (A7) ← (A0)

RRC A   Rotate right through carry flag

Bytes: 1

Cycles: 1

Encoding: | 0 0 0 1 | 0 0 1 1 |

PSW: C, P

Operation: (An) ← (An + 1) n = 0 - 6
                    (A7) ← (C)
                    (C) ← (A0)

SETB <bit>   **Set bit**

SETB C   Set bit

Bytes: 1

Cycles: 1

Encoding: | 1 1 0 1 | 0 0 1 1 |

PSW: C

Operation: (C) ← 1

SETB bit   Set bit

Bytes: 2

Cycles: 1

Encoding: | 1 1 0 1 | 0 0 1 0 | | bit |

PSW: —

Operation: (bit) ← 1

SJMP rel   Short jump

Bytes: 2

Cycles: 2

Encoding: | 1 0 0 0 | 0 0 0 0 | | rel |

PSW: —

Operation: (PC) ← (PC) + 2
                    (PC) ← (PC) + rel

SUBB A,<src-byte>   **Subtract with borrow**

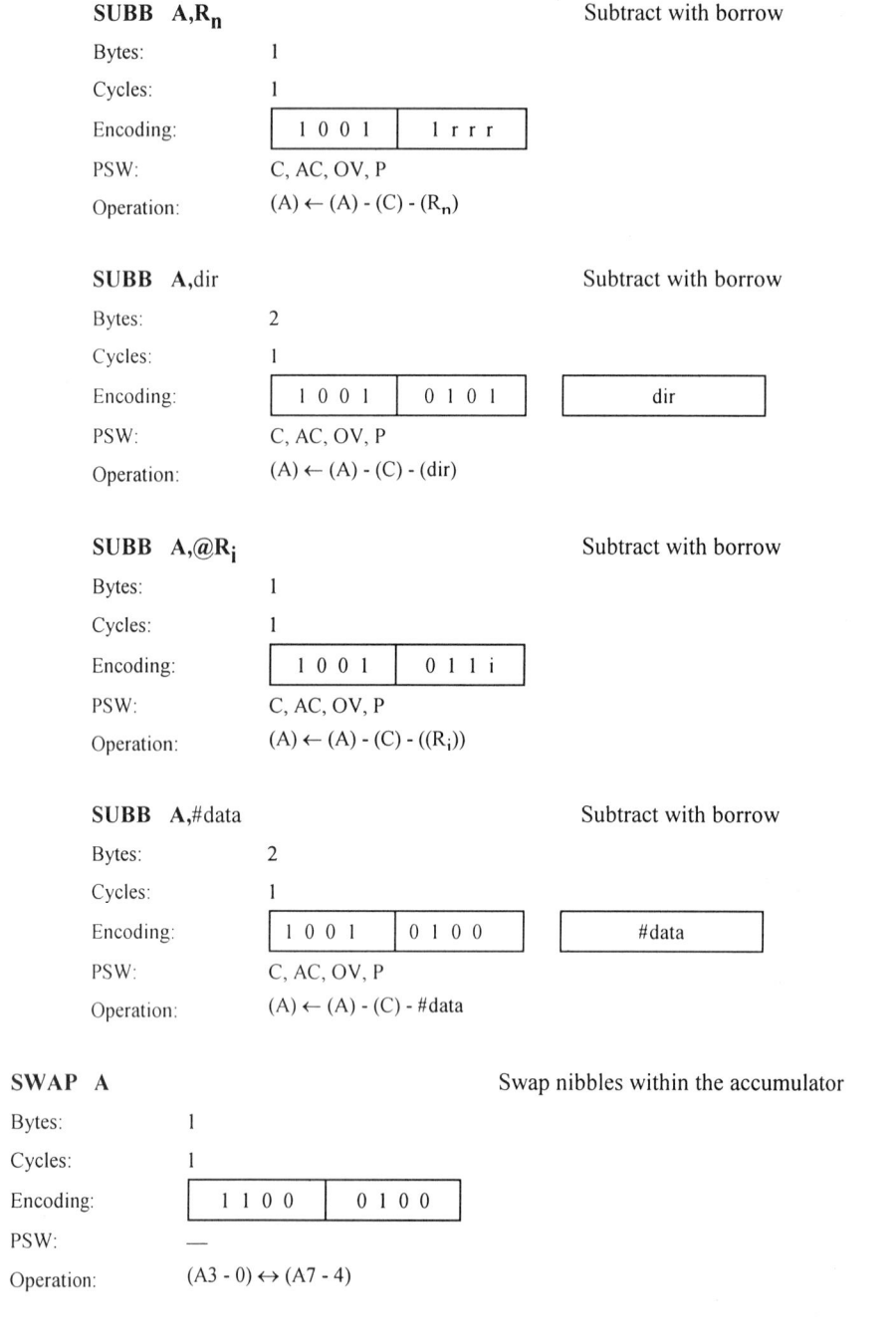

SUBB   A,R$_n$ Subtract with borrow

Bytes: 1

Cycles: 1

Encoding: | 1 0 0 1 | 1 r r r |

PSW: C, AC, OV, P

Operation: (A) ← (A) - (C) - (R$_n$)

SUBB   A,dir   Subtract with borrow

Bytes: 2

Cycles: 1

Encoding: | 1 0 0 1 | 0 1 0 1 | | dir |

PSW: C, AC, OV, P

Operation: (A) ← (A) - (C) - (dir)

SUBB   A,@R$_i$ Subtract with borrow

Bytes: 1

Cycles: 1

Encoding: | 1 0 0 1 | 0 1 1 i |

PSW: C, AC, OV, P

Operation: (A) ← (A) - (C) - ((R$_i$))

SUBB   A,#data   Subtract with borrow

Bytes: 2

Cycles: 1

Encoding: | 1 0 0 1 | 0 1 0 0 | | #data |

PSW: C, AC, OV, P

Operation: (A) ← (A) - (C) - #data

SWAP A   Swap nibbles within the accumulator

Bytes: 1

Cycles: 1

Encoding: | 1 1 0 0 | 0 1 0 0 |

PSW: —

Operation: (A3 - 0) ↔ (A7 - 4)

XCH   A,\<byte\>   **Exchange accumulator with byte variable**

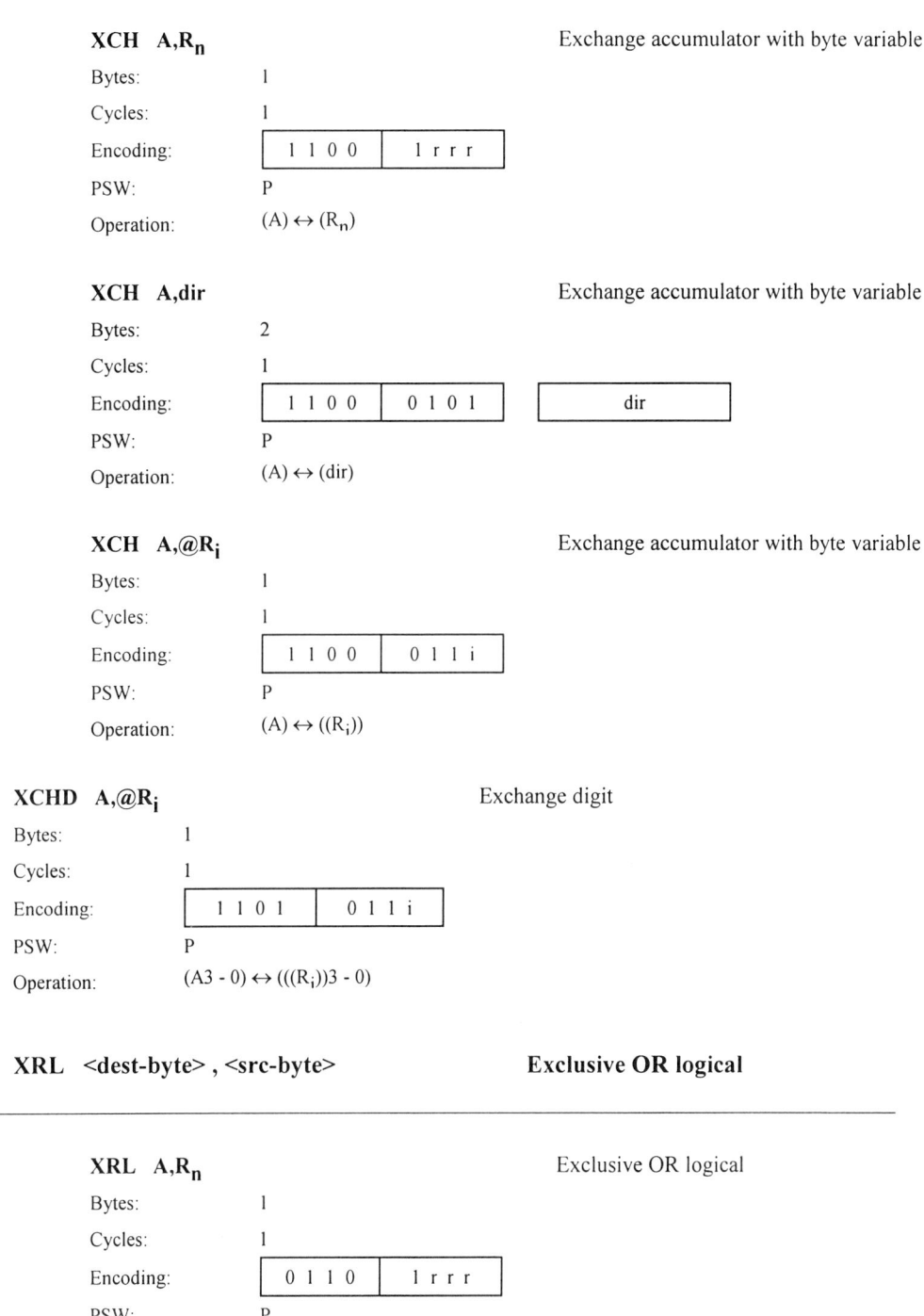

XCH   A,R$_n$ Exchange accumulator with byte variable

Bytes: 1

Cycles: 1

Encoding: | 1 1 0 0 | 1 r r r |

PSW: P

Operation: $(A) \leftrightarrow (R_n)$

XCH   A,dir   Exchange accumulator with byte variable

Bytes: 2

Cycles: 1

Encoding: | 1 1 0 0 | 0 1 0 1 | | dir |

PSW: P

Operation: $(A) \leftrightarrow (dir)$

XCH   A,@R$_i$ Exchange accumulator with byte variable

Bytes: 1

Cycles: 1

Encoding: | 1 1 0 0 | 0 1 1 i |

PSW: P

Operation: $(A) \leftrightarrow ((R_i))$

XCHD   A,@R$_i$ Exchange digit

Bytes: 1

Cycles: 1

Encoding: | 1 1 0 1 | 0 1 1 i |

PSW: P

Operation: $(A3 - 0) \leftrightarrow (((R_i))3 - 0)$

XRL   \<dest-byte\> , \<src-byte\>   **Exclusive OR logical**

XRL   A,R$_n$ Exclusive OR logical

Bytes: 1

Cycles: 1

Encoding: | 0 1 1 0 | 1 r r r |

PSW: P

Operation: $(A) \leftarrow (A) \veebar (R_n)$

XRL   A,dir   Exclusive OR logical

Bytes: 2

Cycles: 1

Encoding: | 0 1 1 0 | 0 1 0 1 | | dir |

PSW: P

Operation: $(A) \leftarrow (A) \veebar (dir)$

XRL   A,@R$_i$ Exclusive OR logical

Bytes: 1

Cycles: 1

Encoding: | 0 1 1 0 | 0 1 1 i |

PSW: P

Operation: $(A) \leftarrow (A) \veebar ((R_i))$

XRL   A,#data   Exclusive OR logical

Bytes: 2

Cycles: 1

Encoding: | 0 1 1 0 | 0 1 0 0 | | #data |

PSW: P

Operation: $(A) \leftarrow (A) \veebar \#data$

XRL   dir,A   Exclusive OR logical

Bytes: 2

Cycles: 1

Encoding: | 0 1 1 0 | 0 0 1 0 | | dir |

PSW: —

Operation: $(dir) \leftarrow (dir) \veebar (A)$

XRL   dir,#data   Exclusive OR logical

Bytes: 3

Cycles: 2

Encoding: | 0 1 1 0 | 0 0 1 1 | | dir | | #data |

PSW: —

Operation: $(dir) \leftarrow (dir) \veebar \#data$

Appendix E

INSTRUCTION SET - SUMMARY

Arithmetic instructions

Mnemonic	Description	Code	Bytes	Cycles
ADD A,R_n	Add	28 - 2F	1	1
ADD A,dir	Add	25	2	1
ADD A,@R_i	Add	26,27	1	1
ADD A,#data	Add	24	2	1
ADDC A,R_n	Add with carry	38-3F	1	1
ADDC A,dir	Add with carry	35	2	1
ADDC A,@R_i	Add with carry	36,37	1	1
ADDC A,#data	Add with carry	34	2	1
SUBB A,R_n	Subtract with borrow	98-9F	1	1
SUBB A,dir	Subtract with borrow	95	2	1
SUBB A,@R_i	Subtract with borrow	96,97	1	1
SUBB A,#data	Subtract with borrow	94	2	1
INC A	Increment	04	1	1
INC R_n	Increment	08-0F	1	1
INC dir	Increment	05	2	1

Mnemonic		Description	Code	Bytes	Cycles
INC	@R$_i$	Increment	06,07	1	1
DEC	A	Decrement	14	1	1
DEC	R$_n$	Decrement	18-1F	1	1
DEC	dir	Decrement	15	2	1
DEC	@R$_i$	Decrement	16,17	1	1
INC	DPTR	Increment	A3	1	2
MUL	AB	Multiply	A4	1	4
DIV	AB	Divide	84	1	4
DA	A	Decimal adjust accumulator	D4	1	1

Logical instructions

Mnemonic		Description	Code	Bytes	Cycles
ANL	A,R$_n$	AND logical	58-5F	1	1
ANL	A,dir	AND logical	55	2	1
ANL	A,@R$_i$	AND logical	56,57	1	1
ANL	A,#data	AND logical	54	2	1
ANL	dir,A	AND logical	52	2	1
ANL	dir,#data	AND logical	53	3	2
ORL	A,R$_n$	OR logical	48-4F	1	1
ORL	A,dir	OR logical	45	2	1
ORL	A,@R$_i$	OR logical	46,47	1	1
ORL	A,#data	OR logical	44	2	1
ORL	dir,A	OR logical	42	2	1
ORL	dir,#data	OR logical	43	3	2
XRL	A,R$_n$	Exclusive OR logical	68-6F	1	1
XRL	A,dir	Exclusive OR logical	65	2	1
XRL	A,@R$_i$	Exclusive OR logical	66,67	1	1
XRL	A,#data	Exclusive OR logical	64	2	1

Mnemonic		Description	Code	Bytes	Cycles
XRL	dir,A	Exclusive OR logical	62	2	1
XRL	dir,#data	Exclusive OR logical	63	3	2
CLR	A	Clear accumulator	E4	1	1
CPL	A	Complement accumulator	F4	1	1
RL	A	Rotate left	23	1	1
RLC	A	Rotate left through carry	33	1	1
RR	A	Rotate right	03	1	1
RRC	A	Rotate right through carry	13	1	1
SWAP	A	Swap nibbles	C4	1	1

Data transfer instructions

Mnemonic		Description	Code	Bytes	Cycles
MOV	A,R_n	Move	E8-EF	1	1
MOV	A,dir	Move	E5	2	1
MOV	A,@R_i	Move	E6,E7	1	1
MOV	A,#data	Move	74	2	1
MOV	R_n,A	Move	F8-FF	1	1
MOV	R_n,dir	Move	A8-AF	2	2
MOV	R_n,#data	Move	78-7F	2	1
MOV	dir,A	Move	F5	2	1
MOV	dir,R_n	Move	88-8F	2	2
MOV	dir,dir	Move	85	3	2
MOV	dir,@R_i	Move	86,87	2	2
MOV	dir,#data	Move	75	3	2
MOV	@R_i,A	Move	F6,F7	1	1
MOV	@R_i,dir	Move	A6,A7	2	2
MOV	@R_i,#data	Move	76,77	2	1

Mnemonic		Description	Code	Bytes	Cycles
MOV	DPTR,#data16	Move	90	3	2
MOVC	A,@A+DPTR	Move code	93	1	2
MOVC	A,@A+PC	Move code	83	1	2
MOVX	A,@R$_i$	Move external	E2,E3	1	2
MOVX	A,@DPTR	Move external	E0	1	2
MOVX	@R$_i$,A	Move external	F2,F3	1	2
MOVX	@DPTR,A	Move external	F0	1	2
PUSH	dir	Push onto stack	C0	2	2
POP	dir	Pop from stack	D0	2	2
XCH	A,R$_n$	Exchange	C8-CF	1	1
XCH	A,dir	Exchange	C5	2	1
XCH	A,@R$_i$	Exchange	C6,C7	1	1
XCHD	A,@R$_i$	Exchange digit	D6,D7	1	1

Boolean instructions

Mnemonic		Description	Code	Bytes	Cycles
CLR	C	Clear carry	C3	1	1
CLR	bit	Clear bit	C2	2	1
SETB	C	Set carry	D3	1	1
SETB	bit	Set bit	D2	2	1
CPL	C	Complement carry	B3	1	1
CPL	bit	Complement bit	B2	2	1
ANL	C,bit	AND logical	82	2	2
ANL	C,/bit	AND logical	B0	2	2
ORL	C,bit	OR logical	72	2	2
ORL	C,/bit	OR logical	A0	2	2
MOV	C,bit	Move carry	A2	2	1
MOV	bit,C	Move bit	92	2	2

Program control instructions

Mnemonic		Description	Code	Bytes	Cycles
ACALL	addr11	Absolute call	◆	2	2
LCALL	addr16	Long call	12	3	2
RET		Return from subroutine	22	1	2
RETI		Return from interrupt	32	1	2
AJMP	addr11	Absolute jump	◆ ◆	2	2
LJMP	addr16	Long jump	02	3	2
SJMP	rel	Short jump	80	2	2
JMP	@A+DPTR	Jump	73	1	2
JZ	rel	Jump if accumulator is zero	60	2	2
JNZ	rel	Jump if accumulator is not zero	70	2	2
JC	rel	Jump if carry	40	2	2
JNC	rel	Jump if no carry	50	2	2
JB	bit,rel	Jump if bit	20	3	2
JNB	bit,rel	Jump if no bit	30	3	2
JBC	bit,rel	Jump if bit and clear	10	3	2
CJNE	A,dir,rel	Compare and jump if not equal	B5	3	2
CJNE	A,#data,rel	Compare and jump if not equal	B4	3	2
CJNE	R_n,#data,rel	Compare and jump if not equal	B8-BF	3	2
CJNE	@R_i,#data,rel	Compare and jump if not equal	B6,B7	3	2
DJNZ	R_n,rel	Decrement and jump if not zero	D8-DF	2	2
DJNZ	dir,rel	Decrement and jump if not zero	D5	3	2
NOP		No operation	00	1	1

◆   11, 31, 51, 71, 91, B1, D1, F1

◆ ◆   01, 21, 41, 61, 81, A1, C1, E1

INDEX